BONINITES

TITLES OF RELATED INTEREST

Carbonatites
K. Bell (ed.)

Cathodoluminescence of geological materials
D. J. Marshall

Chemical fundamentals of geology
R. Gill

Crystal structures and cation sites of the rock-forming minerals
J. R. Smyth & D. L. Bish

The dark side of the Earth
R. Muir Wood

Deformation processes in minerals, ceramics and rocks
D. J. Barber & P. G. Meredith (eds)

Geology and mineral resources of West Africa
J. B. Wright

Igneous petrogenesis
B. M. Wilson

Image interpretation in geology
S. Drury

The inaccessible Earth
G. C. Brown & A. E. Mussett

The interpretation of igneous rocks
K. G. Cox et al.

Introduction to X-ray spectrometry
K. L. Williams

Komatiites
N. Arndt & E. Nisbet (eds)

Mathematics in geology
J. Ferguson

Metamorphism and metamorphic belts
A. K. Miyashiro

Perspectives on a dynamic Earth
T. R. Paton

Petrology of the igneous rocks
F. Hatch et al.

Petrology of the metamorphic rocks
R. Mason

Planetary landscapes
R. Greeley

A practical introduction to optical mineralogy
C. D. Gribble & A. J. Hall

Rheology of the Earth
G. Ranalli

Rutley's elements of mineralogy
C. D. Gribble

Simulating the Earth
J. Holloway & B. Wood

Statistical methods in geology
R. F. Cheeney

Volcanic successions
R. Cas & J. V. Wright

The young Earth
E. G. Nisbet

BONINITES

Edited by

A. J. Crawford

Geology Department, University of Tasmania

London
UNWIN HYMAN
Boston   Sydney   Wellington

© A. J. Crawford, 1989
This book is copyright under the Berne Convention. No reproduction without permission. All rights reserved.

Published by the Academic Division of
Unwin Hyman Ltd
15/17 Broadwick Street, London W1V 1FP, UK

Unwin Hyman Inc.,
8 Winchester Place, Winchester, Mass. 01890, USA

Allen & Unwin (Australia) Ltd,
8 Napier Street, North Sydney, NSW 2060, Australia

Allen & Unwin (New Zealand) Ltd in association with the Port Nicholson Press Ltd,
Compusales Building, 75 Ghuznee Street, Wellington 1, New Zealand

First published in 1989

British Library Cataloguing in Publication Data

Boninites.
1. Boninites
I. Crawford, A. J.
552'.2
ISBN 0-04-445003-6

Library of Congress Cataloging in Publication Data

Boninites/edited by A. J. Crawford.
p. cm.
Includes index.
ISBN 0-04-445003-6 (alk. paper)
1. Boninite. I. Crawford, A. J. (Anthony J.)
QE462.B65B66 1989
552'.2--dc19   88-23549
                                           CIP

Typeset in 10/12 point Times by
Mathematical Composition Setters, Salisbury
and printed in Great Britain at
the University Press, Cambridge

Preface and acknowledgements

Ten years ago, very few petrologists were aware of the existence of boninites, and fewer still would have been able to describe their petrographic or compositional characteristics. Boninites were first described in 1890 from Chichi-jima in the Bonin Islands. They were named a year later by Petersen, and spent the next 75 years in obscurity. In the mid-1960s, Tertiary lavas collected from Cape Vogel in Papua New Guinea by geologists from the Australian Bureau of Mineral Resources were found to contain the unusual pyroxene, clinoenstatite, and to be plagioclase-free. Although not recognized at that stage as boninites, these lavas were described thoroughly and shown to have most unusual compositions, with both high SiO_2 ($> 53\%$) and high MgO ($> 8\%$) contents.

Japanese petrologists brought boninites back into focus in 1977, with a series of papers by Shiraki and colleagues (see Chs 1 and 2 for references) in which it was proposed that the Bonin Island boninites might represent the parent magmas of the island-arc calc-alkaline suite. At the same time, Cameron and colleagues pointed out the occurrence of boninites in some ophiolites, and the similarities, both petrographically and compositionally, between Tertiary boninites and Archaean basaltic komatiites.

In the late 1970s and early 1980s, boninites identical to the type-locality lavas were dredged from the southern Mariana forearc, and somewhat more evolved lavas were drilled from the Mariana forearc at DSDP Site 458. Boninites petrographically and compositionally analogous to Tertiary examples were recognized in Lower Palaeozoic ophiolites in SE Australia and Newfoundland, and, based partly upon the presence of boninites, a supra-subduction-zone origin for the Troodos Ophiolite in Cyprus was generally agreed upon, ruling out pre-existing notions that this ophiolite was a slice of typical, mid-ocean-ridge-generated oceanic crust and upper mantle.

Between 1980 and 1985, detailed petrological, geochemical and isotopic studies of various boninite suites were appearing, and a reasonably consistent model for boninite genesis had developed. This model, in essence, involved depleted supra-subduction-zone harzburgite being invaded and fluxed by slab-derived, LILE-bearing hydrous fluids, and partially melting to yield boninitic primary magmas. Today that model stands little changed, although the fine details of timing and tectonic setting of boninitic magmatism relative to adjacent arc–backarc magmatism, and the nature of the source components, remain controversial.

PREFACE AND ACKNOWLEDGEMENTS

During 1986, I became aware that many colleagues were working enthusiastically on well known boninite occurrences (e.g. Troodos Ophiolite, Cyprus), while at the same time, new and exciting suites were being discovered, e.g. the N Tonga forearc suite described by Falloon et al. in this volume (Ch. 14) and the New Caledonian suites documented by Cameron in this volume (Ch. 12). The stage appeared set for a flood of boninite-related papers, and I proposed to bring them together in an accessible format in this book.

It became obvious at an early stage that an enormous problem existed with terminology and classification of boninite series lavas and related rocks. This prompted the attempt by Crawford, Falloon and Green (Ch. 1) to formalize a major-element-based classification scheme for such rocks, recognizing nevertheless the existence of a continuum of compositions in nature and the sliding compositional boundaries or 'pigeonholes' arbitrarily erected in this scheme. No attempt was made to stretch editorial privilege and ask or encourage other contributors to work within the proposed classification scheme.

In Chapter 2, Tatsumi and Maruyama review the tectono-magmatic evolution of boninites in the SW Japan–Bonin–Mariana region and present a petrogenetic model which has important implications for the entire spectrum of primary magma compositions in arcs. Chapter 3 (Foley and Venturelli), dealing with ultrapotassic rocks with high-MgO, high-SiO_2 affinities, is perhaps, at first glance, out of place in this volume. However, mineralogical (e.g. very Cr-rich chromites) and wholerock chemical features (e.g. low CaO/Al_2O_3, low Ti/Zr) indicate that the sources of such magmas were exceptionally refractory, like boninite sources, but had been enriched in K_2O (as mica) rather than Na_2O, as well as water and other LILE. This may offer important insights into the nature of processes and fluids involved in controlling the development of compositional zoning in sub-arc mantle.

Chapters 4 (Umino and Kushiro) and 5 (Van der Laan, Flower and Koster van Groos) report detailed experimental petrological studies of boninites, and Chapters 6 (Sun, Nesbitt and McCulloch), 7 (Hatton and Sharpe) and 8 (Kuehner) document Precambrian boninites and related rocks. Sun et al. show that basaltic komatiites with compositions transitional to boninites are not derived by hydrous melting of shallow, refractory mantle, but rather are formed by extensive fractionation and crustal contamination of komatiite magmas. They show that a similar origin is unlikely for the parent magmas of the Bushveld Complex ultramafics, a conclusion supported and addressed in more detail by Hatton and Sharpe. Undeformed boninitic dykes approximately 2400 Ma old in the Vestfold Hills, Antarctica, are described by Kuehner, who suggests that these magmas formed by partial melting at ~10 kbar of essentially chondritic source peridotite, leaving harzburgitic residues.

Lower Palaeozoic boninites in ophiolites in Tasmania (Ch. 9, Brown and Jenner) and the Appalachians (Ch. 10, Coish) yield important three-dimensional insights into the temporal and spatial sequence of boninitic magmatism, which are often unavailable from the Tertiary arc–forearc–trench boninite suites. The apparent restriction of Tertiary and Mesozoic boninites to forearc and/or backarc settings suggests that boninites in ophiolites can provide key information on the polarity and timing of plate collisions.

Chapters 11–16 provide detailed petrological and geochemical data for some Mesozoic and Tertiary boninite suites from the W Pacific, the Troodos Ophiolite and the Baja California region. Rogers, MacLeod and Murton (Ch. 11) provide new major- and trace-element and isotopic data for boninites within the Limassol Forest Complex of the Troodos Ophiolite, and suggest an alternative petrogenetic model for such rocks, involving the 'chromatographic column' mantle–magma generation process put forward by Navon and Stolper in 1987. This model is also considered by Rogers and Saunders (Ch. 16) with reference to the relatively lower-Mg#, higher-alkali suite of boninite-like lavas from Baja California. New boninite suites of remarkable diversity are described by Cameron (Ch. 12) from New Caledonia, and Falloon, Green and McCulloch (Ch. 14) from the northern termination of the Tonga Trench. These studies focus attention on the exceptional variation of trace-element and isotopic signatures of boninites of broadly similar major-element chemistry, and offer important clues to the nature and origin of 'enriching' fluids involved in boninite genesis in particular and arc magma genesis in general. Dobson and Tilton (Ch. 15) and Hickey-Vargas (Ch. 13) give much-needed Pb isotope data for Bonin Islands and Mariana forearc boninites, respectively, which serve to emphasize further the decoupling of major-element and LILE and isotopic 'fingerprints' of boninites. The Bonin suite shows clear evidence of slab-derived pelagic sediment involvement in its genesis, whereas probably no sediment was involved in DSDP Hole 458 lavas, which have Pb isotope ratios virtually identical to Pacific Ocean basalts.

This volume should serve as an up-to-date (to June 1988) sourcebook for boninite studies, although because of the significant number of accounts of new boninite suites herein, it is not offered nor was it planned as a review volume (in the style of its predecessor, *Komatiites* (eds Arndt and Nisbet)).

I wish to thank all the contributors to this volume for their support and patience during its preparation; in particular, those authors (mainly post-doctoral fellows and graduate students, comprising exactly half the contributors) who submitted manuscripts almost two years ago and dealt with reviews promptly and efficiently. Each paper was reviewed by myself and one or two additional referees, and I thank the following people for refereeing and for their useful comments to an 'apprentice' editor:

Dr Warrington Cameron	Dr Ian H. Campbell	Dr Trevor Falloon
Professor James E. Gill	Professor David H. Green	Dr Trevor H. Green
Dr Paul R. Hamlyn	Dr A. Lynton Jaques	Dr George A. Jenner
Professor John Longhi	Dr James Luhr	Dr Ian A. Nicholls

PREFACE AND ACKNOWLEDGEMENTS

Dr Richard C. Price   Dr John Sheraton   Dr Shen-su Sun
Dr Rick Varne   Dr Graeme Wheller

Roger Jones of Unwin Hyman has been most helpful and patient, and both he and I owe a special debt of gratitude to the man that invented Macintosh microcomputers, and to Susan Barsdell for intelligently word-processing more than half of the manuscripts herein.

I wish to extend my sincere thanks to Professor David Green for the manner in which he has afforded me total research independence during my time as a post-doctoral fellow in his Department. His guidance, wise counsel, support and great enthusiasm for all things petrological have been of immeasurable benefit to me and to all members of our 'Petrology Group'. I have enjoyed and learnt much from many humorous hours spent discussing different projects with Dr Rick Varne, one of geology's great 'devil's advocates'; his originality and advice are much appreciated. This book was prepared during the tenure of a Queen Elizabeth II Post-Doctoral Fellowship Award, for which I am most grateful.

Thanks are also due to the Director, Bureau of Mineral Resources, Geology and Geophysics, Canberra, for authorization to publish Chapter 6 (for which Crown Copyright is reserved); to *Economic Geology* for permission to reproduce Figures 7.2 and 7.8; and to *Journal of Geophysics Research* for permission to reproduce Figure 2.4.

Finally, I cannot imagine where I would be without the encouragement and support of my parents; and the biggest thank-you of all to my wife, Anne, for everything, and to our babies, Dominic, for coming to grips with the concept of sleep just at the right time, and Timmy, for thoughtfully delaying his arrival until this book was with the publisher.

A. J. Crawford

Hobart
1 June 1988

Contents

Preface and acknowledgements	vii
List of tables	xvii
List of contributors	xxi

1 Classification, petrogenesis and tectonic setting of boninites
A. J. Crawford, T. J. Falloon and D. H. Green

1.1	Classification of boninites	2
1.2	Chemical subdivision of the boninite spectrum	5
1.3	Petrography–mineral chemistry of boninite suites	11
1.4	Phase relationships and their significance	17
1.5	Discussion	21
1.6	Petrogenesis of boninites: a summary	26
1.7	Tectonic setting of boninite generation	29
1.8	A new model for boninite petrogenesis	37
1.9	Directions and problems for further study	41
1.10	Conclusions	42
	Acknowledgements	44
	References	44

2 Boninites and high-Mg andesites: tectonics and petrogenesis
Y. Tatsumi and S. Maruyama

2.1	Introduction	50
2.2	Geological constraints	52
2.3	Petrological constraints	57
2.4	Discussion	63
2.5	Conclusions	68
	Acknowledgements	68
	References	68

3 High-K_2O rocks with high-MgO, high-SiO_2 affinities
S. F. Foley and G. Venturelli

3.1	Introduction	72
3.2	Geological setting and mineralogical features	75
3.3	Wholerock chemistry	78

3.4	Petrogenesis	82
	Acknowledgements	85
	References	85

4 Experimental studies on boninite petrogenesis
S. Umino and I. Kushiro

4.1	Introduction	89
4.2	Petrography and mineralogy of selected boninites	91
4.3	Experimental methods	98
4.4	Experimental results	99
4.5	Conditions of boninite generation and crystallization	105
4.6	Summary	108
	Acknowledgements	109
	References	109

5 Experimental evidence for the origin of boninites: near-liquidus phase relations to 7.5 kbar
S. R. Van der Laan, M. F. J. Flower and A. F. Koster van Groos

5.1	Introduction	113
5.2	The simple system analogue for boninite	114
5.3	Previous experimental work relevant to boninite genesis	115
5.4	Experiments	117
5.5	Results	121
5.6	Discussion	135
5.7	Tectonic implications	143
	Acknowledgements	145
	References	145

6 Geochemistry and petrogenesis of Archaean and early Proterozoic siliceous high-magnesian basalts
S.-S. Sun, R. W. Nesbitt and M. T. McCulloch

6.1	Introduction	149
6.2	Brief description of sample occurrences	151
6.3	Analytical methods and results	152
6.4	Geochemical distinction between Archaean and early Proterozoic SHMB and modern boninites	154
6.5	SHMB: komatiitic parent magmas and the role of crustal contamination	156
6.6	Continental lithosphere as a magma source for SHMB	165
6.7	PGE mineralization in Archaean and early Proterozoic layered complexes	166
6.8	Development of a continental lithosphere modified by subduction processes	167

	6.9	Concluding remarks	169
		Acknowledgements	170
		References	170

7 Significance and origin of boninite-like rocks associated with the Bushveld Complex
C. J. Hatton and M. R. Sharpe

7.1	Introduction	174
7.2	Geological setting and stratigraphy of the eastern Bushveld Complex	175
7.3	The parental magmas to the Bushveld Complex	180
7.4	Field relations of boninites	183
7.5	Wholerock geochemistry	186
7.6	Discussion	189
7.7	Summary	202
	References	203

8 Petrology and geochemistry of early Proterozoic high-Mg dykes from the Vestfold Hills, Antarctica
S. M. Kuehner

8.1	Introduction	209
8.2	Characterization of subgroups	210
8.3	Age relationships of the HMG subgroups	211
8.4	Petrography	212
8.5	Wholerock geochemistry	216
8.6	Rare-earth-element geochemistry	224
8.7	Crustal contamination	225
8.8	Estimation of the depth of magma genesis	225
8.9	Summary	227
	Acknowledgements	229
	References	229

9 Geological setting, petrology and chemistry of Cambrian boninite and low-Ti tholeiite lavas in western Tasmania
A. V. Brown and G. A. Jenner

9.1	Introduction	233
9.2	Regional geological setting	235
9.3	Field relationships of the boninitic and low-Ti lavas	237
9.4	Petrography	239
9.5	Mineral chemistry	241
9.6	Geochemistry	241
9.7	Discussion	255

	9.8	Conclusions	259
		Acknowledgements	260
		References	260

10 Boninitic lavas in Appalachian ophiolites: a review
R. A. Coish

	10.1	Introduction	264
	10.2	Regional setting	266
	10.3	Local geological setting	266
	10.4	Petrography of boninitic lavas	268
	10.5	Mineral chemistry	269
	10.6	Wholerock geochemistry	271
	10.7	Volcanic rocks associated with the boninitic lavas	276
	10.8	Petrogenesis of Appalachian boninites	279
	10.9	Tectonic models for boninite generation	280
	10.10	Appalachian boninites and plate tectonics	281
	10.11	Conclusions	284
		Acknowledgements	284
		References	284

11 Petrogenesis of boninitic lavas from the Limassol Forest Complex, Cyprus
N. W. Rogers, C. J. MacLeod and B. J. Murton

	11.1	Introduction	289
	11.2	Geology	289
	11.3	Petrography	291
	11.4	Results	293
	11.5	Discussion	301
	11.6	Possible re-evaluation of Western Pacific boninites?	309
	11.7	Conclusions	309
		Acknowledgements	310
		References	311

12 Contrasting boninite–tholeiite associations from New Caledonia
W. E. Cameron

	12.1	Introduction	314
	12.2	Massif de Koh	315
	12.3	Formation des Basaltes	322
	12.4	Sr–Nd isotopes and tectonic setting	331
		Acknowledgements	336
		References	336

13 Boninites and tholeiites from DSDP Site 458, Mariana forearc
R. Hickey-Vargas

13.1	Introduction	340
13.2	Summary of section and major- and trace-element data	341
13.3	Isotopic data	345
13.4	Discussion	347
13.5	Origin of the high Zr/Sm and Hf/Sm in boninites	351
13.6	Summary and conclusions	353
	Acknowledgements	354
	References	354

14 Petrogenesis of high-Mg and associated lavas from the north Tonga Trench
T. J. Falloon, D. H. Green and M. T. McCulloch

14.1	Introduction	358
14.2	Petrography and mineral chemistry	359
14.3	Implications of mineral chemistry for magma mixing	375
14.4	Geochemistry	376
14.5	Petrogenesis	379
14.6	Conclusions	391
	Acknowledgements	392
	References	392

15 Th, U and Pb systematics of boninite series volcanic rocks from Chichi-jima, Bonin Islands, Japan
P. F. Dobson and G. R. Tilton

15.1	Introduction	396
15.2	Analytical techniques	398
15.3	Results	398
15.4	Secondary processes affecting Th/U	400
15.5	Primary processes affecting Th/U	402
15.6	Th/U systematics in boninite lavas	406
15.7	Pb isotopic data: implications for sediment involvement in boninite genesis	409
15.8	Conclusions	411
	Acknowledgements	412
	References	412

16 Magnesian andesites from Mexico, Chile and the Aleutian Islands: implications for magmatism associated with ridge–trench collision
G. Rogers and A. D. Saunders

16.1	Introduction	417
16.2	Tectono-magmatic setting	418

16.3	Geochemistry	423
16.4	Comparisons with boninitic magmas	431
16.5	Discussion	434
	Acknowledgements	442
	References	442
Index		446

List of tables

1.1	Representative wholerock analyses of low- and high-Ca boninites. Asterisks indicate Fe calculated as either FeO or Fe_2O_3.	9
2.1	Geological background for production of HMA magmas.	52
3.1	Lamproites and rocks with lamproitic affinity with $SiO_2 > 53\%$ and $MgO > 8\%$.	73
3.2	Main petrographic features of high-MgO, high-SiO_2 ultrapotassic rocks.	75
3.3	Representative analyses of high-SiO_2, high-MgO ultrapotassic rocks from the Mediterranean region.	79
4.1	Wholerock and modal analyses of the Chichi-jima boninites used in melting studies reported in this chapter.	92
4.2	Analyses of run products for experiments on Chichi-jima boninites.	96
5.1	Gel compositions, Mg# and CaO/Al_2O_3 of the starting materials, and Cr, Ni contents of the natural samples.	119
5.2	Experimental conditions and results for compositions (a) Margi, (b) Bonin and (c) CV-46.2.	122
5.3	Summary of best matching natural and experimentally (IHPV) derived mineral phases.	136
5.4	Major-element compositions of fertile (Ib/8) and refractory source compositions (others) and calculated mode options used in mass-balance calculations.	140
5.5	Summary of (a) calculated melt residues and weight fraction of (F) melt produced, and (b) solid fractions entering melt, assuming plagioclase–'lherzolite' and spinel–'lherzolite' mantle assemblages.	141
6.1	Major- and trace-element data of some late Archaean and early Proterozoic SHMB from Western Australia and eastern Antarctica.	153
6.2	Nd isotope and Sm, Nd concentration data of some Archaean and early Proterozoic SHMB, two late Archaean sediments from Western Australia and three early Proterozoic dykes from eastern Antarctica.	154
6.3	Enrichment factors of some trace elements in Kambalda 331/477 relative to an idealized komatiite with 24% MgO.	159

LIST OF TABLES

6.4	Correlation of Pd, Cu, Ti and Y abundances in the primitive mantle, average modern boninites, Archaean komatiites and SHMB from the Eastern Goldfield, Western Australia. Data for average Bushveld SHMB type B4 dykes and estimates for parent magma of the Stillwater Complex are also shown.	163
7.1	Compositions of B1 quench-textured micropyroxenites, ultramafic sills derived from B1 magma, B2 fine-grained marginal rock and B3 fine-grained marginal rock.	187
8.1	Selected wholerock analyses of high-Mg dykes from the Vestfold Hills, Antarctica.	217
8.2	Comparison of calculated 'end-member' primitive liquid compositions with experimental 10 kbar analyses.	226
9.1	Compositions of selected Cr-spinels from Tasmanian Cambrian boninite samples.	242
9.2	Wholerock compositions of Tasmanian Cambrian boninites.	244
9.3	Wholerock compositions of Tasmanian Cambrian low-Ti tholeiites (LOTI).	246
9.4	Average compositions and selected ratios of Tasmanian Cambrian boninites (HMA) and low-Ti tholeiites (LOTI), Victorian Cambrian boninites and W Pacific Tertiary boninites.	250
9.5	Rare-earth-element concentrations (ppm) in selected Tasmanian Cambrian boninite (HMA) and low-Ti tholeiite (LOTI) samples.	254
9.6	Sm–Nd isotopic data for Tasmanian Cambrian boninite (HMA) and low-Ti tholeiite (LOTI) samples.	255
10.1	Representative analyses of chromite in boninitic rocks from the Betts Cove ophiolite.	270
10.2	Representative analyses of boninitic samples from the Betts Cove ophiolite.	271
10.3	Representative analyses of boninitic rocks from (a) lower unit Type II, Thetford Mines and (b) upper lava unit, Thetford Mines.	272
10.4	Representative analyses of boninitic rocks from Pacquet Harbour Group.	274
10.5	Nd isotopes in Betts Cove and Thetford Mines ophiolites.	279
11.1	Major- and trace-element concentrations in six lavas and two dykes from the LFC.	294
11.2	Nd and Sr isotope analyses of six LFC lavas with corrected values assuming an age of 85 Ma.	299
12.1	Microprobe analyses of selected minerals.	318
12.2	Major- and trace-element analyses of metaigneous rocks from the Koh Massif, New Caledonia.	320
12.3	Major- and trace-element analyses of igneous rocks from the Formation des Basaltes, west coast of New Caledonia.	326

LIST OF TABLES

12.4	Sr and Nd isotopic data.	332
13.1	Representative major-element analyses of lavas from DSDP Hole 458, a typical Bonin Is. boninite, a Parece Vela backarc basin basalt and an experimental melt of broadly boninitic composition.	341
13.2	Isotopic data for lavas from DSDP Hole 458.	345
13.3	Calculated mixture of 80% Unit I–III bronzite andesite and 20% Unit IVa tholeiitic andesite compared to Unit IVb bronzite andesite.	349
14.1	Summary of the petrography and mineral chemistry of the dredged lavas from north Tonga.	360
14.2	Representative electron microprobe analyses of groundmass pyroxenes from north Tonga lavas.	362
14.3	Representative electron microprobe analyses of pyroxene phenocrysts and microphenocrysts in north Tonga lavas.	366
14.4	Wholerock and groundmass major-element chemistry of high-Mg lavas and basaltic andesites from north Tonga.	368
14.5	Microprobe analyses of olivine-hosted glass inclusions and calculated parental magma compositions.	370
14.6	Representative electron microprobe analyses of chromites from north Tonga lavas.	371
14.7	Representative major-, trace- and rare-earth-element geochemistry of dredged lavas from north Tonga and Lau Basin.	377
14.8	Sr and Nd isotopes of representative dredged lavas from north Tonga and Lau Basin.	378
15.1	Th, U and Pb data from boninites, Chichi-jima, Japan.	399
15.2	Th, U, Pb and rare-earth-element concentrations in boninites from other studies.	400
15.3	Calculated K_D values (Opx–glass).	404
16.1	Selected wholerock analyses of magnesian Cenozoic lavas, Baja California, Isla Cook (Chile) and the Aleutian Islands.	424

List of contributors

A. V. Brown
Tasmanian Geological Survey,
Department of Mines, PO Box 56,
Rosny Park, Tasmania 7018,
Australia

W. E. Cameron
Department of Geology, Australian
National University, Canberra,
ACT 2601, Australia

R. A. Coish
Geology Department, Middlebury
College, Middlebury, VT 05753,
USA

A. J. Crawford
Geology Department, University of
Tasmania, PO Box 252C, Hobart,
Tasmania 7001, Australia

P. F. Dobson
Division of Geological and
Planetary Sciences, California
Institute of Technology, Pasadena,
CA 91106, USA

T. J. Falloon
Geology Department, University of
Tasmania, PO Box 252C, Hobart,
Tasmania 7001, Australia

M. F. J. Flower
Department of Geological Sciences,
University of Illinois at Chicago,
Box 4348, Chicago, IL 60680, USA

S. F. Foley
Max-Planck-Institut für Chemie,
Saarstrasse 23, D-6500 Mainz,
Federal Republic of Germany

D. H. Green
Geology Department, University of
Tasmania, PO Box 252C, Hobart,
Tasmania 7001, Australia

C. J. Hatton
Institute for Geological Research on
the Bushveld Complex, University
of Pretoria, Hillcrest, Pretoria 0002,
South Africa

R. Hickey-Vargas
Department of Geology, Florida
International University, Miami,
FL 33199, USA

G. A. Jenner
Department of Earth Sciences,
Memorial University of
Newfoundland, St John's,
Newfoundland, Canada, A1B 3X5

A. F. Koster van Groos
Department of Geological Sciences,
University of Illinois at Chicago,
Box 4348, Chicago, IL 60680, USA

S. M. Kuehner
Department of Geophysical
Sciences, The University of Chicago,
Chicago, Illinois, USA

LIST OF CONTRIBUTORS

I. Kushiro
Geological Institute, Faculty of
Science, University of Tokyo,
Bunkyo-ku, Tokyo 113, Japan

M. T. McCulloch
Research School of Earth Sciences,
Australian National University,
Canberra, ACT 2601, Australia

C. J. MacLeod
Department of Earth Sciences, Open
University, Milton Keynes,
MK7 6AA, UK

S. Maruyama
Department of Earth Sciences,
Faculty of Education, Toyama
University, Toyama 930, Japan

B. J. Murton
Department of Earth Sciences, Open
University, Milton Keynes,
MK7 6AA, UK

R. W. Nesbitt
Department of Geology,
The University, Southampton,
SO9 5NH, UK

G. Rogers
Scottish Universities Research and
Reactor Centre, East Kilbride,
Glasgow, UK

N. W. Rogers
Department of Earth Sciences, Open
University, Milton Keynes,
MK7 6AA, UK

A. D. Saunders
Department of Geology,
University of Leicester, Leicester,
LE1 7RH, UK

M. R. Sharpe
Institute for Geological Research on
the Bushveld Complex, University
of Pretoria, Hillcrest, Pretoria 0002,
South Africa

S.-S. Sun
Division of Petrology and
Geochemistry, Bureau of Mineral
Resources, Geology and Geophysics,
PO Box 378, Canberra, ACT 2601,
Australia
and
Research School of Earth Sciences,
Australian National University,
Canberra, ACT 2601, Australia

Y. Tatsumi
Department of Geology and
Mineralogy, Faculty of Science,
Kyoto University, Kyoto 606, Japan

G. R. Tilton
Department of Geological Sciences,
University of California, Santa
Barbara, CA 93106, USA

S. Umino
Department of Geology, Shizuoka
University, Shizuoka, Japan

S. R. Van der Laan
Division of Geological and
Planetary Sciences, California
Institute of Technology, Pasadena,
CA 91125, USA

G. Venturelli
Istituto di Petrografia, Università di
Parma, Viale di Scienze 78, 43100
Parma, Italy

BONINITES

1 Classification, petrogenesis and tectonic setting of boninites

ANTHONY J. CRAWFORD,
TREVOR J. FALLOON & DAVID H. GREEN

Abstract

Based on major-element compositions, boninites are divided into two classes, low-Ca and high-Ca suites. High-Ca boninites, exemplified by the Upper Pillow Lavas of the Troodos Ophiolite in Cyprus, always have $SiO_2 < 56\%$ and $CaO/Al_2O_3 > 0.75$. They generally contain olivine as a phenocryst phase, and may crystallize both high-Ca and low-Ca pyroxenes as phenocrysts or microphenocrysts; their high-CaO nature inhibits clinoenstatite crystallization.

Low-Ca boninites are subdivided into three types. Type 1 are characterized by very low CaO ($< 6\%$), high SiO_2 ($> 58\%$) and total alkalis $> 2\%$; they have low $CaO/Al_2O_3 < 0.5$ and low FeO^* ($< 6\%$); characteristic phenocrysts are olivine up to Fo_{94} and low-Ca pyroxenes, including both ortho- and clinoenstatite. Very refractory Cr-spinel ($Cr\# = 0.80-0.98$) is always present. The best example of Type 1 boninite is the Nepoui suite from New Caledonia.

Type 2 low-Ca boninites also have very low CaO/Al_2O_3 (< 0.55) and low FeO^* ($< 7\%$) but have higher total alkalis and lower SiO_2 contents than Type 1 boninites. They crystallized at lower temperatures (1150–1200°C) than Type 1 boninites, and trace-element data suggest that they may represent lower-degree partial melts of the same refractory harzburgitic sources which yielded Type 1 lavas. They usually carry olivine phenocrysts ± orthopyroxene ± clinopyroxene phenocrysts. The Setouchi (SW Japan) and Baja California suites are the best-documented Type 2 low-Ca boninites.

Type 3 low-Ca boninites have higher CaO/Al_2O_3 values (mainly 0.5–0.75) than other low-Ca suites, and are compositionally intermediate between low-Ca Type 1 and high-Ca boninites. They crystallize olivine and chromite, followed by clinoenstatite and orthopyroxene. As in all boninite suites, plagioclase is restricted to the groundmass of evolved, more slowly cooled samples. The type locality Bonin Island and Cape Vogel suites are representative of this group.

Boninite sources were considerably more refractory than sources for MORB or typical arc basalts, owing to one or more prior melt extraction events. However, invasion of these depleted harzburgites by hydrous fluids, thought to be derived from the subducted slab, led to enrichments in LILE including LREE and Zr, and importantly also in Na_2O and SiO_2. The Nd–Sr isotopic composition of the boninites generated from such enriched refractory peridotite is controlled by the bulk Nd–Sr isotopic composition of the ocean-crust section yielding the hydrous fluids. If this crust is young and relatively sediment-free, hydrous fluids derived from the slab will have Nd–Sr isotopic signatures little different from MORB. If, however, the slab contains a component of recycled ancient crust (e.g. pelagic sediments), the resultant isotopic signatures in boninites generated are much more radiogenic in Sr and have lower ε_{Nd} than the former case.

Two key ingredients for boninite production are a supply of hydrous fluids into refractory peridotite to lower solidus temperatures and permit partial melting, and a mechanism for the maintenance of very high temperatures (1150–1350°C) at shallow levels (< 50 km) in the subduction-zone environment. We suggest that a spectrum of boninitic magmas from initial high-Ca boninites to Type 3, then Type 1, boninites may be generated sequentially from supra-subduction-zone peridotite by continued influx of hydrous fluids, and partial melting of increasingly refractory but increasingly hydrous (and Na-, Si- and LILE-rich) harzburgites. The source of the high temperatures required to permit continued melting in this fashion is either a MORB-source diapir rising beneath the arc during incipient backarc opening, or a subducted active spreading centre. In the latter case, the subducted spreading ridge may be 'resurrected', to yield 'backarc-basin-type' tholeiites overlying the most refractory boninites, but in a forearc setting. Such a scenario is hypothesized to explain the sequence of magma types in several key boninite- and tholeiite-bearing ophiolites.

1.1 Classification of boninites

1.1.1 Introduction

Volcanic suites herein considered as boninitic are those in which the volumetrically dominant lavas either have > 53% SiO_2 and Mg#* > 0.6, or are demonstrably derived from parental magmas meeting these compositional requirements. Magmas with boninitic affinities have been erupted throughout Earth history and are important components of some Archaean greenstone belts, many Palaeozoic and Mesozoic 'ophiolites', forearc regions of certain

---

*Throughout this volume, unless indicated otherwise, Mg# signifies $Mg/(Mg + Fe^{2+})$, where Fe^{2+} is calculated as total Fe.

modern or Tertiary intra-oceanic arcs, and also in several continental-margin convergent plate boundary settings (e.g. Baja California and SW Japan). The economic importance and potential of boninitic rocks are emphasized by the fact that two of the world's largest platinum-group element deposits, the Bushveld and Stillwater Complexes, had parental magmas with boninitic affinities.

A major problem confronting successful reporting of petrological–geochemical studies of boninites and related rocks lies in the inadequate classification and nomenclature presently applied to such lavas. This is highlighted by the plethora of names used in the recent literature for lavas and lava suites which meet the above chemical requirement for boninite. These terms include boninite, transitional boninite, high-Mg andesite, marianite, sanukitoid, low-Ti ophiolitic basalt, komatiitic basalt, basaltic komatiite, siliceous high-Mg basalt, low-Ti quartz tholeiite, and magnesian quartz tholeiite.

Two reasons for the existence of such confused nomenclature for boninites and related lavas may be identified. First, owing to the relatively recent emergence and recognition of boninites as a distinct magma series, existing classification schemes for volcanic rocks are entirely inadequate when applied to boninitic lavas. For example, in the system recommended by the IUGS Subcommission on the Systematics of Igneous Rocks (LeMaitre 1984), Bonin (Ogasawara) Islands (type location) boninites are classified as basaltic andesites. Secondly, within the suites herein considered as boninites and related rocks, there exist pronounced variations in wholerock compositions, modal phenocryst mineralogy and textures. These variations reflect the continuum, in nature, of petrogenetically important intensive variables (P, T, X_{H_2O}, degree of source depletion in 'basaltic' components and so on) operative in the upper-mantle sources of such magmas. For example, in a suprasubduction-zone setting at a nominal temperature of 1300°C and under H_2O-undersaturated conditions, magma segregation at increasing depths from 20 to 100 km would generate a spectrum of primary magma compositions ranging from high Si, high Mg boninitic compositions to low Si, high-Mg picritic or olivine tholeiitic end-members. Classification or assignment of lava compositions falling close to these end-member extremes will be relatively simple, but those falling part-way along the spectrum will generate nomenclatural problems.

Any attempt to subdivide and assign names to primary magmas (or their derivatives following crystal–liquid fractionation) constituting this spectrum of compositions must necessarily erect arbitrary and artificial compositional limits on rock types defined. Ideally, such a classification scheme would successfully relate phenocryst mineralogy and crystallization sequences to wholerock compositions, so that specification of one particular rock name would allow certain petrogenetic conclusions to be drawn about that particular lava, given no further information. In reality, however, numerous factors (e.g. variable cooling rates, crystal fractionation (e.g. sinking of olivine), magma mixing, contamination, post-eruptive alteration) contrive to complicate and reduce the effectiveness of such an ideal scheme, forcing compromise.

As a prelude to providing a workable classification scheme for boninites and related rocks, we have collated petrographic, mineral chemical and major-element chemical data for eight suites which we believe to be representative of the spectrum of natural compositions of boninites and related rocks. Very brief descriptions of the occurrence of each suite are given below; petrographic characteristics and mineral chemistry are discussed further on, following a review of the compositional variability of boninite magmas.

1.1.2 Boninite suites used in the classification

SUITE 1. BONIN (OGASAWARA) ISLANDS

Although first described and named last century (Kikuchi 1890, Petersen 1891), the type location boninites have only recently been carefully documented (Kuroda & Shiraki 1975, Kuroda et al. 1978, Shiraki et al. 1978, 1980, Komatsu 1980, Umino 1985, 1986). On Chichi-jima, boninites and related bronzite andesites and dacites of Eocene age are overlain by Oligocene and Lower Miocene limestones. They are dominantly pillowed flows and volcanic breccias and have been described in detail by Umino (1985, 1986).

SUITE 2. CAPE VOGEL, PAPUA NEW GUINEA

These boninites were described by Dallwitz et al. (1966), Dallwitz (1968), Jenner (1981) and Walker & Cameron (1983). They form part of the poorly outcropping Dabi Volcanics, which also contains basalts transitional to arc tholeiites (Jenner 1981), and are of late Palaeocene age (Walker & McDougall 1982). They form part of a series of allochthonous thrust sheets which include the Papuan ultramafic belt that formed in an intra-oceanic arc but were overthrust onto the northern margin of the Australian Plate during arc–continent collision (Jaques & Robinson 1977).

SUITE 3. MARIANA TRENCH

Primitive boninites of probable Eocene age have been dredged from the inner wall of the Mariana Trench at Dmitri Mendelev Site 1403 south of Guam (Dietrich et al. 1978, Sharaskin et al. 1980, Crawford et al. 1981) and several other locations further north along the trench wall (Bloomer 1987, Bloomer & Hawkins 1987). These are almost certainly correlates of the Bonin Island and DSDP Hole 458–Guam boninite occurrences.

SUITE 4. NEW CALEDONIA

Two distinct suites of boninites occur on New Caledonia, one of Permo-Triassic age, the other early Cretaceous–Middle Tertiary (Cameron 1989). The latter, which outcrop around Nepoui, have been described by Sameshima *et al.* (1983), Campiglio *et al.* (1986) and Cameron (1989) and are unusual high-Si, very low-Ca boninites. The three analyses included in this data set are from Cameron *et al.* (1983), though we note that more extensive new data presented herein by Cameron (1989) conform to the general compositional trends and interpretation we have derived from the three published analyses.

SUITE 5. UPPER PILLOW LAVAS OF THE TROODOS OPHIOLITE, CYPRUS
Despite early controversy regarding the origin and tectonic setting of eruption of Upper Cretaceous lavas forming the volcanic carapace of the Troodos Ophiolite (see review by Robinson *et al.* 1983), recent studies have arrived at some consensus of opinion. These lavas include a lower sequence of evolved arc tholeiitic andesites and dacites overlain by a series of more primitive lavas that includes picrites and boninites (Robinson *et al.* 1983, McCulloch & Cameron 1983, Cameron 1985, Thy *et al.* 1985, Rautenschlein *et al.* 1985, Flower & Levine 1987, Rogers *et al.* 1989), suggesting a supra-subduction-zone setting in a primitive intra-oceanic arc or incipient backarc basin.

SUITE 6. MARIANA FOREARC AND GUAM
Relatively evolved late Eocene–early Oligocene boninite series lavas were drilled from DSDP Hole 458 in the Mariana forearc, where they overlie arc tholeiites (Meijer 1980, Wood *et al.* 1981, Kushiro 1981, Bougault *et al.* 1981, Natland 1981, Hickey & Frey 1982). Similar lavas outcrop on Guam in the late Middle Eocene Facpi Formation and the late Eocene–early Oligocene Alutom Formation (Reagan & Meijer 1984, Hickey-Vargas & Reagan 1987); the latter are associated with arc tholeiite basalts.

SUITE 7. SETOUCHI VOLCANIC BELT, SW JAPAN
An unusual magma series ranging from basalt to high-Si, high-Mg lavas (sanukitoids) occurs along the Seto Inland Sea (Setouchi) region of SW Japan. These lavas have been described by Tatsumi (1981, 1982, 1983), Tatsumi *et al.* (1983), Tatsumi & Ishizaka (1981, 1982a, b), Ishizaka & Carlson (1983) and Tatsumi & Maruyama (1989) and were erupted during a brief magmatic event at approximately 13 Ma.

SUITE 8. BAJA CALIFORNIA
Saunders *et al.* (1987) and Rogers & Saunders (1989) describe a compositionally diverse suite of lavas occurring in Baja California that ranges in age from 12 to 4 Ma and includes a spectrum of rock types from basalt to high-Si, high-Mg lavas with boninitic affinities; the latter were termed 'bajaites' by Saunders *et al.* (1987).

1.2 Chemical subdivision of the boninite spectrum

As a first step towards establishing the existence of compositionally (petrogenetically?) distinct magma groups within the boninite spectrum, we have plotted major-element chemical variations and CaO/Al_2O_3 of the eight selected suites versus Mg# in Figures 1.1 and 1.2. Suites 1 to 6 are characterized by abundant lavas with > 10% MgO, whereas suites 7 and 8 lavas more frequently have < 10% MgO. Important differences between the suites are evident, especially for SiO_2, CaO, FeO^* and Na_2O contents. From Figures 1.1 and 1.2, we distinguish two distinct lineages of boninitic magmas, low-Ca boninites and high-Ca boninites. Low-Ca boninites are further sub-

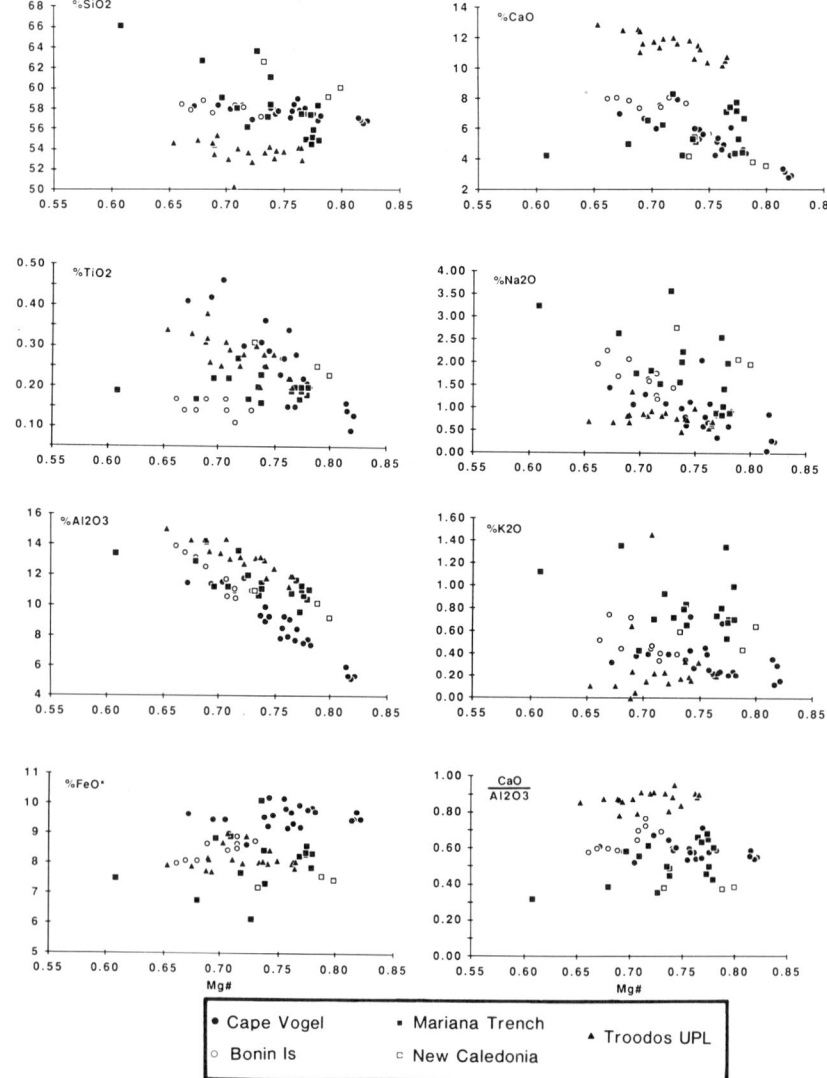

Figure 1.1 Element–Mg# plots for reference boninite suites from Cape Vogel and Bonin Islands (mainly low-Ca Type 3), Mariana Trench (Type 3 and 1), New Caledonia (Type 1) and the high-Ca Troodos Upper Pillow Lavas for reference. Data sources in text.

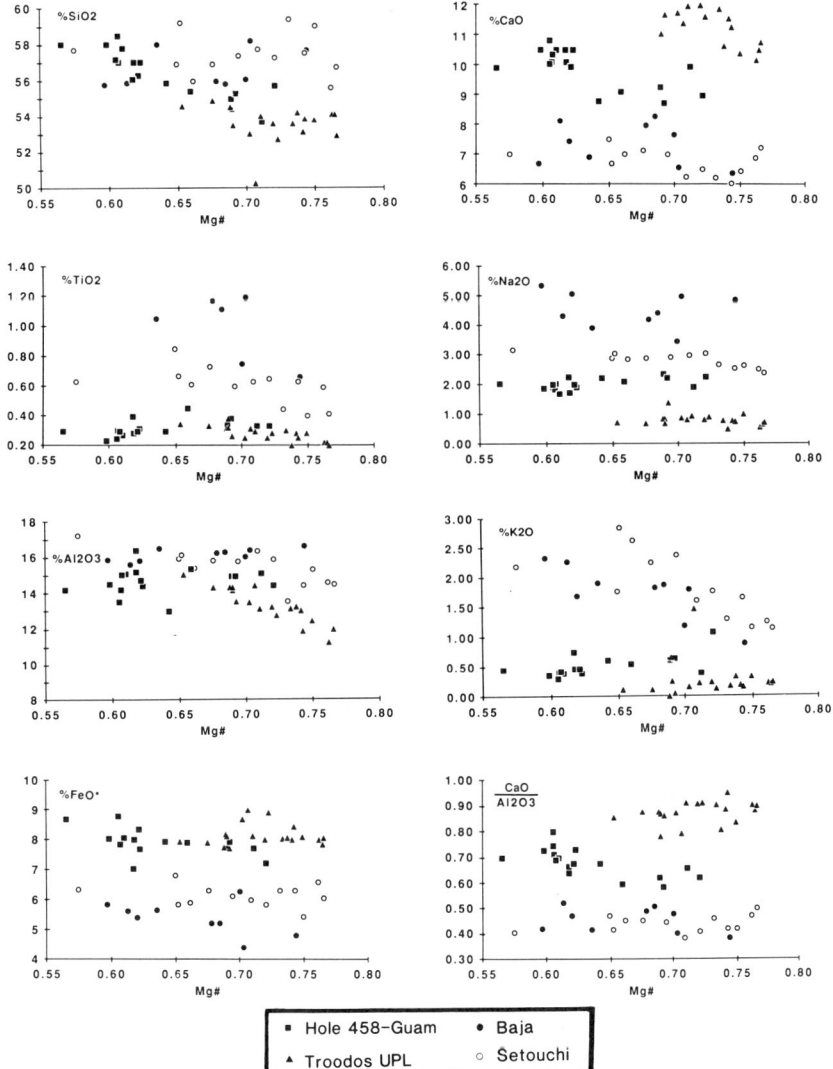

Figure 1.2 Element–Mg# plots for reference boninite suites from Setouchi and Baja (low-Ca Type 2), DSDP Hole 458–Guam (mainly evolved Type 3) and the high-Ca Troodos Upper Pillow Lavas. Data sources in text.

divided into three classes:

Type 1 low-Ca boninites Comprising New Caledonian late Mesozoic–early Tertiary boninites and some boninites from the Mariana Trench, this group has very low CaO/Al_2O_3 (< 0.55) values and low CaO and FeO^* contents (both < 8%), and high Al_2O_3 and alkali contents (Fig. 1.1).

Type 2 low-Ca boninites Included here are the high-Si, high-Mg lavas from the Setouchi area in SW Japan (sanukitoids), and those from Baja California (bajaites). These lavas rarely have more than 10% MgO, but at any Mg# the Setouchi and Baja lavas have slightly lower SiO_2 and higher alkali contents than other boninites, and might be referred to as alkaline low-Ca

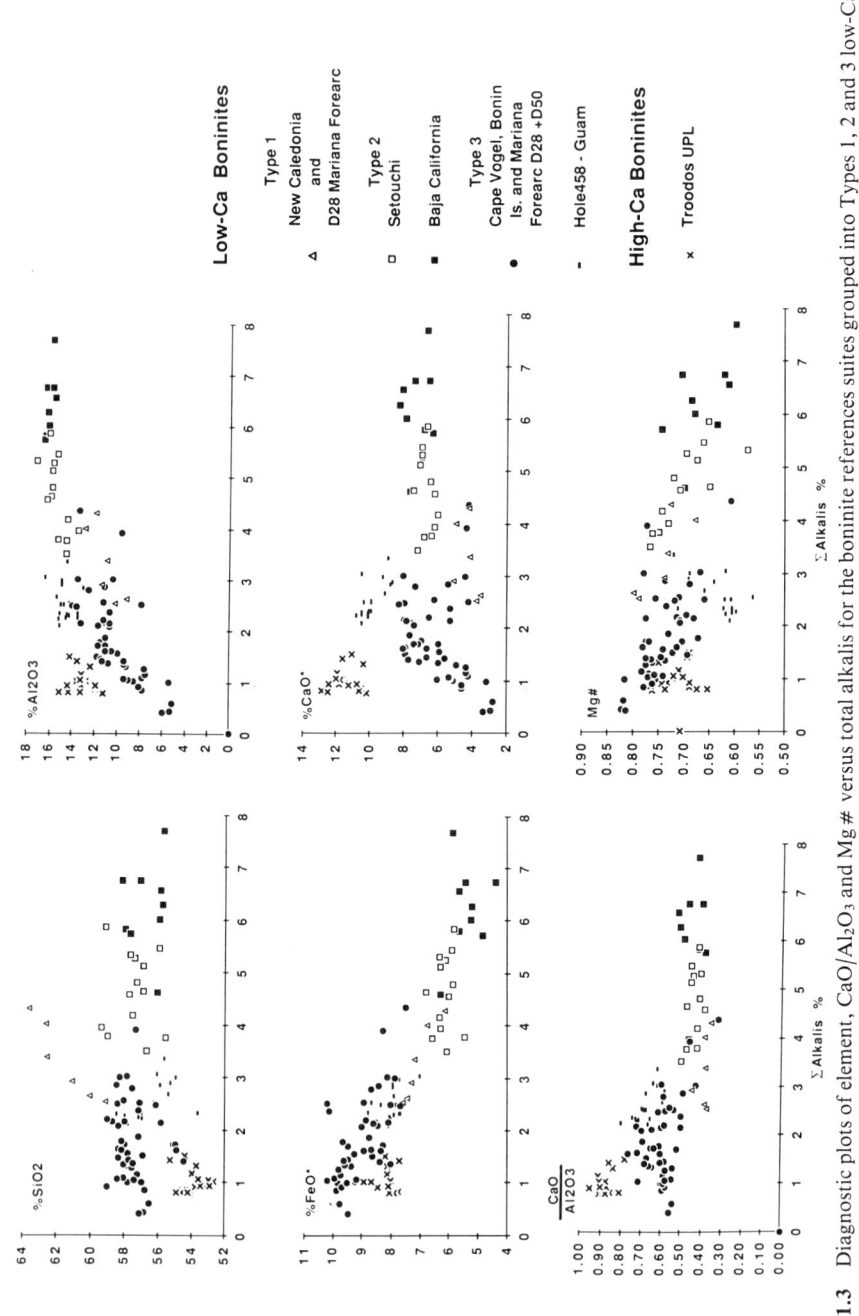

Figure 1.3 Diagnostic plots of element, CaO/Al$_2$O$_3$ and Mg # versus total alkalis for the boninite references suites grouped into Types 1, 2 and 3 low-Ca, and high-Ca magma types. A clear separation is achieved between dominant population of each suite and other suites.

Table 1.1 Representative wholerock analyses of low- and high-Ca boninites. Asterisks indicate Fe calculated as either FeO or Fe_2O_3.

	Low-Ca Type 1				Low-Ca Type 2				Low-Ca Type 3				High-Ca boninites			
	1	2	3	4	5	6	7	8	9	10	11	12	13	14	15	16
SiO_2	54.52	59.64	59.46	57.12	58.50	56.99	57.20	55.80	54.75	56.20	54.41	53.71	51.32	50.19	53.83	54.99
TiO_2	0.23	0.30	0.16	0.18	0.43	0.60	0.65	0.74	0.26	0.13	0.21	0.21	0.22	0.24	0.36	0.30
Al_2O_3	9.36	10.48	11.16	10.18	13.30	15.63	16.50	15.90	13.28	10.57	10.93	7.77	12.42	12.74	9.63	10.40
Fe_2O_3	2.35	1.28	1.37	—	2.50	2.01	5.24*	1.06	2.40	2.02	1.47	2.33	1.73	2.19	—	—
FeO	4.88	5.71	5.90	7.70*	3.95	4.27	—	5.30	5.33	6.23	6.98	7.03	5.85	6.24	9.71*	8.30*
MnO	0.27	0.15	0.13	0.16	0.12	0.14	0.08	0.11	0.13	0.16	0.17	0.17	0.15	0.15	0.19	0.17
MgO	14.62	10.55	11.26	15.25	9.47	7.75	7.80	8.20	10.72	11.19	10.65	18.49	11.13	10.38	16.30	13.38
CaO	3.55	3.97	4.97	4.36	6.13	6.95	6.30	7.61	8.11	7.44	7.98	4.64	10.30	11.11	8.50	9.00
Na_2O	1.93	2.64	2.19	1.96	2.61	2.88	4.80	3.40	1.50	1.54	1.01	0.83	0.97	0.83	1.27	1.12
K_2O	0.41	0.57	0.64	0.98	1.28	2.35	0.88	1.18	0.91	0.40	0.41	0.19	0.31	0.14	0.16	0.42
P_2O_5	0.03	0.05	0.04	0.02	0.13	0.15	0.22	0.20	0.05	0.02	0.03	0.03	0.03	0.02	0.04	0.07
H_2O^+	4.66	3.96	2.92	—	1.37	0.75	—	—	2.95	3.02	3.39	2.65	3.76	4.03	0.24	1.87
H_2O^-	2.96	0.56	—	—	0.22	0.05	—	—	—	0.93	2.05	1.64	1.31	1.06	—	—
CO_2	<0.10	<0.10	0.04	—	—	—	—	—	0.08	<0.10	<0.10	<0.10	0.10	0.24	—	—
total	99.77	99.86	100.24	97.91	100.01	100.52	99.67	99.50	100.47	99.85	99.69	99.70	99.60	100.06	99.99	100.02
Ni	400	194	234	337	184	174	280	150	184	194	134	540	237	279	341	189
Cr	1100	655	722	1310	472	408	322	205	548	695	510	1930	780	650	1294	927
V	105	962	126	119			128	160	164	181	203	136	222	233	225	214
Sc	22	23			22					38	37	24	41	43	44	43
Zr	63	78	50	42			63	166	50	20	25	30	9	6	20	32
Nb	1.5	2	6	4			2	5	4	1	1	1.5	<1	<1	<1	8
Y	7	12	5	6			6	8	10	5	8	4	7	8	10	8
Sr	325	131	154	93			1189	2068	90	61	83	63	136	38	68	159
Rb	6	13	10				6	9	14	9	8	4	7	3	3	8
Ba	54	52	36				280	804	20	30	28	30	50	17	41	105
Th	0.4	0.5					0.6	2.09		0.1	0.3	0.5				
Hf	1.7	1.6					1.69	4.13		0.5	0.6	0.7				

1 and 2, New Caledonia 49 and 52 (Cameron *et al.* 1983); 3 and 4, Mariana Trench 28-1 and 2815 (Bloomer 1987); 5 and 6, Setouchi TG-1 and SD407 (Tatsumi & Ishizaka 1982a); 7 and 8, Baja California J7.2 and J2.1 (Saunders *et al.* 1987); 9, Mariana Trench 50-23 (Bloomer 1987); 10, Chichi-jima 106 (Cameron *et al.* 1983); 11 and 12, Cape Vogel 52.1A and B6 (Cameron *et al.* 1983); 13 and 14, Upper Pillow Lavas, Cyprus, Kapilio 13 and Arakapas 24 (Cameron 1985); 15 and 16, north Tonga Trench 5-25 and 6-3 (Falloon *et al.* 1989).

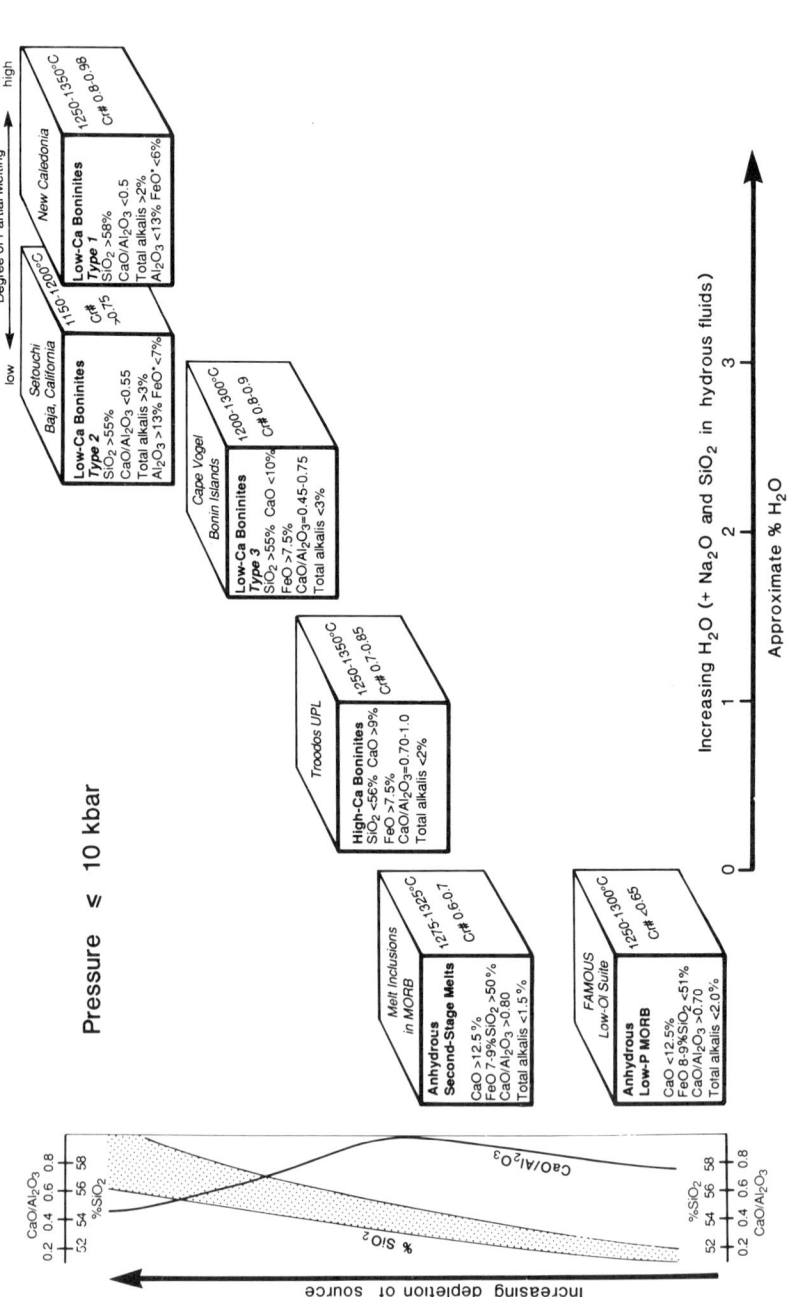

Figure 1.4 Petrogenetic scheme for boninites and related rocks at $P < 10$ kbar, emphasizing the control over primary magma composition exerted by the interplay between extent of depletion of source peridotite and the amount of hydrous fluids required to flux a refractory source and initiate partial melting. Compositional 'pigeonholes' on front faces of boxes are only broad guidelines, and assignment of a suite within this classification should also be done with reference to fields in Figure 1.3. Likewise, range of liquidus temperatures and liquidus chromite Cr # on side faces of boxes is a broad bracketing only, and it is certain that a continuum exists in both these parameters and wholerock compositions between the boxes shown in this figure.

boninites. However, their low CaO/Al_2O_3 values (< 0.55) and low CaO and FeO^* contents (Fig. 1.2) indicate strong similarities with Type 1 boninites.

Type 3 low-Ca boninites Comprising boninites from the type locality on Chichi-jima, from Cape Vogel and the Mariana forearc (DSDP Hole 458 and Guam) and also from the Mariana Trench, these have lower Na_2O and higher CaO and FeO^* contents (Fig. 1.1) than Types 1 or 2 lavas at any Mg# value, and $CaO/Al_2O_3 = 0.5$–0.8. The Cape Vogel and Chichi-jima lavas form a coherent trend, with the latter defining the lower Mg# end of the compositional spread.

Some Mariana Trench lavas (Dredge MT50) fall at slightly lower SiO_2, FeO^* and Na_2O levels and significantly higher CaO, K_2O and Al_2O_3 contents at any Mg# than Chichi-jima or Cape Vogel lavas (Fig. 1.1). Although we include the MT50 lavas in the Type 3 boninite category, their higher CaO and lower Na_2O and SiO_2 levels than Chichi-jima or Cape Vogel lavas suggest that in some respects they are transitional to high-Ca boninites, described below. Hole 458 and Guam lavas plot at the low Mg# end of the Type 3 boninite spectrum, and at least in major-element terms (Fig. 1.2) are compositionally appropriate to represent the evolved, more fractionated end of the lower-SiO_2, higher-CaO part of the Type 3 low-Ca boninite spectrum (see also Fig. 1.3).

High-Ca boninites Our reference suite is the Troodos Upper Pillow Lavas (Figs 1.1 & 1.2); other high-Ca boninite suites (not plotted) include high-Si, high-Mg Proterozoic dykes from the Vestfold Hills, Antarctica (Kuehner 1989) and Greenland (Hall & Hughes 1987), and late Tertiary lavas dredged from the N Tongan forearc (Falloon *et al.* 1987, 1989, Falloon & Crawford 1989). High-Ca boninites have notably lower SiO_2, Na_2O and K_2O contents than low-Ca boninites, and have higher CaO and FeO^* contents, and probably grade with decreasing SiO_2 contents into low-Ti arc basalts such as those from Manam, Papua New Guinea (Johnson *et al.* 1985).

The four boninitic magma types are clearly distinguished from each other in Figures 1.3a–e; representative analyses are listed in Table 1.1. In Figure 1.4, we propose a petrogenetic-based classification for boninites (discussed in detail further on) and suggest broad chemical 'pigeonholes' for each of the suites recognized. Further discussion of the petrogenetic implications of these four classes of boninitic magmas follows a review of their petrographic–mineral chemical characteristics.

1.3 Petrography–mineral chemistry of boninite suites

1.3.1 Introduction

Cameron *et al.* (1979) defined boninite as 'a highly magnesian but relatively siliceous glassy rock containing one or more varieties of pyroxene, some or all of which have a morphology characteristic of rapid growth, accessory magnesiochromite and, commonly, minor amounts of olivine. Laths of amphibole or plagioclase microlites are rare.' As noted by Natland (1981), this definition encounters difficulty in accommodating more slowly cooled, non-glassy lavas with boninite compositions. In addition, Meijer (1980) pointed out that boninites, like any other magma type, may experience extensive fractionation en route to eruption, and he proposed the existence of a 'boninite series'. The most primitive lavas in this series have been termed olivine boninites, and more fractionated lavas range from boninites through boninitic bronzite andesites to boninitic hypersthene dacites (e.g. Bloomer & Hawkins 1987). In this paper, we restrict discussion to lavas with $Mg\# > 0.6$, as these are more informative with respect to source compositions and petrogenetic processes.

1.3.2 Low-Ca boninites

There are no significant differences in petrography and mineral chemistry between primitive Type 1 and Type 3 low-Ca boninite lavas ($Mg\# > 0.70$). Detailed mineralogical data are available for primitive Type 1 low-Ca boninites from Nepoui (New Caledonia) (Cameron *et al.* 1983, Sameshima *et al.* 1983, Campiglio *et al.* 1986, Cameron 1989), and for primitive Type 3 low-Ca boninites from Cape Vogel (Dallwitz *et al.* 1966, Walker & Cameron 1983) and Chichi-jima (Umino 1986). Lavas of both groups carry phenocrysts and microphenocrysts of olivine, chromite, clinoenstatite and orthopyroxene. Plagioclase is conspicuously absent.

Olivine phenocrysts, which often show some evidence of reaction and resorption, range from $Fo_{94.1}$ to Fo_{88}. Boninitic olivines with $Fo_{>92}$ have characteristically high Cr_2O_3 contents between 0.15 and 0.3%. Clinoenstatite ($Wo_{<0.5}En_{92.7-88}$) occurs as phenocrysts generally < 1 mm long, although crystals up to 5 cm long are present in some Cape Vogel lavas; all show characteristic polysynthetic twinning on (1 0 0) developed during inversion from the high-temperature polymorph protoenstatite. Orthopyroxene phenocrysts range from close to $Wo_{>1.5}En_{90}$ to compositions as Fe-rich as Wo_4En_{78}, and always have significantly more CaO and Al_2O_3 (> 0.5% and 0.9% respectively) than coexisting clinoenstatite. Chromite occurs as inclusions in olivine and low-Ca pyroxene, and less frequently as discrete phenocrysts up to 2 mm diameter. They are very Cr-rich and Al-poor magnesiochromites, with $Cr\#$ (i.e. $Cr/(Cr + Al)) = 0.95-0.80$, and $Mg\#$ from 0.83 to 0.20; low $Mg\#$ values result from subsolidus re-equilibration.

Groundmass pyroxenes are typically very strongly zoned from bronzite or magnesian pigeonite cores through subcalcic augite to augite or ferroaugite rims; they form compositional arrays which trend across the pyroxene quadrilateral at relatively constant $Mg\#$ for any individual lava. Pale-green pargasitic amphibole with $Mg\#$ generally from 0.50 to 0.30 occurs as quench needles and overgrowths on groundmass augite, and demonstrates clearly that low-Ca boninitic magmas contain primary water. In slowly cooled, low-Ca boninitic dykes with 10–12% MgO from Heathcote, Victoria, plagioclase occurs only in the intrafasciculate-textured groundmass as blades and sheaves intergrown with similar calcic pyroxene. Textures of primitive low-Ca boninites are shown in Figures 1.5 and 1.6, and on the cover of this book.

Evolved Type 3 boninites, such as those from DSDP Hole 458, have been described in detail by Natland (1981). These form massive and pillowed flows with glass occurring only in quenched flow margins and pillow rims. Microphenocrysts include euhedral bronzite and augite, which are embedded in a groundmass formed by spherulitic clinopyroxene and plagioclase in more rapidly cooled examples, and microlitic strongly zoned clinopyroxenes and spherulitic plagioclase in more slowly cooled lavas. Olivine and clinoenstatite are absent, owing to the evolved, low Mg# nature of the Hole 458 lavas (Natland 1981). Slightly more primitive lavas in this suite from Guam, however, do contain olivine both as phenocrysts and in the groundmass. The higher CaO (Figs 1.2 & 1.3) and low Mg# in these lavas relative to most other Type 3 boninites is responsible for the coexistence of augite phenocrysts with olivine and bronzite. The crystallization sequence indicated is still olivine (+ chromite)→bronzite→clinopyroxene→plagioclase.

The Setouchi and Baja alkaline (Type 2) low-Ca boninites are generally much less glassy and less porphyritic than other low-Ca boninites. They are usually sparsely porphyritic (< 10 modal% phenocrysts) with olivine as the dominant phenocryst phase; compositions of olivine phenocrysts range from

Figure 1.5 Photomicrograph of orthopyroxene (untwinned) and clinoenstatite (twinned) phenocrysts in isotropic glass groundmass charged with strongly zoned pyroxene microlites. Cape Vogel boninite; field of view 3 mm.

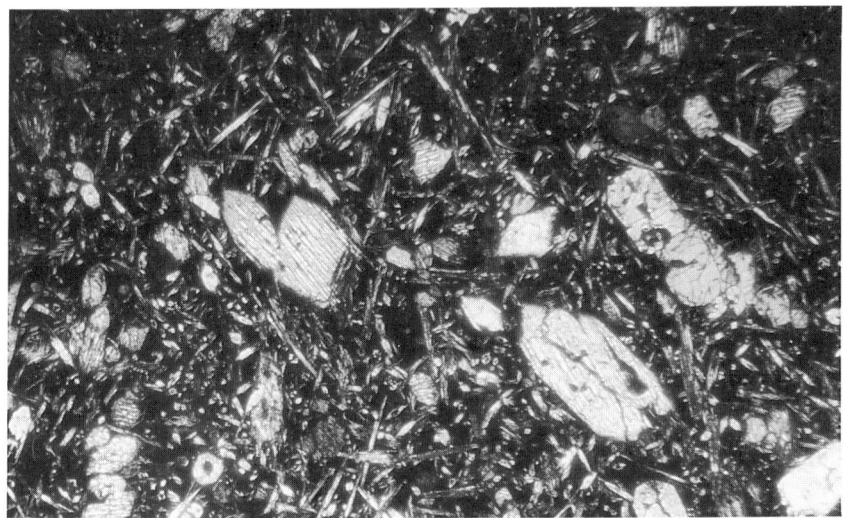

Figure 1.6 Another Cape Vogel boninite showing typical porphyritic texture, with polysynthetically twinned clinoenstatite phenocrysts and orthoenstatite phenocryst in plagioclase-free, glassy, pyroxene-charged groundmass.

Fo_{91-70} for the Setouchi lavas (av. Fo_{85-90}; Tatsumi & Ishizaka 1982b) and Fo_{92-87} for the Baja lavas (Rogers & Saunders 1989). In the Setouchi lavas, olivine may be accompanied by phenocrysts of bronzite (Mg # = 0.89–0.82; Al_2O_3 generally < 1.5%) or augite or both, and in the Baja lavas other phenocryst phases include occasional augite, less abundant bronzite, and rare plagioclase phenocrysts and quartz xenocrysts. Clinoenstatite phenocrysts are unknown in these alkaline low-Ca boninites. Evolved lavas from both areas contain hornblende phenocrysts.

1.3.3 High-Ca boninites

Detailed petrographic and mineral chemical data are available for high-Ca boninites from the Troodos Upper Pillow Lavas (Cameron 1985, Duncan & Green 1987, Flower & Levine 1987) and dredged lavas from the N Tongan forearc (Falloon et al. 1987, 1989). In this compilation, for the Troodos lavas we have used data for only the most refractory Upper Pillow Lavas (i.e. Group III of Cameron (1985) and Group C of Flower & Levine (1987)). These generally vitrophyric lavas have phenocrysts of magnesian olivine, chromite and magnesian orthopyroxene, microphenocrysts of olivine, chromite, orthopyroxene, pigeonite and diopside, and a quenched groundmass composed of clinopyroxene, plagioclase and occasional pargasite microlites in glass. Olivine phenocryst compositions range up to $Fo_{91.5}$, but are mostly in the range Fo_{90-88}; orthopyroxene phenocrysts fall mainly between En_{89} and En_{87} and have low Al_2O_3 contents ($< 1\%$). Clinopyroxene microphenocrysts have $Mg\# < 0.90$ and are often strongly zoned to subcalcic augite rims with lower $Mg\#$ and Wo contents. Complex clinopyroxene microphenocryst–groundmass compositional relationships exist in these lavas, with Mg-pigeonite cores (Wo_{9-14}; $Mg\# \sim 0.88$) to some microphenocrysts being separated by a sharp compositional gap from strongly zoned outer parts of the crystal which range from magnesian diopside to Fe-rich subcalcic augite (Duncan & Green 1987). Chromite in these lavas has $Cr\# = 0.65-0.85$ and $Mg\# = 0.70-0.47$, although most compositions cluster around $Cr\# = 0.72-0.78$ and $Mg\# = 0.65-0.55$.

1.3.4 Two test cases

To test the usefulness of this classification scheme, we selected two well studied suites of high-Mg, relatively high-Si rocks for assignment within the proposed classification. These suites are the following:

(a) Lavas from the Kambalda area of the Archaean Yilgarn Block, W Australia, described in detail by Redman & Keays (1985) and Arndt & Jenner (1986) and called by the former authors 'siliceous high-Mg basalts'. These contain phenocrysts, microphenocrysts and skeletal quench crystals of olivine in a groundmass of spherulitic-to-dendritic quench pyroxenes and less-abundant spherulitic-to-acicular plagioclase. Spherical felsic ocelli are occasionally present in these lavas. Primary olivine has been entirely replaced by secondary phases. We consider that these Kambalda lavas are entirely representative of similar high-Si, high-Mg basalts in other Archaean greenstone belts.
(b) Quench-textured micropyroxenite sills and dykes emplaced in Transvaal Sequence rocks penecontemporaneously with the Bushveld Complex are described in detail by Hatton & Sharpe (1989), and are considered to be the most likely candidates for parent magmas of the lower, ultramafic section of the layered Bushveld Complex (Davies et al. 1980, Cawthorn et al. 1981, Sharpe & Hulbert 1985). These sills are characterized by orthopyroxene phenocrysts ($Mg\# = 0.92-0.78$) and sparse subhedral-to-skeletal olivine phenocrysts (Fo_{90-72}) in a quench-textured groundmass composed of devitrified glass, comb-textured plagioclase and acicular or bladed orthopyroxene microlites. A peridotite sill compositionally related to the micropyroxenite suite has also been included in this data set.

Figure 1.7 shows compositional features of these two 'test' suites which are pertinent to their assignment within the proposed boninite classification scheme. The 'siliceous high-Mg basalts' (herein SHMB) are closest to high-Ca boninites in CaO/Al_2O_3; however, they have much higher TiO_2 ($> 0.55\%$), FeO^* (mainly 10–11%) and total alkali (2–3%) contents and lower SiO_2

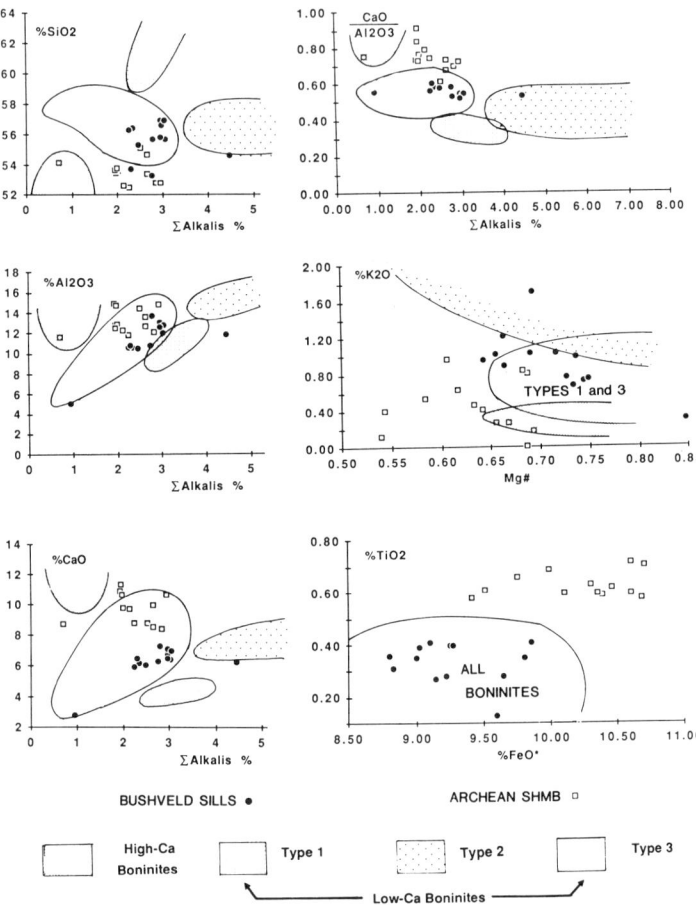

Figure 1.7 Same diagram as Figure 1.3, but showing only outlines of compositional fields for the four boninite magma types defined. Bushveld sills consistently plot in the low-Ca boninite suite, and are quite unlike the Archaean siliceous high-Mg basalts.

contents than the Troodos Upper Pillow Lavas. These features are all in keeping with results from recent detailed trace-element and isotopic studies of these Archaean lavas (Arndt & Jenner 1986; Sun et al. 1989) which concluded that they represent evolved komatiitic lavas extensively modified by combined crustal contamination and fractional crystallization processes.

This leads to the question of whether or not a suite of SHMB demonstrably derived from crustal contamination of peridotitic komatiites should be called boninites, even if they *are* extremely similar in major-element composition to modern high-Ca boninites which are manifestly *not* derived by crustal contamination of komatiites. Does a komatiite-derived basalt become a boninite when it crosses the 53% SiO_2 line at Mg# > 0.60? We prefer to restrict the term 'boninite' to lava suites whose parent magmas had $\geqslant 53\%$ SiO_2, and we note that there are magma suites, particularly in Proterozoic sequences, that evolve to members with $> 53\%$ SiO_2 at Mg# values still > 0.60 (e.g. Hall & Hughes 1987). These suites are probably better referred to as magnesian quartz tholeiites or siliceous high-Mg basalts than boninites. Had the source of any such suite partially melted at lower pressures (< 10 kbar?) or in the presence of more H_2O (if indeed H_2O was involved at all in the petrogenesis of these suites), resulting parent magmas might well have been boninitic. This topic is addressed further in Section 1.5.

The Bushveld sills are clearly a low-Ca boninite suite (Fig. 1.7), with CaO contents $< 7\%$ and CaO/Al_2O_3 values < 0.60; they have higher SiO_2 and K_2O contents and lower FeO^* and TiO_2 contents than SHMB, and, on a major-element basis alone, an origin for these rocks via extensive crustal contamination of peridotitic komatiites would seem to be ruled out. The parent magma of the Bushveld, if these sills are representative, was closer to a low-Ca boninite composition. The notably higher K_2O contents of the Bushveld sills relative to typical Types 1 and 3 low-Ca boninites erupted in forearc settings may be due to either derivation from a mantle source enriched by incorporation of ancient subducted sediments (see Hatton & Sharpe 1989), or some limited extent of crustal contamination accompanying ascent through cratonic crust and intrusion of this suite into the Transvaal Sequence.

1.4 Phase relationships and their significance

In this section we examine the major-element compositional variation displayed by the boninite suites shown in Figures 1.1 and 1.2 within a framework based upon experimental petrology and constraints imposed by factors influencing magma generation in the upper mantle. These factors, discussed by Green *et al.* (1987), include:

(a) $P-T$ variations (i.e. local geotherm) at the site of magma segregation,
(b) the amount and composition of C–O–H fluids at the site of partial melting, and
(c) the bulk composition (fertile or refractory?) of the source peridotite.

For boninitic magmas in general, experimental studies have indicated that they segregated from clinopyroxene-poor lherzolite or, more frequently, harzburgite sources at low pressures, probably less than 10 kbar for realistic H_2O contents in these magmas ($< 5\%$) (Green 1976, Tatsumi 1982, Cawthorn & Davies 1983, Jenner 1983, Umino and Kushiro 1989, Van der Laan *et al.* 1989). Furthermore, with the exception of the very similar Baja and Setouchi suites, the low TiO_2 contents of boninites (Figs 1.1 & 1.2), which result in high CaO/TiO_2 and Al_2O_3/TiO_2 values and low Ti/V and Ti/Sc values for these lavas, suggest derivation from a depleted, refractory magma source (Sun & Nesbitt 1978, Cameron *et al.* 1979, Duncan & Green 1980, Crawford *et al.* 1981, Hickey & Frey 1982, Beccaluva & Serri 1988). Such refractory source peridotites are likely to have been residua from earlier partial melting events which led to a depletion in basaltic components; thus the term 'second-stage melts' has been applied to boninitic magmas by Duncan & Green (1980, 1987) and Hamlyn *et al.* (1985) to convey the notion of prior depletion of the source peridotites of boninites.

Given that pressures are probably < 10 kbar during boninite segregation, the major factors influencing or controlling boninite magma compositions are, therefore, source composition, temperature, and the amount and nature of hydrous fluids at the site of partial melting. Figure 1.4 attempts to summarize the role of these key factors in boninite genesis.

To understand and portray the chemical variation in boninite suites and learn more of their petrogenetic significance, we use two slightly different types of molecular norm-based projections. In the first type, CIPW normative minerals are assigned to four end-members of a modified basalt tetrahedron of Yoder & Tilley (1962), as outlined by Green (1970). Two projections are used to illustrate important compositional variations and implications within the three-dimensional tetrahedron. These are (1) a projection from Di onto the base of the tetrahedron (Jd + CaTs)–Ol–Qtz (Fig. 1.8a), and (2) a projection from Ab + An onto the face Ol–Di–Qtz (Fig. 1.8b). The second type of projection used here to provide information about the source and crystallization conditions of the selected boninite suites is based on Walker *et al.*'s (1979) modification of the CMAS projections of O'Hara (1968) in which O'Hara's system is simplified to a tetrahedron based on plagioclase (Pl), olivine (Ol), quartz (Qtz) and diopside (Di). We use a projection in which compositions are projected from Pl onto the face Di–Ol–Qtz (Fig. 1.9a), and another from Qtz onto the face Pl–Ol–Di (Fig. 1.9b).

In addition to the compositional fields of the 10 boninite suites referred to above, Figures 1.8 and 1.9 also show the compositions of equilibrium partial melts of the depleted Tinaquillo peridotite at 20, 10, 8, 5 and 2 kbar (Jaques & Green 1980, Falloon & Green 1987, Falloon *et al.* 1989), plus the compositional spread of peridotites from the Ronda ultramafic massif; the latter are interpreted to be residues generated during extraction of a MORB-type parent magma (Frey *et al.* 1985).

The most important point to be gained from Figures 1.8a and b is that low-Ca boninites cannot be derived by anhydrous partial melting at any $P–T$ condition of a MORB source- or Tinaquillo-type peridotite. As discussed in Green *et al.* (1987) and Falloon *et al.* (1989), no partial melt composition can lie to the right of an olivine control line extrapolated through the source composition to the olivine apex. This conclusion is independent of the influence of C–O–H fluids on melting relationships (Green *et al.* 1987). The fields defined by low-Ca boninites demand a more refractory source, so that

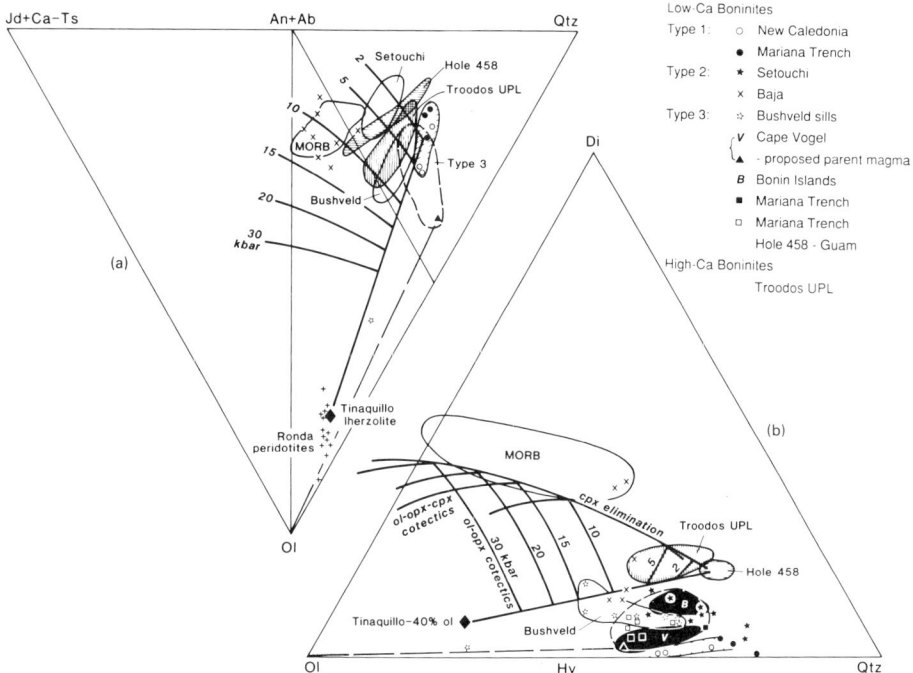

Figure 1.8 Projections from Di (a) and Pl (b) (see text for background) of compositional fields for boninite reference suites. Also shown are the equilibrium melt path compositions for partial melts of the slightly refractory Tinaquillo lherzolite (Falloon & Green 1987, Falloon et al. 1989). Important points (see text) to note from these projections are that the Cape Vogel parent magma estimated by Walker & Cameron (1983) is derived from a source considerably more refractory than the Tinaquillo lherzolite, and more refractory than the majority of the Ronda peridotites. No low-Ca boninites can be generated by anhydrous partial melting of a Tinaquillo lherzolite source at any pressure, but the Troodos high-Ca boninites may be generated from such a source under anhydrous or water-deficient conditions at $P < 8$ kbar.

the olivine control line drawn from the olivine apex through and beyond the source composition will lie well over towards the Ol–Qtz side of both projections. Only the most refractory Ronda spinel harzburgite (2 modal% clinopyroxene) is potentially capable of generating the calculated parent magma of the Cape Vogel boninites according to this projection, but even that sample has $CaO/Al_2O_3 = 0.88$ and olivine ($Fo_{91.6}$) suggesting it was less depleted than low-Ca boninite source peridotites. This is an important point, since some models for boninite genesis derive primitive boninitic magmas from shallow, depleted mantle which has yielded only MORB-type magmas during prior melting events (e.g. Model 2 of Hickey & Frey 1982, Bloomer & Hawkins 1987). Low-Ca boninite source peridotites must be considerably more refractory than any residuum of MORB extraction.

The projection from Di (Fig. 1.8a) does not bring out the compositional difference between Type 3 (e.g. Cape Vogel and Chichi-jima) and Type 1 (e.g. New Caledonia) low-Ca boninites, which plot in broadly overlapping compositional fields. This is due to the enhanced Na_2O contents of Type 1 boninites forcing these lavas to plot at higher Ab + An component than they would if they had similar Na_2O contents to the Type 2 boninites. Note from Figure

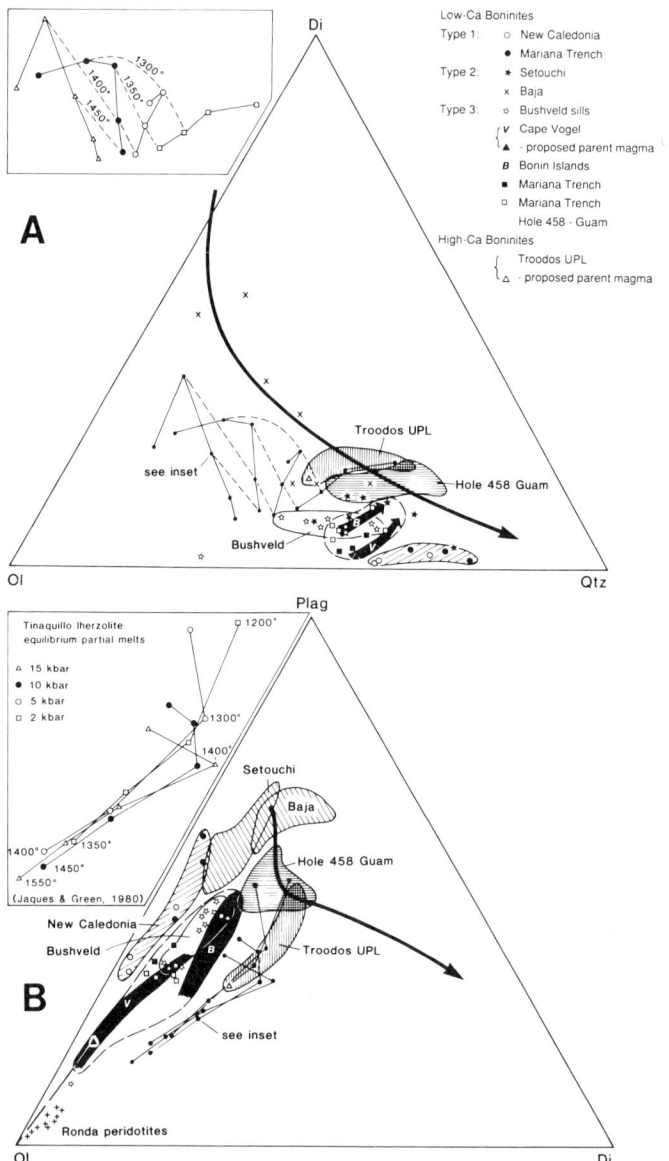

Figure 1.9 Projections after Walker *et al.* (1979) from Pl (A) and Qtz (B) showing fields for reference boninite suites, and 1 atm cotectics (heavy curves). Equilibrium partial melts of Tinaquillo lherzolite (Jaques & Green 1980) at 2, 5, 10 and 15 kbar are also shown. High-Ca boninites from Troodos can be generated by anhydrous or water-deficient partial melting of a Tinaquillo-like source at $P < 10$ kbar. No low-Ca boninites can be generated from Tinaquillo lherzolite sources at any pressure under anhydrous conditions. The equilibrium melts of Tinaquillo lherzolite in (B) fall in a fairly narrow band irrespective of pressure; since the low-Ca boninites fall at notably lower Di component in this figure, compositional differences between the boninite groups defined reflect extent of prior depletion of the sources involved in generating these magmas. Type 1 low-Ca boninites require more refractory sources than Type 3 boninites. See text for more detailed discussion.

1.8b, however, that Type 1 boninites require a source even more depleted in Di than Type 2 boninites, implying that Na_2O in Type 1 low-Ca boninites has been introduced in hydrous fluids which metasomatized the source peridotites before or during partial melting (see later).

High-Ca boninites, defined by the Troodos Upper Pillow Lavas (UPL), plot in Figures 1.8a and b in positions appropriate for low-pressure (< 8 kbar) anhydrous partial melts of Tinaquillo lherzolite or a somewhat more refractory source. This source was notably more Di- (i.e. essentially Cpx-) rich than sources of low-Ca boninites. Experimental studies of a potential Troodos UPL parental magma (Duncan & Green 1987) have shown that it could have segregated from a harzburgitic residuum at 7–8 kbar and 1360°C. However, the observed relationships between pigeonite and orthopyroxene microphenocrysts, and presence of minor quench pargasite in some Troodos UPL, suggested to these authors that 0.5–1% H_2O was present in the primary magmas of this suite.

The projection from Pl (Fig. 1.9A) also shows that low-Ca boninites cannot be derived from anhydrous melting of Tinaquillo lherzolite under any $P-T$ condition, and that the source which yielded low-Ca boninite suites was significantly more refractory than Tinaquillo lherzolite. Furthermore, the source for the Type 1 low-Ca boninites was more refractory than that which yielded the Type 3 low-Ca boninites.

Both these points are shown more clearly on the projection from Qtz (Fig. 1.9B), in which anhydrous partial melts of Tinaquillo lherzolite from 2 to 30 kbar lie within a narrow band displaced at significantly higher Di values than any of the low-Ca boninite suites. Clearly, pressure at the site of magma extraction cannot be responsible for the low Di content of low-Ca boninite suites; rather, the degree of depletion of the source peridotite determines the Di content of the partial melt, and the source for Type 1 low-Ca boninites was more refractory than that for Type 3 low-Ca boninites. The Setouchi and Baja suites plot at higher Pl component than other low-Ca boninites owing to their enhanced Na_2O contents, and also could not have originated by partial melting of a MORB- or Tinaquillo-source peridotite at any pressure. Further aspects of their petrogenesis are discussed in the next section.

1.5 Discussion

1.5.1 High-Ca boninites

Experimental petrological studies (Duncan & Green 1987, Van der Laan et al. 1989) have shown that low-pressure (< 10 kbar) partial melting of a depleted, clinopyroxene-poor lherzolite source (such as residuum following MORB extraction), under either anhydrous conditions or in the presence of a very small amount of water (probably < 0.5% H_2O in source), requires temperatures in the range 1250–1350°C and will generate high-Ca boninite magmas similar to the Troodos Upper Pillow Lavas. At Mg# = 0.75, these have Na_2O contents < 1%, CaO/Al_2O_3 values > 0.7 and CaO/Na_2O values from 4 to 10, and crystallize liquidus olivine and magnesiochromite with maximum Fo and Cr# values of 92 and 0.85 respectively. High CaO contents at high Mg# relative to other boninites inhibit clinoenstatite crystallization but permit crystallization of microphenocrystal calcic pyroxene, although augite appearance invariably follows that of olivine and low-Ca pyroxenes. Plagioclase crystallizes only as a late-stage phase in the groundmass of more slowly cooled lavas.

The residuum following segregation of this magma must be harzburgitic, and have considerably less clinopyroxene and lower CaO/Al_2O_3 than the source of the high-Ca boninites. To partially melt this residuum and yield a further magma batch under anhydrous or very low H_2O content conditions requires prohibitively high temperatures, probably well in excess of 1400°C.

1.5.2 Type 1 and Type 3 low-Ca boninites

Such a refractory source may, however, melt further if its solidus temperature is lowered by influx of relatively large volumes of hydrous fluids. We argue that Types 1 and 3 low-Ca boninites are generated in this manner, with Type 1 lavas being produced from more refractory harzburgitic sources than Type 3 boninites, as indicated by their lower CaO/Al_2O_3 and lower CaO and FeO* contents. If, as we claim, low-Ca boninites are derived by hydrous partial melting of residual harzburgites left after segregation of high-Ca boninitic magma, then the high alkali (Na_2O in particular) contents of the low-Ca boninites relative to Troodos high-Ca boninites requires explanation. Both Na_2O and K_2O contents should decrease further with increased (second- or third-stage) partial melting, to values well below those shown by the high-Ca boninites. In fact, they are notably higher, especially for the Type 1 low-Ca boninites. The high Na_2O Type 1 low-Ca boninites also contain significantly higher SiO_2 contents than Type 3 boninites or high-Ca boninites.

A relatively simple explanation of the high SiO_2 contents of Type 1 low-Ca boninites involves incongruent melting of enstatite. As is obvious from Figures 1.8 and 1.9, and as discussed by Van der Laan *et al.* (1989), most primitive boninite compositions, whether of high-Ca or low-Ca type, plot outside the Alkemade triangles defined by the dominant mantle melting assemblage Ol–Opx–Cpx. It is concluded, therefore, that incongruent melting of enstatite (to yield forsteritic olivine and a more silicic liquid) must have been involved in the generation of boninitic magmas. Van der Laan *et al.* (1989) argue that the extent of this low-pressure, peritectic style of partial melting will be enhanced by increased H_2O contents. Thus, since petrogenesis of low-Ca boninites involves significantly more H_2O than is involved in production of high-Ca boninites, it is to be expected that increased extents of peritectic melting for the more H_2O-rich sources result in production of more SiO_2-rich magmas. By implication, Type 1 low-Ca boninites must have had more H_2O involved in their generation than the less SiO_2-rich Type 3 boninites. Such an explanation seems intuitively reasonable, but does not provide an obvious explanation for the high Na_2O contents of Type 1 boninites.

Experimental studies by Kushiro et al. (1968), Kushiro (1972), Nakamura & Kushiro (1974) and Ryabchikov et al. (1982) demonstrated conclusively that SiO_2 and Na_2O can be effectively transported in the upper mantle by high-temperature aqueous fluids. The solubility of SiO_2 increases with increasing pressure and temperature, and at 15 kbar substantial amounts of SiO_2 component exist in the aqueous fluid coexisting with enstatite and forsterite. At constant temperature, the solubility of Na_2O increases with *decreasing* pressure. We believe that both SiO_2 and the alkalis were introduced into the refractory peridotite sources of boninitic lavas in the hydrous fluids which initiated partial melting. Greater volumes of fluid for a given volume of peridotite are required to produce partial melting of increasingly refractory sources, so that the more refractory Type 1 boninites have higher SiO_2, Na_2O (2–2.5% Na_2O) and lower $CaO/Na_2O < 2.5$ than Type 3 low-Ca boninites (0.7–1.4% and 4–10 respectively) at any Mg#. This interplay between the degree of depletion of a peridotite source and the amount of hydrous fluids required to initiate partial fusion is perhaps the key factor in understanding the compositional variation of boninite suites, and is shown diagrammatically in Figure 1.4. If high-temperature aqueous fluids are capable of controlling to a significant degree the major-element composition of boninites, then the trace-element and isotopic signatures of boninites demand explanation in the same petrogenetic scenario.

1.5.3 *Type 2 (alkali) low-Ca boninites*

As noted above, the Setouchi and Baja Type 2 low-Ca boninites share many important major-element compositional features (high Na_2O, low FeO^* and CaO, and $CaO/Al_2O_3 < 0.55$) with Type 1 low-Ca boninites. This suggests that they are also derived from a very refractory peridotite source. Their SiO_2 contents at Mg# = 0.70–0.75 are variable (56–60%), but are lower than those of Type 1 boninites by at least 2%. In contrast, their Na_2O contents over this same Mg# range are 1% (Setouchi) to 3% (Baja) higher, and their Al_2O_3 contents are nearly 4% higher than in Type 1 low-CaO boninites (Figs 1.1 & 1.3). These differences are unlikely to be due to differences in either the amount or the composition of the aqueous fluids responsible for initiating melting in the refractory mantle sources of these lavas. A more reasonable explanation is that the parent magmas of the Type 2 low-Ca boninites were generated by much lower degrees of partial melting than were involved in production of the Type 1 boninites.

Melting experiments on a near-aphyric Type 2 low-Ca boninite from the Setouchi belt (Tatsumi 1981) under H_2O-undersaturated (8% H_2O) conditions showed that olivine and orthopyroxene can coexist on the liquidus of the composition used (9.6% MgO) at 11 kbar and 1110°C; at lower, more realistic, H_2O contents this composition would be in equilibrium with harzburgite at lower pressures still, but at higher temperatures. These results indicate that the liquidus temperatures of Type 2 low-Ca boninites are at least 50–100°C lower than for Type 1 and Type 3 boninites (> 1225°C; Van der Laan *et al*. 1989, Umino & Kushiro 1989), supporting the hypothesis that Type 2 boninites may have been generated by lower degrees of partial melting than other low-Ca boninites. The relatively high contents of K_2O, TiO_2 and LILE in both suites of Type 2 boninites (Tatsumi & Ishizaka 1982a,b, Saunders *et al*. 1987, Rogers & Saunders 1989) further support this hypothesis.

Some of the compositional variability of the Setouchi and Baja lavas (e.g. in SiO_2, K_2O and LILE) might be explained by limited crustal contamination during ascent, since both suites were erupted through relatively thick continental-type crust, whereas other boninite suites were erupted through thin oceanic-type crust and had less opportunity for pooling, assimilation and fractionation. However, both Ishizaka & Carlson (1983) and Saunders *et al*. (1987) rule out crustal contamination as having played a significant role in the compositional evolution of Setouchi and Baja Type 2 low-Ca boninites. We note, however, that data in Tables 3 and 4 of Ishizaka & Carlson (1983) show that 5% crustal contamination (by Sari Granodiorite, taken to be representative of the lower crust of SW Japan) of a primitive Type 2 boninite magma at the high ε_{Nd}–low $^{87}Sr/^{86}Sr$ end of the Setouchi spectrum can produce a resultant magma with Nd–Sr isotopic ratios still well within the range reported by these authors for Setouchi low-Ca boninites (ε_{Nd} = + 1.8 to − 2.5; initial $^{87}Sr/^{86}Sr$ = 0.70487–0.70537).

1.5.4 *Coexistence of low-Ca and high-Ca boninites*

We have argued above that the major-element compositional spectrum of boninites is generated by a subtle interplay between degree of depletion ('refractoriness') of the source peridotite, the composition and amount of hydrous fluids which initiate melting of otherwise refractory peridotite, and the ambient temperature at the site of partial melting (Fig. 1.4). Individual boninite suites clearly belong dominantly to either high-Ca or low-Ca suites (Fig. 1.3). However, a very important observation to take into account in models for boninite petrogenesis is that, in many boninite sequences, both low-Ca and high-Ca boninites coexist, albeit that one or the other type may be volumetrically far more significant than the other. For example, Umino (1986) showed that whereas our Type 3 low-Ca boninites dominate the Chichi-jima lava pile, sparse high-Ca boninites (his Type 4 petrographic group of

Chichi-jima lavas) do occur in this sequence. These high-Ca boninites have microphenocrysts of calcic pyroxene and bronzite, and as expected show lower SiO_2 and higher CaO (10% versus 7–8% at Mg# ~ 0.72) contents than low-Ca boninites in the same sequence. Similarly at Cape Vogel, while the dominant lavas are Type 3 low-Ca boninites with CaO/Al_2O_3 values of 0.54–0.60, several lavas reported by Dallwitz (1968) and Jenner (1981) have values around 0.66–0.71, and at any Mg# notably lower SiO_2 and higher CaO (again by around 2%) than the dominant lavas in the sequence. These higher-Ca lavas, although still plotting as low-Ca boninites in Figures 1.1 and 1.3, trend strongly towards high-Ca boninite compositions. Also from the Cape Vogel sequence is at least one sample (173 of Walker & McDougall 1982) with significantly higher SiO_2 (64.2%) and lower CaO (4.5%) at Mg# = 0.75 than other Cape Vogel boninites; this sample shows strong affinities to Type 1 low-Ca boninites. This compositional variation within single boninite sequences serves to emphasize the continuum existing in nature between the series defined here.

Relative ages of lower-Ca and higher-Ca boninites where they coexist at Cape Vogel and Chichi-jima are unknown. However, in the Cambrian Heathcote and Mount Wellington greenstone belts in SE Australia, time relationships between coexisting boninite suites can be determined with certainty (Crawford et al. 1984, Crawford & Cameron 1985, Crawford & Keays 1987). In both greenstone belts there is an upward passage from higher-Ca, less refractory boninites at the base of the succession to Type 1 low-Ca, exceptionally refractory clinoenstatite-bearing lavas at the top. The latter have microphenocrysts of clinoenstatite growing around reacted and resorbed phenocrysts of olivine (Fo_{94}), and chromite (Cr# = 0.97) occurs as phenocrysts and inclusions in olivine and clinoenstatite (Crawford 1980). Using the olivine–chromite geothermometer of O'Neill & Wall (1987), these highly refractory lavas had liquidus temperatures from 1330 to 1350°C, whereas those lower in the sequence probably crystallized around 1200–1250°C. This is compatible with a petrogenetic model in which successive batches of progressively lower-Ca, more refractory boninites are derived from increasingly refractory harzburgitic sources following repeated or continuous influx of solidus-lowering hydrous fluids. In both the Heathcote and Mount Wellington greenstone belts, the uppermost, most refractory boninites have the lowest initial ε_{Nd} values and strongest LREE enrichment (Nelson et al. 1984). If the hydrous fluids initiating melting are also the LREE transporting agent (Hickey & Frey 1982, Cameron et al. 1983), then the strongest LREE enrichment and lowest ε_{Nd} in the most refractory (in terms of TiO_2 or HREE levels) lavas at the top of the sequence indicate that greater water/rock ratios were required to melt increasingly refractory residues of successive melting events. This cannot happen within a single diapir via sequential or dynamic partial melting, since the majority of LILE will be removed from the diapir during the first major melt extraction. Subsequent melts will be increasingly depleted in LILE, the opposite of that observed in the boninite sequences described above.

1.6 Petrogenesis of boninites: a summary

1.6.1 Overview

In summary, we believe that boninites are derived from source peridotites more refractory than residua from MORB generation. To melt this depleted peridotite requires either higher temperatures than those required for the prior melting event, or the influx of solidus-lowering hydrous fluids. If only very small volumes of hydrous fluids invade this peridotite, then high-Ca boninites such as the Troodos Upper Pillow Lavas suite may be generated at $P < 10$ kbar and $1250-1350°C$. The residuum from this melting event will not melt further without input of larger volumes of hydrous fluids (and extra heat?) than were involved in high-Ca boninite production.

If both high temperatures and a supply of hydrous fluids to the site of partial melting can be maintained, then further partial melting (at pressures probably from 2 to 8 kbar) will generate first Type 3 low-Ca boninites and then Type 1 low-Ca boninites. Further access of hydrous fluid is required to generate the latter from the very refractory residues of the Type 3 boninite-producing melting event. The volume of Type 1 low-Ca boninites generated would be expected to be very small relative to volumes of high-Ca and Type 3 low-Ca boninites.

Partial melting of a very refractory peridotite source at higher pressures (?8–15 kbar) and lower temperatures ($1100-1150°C$) will generate lower-degree Type 2 low-Ca boninite partial melts, such as the Baja and Setouchi boninites. Crustal thicknesses under both Baja California and SW Japan demand upper-mantle magma segregation at pressures of at least 10 kbar.

1.6.2 Identity of the enriched and depleted source components in boninite genesis

All published boninite petrogenetic schemes agree that the source of these magmas must be a depleted, refractory peridotite which is enriched in LILE and some other elements (e.g. Si, Na, Zr) by a hydrous fluid prior to or during partial melting which generated the boninitic magma (e.g. Sun & Nesbitt 1978, Crawford *et al.* 1981, Hickey & Frey 1982, Tatsumi 1982, Cameron *et al.* 1983, Bloomer 1987, Beccaluva & Serri 1988).

THE DEPLETED COMPONENT
From considerations of major-element chemistry and phase relationships of

eight suites of boninites, we have shown that the source peridotites of high-Ca boninites may be refractory residua from prior melting events involving extraction of MORB parent magmas. Clinopyroxene was present in these sources prior to melting (see also Beccaluva & Serri 1988), as indicated by the relatively high CaO/Al_2O_3 (0.7–1.0) and high Sc contents (av. 40 ppm at Mg# = 0.70 for Troodos Upper Pillow Lavas) of high-Ca boninites; it was, however, eliminated during generation of the high-Ca boninite parent magma. Low-Ca boninites have low to very low CaO/Al_2O_3, ranging from 0.45–0.7 for Type 3 lavas to 0.28–0.5 for Type 1 low-Ca boninites. This suggests only a very minor role for clinopyroxene in genesis of the Type 3 low-Ca boninites (Sc = 30–35 ppm at Mg# = 0.70), and that clinopyroxene might have been absent from the sources of Type 1 low-Ca boninites (Sc = 20–25 ppm at Mg# 0.70; Cameron 1989). Note that the Setouchi Type 2 low-Ca boninites have an average Sc abundance of 22 ± 4 ppm (Tatsumi & Ishizaka 1982b); this, together with their low CaO contents and low CaO/Al_2O_3 values, supports the above model that they are derived from very refractory, clinopyroxene-free harzburgitic sources similar to sources of type 1 low-Ca boninites.

THE ENRICHED COMPONENT(S)

A detailed discussion of the nature and origin of the one or more enriched components involved in boninite genesis is beyond the scope of this chapter. We wish to stress several points, however, which must be taken into account in attempting to identify these components and their role in boninite petrogenesis.

(1) We are convinced that significant amounts of at least Na_2O and SiO_2 were transported into low-Ca boninite sources by hydrous fluids which initiated partial melting. If high-temperature aqueous fluids dissolve and transport SiO_2, it is highly likely that they will also transport LILE, including REE (Tatsumi et al. 1986), which may reside largely on grain boundaries in upper-mantle peridotites (e.g. Frey & Green 1974). Kyser et al. (1986) and Dobson & O'Neil (1987) have shown that H_2O involved in boninite genesis is probably derived from altered, subducted oceanic lithosphere. There is strong evidence in some suites (e.g. Pb isotopes in Chichi-jima and Victorian Cambrian (Howqua) boninites; Dobson & Tilton 1989; B. L. Gulson pers. comm. 1987) of some LILE enrichment derived from an ancient crustal component (?sediments) in the dehydrating slab. As shown by Nelson et al. (1984), this subducted sedimentary component must have had ε_{Nd} values at least as low as -8.

However, detailed trace-element and isotopic studies of boninites (Jenner 1981, Hickey & Frey 1982, Cameron et al. 1983, McCulloch & Cameron 1983, Sharaskin et al. 1983, Bloomer 1987, Hickey-Vargas & Reagan 1987, Beccaluva & Serri 1988, Cameron 1989, Fallon et al. 1989, Hickey-Vargas 1989, Dobson & Tilton 1989) have all suggested that the enriched component is unlikely to be simply fluids derived by dehydration of altered subducted ocean crust (including sediments). There is strong evidence for a second enriching component, very similar to that responsible for enrichments seen in ocean island basalts (OIB) and in spinel lherzolite nodules in alkali basalts. The nature of this second enriching component is poorly understood. Jenner (1981), Hickey & Frey (1982), Hickey-Vargas & Reagan (1987) and Beccaluva & Serri (1988) suggested that it may be silicate partial melts derived from OIB source mantle, which invaded refractory peridotite in the mantle wedge presumably before initiation of subduction; subsequently, this mantle with OIB-type enrichment was further modified by hydrous fluids derived from the slab.

(2) Clues to the origin of the enriched component(s) in boninite genesis come from consideration of the unusual compositional features of the Nepoui (New Caledonia) Type 1 low-Ca boninites. These were clearly derived from a highly refractory source peridotite, as indicated by their very low Sc, CaO and FeO* contents and low CaO/Al_2O_3 values (< 0.5), yet they have higher SiO_2 and Na_2O contents, and stronger LREE enrichment ($(La/Sm)_N \sim 2$) than any Type 3 low-Ca boninites or high-Ca boninites (Cameron et al. 1983, Cameron 1989). Despite strong LREE enrichment, Nepoui lavas have MORB-type ε_{Nd} values ($+8$ to $+10$) and low initial $^{87}Sr/^{86}Sr$ (0.7034–0.7035). This suggests that the hydrous fluids which introduced SiO_2, Na_2O and LILE (including REE and Zr) and initiated partial melting were derived almost entirely from a young, subducted MORB source modified only by limited interaction (as indicated by slightly radiogenic Sr relative to MORB) with sea water. There is no isotopic evidence for the existence of the hypothetical OIB-type enriched component in the source of these lavas. Therefore, it is important to note that the component in boninite genesis which enriches LREE (and Zr, leading to the characteristic low Ti/Zr values (< 50) of low-Ca boninites) *need not* be assigned an OIB enrichment-type origin. Apparently, high-temperature aqueous fluids transporting SiO_2 and Na_2O from the slab (and near-slab peridotites in the mantle wedge) can effectively extract and move LREE and Zr from basalts in subducted oceanic crust, and introduce these elements into the refractory boninite source.

We suggest, therefore, that two components may enrich the refractory mantle source of low-Ca boninites. One of these is ubiquitous, being derived via extraction by slab-derived, superheated aqueous fluids of Si, Na, LILE including LREE, and Zr from subducted MORB. These are transported upwards into the refractory mantle wedge and initiate partial melting if ambient wedge temperatures exceed the solidus. If the oceanic crust from which these hydrous fluids were derived and equilibrated was typical MORB, fluids inherited ε_{Nd} values from $+8$ to $+10$. The second enriched component in boninite genesis had significantly lower $\varepsilon_{Nd}(< -8)$ and a Pb isotopic signature suggestive of derivation from an ancient sedimentary component in the slab; it also was probably LREE-enriched and had a relatively low Ti/Zr

(< 50?). Involvement of the latter component in the genesis of some boninite suites, such as the New Caledonian (Nepoui) and Guam–Hole 458 suites (Meijer & Hanan 1981, Hickey-Vargas 1989) was insignificant, whereas it was an important component in the genesis of other suites, such as the Chichi-jima and Howqua (Victorian Cambrian) suites.

Finally, we note that Hole 458 and Guam boninites have $^{206}Pb/^{204}Pb$ higher than MORB, and closer to values for NE Pacific OIB (Hickey-Vargas & Reagan 1987). However, as Bloomer (1987) has pointed out, Pb isotope data for altered Jurassic MORB in the W Pacific (Meijer 1976) are very similar to the Mariana forearc–Guam boninite Pb isotope data; since it is precisely this crust that is being subducted around the W Pacific rim, pre-subduction OIB-type source enrichments in upper mantle eventually trapped in the wedge may not be necessary.

1.7 Tectonic setting of boninite generation

1.7.1 Physical conditions of boninite generation

Melting studies of boninites summarized earlier in this chapter indicate that magmas with these high-SiO_2, high-MgO characteristics can only be extracted from peridotitic upper mantle at very shallow levels, certainly shallower than 50 km and probably shallower than 30 km. Requisite temperatures are $> 1100°C$ for Type 2 low-Ca boninites, and probably $> 1200°C$ for other low-Ca and high-Ca boninites. Water is a key ingredient in low-Ca boninite genesis, and a small amount of water (< 0.5%) is probably also present in the sources of high-Ca boninites. Under anhydrous conditions, magmas with high SiO_2 and high MgO contents approaching boninitic compositions can only be generated at $P < 5$ kbar, corresponding to crustal depths for island arc or continental settings.

1.7.2 Tectonic scenarios for boninite generation

Foldbelts are produced by continent–continent, continent–arc or arc–arc collisions, at the final phase of a Wilson cycle of ocean-basin opening and closing. In theory, boninites incorporated in ophiolites in any particular foldbelt could have been generated during any one of the preceding three phases of the Wilson cycle which produced that foldbelt; that is, during (i) initial continental rifting and rupturing, (ii) ocean-floor spreading at a mid-ocean-ridge spreading centre, or (iii) subduction-related magmatism (including arc, forearc and backarc settings) as the ocean is closing.

Below, we construct idealized but actualistic scenarios for each of these pre-collisional stages of the Wilson cycle. For each stage, we determine whether the prerequisite conditions for boninite generation might be attained, and we predict the stratigraphic sequence of lavas and associated rocks in which the boninites should occur, if at all.

1.7.3 Stage 1: Continental rifting

Magmatism accompanying continental rifting results from partial melting of one or more diapirs of MORB-source mantle rising essentially adiabatically beneath gradually attenuating continental crust. Earliest erupted lavas tapped from the diapir(s) at relatively deep levels (100–60 km) may be slightly alkaline, or more frequently, T-type (transitional) MORB compositions (Fodor & Vetter 1984). Magmas subsequently generated by partial melting of the diapirs at shallower levels (60–40 km) approach normal (N-) MORB compositions, until the continental crust ruptures, and a steady-state spreading centre is established.

Hydrous fluids are unlikely to be present in the source areas of these anhydrous, tholeiitic rift-related magmas. Therefore, relatively siliceous high-Mg magmas could be produced only if the diapirs which yielded MORB-type lavas by partial melting at deeper levels were able to continue ascending and maintain their heat to levels perhaps as shallow as 10–15 km (Duncan & Green 1980). In this situation of second-stage melting, relatively minor volumes of strongly depleted magnesian quartz tholeiites (51–53% SiO_2) might be generated, and residual refractory harzburgite will freeze close to the moho. Such lavas are compositionally transitional between tholeiitic basalts and boninites. If diapirism associated with rifting progresses to a steady state, then new, relatively fertile diapirs of MORB-source mantle will continuously supply large volumes of MORB, and the chance for the generation or preservation of these second-stage melts will be very limited. We suggest, therefore, that high-Si, high-Mg magmas will only be generated in rifts where supply of MORB-source diapirs is aborted, and spreading jumps to a more favourable, successful site.

The idealized sequence of magmatic events accompanying and driving continental rifting predicts a stratigraphic pile in which early alkaline and transitional lavas are overlain by more abundant T- to N-MORB (Fig. 1.10a), which may be overlain, in turn, by volumetrically subordinate strongly depleted boninitic basalts (Fig. 1.10b). Chemically, these lavas will approach high-Ca boninites; Na_2O and TiO_2 contents will be very low, and LREE in the parent magmas will be highly depleted. Any LILE enrichment will be limited to the earliest-erupted second-stage melt lavas, which have the most opportunity to digest continental crust. Subsequent lavas are likely to ascend through conduits chemically 'insulated' by earlier magmas, and may erupt with unmodified LILE contents.

Lavas produced during the transitional rifting stage to ocean opening will be most likely covered by thick sedimentary prisms; they can only be studied if deep continental-margin drilling has reached basement (e.g. Fodor & Vetter

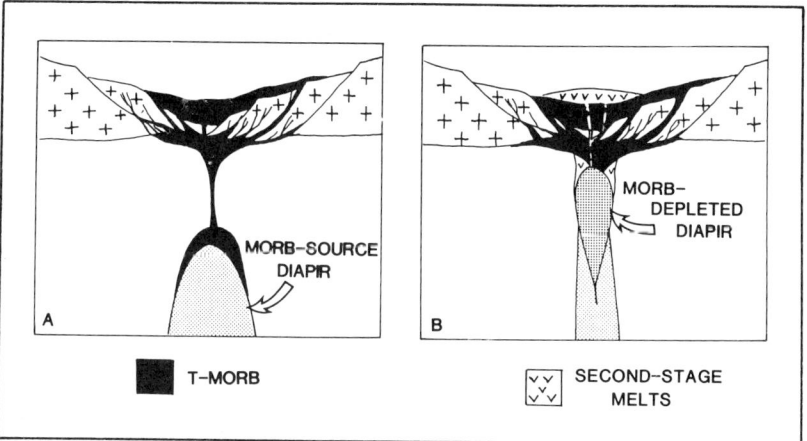

Figure 1.10 Hypothetical scenario to generate second-stage metals in an intracontinental rift. (A) Ascent and partial melting of a MORB diapir beneath an actively attenuating continental rift may generate voluminous T-MORB. If regional plate kinematics or local crustal mechanical factors are unfavourable for the development of a steady-state spreading centre, the ridge may jump elsewhere in the crust. (B) An 'unsupported' diapir may continue to rise rapidly and partially melt a second time to yield magnesian quartz tholeiites with very low TiO_2 contents and strongly LREE-depleted REE patterns; these second-stage melts are transitional to high-Ca boninites.

1984), or if the passive margin produced during successful rifting eventually collides with an arc or continent, to be uplifted and exposed during orogenic deformation (Crawford & Berry 1989).

Lava piles representing the rift-to-ocean stage of a Wilson cycle are known in early Proterozoic (Canada: Francis *et al.* 1983), late Proterozoic–Palaeozoic (Newfoundland: Strong & Dostal 1980; Tasmania: Brown & Waldron 1982, Crawford & Berry 1989; Greenland: Kalsbeek & Jepsen 1984) and Mesozoic (Bertrand *et al.* 1982) foldbelts. In each of these occurrences, the lava sequence is as predicted. Early transitional or alkaline lavas pass upwards into a tholeiite pile which includes lower T-type MORB overlain by more depleted tholeiites approaching or attaining N-MORB compositions.

Second-stage melts compositionally transitional to boninite with strongly LREE-depleted REE patterns are present at several localities in late Proterozoic rift sequences in W Tasmania. On King Island, least altered picrites (underlain by T-type MORB tholeiites) with 15–20% MgO contain around 50% SiO_2 (cf. 46% for Baffin picrites; Francis 1985), have 0.2–0.3% TiO_2 and $(La/Sm)_N \sim 0.4$ (Brown & Waldron 1982) and are clearly second-stage melt lavas. Further south, at Double Cove, T-MORB are overlain by highly LREE-depleted basalts with around 52% SiO_2 and 0.3–0.6% TiO_2 (Crawford & Berry 1989). In the mid-Proterozoic Cape Smith Foldbelt in Canada (Francis *et al.* 1983), similar but somewhat less depleted lavas constitute the Chukotat Group, and overlie transitional alkaline to T-MORB tholeiites of the Povungnituk Group. Francis *et al.* (1983) proposed that the Chukotat lavas may have been generated by further melting upon continued adiabatic ascent of the diapir(s) which yielded the Povungnituk lavas during partial melting at deeper levels. The high Ni and Cr, and very low LREE contents of the Tasmanian and Chukotat tholeiites rule out the possibility that the relatively high-Si character of these magnesian lavas was produced by crustal contamination of more typical (i.e. not second-stage melt) tholeiitic magmas.

In summary, it is possible that basalts transitional to high-Ca boninites could be generated in a continental rift during aborted, earliest stages of ocean opening. In such tensional settings, subordinate volumes of depleted magnesian quartz tholeiites (SiO_2 of 50–53%) may be generated as a final melt increment extracted from a diapir which continued to rise to shallow levels after yielding earlier MORB-type basalts during partial melting deeper in the upper mantle (see Fig. 1.12b). During the ocean closing phase of the Wilson cycle, it is inevitable (except perhaps during arc–arc collisions) that a passive margin bearing a rift-generated basalt pile (which may include some 'second-stage' depleted magnesian quartz tholeiites) will collide with an arc or continental block, generating a foldbelt. Taking into account geochemical evidence and the stratigraphic sequence of lavas and associated sediments, there is little likelihood that depleted magnesian quartz tholeiites in such rift-produced lava piles incorporated within a foldbelt will be misidentified as boninite products of convergent plate boundary magmatism.

1.7.4 Stage 2: Ocean-floor spreading

During the ocean-floor spreading phase of the Wilson cycle, steady-state ascent of MORB-source diapirs from deeper in the mantle provides a near-continuous supply of MORB to spreading centres, with magma segregation occurring mainly at depths exceeding 50 km (Green *et al.* 1979). Continued upward ascent and second-stage melting of these diapirs is likely to be very limited, since new diapirs are regularly arriving at depth beneath the spreading centre. However, MORB with SiO_2 contents in excess of 50%, and with relatively low Na_2O and TiO_2 contents are known from the ocean basins (see Falloon & Green 1988), and may represent low-P (~10 kbar) melts.

To explain the presence of very calcic plagioclase megacrysts and xenocrysts, and also high-Mg#, high Ca/Na melt inclusions recorded from some MORB, Duncan & Green (1980, 1987) proposed that low-pressure (< 10 kbar) second-stage melting of diapirs which had already yielded picritic MORB might occur beneath spreading centres. Continued adiabatic ascent of the MORB-depleted diapirs might occasionally occur, leading to a further ~10% partial melting and production of magnesian quartz tholeiites with high CaO/Na_2O, low TiO_2 and very depleted LILE contents. Such magmas are rarely preserved, probably due to mixing with more typical MORB in crustal magma chambers. However, the recent discovery (Danjushevsky *et al.* 1987) in some MORB of magnesian olivine-hosted melt inclusions with 52.8% SiO_2,

0.3%TiO_2 and $CaO/Na_2O = 12.6$ demonstrates that second-stage melts transitional to boninite compositions do occur in the ocean basins, as proposed by Duncan & Green (1980).

Although we doubt that major ocean-basin crust can be incorporated in foldbelts during plate collisions, there is no reason why the process described above for generations of second-stage melts cannot also take place in backarc-basin spreading centres. Even so, their rare, very limited occurrence in a pile of volumetrically dominant MORB-type lavas will render second-stage melts (with compositions transitional to high-Ca boninites) generated at mid-ocean-ridge or backarc-basin spreading centres immediately identifiable as such; their misidentification as subduction-related boninites is most unlikely.

1.7.5 Stage 3: Subduction-related magmatism

A consensus of opinion exists that both high-Ca and low-Ca boninites are products of partial melting above a subducted slab of oceanic lithosphere at convergent plate boundaries. However, the apparent absence of boninitic lavas in many well studied island arcs, such as the Aleutians, Vanuatu, Sunda and Lesser Antilles arcs, indicates that the requisite conditions for boninite generation are probably not met with at all convergent plate boundaries (Crawford et al. 1981). The missing ingredient in such settings is almost certainly the requirement of very high temperatures ($> 1100°C$) at very shallow levels in the mantle wedge (< 50 km). All geophysical models for the thermal structure of subduction zones in island arc settings suggest that ambient temperatures in the forearc wedge at levels shallower than 50 km should be $< 800°C$, and probably $< 500°C$ (e.g. Hsui & Toksoz 1979). To generate boninites in forearcs, where they occur in the Bonin–Mariana arc and the N Tonga arc, a mechanism capable of raising the shallow upper-mantle temperature to $> 1100°C$ must be identified.

For the Bonin–Mariana setting, two different models have been proposed to provide the extra heat required to generate boninites at depths shallower than 50 km in the forearc mantle wedge. The first of these, suggested by Cameron et al. (1979) and expanded by Meijer (1980), envisages boninites as a magma type generated during the earliest stages of initiation of a subduction zone. Meijer's (1980) model, adopted by Hickey & Frey (1982), Hawkins et al. (1984) and Hickey-Vargas & Reagan (1987) is based largely on DSDP Hole 458 in the Mariana forearc (late Eocene–basal Oligocene boninitic lavas overlying arc tholeiites) and on Guam (late Middle Eocene boninites interbedded with and succeeded by arc tholeiites). The model suggests that the change in spreading direction of the Pacific plate at $\sim 44-42$ Ma resulted in initiation of a subduction zone along a former major N–S transform now broadly marked by the Palau–Kyushu Ridge. West of the new trench lay an active spreading centre segment of the West Philippine Sea, orientated approximately NW–SE

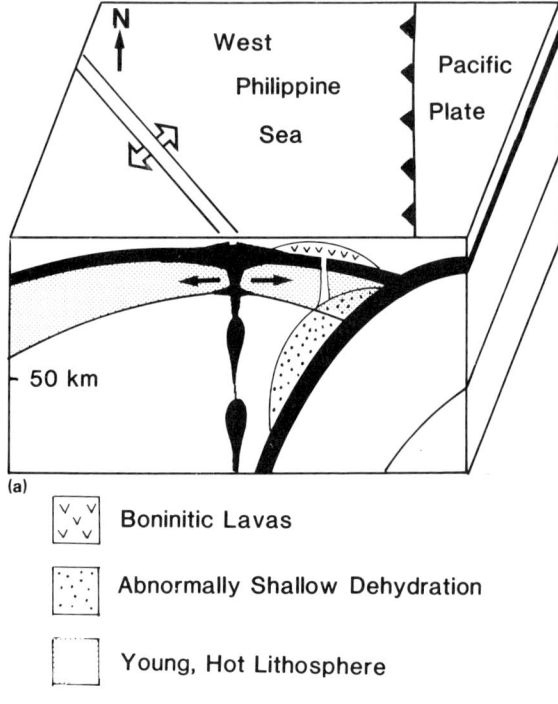

Boninitic Lavas

Abnormally Shallow Dehydration

Young, Hot Lithosphere

(Seno & Maruyama 1984). Subduction of old, cold oceanic crust into this young, hot subspreading centre upper mantle resulted in shallower-than-normal dehydration of altered oceanic crust and ingress of water into peridotites which had yielded MORB parent magmas only a short time earlier (Fig. 1.11a). Arc tholeiitic magmas were generated by shallow melting of this relatively refractory sub-oceanic upper mantle, and boninitic magmas were generated from the residuum of the arc tholeiite-producing melting event.

To provide a source for the extra heat required to generate boninites in this setting *after* arc tholeiites (in DSDP Hole 458), Hickey & Frey (1982) adopted the mechanism central to Model 2, suggested by Crawford et al. (1981). This model involves heat conduction from an ascending MORB-source mantle diapir in the earliest stages of backarc spreading. This diapir, which eventually supplies backarc-basin basalts that rift the arc in two, must initially pass through hydrous, depleted sub-oceanic lithosphere within the metasomatic halo of the subducting slab (Fig. 1.11b). Boninites are produced by contact

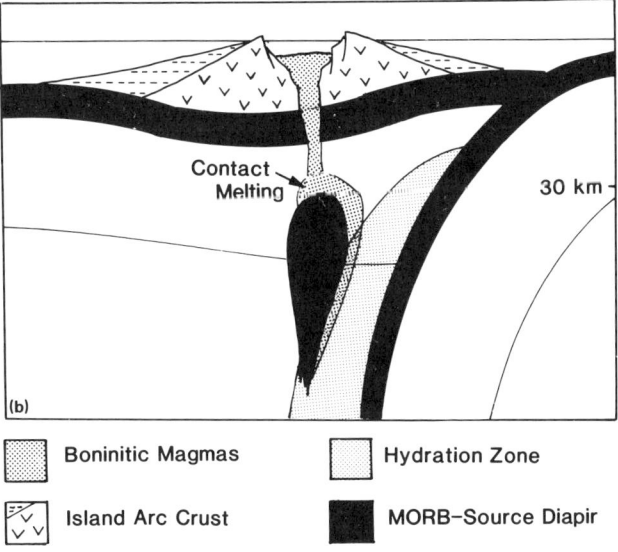

Figure 1.11 (a) Model for boninite genesis after Meijer (1980), in which subduction of old, cold oceanic crust into young, hot sub-Philippine Sea lithosphere initiates shallow dehydration of the slab, and boninite generation occurs along that section of the new plate boundary still within the region of elevated geotherms related to the West Philippine Sea spreading centre. (b) Model for boninite genesis after Crawford *et al.* (1981), in which, during the earliest stages of backarc rifting and opening, a MORB-source diapir intrudes hydrous sub-arc peridotite and initiates limited contact melting, generating boninites. Eruptions of basalts transitional between MORB and arc tholeiites, and derived from the MORB-source diapir, may cover the boninites.

melting of this hydrated sub-arc peridotite and are volumetrically minor relative to subsequent eruptions of backarc basalts. Boninites are then buried by voluminous eruptions of early backarc-basin basalts and sedimentary aprons washing off the sundering arc fragments. The major difference between this model and that of Meijer (1980) is that boninites are not generated in the earliest phase of subduction, but immediately precede backarc-basin opening; the latter event may occur tens of millions of years after the initiation of subduction.

Further consideration of Models 1 and 2, strongly influenced by new observations and data from both modern (e.g. Crawford *et al.* 1986, Falloon *et al.* 1989) settings and ophiolites (Crawford & Keays 1987, Flower & Levine 1987), suggests that both models encounter major problems and require significant modification before they can adequately account for the petrogenesis of boninites.

Model 1 has several important shortcomings:

(a) Arc tholeiites clearly occur *below* boninites in Hole 458, and occur interbedded with and after boninites on Guam. This suggests that 'normal' arc magmatism was active early in history of this subduction zone, and boninites here cannot be a peculiar feature accompanying the initiation of subduction. A return to 'normal' arc magma generation followed boninite eruption.

(b) If the heat source for boninite generation in the Mariana forearc was hot upper mantle below the southern end of the West Philippine Sea spreading centre, as suggested by Meijer (1980), it is unlikely that the zone of anomalously hot upper mantle extended more than a few hundred kilometres either side of the spreading centre, which trended NW away from the new convergent plate boundary (Fig. 1.11a). However, boninites of similar age (middle–late Eocene) occur in the Bonin Islands some 1500 km further north, and could not have been generated in this petrogenetic scenario.

(c) There is evidence from the northern part of the Palau–Kyushu Ridge that arc lavas were erupting along the Palau–Kyushu Ridge as early as 48 Ma (Seno 1984), indicating that boninites in the Mariana–Bonin region were not the first manifestations of arc magmatism following initiation of subduction.

Model 2 also has its problems, although these appear less severe than those noted above for Model 1:

(a) It predicts that boninitic magmatism is a likely precursor of backarc-basin spreading and basalt generation. However, although it is possible that they are buried by sediments and backarc basalts, boninites remain unrecorded from many arc–backarc basin settings, such as the Manus Basin, N Fiji Basin, Japan Sea, S Lau Basin–Havre Trough and Scotia Sea.

(b) If this contact melting mechanism operates as a MORB-source diapir rises beneath the arc, then the first boninitic melts should be the most LILE- and H_2O-enriched, and subsequent melts should be depleted in LILE (if indeed the anhydrous residue of the first boninite-producing melting event can be melted further). Well exposed boninite piles in Cambrian ophiolites in SE Australia show the opposite, with the most LILE-enriched boninites at the top of the pile and high-Ca boninites at the base.

Although we do not believe that Model 2 adequately accounts for the petrogenesis of the best-studied modern and ancient low-Ca boninite suites, it is hard to envisage that MORB-source diapirism ($\sim 1400°C$) initiating backarc spreading *can avoid* causing some degree of contact melting of hydrated sub-arc peridotites (solidus temperature ~ 1000–1100°C). Limited volumes of magmas compositionally transitional between arc tholeiites and high-Ca boninites might be generated in this fashion. High-Ca boninites of probable late Tertiary age occur in a complex tectonic setting at the northern termination of the Tonga Trench (Falloon *et al.* 1987, Falloon *et al.* 1989). One possible tectonic scenario for generation of these boninites is that the spreading centre at the very north end of the Lau Basin bifurcates, with one branch heading off in a NW direction and the other trending NE, propagating into and transecting the forearc of the N Tongan arc. Diapirs supplying MORB to this postulated NE-trending spreading centre must pass through shallow, refractory and probably hydrous subforearc upper mantle, and could have been the heat source necessary to generate these high-Ca boninites from such mantle at depths probably less than 50 km.

1.8 A new model for boninite petrogenesis

We consider that neither Models 1 nor 2 above for the generation of boninites can be applied successfully to the type location Bonin–Mariana boninites, or to the well documented Cambrian boninite occurrences in SE Australia. Key observations which must be explained in a realistic petrogenetic model for these boninite occurrences include the following:

(a) Boninites were erupted in a relatively shortlived event in middle or late Eocene along more than 1500 km of the convergent plate boundary on the eastern margin of the West Philippine Sea, and were preceded by 'normal' arc magmatism.

(b) In the Cambrian boninite piles, lavas become more refractory upwards in terms of 'basaltic' components, but at the same time more enriched in LILE and with decreasing ε_{Nd} (Nelson *et al.* 1984, Crawford & Cameron 1985). The uppermost boninites are exceptionally depleted lavas (chromite $Cr\# > 0.95$) with pronounced LREE enrichment and low ε_{Nd} (~ -8).

(c) Also in both the Victorian and Newfoundland (Betts Cove) Lower Palaeozoic ophiolites, tholeiites chemically most similar to early backarc-basin tholeiites occur interbedded in and/or overlying the uppermost boninites (Crawford & Keays 1987, Coish & Church 1979, Coish *et al.* 1982, Coish 1989). The tholeiites are derived from a very different source than the boninites.

(d) In the Chichi-jima, Mariana forearc, Cape Vogel and Victorian Cambrian boninite sequences, a range of boninite compositions is recorded, from less refractory high-Ca boninites (or lavas transitional from high-Ca to low-Ca boninites) to refractory low-Ca boninites.

CLASSIFICATION, PETROGENESIS AND TECTONIC SETTING

We suggest that the most likely mechanism for generating boninites, and low-Ca boninites in particular, is subduction of an active spreading centre subparallel to a trench fronting an intra-oceanic arc (Fig. 1.12a). As the hot lithosphere on either side of the spreading centre approaches the trench, the dip of the slab probably decreases and isotherms in the mantle wedge rise, causing partial melting of depleted subforearc oceanic lithosphere and genera-

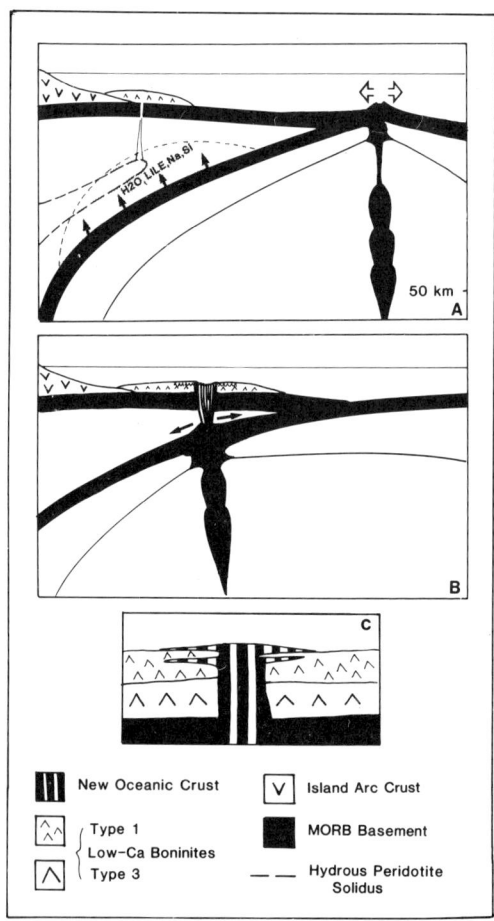

Figure 1.12 An alternative model for boninite generation involving subduction of an active spreading centre (A) accompanied by rising isotherms beneath the forearc and generation of first boninites from refractory, hydrous sub-arc mantle. Continued descent of the ridge, which remains active and hot, generates more refractory boninites, but only if the supply of water to the site of partial melting is maintained. In a tensional forearc setting, tension on the base of the thin mantle wedge may initiate rifting within the forearc (B) and 'resurrection' of the spreading centre. Backarc-basin-type basalts may erupt on the uppermost, highly refractory boninites (C). See text for detailed discussion.

tion of high-Ca boninites. Arc magmatism on the adjacent arc may shut off due to the change in dip of the slab, and possible interruption of the supply of hydrous fluids from the slab to the site of 'normal' arc magma generation at depths of ~100 km.

Harzburgite residues following high-Ca boninite extraction require further influx of hydrous fluids from the slab, and probably increasing temperature, to partially melt. Hydrous fluids passing up into the wedge can transport in (and extract from surrounding upper mantle) large amounts of 'enriched' component(s), which are a key factor in boninite geochemistry. The final boninites (Type 1 low-Ca boninites) which can be extracted before the residual mantle becomes infusible have the decoupled signature of highly refractory (low TiO_2, CaO, CaO/Al_2O_3, Sc and very low HREE), and Na-, Si- and LILE-enriched source components.

The increase in temperature required to generate low-Ca boninites in this setting is probably due to arrival of the subducted spreading centre beneath the forearc. We argue that MORB-type magmatism, derived from the spreading centre as it commences descent along the now very shallowly dipping subduction zone, need not necessarily cease, thereby maintaining high temperatures at shallow levels. Diapirs rising from around 150 km to feed the spreading centre are unlikely to 'see' the thin mantle wedge sliding above the recently subducted spreading centre, so that hot (~$1400°C$) lithosphere below the spreading centre will come into close proximity with multiply depleted peridotite at very shallow levels, leading to further boninite generation. MORB-type magmas in shallow magma chambers below the subducted spreading centre may erupt contemporaneously with and after highly depleted Type 1 low-Ca boninites, as at Howqua in Victoria (Crawford & Keays 1987).

If the forearc is under tensional stress, as in the modern Mariana–Bonin and Tongan arcs, continued spreading on the subducted spreading centre might lead to splitting of the forearc sliver of mantle wedge (Fig. 1.12b), leading to a 'resurrection' of the spreading centre, initiation of 'backarc-type' spreading in a forearc setting, and the cessation of boninitic magmatism. This emergence of a subducted spreading centre has been hypothesized to occur, and termed 'eduction' by Dixon & Farrar (1980). The thick tholeiite pile above boninites in the Victorian Cambrian greenstone belts and in the Betts Cove ophiolite in Newfoundland might have been produced in this manner (Crawford & Keays, 1987, Coish 1989).

Following the prescient hypothesis of Seno (1984), we believe that subduction of an active spreading centre around 48–43 Ma is the most probable explanation for the extensive occurrences of boninites along the Bonin–Mariana forearc. Seno (1984) found that the most simple way to account for regional geophysical–tectonic relationships in the central W Pacific was to hypothesize the former existence of a broadly NW–SE orientated spreading centre which was subducted at a NW–SE trending trench (roughly at the site of the Palau–Kyushu Ridge) between 48 and 43 Ma. Figure 1.13 shows his

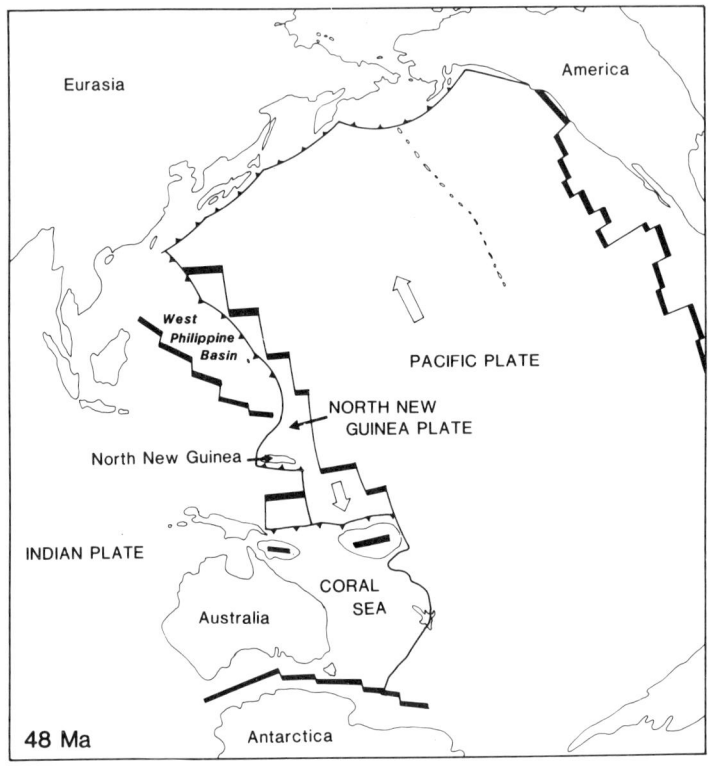

Figure 1.13 Reconstruction of the W Pacific at 48 Ma after Seno (1984), showing the spreading centre forming the eastern margin of the North New Guinea Plate about to be subducted along several thousands of kilometres of plate boundary. We argue that this subducted spreading centre provided the heat necessary to generate the Mariana–Bonin boninites from shallow, hydrous and refractory subforearc mantle. Subduction of parts of the same spreading system some 10–20 Ma earlier may have generated the Cape Vogel (50 Ma) and New Caledonian (early Tertiary) boninites.

reconstruction at 48 Ma. Subduction of more southerly segments of the same spreading centre around 53 Ma may well have been responsible for the generation of the Cape Vogel low-Ca boninites, and possibly also those in the latest Cretaceous–early Eocene in New Caledonia.

As described by Tatsumi & Maruyama (1989), Saunders *et al.* (1987) and Rogers & Saunders (1989), a broadly similar petrogenetic scenario may explain the origin of the Setouchi and Baja Type 2 low-Ca boninites, and similar lavas in S Chile (Isla Cook). Very young, hot oceanic lithosphere was subducted beneath both the former regions immediately prior to boninite generation, which occurred during a fairly brief (1 Ma duration for Setouchi) magmatic event. Magma segregation in the uppermost mantle (30–40 km) beneath these regions generated parental low-Ca boninites probably more alkali- and LILE-rich than Types 1 and 3 boninites; this may be partly due to lower amounts of water and lower degrees of partial melting at 10–15 kbar relative to Types 1 and 3 magmas generated by melting in the wedge at pressures ≤ 8 kbar. Type 2 boninites at these localities are rarely more magnesian than 10% MgO, probably due to the greater opportunity for pooling and fractionation in the thick arc crust of SW Japan and central Mexico relative to the very thin ocean-type crustal thicknesses in the forearcs of the intra-oceanic arcs such as the Bonin–Mariana arc. Boninites constitute only one part of the compositional spectrum of magmas erupted in both regions following subduction of this hot lithosphere. Basalts, derived at higher pressures and from different sources than the boninites, are an important component of both the Baja and Setouchi suites, and deserve further study.

1.9 Directions and problems for further study

Detailed studies of boninitic rocks are still comparatively few relative to those on other unusual magma suites in the geological record (e.g. kimberlites, carbonatites and komatiites). Despite reasonable consensus regarding their origin and tectonic significance, a number of outstanding problems still confront the interpretation and understanding of boninites. Among these problems are:

(a) Boninites fractionate to bronzite andesites and dacitic compositions on Chichi-jima. Are these highly evolved boninites compositionally distinctive, and how closely do they approach more typical calc-alkaline andesites and dacites?
(b) What is the exact identity and origin of the enriched component(s) in boninite genesis? Are these same enriched components involved in typical arc basalt genesis at depths closer to 100 km?
(c) How effectively can superheated aqueous fluids extract and transport LILE, REE, high-field-strength elements such as Zr, Nb and Y, and radiogenic Sr through the upper mantle?
(d) What is the nature of the basement upon which the type location Bonin Island boninites were erupted? Planned basement drilling in several W Pacific arcs and forearcs during the next few years, using the JOIDES Resolution, should clarify temporal–spatial relationships of boninite suites in these essentially *in situ* settings, as should more detailed studies of boninite-bearing ophiolites. Also, it should be possible to examine modern plate-tectonic settings and predict locations where boninitic magmatism might be presently or formerly active. One such location is the southern end of the N Fiji Basin, where an active N–S orientated spreading centre segment in the N Fiji Basin abuts the Hunter Fracture Zone (extending from S Vanuatu to Fiji). The latter, although poorly understood, may delineate a site of former (or incipient) subduction.

Juxtaposition of hot subspreading centre lithosphere and old, cold and probably hydrated sub-arc upper mantle provides the necessary ingredients for boninite generation along that part of the Hunter Fracture Zone close to the end of the spreading centre.
(e) Refinement of experimental petrological techniques for H_2O-undersaturated melting of peridotite at high temperatures and low pressures, aimed at avoiding quench problems, will further constrain relationships between source depletion and fluids (compositions, volumes, etc.) involved in boninite petrogenesis.
(f) More extensive PGE data are required for modern and ancient boninitic rocks, to aid in understanding their petrogenesis and their potential as sources for precious-metal mineralization.
(g) Most boninite suites occur in close association with tholeiitic basalts which seem to range between arc tholeiite and backarc-basin tholeiite compositions. Further documentation of these tholeiites is required to determine their tectonic settings of eruption; a better understanding of their petrogenesis will further constrain models for boninite genesis.

1.10 Conclusions

We suggest a major-element chemistry-based classification scheme for boninites and related rocks which subdivides these into high-Ca suites ($CaO/Al_2O_3 = 0.7-1.0$ at $Mg\# > 0.65$), typified by the Troodos Upper Pillow Lavas, and compositionally more diverse low-Ca suites. The latter are further broken down into three groups: Types 1 and 2 are generated from high and low degrees, respectively, of partial melting of very depleted, probably clinopyroxene-free harzburgite, and consequently have very low CaO/Al_2O_3 values (< 0.55), and low CaO and FeO^* contents. Type 1 low-Ca boninites are best represented by the late Cretaceous–early Tertiary lavas from Nepoui, in New Caledonia, and Type 2 low-Ca boninites, with total alkali contents $> 3\%$, are known from the Setouchi region of SW Japan and Baja California. Type 3 low-Ca boninites, which include the type location suite from Chichi-jima, have somewhat higher CaO and FeO^* contents, and higher CaO/Al_2O_3 (0.55–0.7) than Types 1 and 2 suites, and were derived from slightly less refractory harzburgite sources.

Refractory sources of low-Ca boninites were metasomatized by multi-source, LILE-enriched hydrous fluids prior to or during boninite generation. We concur with previous studies of boninites in suggesting that the hydrous fluids involved in boninite genesis were derived from slab dehydration. Boninite radiogenic isotope signatures were controlled by those of the bulk subducted crustal section with which they equilibrated. Where this subducted crust was young, fresh and virtually sediment-free MORB, such as for the

Nepoui boninites, fluid ε_{Nd} values are MORB-like. If significant volumes of old, continent-derived sediments were also in the subducted slab, ε_{Nd} values as unradiogenic as -8 were carried into the mantle wedge with the fluids initiating boninite generation from refractory peridotite.

High-Ca boninites, such as those in the Troodos Upper Pillow Lavas and in the N Tongan forearc, may be generated from depleted sub-oceanic lithosphere by contact melting adjacent to rising MORB-source diapirs. These $1400°C$ diapirs, which eventually lead to arc rifting and backarc-basin formation if their ascent is not aborted by regional changes in plate motion (e.g. arc–continent collision), must pass through and locally partially melt shallow, hydrated sub-arc mantle. High-Ca boninites in this idealized scenario must be preceded by 'normal' arc lavas, and possibly followed by backarc-basin-type basalts.

Low-Ca boninites are generated from more refractory sources than high-Ca boninites, and consequently hydrous fluids are required to initiate melting of such highly depleted sources. These fluids carried significant amounts of Si, Na, LILE (including LREE) and possibly Al. Type 3 low-Ca boninites (represented by the Cape Vogel and Chichi-jima suites) are derived from somewhat less refractory harzburgites than Type 1 low-Ca boninites, as reflected in their respective CaO, FeO*, Sc and HREE levels. Greater access of incompatible elements is required for Type 1 genesis and is reflected in their higher Na, Si and LILE contents. Temperatures required to generate such melts at depths in the upper mantle less than 30 km are probably $> 1200°C$. We suggest subduction of an active spreading centre subparallel to the trench may provide the necessary heat for low-Ca boninite generation, which is predicted to be limited to the forearc region of oceanic arcs. Boninites erupted in this setting are in an optimum location for subsequent incorporation into foldbelts during arc–continent collision. The Cape Vogel, New Caledonian and Victorian Cambrian boninite suites are all thought to have been overthrust onto thinned passive margin continental crust during such arc–continent collisions.

Type 2 low-Ca boninites, such as the Setouchi and Baja California suites, are generated from depleted mantle similar to the refractory sources of Type 1 boninites. However, their much higher contents of alkalis, P_2O_5 and LILE suggest they were generated via significantly lower degrees of partial melting of a Type 1 source. Type 2 boninites occur where a spreading centre or very young, hot oceanic lithosphere has attempted to underthrust an active continental margin. Greater crustal thicknesses from source to surface result in Type 2 low-Ca boninites being generally more evolved than Type 1 and 3 boninites erupted through thin forearc crust.

Finally, we note that boninitic magmas were erupted in the Precambrian, as early as 2050 Ma (e.g. parent magmas to ultramafic section of the Bushveld Complex). Siliceous high-Mg basalts in Archaean greenstone belts are compositionally transitional between tholeiitic basalts and high-Ca boninites;

however, they are probably generated by extensive fractionation and contamination of peridotitic komatiite magmas, and are best not classified with boninite series magmas.

Acknowledgements

The ideas expressed in this paper have arisen over the last eight years as a result of considerable input and stimulus from a number of people; we especially acknowledge discussions with Drs R. A. Duncan, G. A. Jenner, W. E. Cameron, S.-S. Sun, L. Beccaluva, G. Serri and R. Varne. June Pongratz is thanked for her help with the drafting.

References

Arndt, N. T. & G. A. Jenner 1986. Crustally contaminated komatiites and basalts from Kambalda, Western Australia. *Chem. Geol.* **56**, 229–55.

Beccaluva, L. & G. Serri 1988. Boninitic and low-Ti subduction-related lavas from intraoceanic arc–backarc systems and low-Ti ophiolites: a reappraisal of their petrogenesis and original tectonic setting. *Tectonophysics* **146**, 291–315.

Bertrand, H., J. Dostal & C. Dupuy 1982. Geochemistry of early Mesozoic tholeiites from Morocco. *Earth Planet. Sci. Lett.* **58**, 225–39.

Bloomer, S. H. 1987. Geochemical characteristics of boninite- and tholeiite-series volcanic rocks from the Mariana forearc and the role of an incompatible element-enriched fluid in arc petrogenesis. Geol. Soc. Am. Spec. Paper no. 215, 151–64.

Bloomer, S. H. & J. W. Hawkins 1987. Petrology and geochemistry of boninite series volcanic rocks from the Mariana Trench. *Contrib. Mineral. Petrol.* **97**, 361–77.

Bougault, H., R. C. Maury, M. El Azzouzi, J.-L. Joron, J. Cotten & M. Treuil 1981. Tholeiites, basaltic andesites, and andesites from Leg 60 sites: geochemistry, mineralogy, and low partition coefficient elements. Init. Rep. DSDP Leg 60, 657–77.

Brown, A. V. & H. M. Waldron 1982. *Preliminary review of the Eocambrian–Cambrian basaltic association and tectonic setting within western and north-western Tasmania.* Dept of Mines, Tasmania, Rep. no. 1982/10.

Cameron, W. E. 1985. Petrology and origin of primitive lavas from the Troodos ophiolite, Cyprus. *Contrib. Mineral. Petrol.* **89**, 256–62.

Cameron, W. E. 1989. Contrasting boninite–tholeiite associations from New Caledonia (this volume).

Cameron, W. E., M. T. McCulloch & D. A. Walker 1983. Boninite petrogenesis: chemical and Nd–Sr isotopic constraints. *Earth Planet. Sci. Lett.* **65**, 75–89.

Cameron, W. E., E. G. Nisbet & V. J. Dietrich 1979. Boninites, komatiites and ophiolitic basalts. *Nature* **280**, 550–3.

Campiglio, C., C. Marion & M. Vannier 1986. Etude d'une boninite à olivine de Nouvelle-Calédonie: pétrographie et chimisme des phases. *Bull. Mineral.* **109**, 423–40.

Cawthorn, R. G. & G. Davies 1983. Experimental data at 3 kbar pressure on parental magma to the Bushveld Complex. *Contrib. Mineral. Petrol.* **83**, 128–35.

Cawthorn, R. G., G. Davies, A. Clubley-Armstrong & T. S. McCarthy 1981. Sills associated with the Bushveld Complex, South Africa: an estimate of the parental magma composition. *Lithos* **14**, 1–15.

Coish, R. A. 1989. Boninitic lavas in Appalachian ophiolites: a review (this volume).

Coish, R. A. & W. R. Church 1979. Igneous geochemistry of mafic rocks in the Betts Cove Ophiolite, Newfoundland. *Contrib. Mineral. Petrol.* **70**, 29–39.

Coish, R. A., R. Hickey & F. A. Frey 1982. Rare earth element geochemistry of the Betts Cove ophiolite, Newfoundland: complexities in ophiolite formation. *Geochim. Cosmochim. Acta* **46**, 2117–34.

Crawford, A. J. 1980. A clinoenstatite-bearing cumulate olivine pyroxenite from Howqua, Victoria. *Contrib. Mineral. Petrol.* **75**, 353–67.

Crawford, A. J. & R. F. Berry 1989. Tectonic implications of late Proterozoic–early Palaeozoic igneous rock series in western Tasmania, (in preparation).

Crawford, A. J. & W. E. Cameron 1985. Petrology and geochemistry of Cambrian boninites and low-Ti andesites from Heathcote, Victoria. *Contrib. Mineral. Petrol.* **91**, 93–104.

Crawford, A. J. & R. R. Keays 1987. Petrogenesis of Victorian Cambrian tholeiites and implications for the origin of associated boninites. *J. Petrol.* **28**, 1075–109.

Crawford, A. J., L. Beccaluva & G. Serri 1981. Tectono-magmatic evolution of the West Philippine–Mariana region and the origin of boninites. *Earth Planet. Sci. Lett.* **54**, 346–56.

Crawford, A. J., L. Beccaluva, G. Serri & J. Dostal 1986. Petrology, geochemistry and tectonic implications of volcanics dredged from the intersection of the Yap and Mariana trenches. *Earth Planet. Sci. Lett.* **80**, 265–80.

Crawford, A. J., W. E. Cameron & R. R. Keays 1984. The association boninite–low-Ti andesite–tholeiite in the Heathcote Greenstone Belt, Victoria: ensimatic setting for the early Lachlan Foldbelt. *Austr. J. Earth Sci.* **31**, 197–208.

Dallwitz, W. B. 1968. Chemical composition and genesis of clinoenstatite-bearing volcanic rocks from Cape Vogel, Papua: a discussion. In *23rd Int. Geol. Congr.* **2**, 229–42.

Dallwitz, W. B., D. H. Green & J. E. Thompson 1966. Clinoenstatite in a volcanic rock from the Cape Vogel area, Papua. *J. Petrol.* **7**, 375–403.

Danjushevsky, L. V., A. V. Sobolev & L. V. Dmitriev 1987. Orthopyroxene-bearing low-Ti tholeiites; a new type of ocean ridge tholeiite. *Dokl. Akad. Nauk SSSR* **292**, 1449–52.

Davies, G., R. G. Cawthorn, J. M. Barton & M. Morton 1980. Parental magma to the Bushveld Complex. *Nature* **287**, 33–5.

Dietrich, V., R. Emmermann, R. Oberhansli & H. Puchelt 1978. Geochemistry of basaltic and gabbroic rocks from the West Mariana Basin and Mariana trench. *Earth Planet. Sci. Lett.* **39**, 127–44.

Dixon, J. M. & E. Farrar 1980. Ridge subduction, eduction and the Neogene tectonics of SW North America. *Tectonophysics* **67**, 81–99.

Dobson, P. F. & J. R. O'Neil 1987. Stable isotope compositions and water contents of boninite series volcanic rocks from Chichi-jima, Bonin islands, Japan. *Earth Planet. Sci. Lett.* **82**, 75–86.

Dobson, P. F. & G. R. Tilton 1989. Th, U and Pb systematics of boninite series volcanic rocks from Chichi-jima, Bonin Islands, Japan (this volume).

Duncan, R. A. & D. H. Green 1980. Role of multistage melting in the formation of oceanic crust. *Geology* **8**, 22–6.

Duncan, R. A. & D. H. Green 1987. The genesis of refractory melts in the formation of oceanic crust. *Contrib. Mineral. Petrol.* **96**, 326–42.

Falloon, T. J. & A. J. Crawford 1989. Petrogenesis of a primitive arc tholeiite suite from the N Tonga forearc. In preparation.

Falloon, T. J. & D. H. Green 1986. Glass inclusions in magnesian olivine phenocrysts from Tonga: evidence for highly refractory parental magmas in the Tonga arc. *Earth Planet. Sci. Lett.* **81**, 95–103.

Falloon, T. J. & D. H. Green 1987. Anhydrous partial melting of MORB pyrolite and other peridotite compositions at 10 kbar: implications for the origin of primitive MORB glasses. *Mineral. Petrol.* **37**, 181–219.

Falloon, T. J. & D. H. Green 1988. Anhydrous partial melting of peridotite from 8 to 35 kbars and the petrogenesis of MORB. *J. Petrol.* Special Lithosphere Issue, 379–414.

Falloon, T. J., D. H. Green & A. J. Crawford 1987. Dredged igneous rocks from the northern termination of the Tofua magmatic arc, Tonga and adjacent Lau Basin. *Austr. J. Earth Sci.* **34**, 487–506.

Falloon, T. J., D. H. Green, C. J. Hatton & K. L. Harris 1988. The anhydrous partial melting of a fertile and depleted peridotite from 2–30 kbar. *J. Petrol.* (in press).

Falloon, T. J., D. H. Green & M. T. McCulloch 1989. Petrogenesis of high-Mg and associated lavas from the north Tonga Trench (this volume).

Flower, M. F. J. & H. Levine 1987. Petrogenesis of a tholeiite–boninite sequence from Ayios Mamas, Troodos Ophiolite: evidence for splitting of a volcanic arc. *Contrib. Mineral. Petrol.* **97**, 509–24.

Fodor, R. V. & S. K. Vetter 1984. Rift-zone magmatism: petrology of basaltic rocks transitional from CFB to MORB, southeastern Brazil. *Contrib. Mineral. Petrol.* **88**, 307–21.

Francis, D. 1985. The Baffin Bay lavas and the value of picrites as analogues of primary magmas. *Contrib. Mineral. Petrol.* **89**, 144–54.

Francis, D., J. Ludden & A. Hynes 1983. Magma evolution in a Proterozoic rifting environment. *J. Petrol.* **24**, 556–82.

Frey, F. A. & D. H. Green 1974. The mineralogy, geochemistry and origin of lherzolite inclusions in Victorian basanites. *Geochim. Cosmochim. Acta* **38**, 1023–59.

Frey, F. A., C. J. Suen & H. W. Stockman 1985. The Ronda high temperature peridotite: geochemistry and petrogenesis. *Geochim. Cosmochim. Acta* **49**, 2469–91.

Green, D. H. 1970. The origin of basaltic and nephelinitic magmas. *Trans. Leics. Lit. Phil. Soc.* **64**, 28–54.

Green, D. H. 1976. Equilibrium testing of 'equilibrium' partial melting of peridotite under water-saturated, high-pressure conditions. *Can. Mineral.* **14**, 255–68.

Green, D. H., T. J. Falloon & W. R. Taylor 1987. Mantle derived magmas – role of variable source peridotite and variable C–H–O fluid compositions. In *Magmatic processes: physiochemical principles*, B. O. Mysen (ed.), 139–54. Geochem. Soc. Spec. Publ. no. 1.

Green, D. H., W. D. Hibberson & A. L. Jaques 1979. Petrogenesis of mid-ocean ridge basalts. In *The Earth: its origin, structure and evolution*, M. W. McElhinny (ed.), 265–99. London: Academic Press.

Hall, R. P & D. J. Hughes 1987. Noritic dykes of southern West Greenland: early Proterozoic boninitic magmatism. *Contrib. Mineral. Petrol.* **97**, 169–82.

Hamlyn, P. R., R. R. Keays, W. E. Cameron, A. J. Crawford & H. M. Waldron 1985. Precious metals in magnesian low-Ti lavas: implications for metallogenesis and sulfur saturation in primary magmas. *Geochim. Cosmochim. Acta* **49**, 1797–811.

Hatton, C. J. & M. R. Sharpe 1989. Significance and origin of boninite-like rocks associated with the Bushveld Complex (this volume).

Hawkins, J. W., S. H. Bloomer, C. A Evans & J. T. Melchior 1984. Evolution of intra-oceanic arc–trench systems. *Tectonophysics* **102**, 175–205.

Hickey-Vargas, R. L. 1989. Boninites and tholeiites from DSDP Site 458, Mariana forearc (this volume).

Hickey, R. & F. A. Frey 1982. Geochemical characteristics of boninite series volcanics: implications for their source. *Geochim. Cosmochim. Acta* **46**, 2099–115.

Hickey-Vargas, R. & M. K. Reagan 1987. Temporal variation of isotope and rare earth element abundances in volcanic rocks from Guam: implications for the evolution of the Mariana Arc. *Contrib. Mineral. Petrol.* **97**, 497–508.

Hsui, A. T. & M. N. Toksoz 1979. The evolution of thermal structures beneath a subduction zone. *Tectonophysics* **60**, 43–60.

Ishizaka, I. & R. W. Carlson 1983. Nd–Sr systematics of the Setouchi volcanic rocks, SW Japan: a clue to the origin of orogenic andesite. *Earth Planet. Sci. Lett.* **64**, 327–40.

Jaques, A. L. & D. H. Green 1980. Anhydrous melting of peridotite at 0–15 kbar pressure and the genesis of tholeiitic basalts. *Contrib. Mineral. Petrol.* **73**, 287–310.

Jaques, A. L. & G. P. Robinson 1977. The continent–island arc collision in northern Papua New Guinea. *BMR J. Austr. Geol. Geophys.* **2**, 289–303.

Jenner, G. A. 1981. Geochemistry of high-Mg andesites from Cape Vogel, Papua New Guinea. *Chem. Geol.* **33**, 307–32.

Jenner, G. A. 1983. Petrogenesis of high-Mg andesites: an experimental and geochemical study with emphasis on high-Mg andesites from Cape Vogel, PNG. Ph.D. Thesis, Univ. of Tasmania, Hobart.

Johnson, R. W., A. L. Jaques, R. L. Hickey, C. O. McKee & B. W. Chappell 1985. Manam Island, Papua New Guinea: petrology and geochemistry of a low TiO_2 basaltic island-arc volcano. *J. Petrol.* **26**, 283–323.

Kalsbeek, F. & H. F. Jepsen 1984. The late Proterozoic Zig-Zag Dal basalt formation of eastern North Greenland. *J. Petrol.* **25**, 644–64.

Kikuchi, Y. 1890. On pyroxene components in certain volcanic rocks from Bonin Island. *J. Coll. Sci. Imp. Univ. Japan* **3**, 67–89.

Komatsu, M. 1980. Clinoenstatite in volcanic rocks from the Bonin Islands. *Contrib. Mineral. Petrol.* **74**, 329–38.

Kuehner, S. M. 1989. Petrology and geochemistry of early Proterozoic high-Mg dykes from the Vestfold Hills, Antarctica (this volume).

Kuroda, N. & K. Shiraki 1975. Boninite and related rocks of Chichi-jima, Bonin Island, Japan. *Rep. Fac. Sci. Shizuoka Univ.* **10**, 145–55.

Kuroda, N., K. Shiraki & H. Urano 1978. Boninite as a possible calc-alkaline primary magma. *Bull. Volcanol.* **41**, 563–75.

Kushiro, I. 1972. Effect of water on the composition of magmas formed at high pressures. *J. Petrol.* **13**, 311–34.

Kushiro, I. 1981. Petrology of high-MgO bronzite andesite resembling boninite from site 458 near the Mariana Trench. *Init. Rep. DSDP Leg 60*, 731–3.

Kushiro, I., H. S. Yoder & M. Nishikawa 1968. Effect of water on the melting of enstatite. *Geol. Soc. Am. Bull.* **79**, 1685–92.

Kyser, T. K., W. E. Cameron & E. G. Nisbet 1986. Boninite petrogenesis and alteration history: constraints from stable isotope compositions of boninites from Cape Vogel, New Caledonia and Cyprus. *Contrib. Mineral. Petrol.* **93**, 222–6.

LeMaitre, R. W. 1984. A proposal by the IUGS Subcommission on the Systematics of Igneous Rocks for a chemical classification of volcanic rocks based on the total alkali–silica diagram. *Austr. J. Earth Sci.* **31**, 243–56.

McCulloch. M. T. & W. E. Cameron 1983. Nd–Sr isotopic study of primitive lavas from the Troodos Ophiolite, Cyprus. evidence for a subduction-related setting. *Geology* **11**, 727–31.

Meijer, A. 1976. Pb and Sr isotopic data bearing on the origin of volcanic rocks from the Mariana island arc system. *Geol. Soc. Am. Bull.* **87**, 1358–69.

Meijer, A. 1980. Primitive arc volcanism and a boninite series: examples from western Pacific island arcs. In *Tectonic and geologic evolution of southwest Asian seas and islands*, D. E. Hayes (ed.), 269–82. Am. Geophys. Union. Monogr. no. 23.

Meijer, A. & B. Hanan 1981. Pb isotopic composition of boninite and related rocks from the Mariana and Bonin fore-arc regions. *Eos* **62**, 408.

Nakamura, Y. & I. Kushiro 1974. Composition of gas phase in Mg_2SiO_4–SiO_2–H_2O at 15 kbar. *Carnegie Inst. Washington Yearb.* **73**, 255–8.

Natland, J. H. 1981. Crystal morphologies and pyroxene compositions in boninites and tholeiitic basalts from Deep Sea Drilling Project Holes 458 and 459 in the Mariana forearc region. *Init. Rep. DSDP Leg 60*, 681–709.

Nelson, D. R., A. J. Crawford & M. T. McCulloch 1984. Nd–Sr systematics in Cambrian boninites and tholeiites from Victoria, Australia. *Contrib. Mineral. Petrol.* **88**, 169–77.

O'Hara, M. J. 1968. Are ocean floor basalts primary magmas? *Nature* **220**, 683–6.

O'Neill, H. St C. & V. J. Wall 1987. The olivine–spinel oxygen geobarometer, the nickel precipitation curve and the oxygen fugacity of the Earth's mantle. *J. Petrol.* **28**, 1169–91.

Petersen, J. 1891. Der boninit von Peel Island. *Jahrb. Hamburg Wiss. Anst.* **8**, 341–9.

Rautenschlein, M., G. A. Jenner, J. Hertogen, A. W. Hoffman, R. Kerrich, H.-U. Schmincke & W. M. White 1985. Isotopic and trace element composition of volcanic glasses from the Akaki Canyon, Cyprus: implications for the origin of the Troodos Ophiolite. *Earth Planet. Sci. Lett.* **75**, 369–83.

Reagan, M. K. & A. Meijer 1984. Geology and geochemistry of early arc-volcanic rocks from Guam. *Geol. Soc. Am. Bull.* **95**, 701–13.

Redman, B. A. & R. R. Keays 1985. Archaean basic volcanism in the Eastern Goldfields Province, Yilgarn Block, Western Australia. *Precambrian Res.* **30**, 113–52.

Robinson, P. T., W. G. Melson, T. O'Hearn & H.-U. Schmincke 1983. Volcanic glass compositions of the Troodos ophiolite, Cyprus. *Geology* **11**, 400–4.

Rogers, G. & A. D. Saunders 1989. Magnesian andesites from Mexico, Chile and the Aleutian Islands: implications for magmatism associated with ridge–trench collision (this volume).

Rogers, N. W., C. J. MacLeod & B. J. Murton 1989. Petrogenesis of boninitic lavas from the Limassol Forest Complex, Cyprus (this volume).

Ryabchikov, I. D., W. Schreyer & K. Abraham 1982. Compositions of aqueous fluids in equilibrium with pyroxenes and olivines at mantle pressures and temperatures. *Contrib. Mineral. Petrol.* **79**, 80–4.

Sameshima, T., J.-P. Paris, P. M. Black & R. F. Heming 1983. Clinoenstatite-bearing lava from Népoui, New Caledonia. *Am. Mineral.* **68**, 1076–82.

Saunders, A. D., G. Rogers, G. F. Marriner, D. J. Terrell & S. P. Verma 1987. Geochemistry of Cenozoic rocks, Baja California, Mexico: implications for the petrogenesis of post-subduction magmas. *J. Volcanol. Geotherm. Res.* **32**, 223–45.

Seno, T. 1984. Was there a North New Guinea plate? *Rep. Geol. Surv. Japan* **263**, 29–42.

Seno, T. & S. Maruyama 1984. Paleogeographic reconstruction and origin of the Philippine Sea. *Tectonophysics* **102**, 53–84.

Sharaskin, A. Y., N. L. Dobretsov & N. V. Sobolev 1980. Marianites: the clinoenstatite-bearing pillow lavas associated with the ophiolite assemblage of the Mariana Trench. In *Proc. Int. Ophiolite Symp.*, A. Panayiotou (ed.), 473–9. Cyprus.

Sharaskin, A. Y., S. F. Karpenko, A. V. Ijalikov, S. K. Zlobin & Y. A. Balashov 1983. Correlated $^{144}Nd/^{143}Nd$ and $^{87}Sr/^{86}Sr$ data on boninites from the Mariana and Tonga arcs. *Ofioliti* **8**, 431–8.

Sharpe, M. R. & L. J. Hulbert 1985. Ultramafic sills beneath the eastern Bushveld Complex: mobilized suspensions of early Lower Zone cumulates in a parental magma with boninitic affinities. *Econ. Geol.* **80**, 849–71.

Shiraki, K., N. Kuroda, S. Maruyama & H. Urano 1978. Evolution of the Tertiary volcanic rocks in the Izu Mariana Arc. *Bull. Volcanol.* **41**, 548–62.

Shiraki, K., N. Kuroda, H. Urano & S. Maruyama 1980. Clinoenstatite in boninites from the Bonin Islands, Japan. *Nature* **285**, 31–2.

Strong, D. F. & J. Dostal 1980. Dynamic melting of Proterozoic upper mantle: evidence from rare earth elements in oceanic crust of eastern Newfoundland. *Contrib. Mineral. Petrol.* **72**, 165–73.

Sun, S.-S. & R. W. Nesbitt 1978. Geochemical regularities and genetic significance of ophiolitic basalts. *Geology* **6**, 689–93.

Sun, S.-S., R. W. Nesbitt & M. T. McCulloch 1989. Geochemistry and petrogenesis of Archaean and early Proterozoic siliceous high-magnesian basalts (this volume).

Tatsumi, Y. 1981. Melting experiments on a high-magnesian andesite. *Earth Planet. Sci. Lett.* **54**, 357–65.

Tatsumi, Y. 1982. Origin of high-magnesian andesites in the Setouchi volcanic belt, southwest Japan, II. Melting phase relations at high pressures. *Earth Planet. Sci. Lett.* **60**, 305–17.

Tatsumi, Y. 1983. High magnesian andesites in the Setouchi volcanic belt, southwest Japan and their possible relation to the evolutionary history of the Shikoku Inter-arc Basin. *Am. Geophys. Union Geodyn. Ser.* **11**, 331–41.

Tatsumi, Y. & K. Ishizaka 1981. Existence of andesitic primary magma: an example from southwest Japan. *Earth Planet. Sci. Lett.* **53**, 124–30.

Tatsumi, Y. & K. Ishizaka 1982a. High magnesian andesite and basalt from Shodo-shima island, southwest Japan, and their bearing on the genesis of calc-alkaline andesites. *Lithos* **15**, 161–72.

Tatsumi, Y. & K. Ishizaka 1982b. Origin of high-magnesian andesites in the Setouchi volcanic belt, southwest Japan, I. Petrographical and chemical characteristics. *Earth Planet. Sci. Lett.* **60**, 293–304.

Tatsumi, Y. & S. Maruyama 1989. Boninites and high-Mg andesites: tectonics and petrogenesis (this volume).

Tatsumi, Y., D. L. Hamilton & R. W. Nesbitt 1986. Chemical characteristics of fluid phase released from a subducted lithosphere and origin of arc magmas: evidence from high-pressure experiments and natural rocks. *J. Volcanol. Geotherm. Res.* **29**, 293–309.

Tatsumi, Y., M. Sakuyama, H. Fukuyama & I. Kushiro 1983. Generation of arc basalt magmas and thermal structure of the mantle wedge in subduction zones. *J. Geophys. Res.* **88**, 5815–25.

Thy, P., C. K. Brooks & J. N. Walsh 1985. Tectonic and petrogenetic implications of major and rare earth element chemistry of Troodos glasses, Cyprus. *Lithos* **18**, 165–78.

Umino, S. 1985. Volcanic geology of Chichijima, the Bonin Islands (Ogasawara Islands). *J. Geol. Soc. Japan* **91**, 505–23.

Umino, S. 1986. Magma mixing in boninite sequence of Chichijima, Bonin Islands. *J. Volcanol. Geotherm. Res.* **29**, 125–57.

Umino, S. & I. Kushiro 1989. Experimental studies on boninite petrogenesis (this volume).

Van der Laan, S. R., M. F. J. Flower & A. F. Koster van Groos 1989. Experimental evidence for the origin of boninites: near-liquidus phase relations to 7.5 kbar (this volume).

Walker, D. A. & W. E. Cameron 1983. Boninite primary magmas: evidence from the Cape Vogel Peninsula, PNG. *Contrib. Mineral. Petrol.* **83**, 150–8.

Walker, D. A. & I. McDougall 1982. $^{40}Ar/^{39}Ar$ and K–Ar dating of altered glassy volcanic rocks: the Dabi Volcanics, PNG. *Geochim. Cosmochim. Acta* **46**, 2181–90.

Walker, D., T. Shibata & S. E. Delong 1979. Abyssal tholeiites from the Oceanographer Fracture Zone. II: Phase equilibria and mixing. *Contrib. Mineral. Petrol.* **70**, 111–25.

Wood, D. A., N. G. Marsh, J. Tarney, J.-L. Joron, P. Fryer & M. Treuil 1981. Geochemistry of igneous rocks recovered from a transect across the Mariana Trough, arc, fore-arc, and trench, Sites 453 through 461, DSDP Leg 60. *Init. Rep. DSDP Leg 60*, 611–45.

Yoder, H. S. & C. E. Tilley 1962. Origin of basalt magmas: an experimental study of natural and synthetic rock systems. *J. Petrol.* **3**, 342–532.

2 Boninites and high-Mg andesites: tectonics and petrogenesis

YOSHIYUKI TATSUMI &
SHIGENORI MARUYAMA

Abstract

Tectonic and petrological constraints on the genesis of high-Mg andesite magmas have been examined, as deduced from two well studied areas: the Setouchi volcanic belt, SW Japan, and the Bonin Islands. In both localities, very young oceanic lithosphere (the Shikoku Basin and the North New Guinea plate, respectively) was subducted beneath lithosphere which had experienced backarc opening only slightly earlier (the Japan Sea and the West Philippine Basin, respectively).

During subduction of oceanic lithosphere in a normal arc–trench system, in which basaltic magmas are the only primary magmas generated, some hydrous fluids expelled from the dehydrating slab are fixed in a hydrous peridotite column built in the near-trench, subforearc mantle wedge; most of these hydrous fluids, however, are transported convectively along the base of the mantle wedge until they reach depths around 110 km, where primary basaltic magmas are generated and ascend to form arc volcanoes. In contrast, in subduction zones characterized by high geothermal gradients, owing to both recent backarc opening in the upper plate and subduction of young, hot oceanic lithosphere, the solidus temperature of hydrous peridotite can be attained in the hydrous column of subforearc mantle wedge peridotite, producing high-Mg andesite magmas in the near-trench region. The Mexican volcanic belt and the Papua New Guinea region (south of New Britain) are two modern subduction-related settings in which the above tectonic conditions may be met with, and where high-Mg andesite magmas may be produced.

2.1 Introduction

An origin for orogenic andesites via direct partial melting of hydrous peridotite was proposed, based on experimental evidence in simple systems, by O'Hara (1965) and Kushiro (1972), who showed that partial melts of hydrous peridotites are enriched in SiO_2 relative to partial melts of anhydrous peridotites. Mysen & Boettcher (1975) also found that partial melting of a natural hydrous peridotite can produce a broadly andesitic liquid at upper-mantle pressures. These experimentally produced 'andesites' are relatively Mg-rich and compositionally distinct from the andesites in most island arcs. However, petrological studies of some unusual W Pacific arc-related lavas have suggested that direct partial melting of hydrous upper-mantle peridotite does operate in nature, producing Mg-rich 'andesitic' magmas including boninites from the Bonin Islands (Kuroda et al. 1978) and sanukitoids from SW Japan (Tatsumi & Ishizaka 1981, 1982a,b). Where field relationships are known, these high-Mg andesites (abbreviated as HMA hereafter) are found in a near-trench environment, where, under normal convergent boundary thermal conditions, typical arc magmas cannot be produced because the solidus temperature of hydrous peridotite is not attained in the mantle wedge at these shallow levels.

In order to understand the unusual tectonic conditions responsible for the abnormally high temperatures required to generate HMA magmas at shallow levels in the subforearc upper mantle, several models have been proposed:

(a) Subduction of oceanic lithosphere beneath young, hot lithosphere formed by sea-floor spreading processes (Shiraki et al. 1978, Meijer 1980); under such conditions, slab-derived H_2O could cause hydrous partial melting of the abnormally hot mantle wedge.
(b) Injection of MORB-source mantle diapirs into the hydrous mantle wedge (Crawford et al. 1981); in this model, HMA magmas could be generated as heat is supplied from the ascending diapir which eventually causes arc rifting and backarc spreading in the region.
(c) Initial subduction of a young, hot oceanic lithosphere (Tatsumi 1982, Crawford et al. 1986); in this model, the mantle wedge is not cooled by the subducted slab during the early stages of subduction, but rather is heated enough to exceed the hydrous peridotite solidus.

In all these models, however, processes of hydration of the near-trench mantle wedge, which are essential to an understanding of HMA genesis, have not been fully taken into account. Recently, Tatsumi (1986, 1989) has proposed a mechanism which may be responsible for controlling the formation of the volcanic front in normal subduction zones, and discussed slab dehydration and wedge hydration processes relevant to forearc and arc magma genesis. In this paper, we review geological, petrological and tectonic constraints on the petrogenesis of HMA magmas, as deduced from two well studied areas, the Setouchi volcanic belt (SW Japan) and the Bonin (Ogasawara) Islands in the

W Pacific region; we enlarge upon the concepts presented in Tatsumi (1986, 1988) and propose a generalized petrological model for HMA magma genesis.

2.2 Geological constraints

2.2.1 Key tectono-magmatic features of HMA occurrences

Geological backgrounds for the Setouchi volcanic belt and the Bonin Islands are summarized in Table 2.1. The following observations should be taken into account in any tectonic model for production of these HMA magmas:

(a) palaeotectonic environments of HMA magma production appear to be closely associated with the formation of marginal or backarc basins,
(b) marginal basin formation preceded HMA magmatism,
(c) HMA volcanism was shortlived,
(d) the subducted oceanic lithosphere beneath the region was necessarily very young and hot, and
(e) basaltic and HMA magmas appear to have been generated simultaneously.

2.2.2 The Setouchi volcanic belt

The Setouchi volcanic belt extends for about 600 km along the SW Japan arc (Fig.2.1). Radiometric dating indicates that volcanism in the Setouchi volcanic belt was shortlived, occurring at 13 ± 1 Ma (Tatsumi *et al.* 1980, Tatsumi 1983a). The volcanic belt can be divided into five subprovinces, of which HMA occur only in the central three (Fig. 2.1). Although basalts in the Setouchi belt are volumetrically subordinate to HMA and normal calc-alkaline andesites, both HMA and basalts occur close together in time and space (Tatsumi 1983b).

Magma generation in the Setouchi volcanic belt is related to subduction of

Table 2.1 Geological background for production of HMA magmas.

	Setouchi volcanic belt	Bonin Is.
Age of volcanism	13 Ma	47–48 Ma
Backarc basin	Japan Sea	W Philippine Sea
Age of backarc basin	15 ± 1 Ma	49–59 Ma
Subducting plate	Philippine Sea	N New Guinea Plate
Age of slab	25–17 Ma	~50 Ma
Related volcanics	Ol-tholeiites	Ol-tholeiites
	CA-andesites	CA-andesites

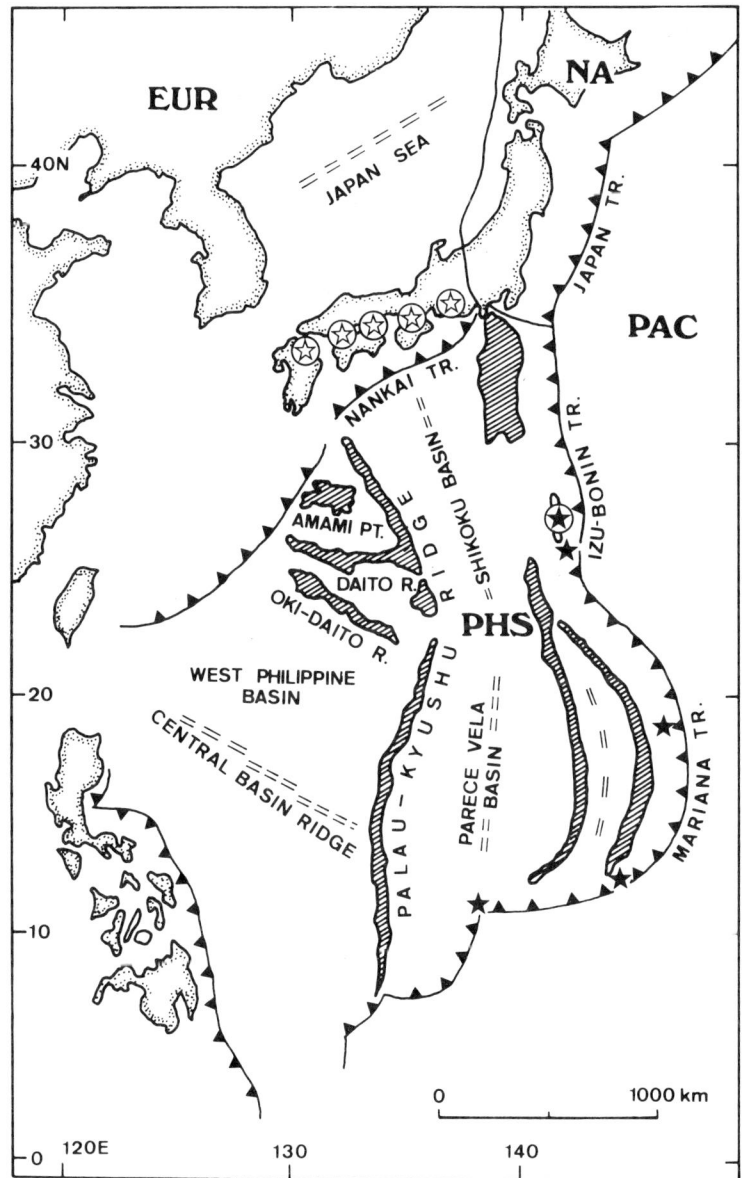

Figure 2.1 Localities of HMA and related rocks in the Setouchi volcanic belt (open stars) and the Bonin Islands (filled star in circle). HMA (boninites) have also been found along the Mariana Trench (filled stars). EUR, Eurasian plate; PAC, Pacific plate; PHS, Philippine Sea plate. Hatched areas indicate volcanic ridges and remnant arcs.

the Philippine Sea plate beneath the Eurasian plate. It has been suggested that the Shikoku Basin, located south of the Setouchi volcanic belt (Fig. 2.1), was formed by sea-floor spreading behind the Palau–Kyushu arc at 25–17 Ma (Kobayashi & Nakada 1978). DSDP drilling has shown that off-ridge volcanism took place in the Shikoku Basin around 15 Ma (Klein *et al.* 1978, Klein & Kobayashi 1980). These data strongly suggest that the Setouchi volcanic belt formed by processes resulting from subduction of newly borne and consequently relatively hot Shikoku Basin oceanic lithosphere.

Recent palaeomagnetic studies of Cretaceous to Recent rocks from SW Japan (Otofuji *et al.* 1985) have indicated that the SW Japan arc rotated clockwise through 40–50° at 15 ± 1 Ma (Fig. 2.2). This rotation can be correlated with spreading of the Japan Sea backarc basin, which, significantly, was followed shortly after by Setouchi belt volcanism. Setouchi belt lavas, at 13 ± 1 Ma, are magnetized parallel to the modern field, but strata conformably overlain by Setouchi lavas (Tatsumi 1983b) show clockwise diffractions of about 40°. During this Middle Miocene episode of HMA and backarc-basin magmatism, the Shikoku Basin lithospheric plate was young and hot, and consequently not easily subducted. Attempted subduction of this hot plate could have resulted from obduction of the SW Japan arc onto the Shikoku Basin plate as a result of opening of the Japan Sea backarc basin.

Some felsic-to-intermediate volcano-plutonic complexes were produced even closer to the trench than the Setouchi volcanic belt during the Middle Miocene (14 Ma; Shibata 1978). Palaeomagnetic studies of these rocks (Otofuji *et al.* 1985) demonstrated that this magmatism preceded backarc spreading (Fig.2.2); it may have been related to northward subduction of the Shikoku Basin plate.

2.2.3 Bonin Islands

The Bonin Islands lie astride the Bonin Ridge, which forms a forearc high facing the Bonin Trench, at which Jurassic Pacific oceanic crust is being subducted (Fig.2.1). They are composed of three island groups, from north to south, Muko-jima, Chichi-jima and Haha-jima. The former two consist of HMA (boninites) and their fractionation products, while the latter is characterized by basalts (Kuroda *et al.* 1983). Along the inner wall of the Bonin Trench in this region, HMA, tholeiitic basalts, more felsic volcanics and harzburgite–dunite have been dredged (Ishii 1985), and similar rock suites, including HMA, have also been dredged and drilled further south along the Mariana forearc (Fig. 2.1; Dietrich *et al.* 1978, Crawford *et al.* 1981, Hawkins *et al.* 1984, Bloomer & Hawkins 1987). Ages for Bonin–Mariana HMA volcanism were first given by Tsunakawa (1983), who reported K–Ar dates ranging from 43 to 4 Ma; this wide range of K–Ar dates is probably due to alteration of some dated lavas. Dobson (1986) gave a K–Ar age of 47.5 Ma for fresh glass separated from a Bonin Islands boninite, and found radiolarians

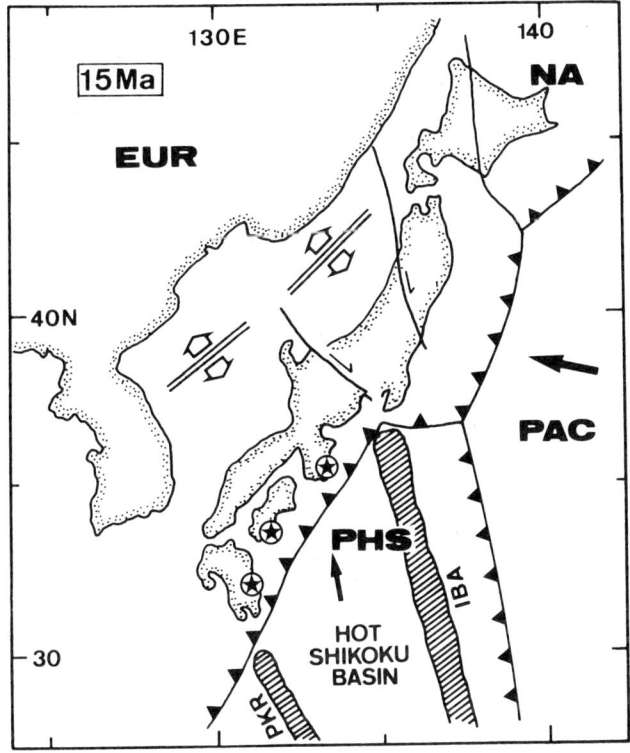

Figure 2.2 Palaeogeography of the Japanese Islands at 15 Ma. The backarc basin was opening to form the Japan Sea. The young Philippine Sea plate (Shikoku Basin between the Palau–Kyushu Ridge (PKR) and the Izu–Bonin arc (IBA)) was subducted to produce felsic-to-intermediate magmas along the region between the Setouchi volcanic belt and the trench (cf. Fig. 2.1). The leading edge of the subducting Philippine Sea plate has not yet arrived beneath the Setouchi volcanic belt.

in sediments interlayered with Chichi-jima boninites which indicate ages of 42.3–47 Ma.

At the time of HMA generation, the Bonin Islands are estimated to have been between 5°N and 5°S of the Equator (Fig. 2.3; Kodama *et al.* 1983, Keating *et al.* 1983). The West Philippine Basin, which was formed at the Central Basin spreading system, was situated just west of the proto-Bonin Islands at 47–48 Ma (Fig. 2.3) according to the reconstruction of the Philippine Sea by Seno & Maruyama (1984). Radiometric ages of opening of

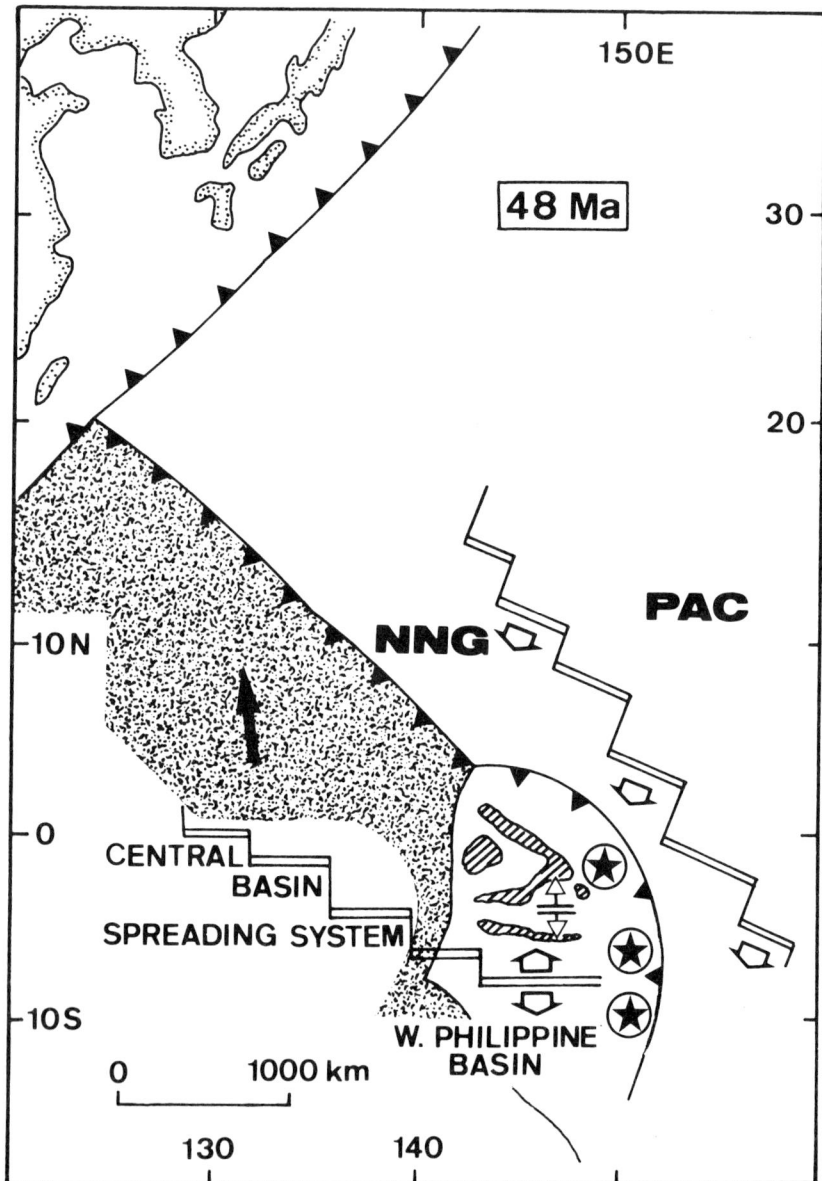

Figure 2.3 Palaeogeography of the Bonin Islands at 48 Ma. The West Philippine Basin formed 50–60 Ma behind the Bonin Islands. The N New Guinea plate (NNG) formed along the spreading system located just NE of the Bonin Islands, and was subducted beneath the young, hot West Philippine plate to generate HMA magmas (stars). Shaded area is that part of Philippine Sea floor subducted since 48 Ma ago.

the West Philippine Basin, a marginal basin situated immediately behind the Bonin Islands, range from 49 to 59 Ma (Seno & Maruyama 1984), indicating that formation of this marginal basin preceded HMA generation on the Bonin Islands (47–48 Ma).

On the basis of geological and geophysical evidence, Seno (1984) suggested that a now-consumed plate, the North New Guinea Plate, existed in the W Pacific during the early Tertiary. The ridge system for this plate was located just east of the Bonin Islands at this time, so that it is probable that recently generated, hot oceanic lithosphere was being subducted beneath the Bonin arc during early Tertiary, at the time of HMA production (Fig. 2.3).

The Palau–Kyushu Ridge, Amami Plateau and Oki–Daito Ridge are all remnant arcs (Fig. 2.1; Seno & Maruyama 1984), of which the Palau–Kyushu Ridge is the youngest (33–49 Ma; Seno & Maruyama 1984, Shiki *et al.* 1985) and the Amami Plateau is the oldest (85–75 Ma; Shiki *et al.* 1985). The absence of radiometric dates between 75 and 49 Ma suggests that no active subduction zone existed in this region before the HMA magmatic event at 47–48 Ma

2.3 Petrological constraints

2.3.1 Basic concepts in subduction-zone magma genesis

HMA magmatism appears to be restricted, spatially and temporally, to discrete magmatic events dependent on the attainment of certain prerequisite physical conditions; HMA magmas are not produced in all subduction zones. To understand the petrogenesis of HMA magmas, therefore, some knowledge of the mechanism of magma generation in normal subduction zones is required.

Figure 2.4 illustrates a plausible model for generation of subduction-zone basalt magmas (Tatsumi 1989). It should be stressed that in this model hydrous phases in subducted oceanic lithosphere totally decompose at levels shallower than 80 km, controlled by the stability limit of amphibole in the basalt layer of the slab. As the volcanic front (which marks the trenchward boundary of a volcanic arc) overlies the downgoing slab at an average depth of 110 km (Tatsumi 1986) in most of subduction zones, it is probable that the subducted slab can supply H_2O to the mantle wedge only beneath the forearc region. Recent experimental studies (Tatsumi *et al.* 1986, Tatsumi & Nakamura 1986, Tatsumi & Isoyama 1988) have shown that the slab-derived hydrous fluid phase is enriched in LILE (eg. Cs, Rb, K, Ba); also, at least some of the Be in the subducted oceanic crust can be transported by this hydrous fluid phase, implying that *partial melting* of the subducted slab is not a necessary prerequisite for the relocation of crustal-derived Be into subduction-zone lavas (Tera *et al.* 1985). Such fluids released from the slab react with overlying mantle wedge peridotite to form enriched hydrous peridotite. The mineral assemblage in the hydrous peridotite depends on ambient temperature in the region; at temperatures below 600°C, serpentine, amphibole, chlorite and phlogopite must be stabilized in the peridotite; at higher temperatures, serpentine is not stable (Kitahara *et al.* 1971, Tatsumi & Nakamura 1986), so that the hydrous mineral assemblage must be amphibole + chlorite + phlogopite (Fig. 2.4). Amphibole crystallizing in the hydrous peridotite is Mg-rich pargasite, which is stable at higher pressures than amphibole in a basalt system or in the slab (Lambert & Wyllie 1972, Millhollen *et al.* 1974).

The enriched and hydrated peridotite formed beneath the forearc is dragged down parallel to the upper surface of the slab by induced convection in the mantle wedge (Tatsumi 1986, 1989). The effective operation of this process is demonstrated by the following observations:

(a) Subduction-zone magmas are more H_2O-rich than magmas in other tectonic settings (Yoder 1969, Sakuyama 1983), although the subducted slab is essentially anhydrous beneath arc volcanoes.
(b) Geochemical characteristics of the enriched, hydrated peridotite are considered to be similar to those of subduction-zone magmas (Tatsumi *et al.* 1986).
(c) Seismological study of upper mantle beneath the NE Japan volcanic arc (Matsuzawa *et al.* 1986) demands the existence near the surface of the downgoing slab of a layer, less than 10 km thick, with a seismic velocity lower than that of overlying mantle wedge; the low-velocity layer may be composed of hydrated peridotite dragged downwards on the slab.

As the hydrated peridotite is dragged downwards, serpentine decomposes. Serpentine contains a large amount of H_2O ($\sim 13\%$) and is produced from olivine and orthopyroxene, the major constituents of peridotite. Although chlorite also contains around 13% H_2O, much less chlorite than serpentine will crystallize, since chlorite is an aluminous phase. The assemblage amphibole–chlorite–phlogopite, therefore, probably traps less H_2O in the slab than does serpentine. Through decomposition of serpentine in the down-dragged hydrated layer, a column of hydrous peridotite without serpentine is built in the forearc mantle wedge above the 600°C geotherm in the down-dragged layer (Fig. 2.4). In normal subduction zones, the solidus temperature of hydrous peridotite (~ 1000°C; Kushiro *et al.* 1968) is not attained in the hydrous peridotite column formed beneath the forearc, so that no magmas are generated in this region. Given, however, higher geothermal gradients in such regions, HMA magmas may form by partial melting of the hydrous column and be erupted in near-trench locations, as discussed in a later section.

Amphibole and chlorite in hydrated peridotite being dragged down along the subduction zone become unstable virtually independent of temperature, at pressures around 35 kbar, equivalent to the depth (110 km) of the slab beneath the volcanic front (Millhollen *et al.* 1974, Goto *et al.* 1989). Any H_2O released

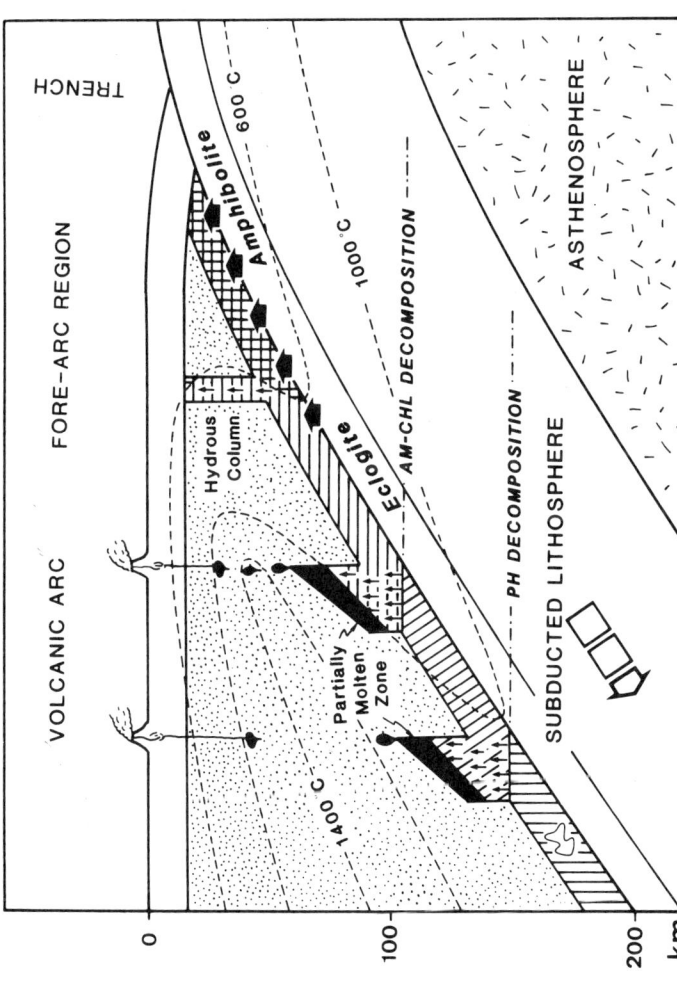

Figure 2.4 A model for migration of H$_2$O and the genesis of basaltic magmas in normal subduction zones (Tatsumi 1989). Hydrous mineral assemblages in the down-dragged peridotite layer at the bottom of the mantle wedge change from serpentine + amphibole (Am) + chlorite (Ch) + phlogopite (Ph), shown with 'grid' hatching, via Am + Ch + Ph (horizontal lines) to Ph (oblique lines). Magma production beneath a volcanic arc is governed by dehydration of Am, Ch and Ph in the down-dragged hydrated peridotite. The solidus temperature of hydrous peridotite (1000°C) is not attained in the forearc mantle wedge, in which the hydrous column is built by decomposition of serpentine in the down-dragged hydrated peridotite.

from amphibole and chlorite migrates upwards to reform amphibole–chlorite peridotite in the higher-temperature and lower-pressure region of the mantle wedge. When the hydrous front reaches a level with a temperature of 1000°C, the solidus temperature of wet peridotite, an initial melt is produced by partial melting of amphibole–chlorite peridotite. Further H_2O supplied to the partially molten zone will expand the partially molten zone upwards. In this zone, the degree of partial melting becomes larger upwards, since the zone is expanding towards the lower-pressure and high-temperature region. A mantle diapir composed of partially molten peridotite rises from the top of the zone, passes through the high-temperature region ($>1400°C$), and stops rising at about 10 kbar (~ 30 km) to release a primary magma for arc volcanism along the volcanic front (Tatsumi et al. 1983). Some H_2O released by decomposition of amphibole and chlorite at 35 kbar must be fixed in phlogopite, which becomes the hydrous phase in peridotite dragged to deeper levels. Preliminary high-pressure experiments (Sudo & Tatsumi, unpublished) indicate that phlogopite in the presence of clinopyroxene breaks down around 60 kbar at 1000°C, perhaps suggesting a role for phlogopite in the formation of magmas on the backarc side of volcanic arcs (Fig. 2.4), where the depth of magma segregation from a mantle diapir is likely to be greater (Tatsumi et al. 1983).

2.3.2 *Experimental petrology studies of HMA magmas*

Some Setouchi and Bonin Islands HMA are unfractionated or primary magmas which have Mg# values appropriate for liquids in equilibrium with mantle peridotite (Kuroda et al. 1978, Tatsumi & Ishizaka 1981, 1982a,b). Melting experiments at high pressures were carried out on several of these HMA to determine $P-T-X_{H_2O}$ conditions for segregation of HMA magmas from mantle peridotite. Tatsumi (1981, 1982) determined phase relations at high pressures and temperatures on two Setouchi HMA, TGI (a bronzite olivine HMA) and SD261 (an augite olivine HMA); both contain less than 6 modal% phenocrysts and are candidates for primary magmas. The results of melting experiments are summarized in Figure 2.5. SD261 and TGI are in equilibrium with lherzolite and harzburgite assemblages, respectively, under both H_2O-saturated ($>15\%$ H_2O in the melt) and undersaturated (7–8% in the melt) conditions. It follows that these HMA may have been produced from peridotite sources leaving different residual phases after magma segregation. Those HMA magmas with lower H_2O contents may equilibrate with peridotite at lower pressures and higher temperatures. However, as the depth to moho beneath the Setouchi region is 30–40 km (~ 10 kbar; Yoshii et al. 1974), HMA magmas formed in the upper mantle were probably generated at pressures greater than 10 kbar. Differences in the temperature of multiple saturation and in residual phase assemblages suggest that the TGI magma formed by higher degree of partial melting than the SD261 magma; this is responsible for the

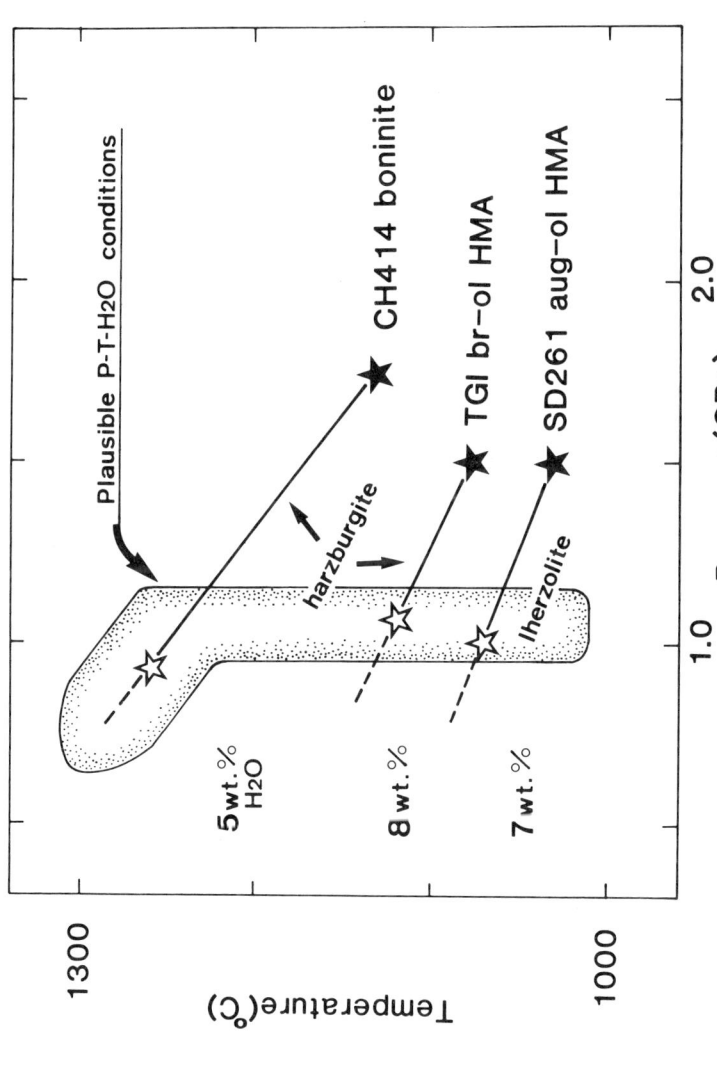

Figure 2.5 Points of multiple saturation for three HMA compositions under H_2O-saturated (filled stars) and H_2O-undersaturated (open stars) conditions. CH414 and TGI coexist with olivine + orthopyroxene and SD261 with olivine + orthopyroxene + clinopyroxene, suggesting generation via different degrees of partial melting (see text). The dotted area indicates plausible P–T conditions for HMA magma generation.

different phenocryst assemblages of these HMA (bronzite + olivine vs. augite + olivine) as discussed by Tatsumi (1982).

High-pressure melting experiments on an HMA CH414 (boninite) from the Bonin Is. which contains 7 modal% of olivine and 12 modal% of bronzite phenocrysts showed that this composition coexists with harzburgite at pressures of 8 and 17 kbar (Fig. 2.5) with 5% and 15% of H_2O, respectively (Umino & Kushiro 1989). Protoenstatite, which is not a phenocryst phase in the HMA studied, appeared as a near-liquidus phase at pressures less than 3 kbar, suggesting that the magma crystallized at deeper levels. As shown in Figure 2.5, this Bonin Is. HMA was generated at higher temperatures and hence by higher degrees of partial melting than Setouchi HMA magmas; this is supported by the fact that clinoenstatite phenocrysts which characterize some HMA in the Bonin Is. are not found in Setouchi HMA.

Since most HMA and their derivative lavas contain no vesicles, and many Setouchi HMA are also quite phenocryst-poor (< 10 modal%), HMA magmas probably form under H_2O-undersaturated conditions. Nevertheless, H_2O contents of HMA magmas are much greater (> 7% in Setouchi HMA) than those in typical subduction-zone basaltic magmas (< 3%; Sakuyama 1983); unusually hydrous conditions are necessary to produce HMA magmas.

Water in HMA magmas must be supplied from downgoing oceanic lithosphere. Dobson & O'Neil (1987) and Dobson (1986) have shown that the δD values for Bonin Is. HMA are about −53 permil; they favoured a subducted slab source of H_2O in these rocks. Tatsumi *et al.* (1986) demonstrated that the mantle sources of Setouchi and Bonin Is. HMA magmas both had geochemical characteristics very similar to those of slab-derived fluids, as determined experimentally (Tatsumi *et al.* 1986, Tatsumi & Nakamura 1986), also supporting a slab origin for H_2O in HMA magmas.

2.3.3 Formation of HMA magmas in the forearc wedge

Release of slab-derived hydrous fluids into the down-dragged wedge peridotite forms a column of hydrous peridotite with chlorite, amphibole and phlogopite in the forearc wedge (Fig. 2.4). Under normal thermal conditions in the sub-arc mantle, the solidus temperature of hydrous peridotite (∼ 1000°C) is not attained beneath the forearc. If, however, an anomalously high thermal gradient develops in the shallow subforearc mantle wedge (due to a number of possible factors discussed later), magmas may be produced by partial melting of the hydrous column. Low-pressure partial melts of peridotite are more SiO_2-rich than basaltic magmas formed at higher pressure by similar degrees of melting (e.g. Kushiro 1969, Takahashi & Kushiro 1983), so that these forearc magmas (HMA) are more siliceous than those generated beneath normal volcanic arcs.

2.4 Discussion

Below we propose a petrogenetic model for HMA generation in the Setouchi and Bonin Is. regions (Fig. 2.6), constrained by both geological and petrological observations.

2.4.1 Initial subduction of young oceanic lithosphere and backarc opening

Palaeomagnetic and K–Ar studies (Otofuji et al. 1985, Shibata 1978) have shown that felsic-to-intermediate magmatism occurred along the southern limit of SW Japan around 15 Ma, before the opening of the Japan Sea backarc basin, suggesting that part of the Philippine Sea plate was subducting beneath this area at this time. However, the absence of subduction-related volcanism in the Setouchi region at this time implies that the leading edge of the subducted Philippine Sea plate had not yet arrived beneath this region. It should also be stressed that the subducting Philippine Sea plate was young (25–17 Ma; Kobayashi & Nakada 1978) and hot enough to cause off-ridge volcanism within the Shikoku Basin. At 15 ± 1 Ma (Otofuji et al. 1985), the Japan Sea formed by backarc spreading involving injection of hot MORB-source asthenosphere (Nohda et al. 1988) into the backarc side of SW Japan, raising ambient upper-mantle temperatures (Fig. 2.6). A high thermal gradient was produced in this region by both the initial subduction of hot oceanic lithosphere and the injection of hot asthenosphere into the mantle wedge. The upper mantle beneath SW Japan which was intruded by MORB-source diapirs during Japan Sea opening was relatively fertile subcontinental lithosphere (Nohda et al. 1988), so that Setouchi HMA were notably enriched in Al, Ca and LILE relative to Bonin HMA (Tatsumi & Ishizaka 1982a).

In the case of the Bonin Is., newly borne oceanic lithosphere created at the spreading centre on the North New Guinea plate probably abutted a transform fault in the West Philippine Basin at 50–60 Ma (Fig. 2.6). The formation of the marginal basin just before production of HMA magmas again suggests that the geothermal gradient in this region was anomalously high at that time. The highly depleted geochemical nature of Bonin Is. HMA demands a mantle source which had undergone previous melting episodes, perhaps involving extraction of MORB and/or arc basalt magmas.

2.4.2 Production of HMA magmas

Initiation of backarc spreading causes the trench to retreat oceanward or the arc to be obducted onto the new oceanic plate, resulting in subduction of young hot oceanic lithosphere (Fig. 2.6). Downgoing lithosphere releases LILE-enriched hydrous fluids which migrate upwards to hydrate peridotite in the mantle wedge. When this hydrous peridotite is dragged downwards to the

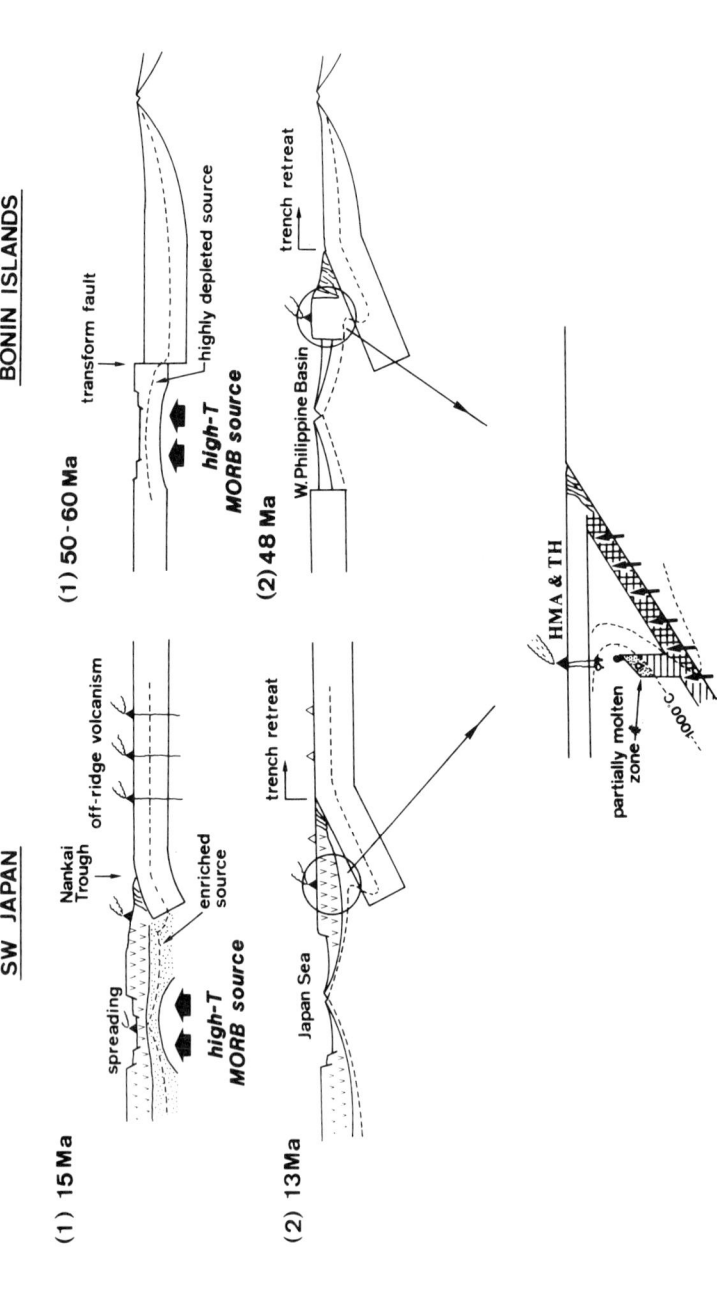

Figure 2.6 Petrogenetic model for production of HMA magmas in the Setouchi and Bonin Is. areas (see text for detailed discussion). Broken curve indicates 1000°C isotherm, the solidus temperature of hydrous peridotite.

region with temperature around 600°C, serpentine in the peridotite decomposes, forming a hydrous column above this region. These processes are thought to occur in the forearc wedge in all active subduction zones. However, both injection of hot, MORB-source mantle beneath the backarc region (Crawford et al. 1981) and subduction of hot lithosphere can contribute to raising the ambient wedge temperature to the solidus temperature of hydrous peridotite (1000°C). When the upward-migrating H_2O front (stabilizing chlorite, amphibole and phlogopite in wedge peridotite in its wake) reaches a level with temperatures of 1000°C and higher, HMA magmas are formed.

Further addition of H_2O to the partially molten zone results in its upward growth towards the higher-temperature region (Fig. 2.6). As the solidus temperature of hydrous peridotite is temperature-dependent at pressures between 10 and 40 kbar (Kushiro et al. 1968), the degree of partial melting increases with increasing temperature in the partially molten column.

Although partial melt in every diapir is H_2O-saturated as it starts rising, the total amount of H_2O in any diapir is larger in a diapir with a higher degree of partial melting. Therefore, a diapir initially with a larger degree of partial melting, produced at a shallower level, will produce an HMA magma and that with lower degree will produce a basalt magma. A mantle diapir with relatively high degree of partial melting should rise fast and nearly adiabatically, whereas a mantle diapir with a smaller degree of melting should rise more slowly, absorbing heat from surrounding higher-temperature upper mantle and producing higher-temperature basaltic melt. Experimental work has suggested that the depth of magma segregation of both HMA and basalt was about 30 km in the Setouchi volcanic belt (Tatsumi 1981, 1982).

Gradual cooling of asthenosphere, which yielded backarc-basin basalt magmas, and continued subduction of oceanic crust necessarily cooler than the mantle wedge both combine to cool the forearc wedge to below the solidus temperature of hydrous peridotite, terminating HMA production. If oceanic lithosphere continues to be subducted, down-dragged and hydrated peridotite releases H_2O at depths around 110 km to produce normal arc magmas along the volcanic front. However, for both the Setouchi and Bonin Is. regions, the subduction velocity of hot, young oceanic lithosphere would have been very low, so that subduction effectively did not operate after obduction of the arc-side plate caused by the backarc spreading. This may be responsible for the absence of any volcanism in these regions after production of near-trench HMA magmas.

2.4.3 *Modern analogues for eruption settings of HMA magmas*

We will now discuss present subduction-zone settings where HMA may occur or may erupt in the near future. Although no subduction zone completely satisfies the geotectonic conditions for production of HMA magmas suggested for the Setouchi and Bonin Is. regions (initial subduction of hot oceanic lithosphere into the mantle wedge heated by the backarc-basin formation), two areas should be noted, central Mexico and the Melanesian region east of New Guinea.

Subduction of newly generated oceanic crust (of the Rivera and the Cocos plates) beneath the North American plate (Fig. 2.7) has produced the Mexican volcanic belt, which is not parallel to, but makes an angle of 15° with, the Middle American Trench, so that the distance between trench and volcanic front decreases westwards. Deep earthquakes (> 150 km) have not been observed beneath active volcanoes in the volcanic belt (Nixon 1982).

Luhr & Carmichael (1980) emphasized the existence of a graben in the western end of the Mexican volcanic belt (Fig. 2.7), and suggested that a segment of the East Pacific Rise is in the process of jumping eastwards, to beneath the graben. Alternatively, we propose that normal backarc spreading is commencing along the graben, associated with subduction of the Rivera plate. If this is the case, HMA might be expected to be generated in this region, since the arc–trench system satisfies the necessary conditions described above for HMA magma production. Bloomfield (1975) reported late-Quaternary 'andesites' rich in MgO (> 7% at 55% SiO_2) from the Mexican volcanic belt. Petrographic (absence of plagioclase phenocrysts) and compositional ($FeO^*/MgO < 1$) characteristics of these andesites suggest that HMA volcanism may be presently active in this region.

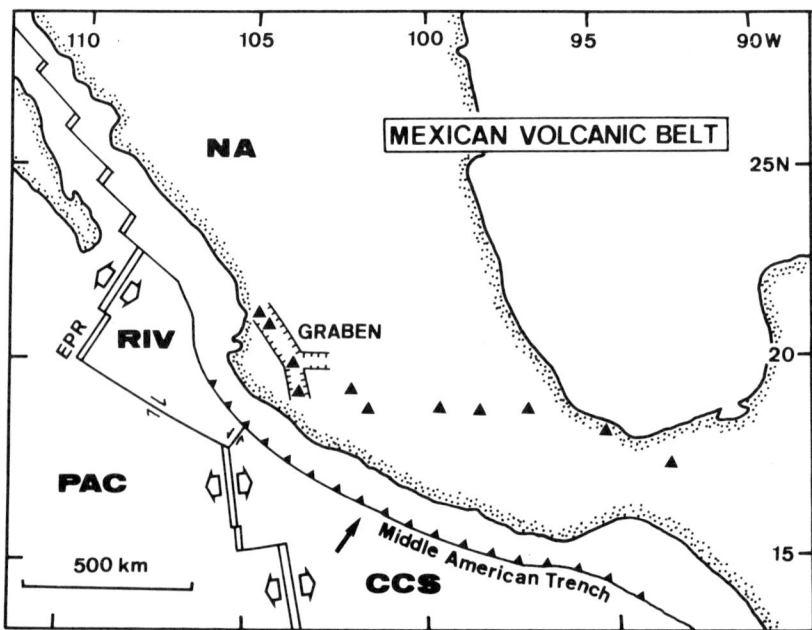

Figure 2.7 Tectonic setting of the Mexican volcanic belt. NA, North American plate; PAC, Pacific plate; CCS, Cocos plate; RIV, Rivera plate; EPR, East Pacific Rise. Triangles mark locations of active volcanoes.

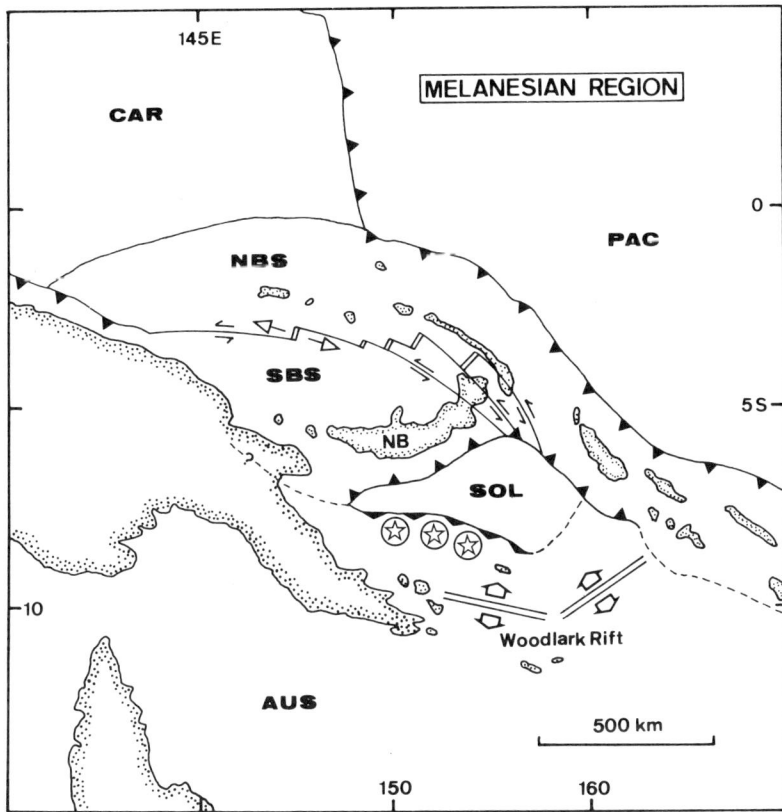

Figure 2.8 Regional tectonic setting of the E Papua New Guinea (Melanesian) region. NB, New Britain arc; CAR, Caroline plate; PAC, Pacific plate; NBS, North Bismark plate; SBS, South Bismark Sea plate; SOL, Solomon Sea plate; AUS, Australian plate. HMA may be produced along the southern boundary of SOL (open stars).

Complex plate interactions govern the tectonics and volcanism in the Melanesian (E Papua New Guinea– New Britain) region (Fig. 2.8), where two active spreading systems, the Manus Rift and the Woodlark Rift, are separated by the Solomon Sea microplate. The latter descends beneath the South Bismark plate to a depth of about 500 km (Johnson 1977), to form Quaternary volcanoes of the New Britain arc. No clear seismic plane has been observed along the southern margin of the Solomon Sea plate (Honza *et al.* 1987). However, it is possible that southward subduction will take place in the future, so that oceanic lithosphere produced at the active Woodlark Rift will obduct onto the Solomon Sea microplate.

Although the Solomon Sea plate is probably composed of oceanic crust of Tertiary age (Hamilton 1979), two factors may contribute to the formation of HMA magmas along the boundary between the Solomon Sea plate and the newly borne plate at the Woodlark Ridge: (a) the Woodlark Rift is presently active, so that hot asthenosphere must be rising to 'overheat' this region, and (b) initial subduction of oceanic lithosphere, even if it is relatively old and cold, would not effectively cool the near-trench mantle wedge. In contrast, in the New Britain arc, behind which backarc opening is also taking place, HMA magmas are not likely to be generated because of the existence of mature slab beneath the arc.

2.5 Conclusions

Both the Setouchi volcanic belt and the Bonin Islands, where HMA magmas were produced in anomalously near-trench settings, were characterized by peculiar and specific tectonic settings at the time of HMA generation; the subducted lithosphere was young and hot, and backarc opening took place just before the HMA magmatism.

HMA magmas can be generated by partial melting of a hydrous peridotite column in the near-trench mantle wedge formed by dehydration of serpentine in down-dragged hydrated peridotite at the base of the wedge, under abnormally high thermal gradients associated with the above tectonic scenario. In normal subduction zones, however, the solidus temperature of hydrous peridotite is not attained in the forearc mantle wedge.

Two areas, the Mexican volcanic belt and the Melanesian (E Papua–Solomon Sea) region, are modern settings which satisfy the above specific tectonic conditions for HMA magma generation, and HMA magmas may be presently or in the near future erupted in these regions.

Acknowledgements

The authors are deeply indebted to Drs A. J. Crawford and G. A. Jenner for helpful comments and criticism on this manuscript. This study was in part sponsored by Grant-in-Aid for Special Project Research of Ministry of Education (No. 62103007) and Inamori Foundation to Y. T., and by a Grant-in-Aid for Scientific Research of Ministry of Education (No. 61540588) to S. M.

References

Bloomer, S. H. & J. W. Hawkins 1987. Petrology and geochemistry of boninite series volcanic rocks from the Mariana Trench. *Contrib. Mineral. Petrol.* **97**, 361–77.

Bloomfield, K. 1975. A late-Quaternary monogenetic volcano field in Central Mexico. *Geol. Rundsch.* **64**, 476–97.

Crawford, A. J., L. Beccaluva & G. Serri 1981. Tectono-magmatic evolution of the West Philippine–Mariana region and the origin of boninites. *Earth Planet. Sci. Lett.* **54**, 346–56.

Crawford, A. J., L. Beccaluva, G. Serri & J. Dostal 1986. Petrology, geochemistry and tectonic implications of volcanics dredged from the intersection of the Yap and Mariana trenches. *Earth Planet. Sci. Lett.* **80**, 265–80.

Dietrich, V., R. Emmermann, R. Oberhansli & H. Puchelt 1978. Geochemistry of basaltic and gabbroic rocks from the West Mariana Basin and Mariana trench. *Earth Planet. Sci. Lett.* **39**, 127–44.

Dobson, P. F. 1986. The petrogenesis of boninite: a field, petrologic and geochemical study of volcanic rocks of Chichijima, Bonin Islands, Japan. Ph.D. Thesis, Stanford University.

Dobson, P. F. & J. R. O'Neil 1987. Stable isotope compositions and water contents of boninite series volcanic rocks from Chichi-jima, Bonin Islands, Japan. *Earth Planet. Sci. Lett.* **82**, 75–86.

Goto, A., A. Sudo & Y. Tatsumi 1989. Stability of chlorite in the upper mantle. *Am. Mineral.* (submitted).

Hamilton, W. 1979. *Tectonics of the Indonesian region.* U.S. Geol. Surv. Prof. Paper no. 1078.

Hawkins, J. W., S. H. Bloomer, C. A. Evans & J. T. Melchior 1984. Evolution of intro-oceanic arc–trench systems. *Tectonophysics* **102**, 175–205.

Honza, E., H. L. Davies, J. B. Keene & D. L. Tiffin 1987. Plate boundaries and evolution of the Solomon sea region. *Geo-Marine Lett.* **7**, 161–8.

Ishii, T. 1985. Dredged samples from the Ogasawara forearc seamount or 'Ogasawara Paleoland' 'forearc ophiolite'. In *Formation of active ocean margins*, N. Nasu (ed.), 307–42. Tokyo: Terrapub.

Johnson, R. W. 1977. *Distribution and major-element chemistry of late Cenozoic volcanoes at the southern margin of the Bismark Sea, Papua New Guinea.* Bur. Min. Res. Austr. Rep. no. 188.

Keating, B. H., K. Kodama & C. E. Helsley 1983. Paleomagnetic studies of the Bonin and Mariana island arcs. in *Arc volcanism: physics and tectonics*, D. Shimozuru and I. Yokoyama (eds), 243–59. Tokyo: Terrapub.

Kitahara, S., S. Takenouchi & G. C. Kennedy 1971. Phase relations in the system $MgO-SiO_2-H_2O$ at high temperatures and pressures. *Am. J. Sci.* **264**, 223–33.

Klein, G. D. & K. Kobayashi 1980. Geological summary of the north Philippine Sea, based on Deep Sea Drilling Project Leg 58 results. Init. Rep. DSDP Leg 58, 951–61.

Klein, G. D., et al. 1978. Off-ridge volcanism and seafloor spreading in the Shikoku Basin. *Nature* **27**, 746–8.

Kobayashi, K. & M. Nakada 1978. Magnetic anomalies and tectonic evolution of the Shikoku Inter-Arc Basin. *J. Phys. Earth* **26**, 391–402.

Kodama, K., B. H. Keating & C. E. Helsley 1983. Paleomagnetism of the Bonin Islands and its tectonic significance. *Tectonophysics* **95**, 25–42.

Kuroda, N., K. Shiraki & H. Urano 1978. Boninite as a possible calc-alkalic primary magma. *Bull. Volcanol.* **41**, 563–75.

Kuroda, N., H. Urano & K. Shiraki 1983. Least differentiated basalt from Muko-Jima in the Haha-Jima Group, Bonin Islands. *J. Geol. Soc. Japan* **89**, 303–6.

Kushiro, I. 1969. The system forsterite–diopside–silica with and without water at high pressures. *Am. J. Sci.* **267A**, 269–94.

Kushiro, I. 1972. Effect of water on the composition of magma formed at high pressures. *J. Petrol.* **33**, 311–34.

Kushiro, I., S. Shono & S.-I. Akimoto 1968. Melting of a peridotite nodule at high pressures and high water pressures. *J. Geophys. Res.* **73**, 6023–9.

Lambert, I. B. & P. J. Wyllie 1972. Melting of gabbro (quartz eclogite) with excess water to 35 kilobars with geological applications. *J. Geol.* **80**, 693–720.

Luhr, J. F. & I. S. E. Carmichael 1980. The Colima volcanic complex, Mexico, I. Post-caldera andesite from Volcan Colima. *Contrib. Mineral. Petrol.* **71**, 343–72.

Matsuzawa, T., M. Umino, A. Hagesawa & A. Takagi 1986. Upper mantle velocity structure estimated from PS-converted wave beneath north-eastern Japan Arc. *Geophys. J. R. Astr. Soc.* **86**, 767–87.

Meijer, A. 1980. Primitive arc volcanism and a boninite series: examples from western Pacific island arcs. In *Tectonic and geologic evolution of southeast Asian seas and islands*, D. E. Hayes (ed.), 269–82. Am. Geophys. Union Monogr. no. 23.

Millhollen, G. K., A. J. Irving & P. J. Wyllie 1974. Melting interval of peridotite with 5.7 per cent water to 30 kbars. *J. Geol.* **82**, 575–87.

Mysen, B. O. & A. L. Boettcher 1975. Melting of a hydrous mantle, II: Geochemistry of crystals and liquids formed by anatexis of mantle peridotite at high pressures and high temperatures as a function of controlled activities of water, hydrogen and carbon dioxide. *J. Petrol.* **16**, 549–90.

Nixon, G. T. 1982. The relationship between Quaternary volcanism in central Mexico and the seismicity and structure of subducted lithosphere. *Geol. Soc. Am. Bull.* **93**, 514–23.

Nohda, S., Y. Tatsumi, Y. Otofuji & K. Ishizaka 1988. Asthenospheric injection and backarc opening: isotopic evidence from northeast Japan. *Chem. Geol.* **68**, 317–27.

O'Hara, M. J. 1965. Primary magmas and the origin of basalts. *Scot. J. Geol.* **1**, 19–40.

Otofuji, Y., A. Hayashida & M. Torii 1985. When was the Japan Sea opened?: paleomagnetic evidence from southwest Japan. In *Formation of active ocean margins*, N. Nasu (ed.), 551–66. Tokyo: Terrapub.

Sakuyama, M. 1983. Petrology of arc volcanic rocks and their origin by mantle diapir. *J. Volcanol. Geotherm. Res.* **18**, 297–320.

Seno, T. 1984. Was there a North New Guinea plate? *Rep. Geol. Surv. Japan* **263**, 29–42.

Seno, T & S. Maruyama 1984. Paleogeographic reconstruction and origin of the Philippine Sea. *Tectonophysics* **102**, 53–84.

Shibata, K. 1978. Contemporaneity of Tertiary granites in the Outer Zone of Southwest Japan. *Bull. Geol. Surv. Japan* **29**, 551–4.

Shiki, T., A. Mizuno & K. Kobayashi 1985. Data listing of the bottom materials dredged and cored from the northern Philippine Sea. In *Geology of the northeastern Philippine Sea*, T. Shiki (ed.), 23–41. Tokyo: Tokai Univ. Press.

Shiraki, K., N. Kuroda, S. Maruyama & H. Urano 1978. Evolution of the Tertiary volcanic rocks in the Izu–Mariana arc. *Bull. Volcanol.* **41**, 548–62.

Takahashi, E. & I. Kushiro 1983. Melting of a dry peridotite at high pressures and basalt magma genesis. *Am. Mineral.* **68**, 859–79.

Tatsumi, Y. 1981. Melting experiments on a high-magnesian andesite. *Earth Planet. Sci. Lett.* **54**, 357–65.

Tatsumi, Y. 1982. Origin of high-magnesian andesites in the Setouchi volcanic belt, southwest Japan, II. Melting phase relations at high pressures. *Earth Planet. Sci. Lett.* **60**, 305–17.

Tatsumi, Y. 1983a. High magnesian andesites in the Setouchi volcanic belt, southwest Japan and their possible relation to the evolutionary history of the Shikoku Inter-arc Basin. *Am. Geophys. Union Geodyn. Ser.* **11**, 331–41.

Tatsumi, Y. 1983b. Volcanic geology of Shodo-shima island, Kagawa Prefecture, southwest Japan, and its bearing on paleoenvironment of the Seto Inland Sea area. *J. Geol. Soc. Japan* **89**, 693–706.

Tatsumi, Y. 1986. Formation of volcanic front in subduction zones. *Geophys. Res. Lett.* **13**, 717–20.

Tatsumi, Y. 1989. Migration of fluid phases and genesis of basalt magmas in subduction zones. *J. Geophys. Res.* (in press).

Tatsumi, Y. & K. Ishizaka 1981. Existence of andesitic primary magma: an example from southwest Japan. *Earth Planet. Sci. Lett.* **53**, 124–30.

Tatsumi, Y. & K. Ishizaka 1982a. High magnesian andesite and basalt from Shodo-shima island, southwest Japan, and their bearing on the genesis of calc-alkaline andesites. *Lithos* **15**, 161–172.

Tatsumi, Y. & K. Ishizaka 1982b. Origin of high magnesian andesites in the Setouchi volcanic belt, southwest Japan, I. Petrographical and chemical characteristics. *Earth Planet Sci. Lett.* **60**, 293–304.

Tatsumi, Y. & H. Isoyama 1988. Transportation of beryllium with H_2O at high pressures and implication for magma genesis in subduction zones. *Geophys. Res. Lett.* **15**, 180–3.

Tatsumi, Y. & N. Nakamura 1986. Composition of aqueous fluid from serpentinite in the subducted lithosphere. *Geochem. J.* **20**, 191–6.

Tatsumi, Y., D. L. Hamilton & R. W. Nesbitt 1986. Chemical characteristics of fluid phase released from a subducted lithosphere and origin of arc magmas: evidence from high-pressure experiments and natural rocks. *J. Volcanol. Geotherm. Res.* **29**, 293–309.

Tatsumi, Y., M. Sakuyama, H. Fukuyama & I. Kushiro 1983. Generation of arc basalt magmas and thermal structure of the mantle wedge in subduction zones. *J. Geophys. Res.* **88**, 5815–25.

Tatsumi, Y., M. Torii & K. Ishizaka 1980. On the age of the volcanic activity and the distribution of the Setouchi volcanic rocks. *Bull. Volcanol. Soc. Japan* **25**, 171–9.

Tera, F., L. Brown, J. Morris & I. S. Sacks 1985. Sediment incorporation in island-arc magmas: inferences from ^{10}Be. *Geochim. Cosmochim. Acta* **50**, 535–50.

Tsunakawa, H. 1983. K–Ar dating on volcanic rocks in the Bonin islands and its tectonic implication. *Tectonophysics* **95**, 221–32.

Umino, S. & I. Kushiro 1989. Experimental studies on boninite petrogenesis (this volume).

Yoder, H. S. 1969. Calc-alkalic andesites: experimental data bearing on the origin of their assumed characteristics. *Proc. Andesite Conf.*, A. R. McBirney (ed.), 77–89. Dept Geol. Mineral. Indust. Oregon, Bull. no. 65.

Yoshii, T., Y. Sasaki, H. Okada, S. Asano, I. Muramatu, M. Hashizume & T. Moriya 1974. The third Kurayoshi explosion and the crustal structure in the western part of Japan. *J. Phys. Earth* **22**, 109–21.

3 High-K_2O rocks with high-MgO, high-SiO_2 affinities

S.F. FOLEY & G. VENTURELLI

Abstract

Ultrapotassic igneous rocks with $SiO_2 > 53\%$ and $MgO > 8\%$ occur predominantly in the W Mediterranean region with localities in SE Spain, the Italian Alps, Tuscany and Algeria. Other isolated occurrences include the Bohemian Massif, Turkey, and the Leucite Hills and Navajo regions of the USA.

The most Mg-rich rocks in the W Mediterranean region are believed to have a similar origin to boninitic rocks in that they are derived by partial melting of strongly depleted peridotitic mantle source rocks at shallow depths in the presence of H_2O. However, in the case of the ultrapotassic rocks, an enrichment event introduced large amounts of LILE into the source region after the major depletion event, but before the melting event which produced the ultrapotassic rocks. This enriching event may be similar in nature to enrichments seen in many boninitic rocks, but introduced much greater volumes of melt/fluid, so that a hydrous phase (probably mica) was stabilized in the source, and formed part of the later melting assemblage. The tectonic setting for partial melting in the W Mediterranean is not directly comparable to the forearc setting of many boninitic rocks, but is post-collisional, involving an anomalous, hydrous mineral-bearing source derived from a previously depleted subcontinental mantle which was later enriched during Andean-type subduction-related processes. High-MgO, high-SiO_2, ultrapotassic rocks from other regions are probably derived by fractional crystallization from mafic alkaline lamprophyric melts, possibly incorporating a minor crustal component in the source.

3.1 Introduction

Ultrapotassic rocks, as recently defined by Foley *et al.* (1987), are rocks with $> 3\%$ K_2O, $> 3\%$ MgO and $K_2O/Na_2O > 2$. These rocks are not abundant, but are widespread, and a large majority are thought to be derived by varying amounts of differentiation from partial melts of upper mantle. Chemically, they are very variable, with SiO_2 ranging from $\sim 30\%$ to $> 70\%$, and MgO ranging up to 28%, and they are enriched in LILE, which broadly correlate with K_2O content. Ultrapotassic rocks form three major groups which tend to have characteristic tectonic settings: continental regions (Group I = lamproites), continental rifts (Group II = kamafugites) and island arcs (Group III). Other rocks were treated as transitional (Group IV) by Foley et al. (1987) because they cannot be consistently placed in any of the major groups. As a result of their unusual chemistry, the ultrapotassic rocks frequently exhibit peculiar mineralogical features such as abundant mica and leucite, and may contain rarer minerals such as priderite, wadeite, richterite, kalsilite and melilite. The mineralogy is described in reviews by Bergman (1987), Mitchell (1985), Gupta & Yagi (1980) and Rock (1984).

Ultrapotassic rocks containing $> 53\%$ SiO_2 and 8% MgO occur very rarely, including only 68 of more than 800 analyses considered by Foley et al. (1987). The locations and references to previous descriptions are given in Table 3.1. These occurrences are restricted to lamproites (Group I) and some transitional rocks with lamproitic affinity.

Lamproites mostly occur in stable continental tectonic settings, but other than two rocks from the Leucite Hills, USA, the rocks treated in this paper

Table 3.1 Lamproites (L) and rocks with lamproitic affinity (LA) with $SiO_2 > 53\%$ and $MgO > 8\%$.

Type	Locality	Age	Analyses	References
L	Betic Cordillera, SE Spain	8.6–5 Ma	50	Fuster et al. (1967), Lopez Ruiz & Rodriguez Badiola (1980), Nixon et al. (1984), Venturelli et al. (1984a, 1988), Wagner & Velde (1986)
L	Northwestern Alps, Italy	31 Ma	5	De Marco (1958), Dal Piaz et al. (1979), Venturelli et al. (1984b), Wagner & Velde (1985)
LA	Orciatico, Tuscany, Italy	4.1 Ma	3	Barberi & Innocenti (1967), Wagner & Velde (1986)
LA	Bohemian Pluton, Czechoslovakia	~331 Ma	3	Nemec (1974), Holub (1977)
LA	Erzgebirge, DDR	~355 Ma	1	Kramer (1976)
LA	Afyon, Turkey	15–9 Ma	1	Keller (1983)
L	Koudiat el Anzazza, Algeria	~11 Ma	2	Vila et al. (1974)
L	Leucite Hills, Wyoming, USA	1.1 Ma	2	Carmichael (1967), Kuehner et al. (1981)
LA	Navajo volcanic field, Arizona, USA	31–25 Ma	2	Roden & Smith (1979), Ehrenberg (1982)

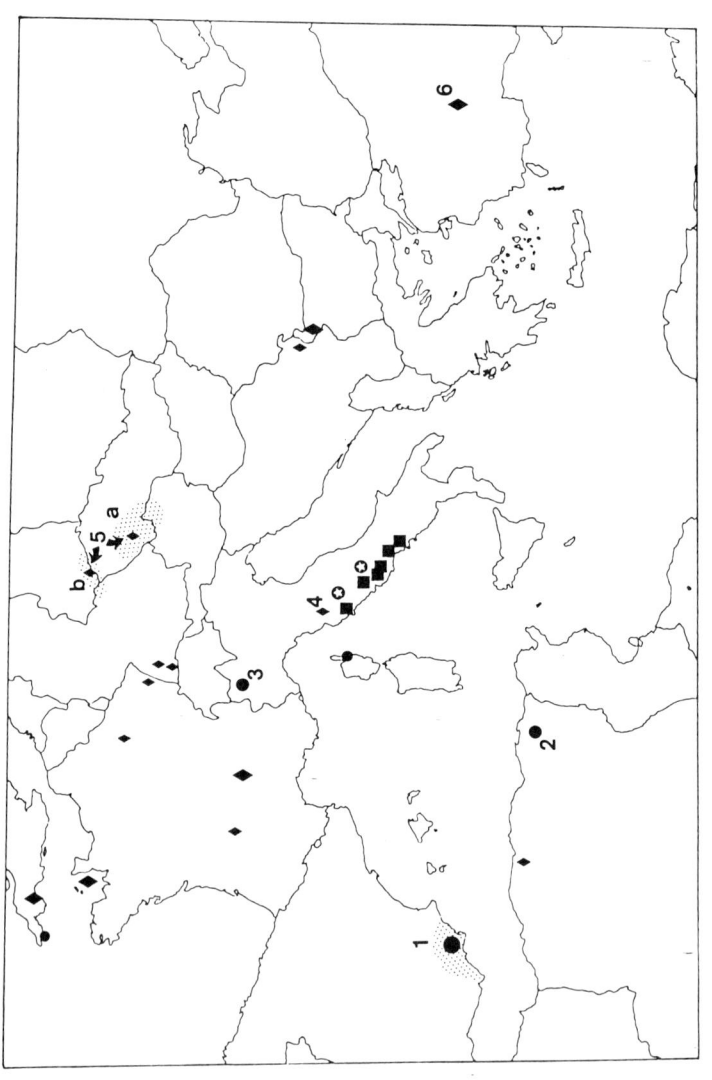

Figure 3.1 Location map of southern European and Mediterranean ultrapotassic rocks with political boundaries for ease of reference. High-MgO, high-SiO$_2$ localities are numbered: 1 = Betic Cordillera; 2 = Koudiat el Anzazza; 3 = northwestern Italian Alps; 4 = Orciatico, Tuscany; 5 = Bohemian Pluton (a) and Erzgebirge (b); 6 = Afyon, Turkey. Symbols designate ultrapotassic rock groupings: full circle, Group I (lamproites); asterisk, Group II (kamafugites); full square, Group III (Al-rich); full diamond, Group IV (transitional). For details of other localities, see Foley *et al.* (1987).

with $SiO_2 > 53\%$ and $MgO > 8\%$ are from orogenic or immediately post-orogenic settings. Two rocks from Western Australia fall into the chemical 'region of interest' but are not included because their compositions are considered to be due to mixing with crustal materials (Jaques *et al.* 1986).

Most rocks discussed here come from the Mediterranean region, and predominantly from the W Mediterranean (Fig. 3.1; references listed in Table 3.1), encompassing several localities in the Betic Cordillera of Spain (e.g. Fortuna, Calasparra, Cancarix, Las Minas de Hellin), the NW Alps of Italy, Orciatico in Tuscany, and Koudiat el Anzazza in Algeria.

Simple comparisons of the number of analyses between areas are misleading because of the large number of Spanish rocks analysed, but it should be noted that the majority of potassic rocks from localities in this W Mediterranean region have high SiO_2 and MgO, whereas other localities listed in Table 3.1 have just one or two examples from a large number of analyses. Later sections of this paper, particularly concerning petrogenesis, therefore concentrate on the W Mediterranean localities and discuss reasons for this geographical distribution.

3.2 Geological setting and mineralogical features

The main mineralogical and petrographic features of rocks from the Betic Cordillera, NW Alps, Orciatico and Koudiat el Anzazza are summarized in Table 3.2. Below we comment on some specific additional features such as

Table 3.2 Main petrographic features of high-MgO, high-SiO_2 ultrapotassic rocks.

Locality	Large crystals (euhedral/subhedral)	Small crystals and groundmass (euhedral/subhedral)
Betic Cordillera, SE Spain		
Fortuna	Ol, Phl	Kf, Phl, Gl, Ap, Opx, Cpx, Amph, Il, CS, Psb
Cancarix	Ol, Phl	Kf, Phl, Ap, Opx, Cpx, Amph, CS, Il
Las Minas de Hellin	Ol, Phl	Kf, Phl, Ap, Cpx, Amph, CS
Calasparra	Ol, Phl	Kf, Phl, Ap, Cpx, Amph, CS
Northwestern Alps, Italy		
Gressoney	Phl, Cpx, Ol	Kf, Amph, P/B, Ap, Sph, Ab, Ep, Cal
Champoluc	Phl	Kf, Amph, Cpx, Sph, Ap, Op
Plan d'Albard	Phl, Ol	Kf, Phl, Cpx, Amph, Ab, Psb, Rut, Sph, CS, ZS
Orciatico	Ol, Cpx, Phl	Kf, Phl, Gl, Ap, Il, Mt, Amph, CS
Koudiat el Anzazza	Ol	Kf, Gl, Ol, Cpx, Psb

Ol = olivine, Phl = phlogopite, Opx = orthopyroxene, Ap = apatite, Amph = Na-K-rich amphibole, Mt = Fe-Ti-spinel, P/B = phlogopite/biotite, Sph = sphene, Ab = albite, Ep = epidote, Psb = pseudobrookite, Rut = rutile, CS = Cr-spinel, Gl = glass, ZS = Cr-Zn-spinel, Op = opaques, Kf = K-feldspar, Il = ilmenite, Cal = calcite.

mineralogical compositions and geological setting, and briefly describe the rock localities.

3.2.1 Betic Cordillera, Spain

Ultrapotassic rocks occur over a relatively wide area of both the Northern (External) and Southern (Internal or Betic) zones of the Betic Cordillera, and form the youngest part of a sequence of Miocene (17–6 Ma) calc-alkaline, shoshonitic and ultrapotassic volcanics, of which the calc-alkaline and shoshonitic members are restricted to the Southern Zone (Lopez Ruiz & Rodriguez Badiola 1980, Venturelli et al. 1984a). Radiometric ages for the potassic rocks range from 7.5 Ma for Sierra de las Cabras to 5.7 Ma for Calasparra (Nobel et al. 1981, Bellon et al. 1983).

The sequence of crystallization in the Spanish lamproites is generally olivine ($Fo_{91.5}$ to Fo_{87}) → phlogopite → clinopyroxene → sanidine and K-rich amphibole. Orthopyroxene ($Mg_{81.7-79}$) occurs at Fortuna, and occasionally in other areas such as Cancarix, where it often occurs as a reaction product mantling olivine. Primary orthopyroxene has not been found in other lamproite localities around the world, with the possible exception of the Kef Hahouner (Algeria) rock, which may be lamproitic (see later section). Phlogopite usually exhibits increasing Fe and Ti towards the rims, has Mg_{94-75} and can be fluorine-rich (up to 4.4%). Clinopyroxene has low Al_2O_3 typical of lamproites and may contain significant Fe_2O_3 (up to 1.2%). Sanidine contains up to 2.2% Fe_2O_3. Amphibole is pale green in colour and is rich in alkalis with a high K/Na (0.31 to 0.63 atomic), although K-richterites in lamproites from other regions often have higher K/Na (e.g. Jaques et al. 1986). Amphibole is also Al-poor and frequently exhibits decreasing Mg# towards the rim. Cr-spinel is frequently associated with olivine, and may be rimmed by olivine or phlogopite, or be present as small, isolated phenocrysts. Early crystallizing Cr-spinels may be extremely Cr-rich (> 60% Cr_2O_3; Foley 1985, Venturelli et al. 1988), indicating a very refractory source for these rocks. Small ultramafic xenoliths and xenocrysts of olivine and orthopyroxene (both Mg_{90}) and Cr-Al-spinel derived from ultramafic rocks are widespread. Other xenocrysts–xenoliths include metamorphic rocks, 'granitic' rocks, limestones and partly resorbed biotite. A more detailed description of these rocks, including mineral analyses, may be found in Fuster et al. (1967), Lopez Ruiz & Rodriguez Badiola (1980) and Venturelli et al. (1984a, 1988).

3.2.2 Northwestern Alps, Italy

The ultrapotassic rocks at Champoluc intrude calc-schists of the Combin unit (Piedmonte Ophiolite Nappe) and paraschists of the Sesia–Lanzo Zone. A dyke cutting the calc-schists has an age of 31.6 Ma (Dal Piaz et al. 1973). The dyke at Plan d'Albard, close to Bard, intruded fine-gained gneiss and mica-schists of the Sesia–Lanzo Zone, and the Grassoney outcrop occurs in eclogitic mica-schists (De Marco 1958, Venturelli *et al.* 1984b).

Detailed mineral compositional data are available only for the dyke from Plan d'Albard (Wagner & Velde 1986). Olivine containing spinel is completely altered. Olivine also crystallized early in the Gressoney rocks, where it occurs together with clinopyroxene as inclusions in mica. Phlogopite in the Plan d'Albard rock has Mg# = 0.94 –0.62; some large crystals have dark reddish brown cores containing low MgO (12–16%) and high TiO_2 (3–5%), FeO (≈ 20%) and Al_2O_3 (≈ 15%). Amphibole is rich in Na_2O and K_2O (Ca-rich arfvedsonite) but has even lower K/Na (0.18) than the Spanish rocks. K-feldspar contains 1–2% Fe_2O_3. A type of Cr-spinel occurs which is rich in ZnO (up to 8.7%) and has very low MgO; the origin of this spinel is unknown, but zincian spinels have been found as inclusions in diamond (Meyer & Boyd 1972, Arai 1986).

3.2.3 Orciatico, Tuscany, Italy

The ultrapotassic rocks constitute a small subvolcanic body emplaced in clay-marly sediments of early Pliocene age. Olivine is usually altered and contains inclusions of Cr-rich spinel. Clinopyroxene may be surrounded by phlogopite flakes. Mineral compositions are available for only one sample, details of which are given by Wagner & Velde (1986). Mica is rich in BaO (1.5%) and TiO_2(6%), and poor in Al_2O_3(11%). K-feldspars may also contain considerable BaO (0.9%). Amphibole has a relatively low K/Na ~ 0.2.

3.2.4 Algeria

In the Koudiat el Anzazza dyke, olivine (Mg_{91-84}) occurs as large or skeletal phenocrysts and in the groundmass. The groundmass contains sanidine, often acicular, and glass of alkali-trachytic composition. This dyke was emplaced in the early Cretaceous Guerrouch Flysch, probably during the Miocene: a nearby occurrence south of Kef Hahouner gives an age of 11 Ma (Bellon *et al.* 1975, Lepvrier & Velde 1976). The Kef Hahouner rock is apparently related petrologically, but no wholerock analysis is available (Raoult & Velde 1971).

3.2.5 Other regions

The rare Leucite Hills (USA) rocks with high MgO and SiO_2 fall into the group classified as olivine orendites by Kuehner *et al.* (1981). They bear the normal mineralogical characteristics of lamproites, with early crystallizing Cr-spinels and low-Al_2O_3 (less than 0.2%) clinopyroxenes. Phlogopites have very high Mg#, low Al_2O_3 and considerable BaO (0.3%); TiO_2 contents of micas are moderate (around 2%) reflecting the intermediate TiO_2 characteristic of the Leucite Hills rocks (Fig. 7 of Foley *et al.* 1987). Leucites are Fe_2O_3-rich (up to 2.7%), indicating high f_{O_2} at the time of crystallization (Kuehner *et al.* 1981,

Foley 1985). The olivine orendites were considered by Kuehner et al. (1981) to be the only rocks in the Leucite Hills suite which do *not* represent near-primary compositions, being related to orendites by crystal fractionation.

The Navajo (USA) rocks (Roden & Smith 1979) contain phenocrysts of diopside, phlogopite and sanidine in a groundmass of the same minerals plus Fe-Ti oxides. They occur in a region of alkaline lamprophyres (minettes) and 'kimberlitic' rocks occurring as dykes and diatremes. On the basis of geochemical, mineralogical and experimental studies, the felsic minettes are thought to originate from mafic minettes by crystal fractionation, possibly involving crustal contamination (Roden 1981, Esperanca & Holloway 1986a, Alibert et al. 1986).

The Afyon (Turkey) rock considered here is part of a suite of K-rich latites, trachytes and leucite-bearing rocks (Keller & Villari 1972) associated with the late Tertiary continent–continent collision along the Tauride foldbelt (Keller 1983). The ultrapotassic rock is the most magnesian of the potassic rocks and is considered by Keller (1983) to represent a near-primary mantle melt composition. Olivines are Fo_{94-90}, and the wholerock has $^{87}Sr/^{86}Sr =$ 0.7094 which Keller argues is not due to crustal contamination.

The occurrences in Czechoslovakia and East Germany are extreme members of a large province of lamprophyric and related rocks of Hyercynian age. The durbachites form stocks rather than dykes and comprise melanocratic granite and quartz syenite rocks which are chemically akin to lamprophyres (Holub 1977). The most melanocratic approach 'amphibole biotitites' with only minor K-feldspar. Amongst the lamprophyre dykes around the Central Bohemian Pluton, a number have $SiO_2 > 53\%$ and $MgO > 8\%$, but only two are also ultrapotassic (Nemec 1974), with most having approximately equal K_2O and Na_2O contents.These include members of both the calc-alkaline and minette series, which are considered by Nemec (1974) to have distinct parageneses. Related rocks of the Central Bohemian Pluton have ages around 330 Ma (van Breemen et al. 1982), whereas the W Erzgebirge rock gives mineral and wholerock ages of 350–360 Ma (Kramer 1976).

3.3 Wholerock chemistry

Representative analyses for the localities in the W Mediterranean region are given in Table 3.3. Most of the chemical features are characteristic of lamproites rather than other ultrapotassic rocks, although many trace elements show characteristics intermediate between lamproites and island-arc ultrapotassic rocks. Ultrapotassic rocks in general have high MgO, Ni and Cr, and also high contents of LILE. Lamproites form a subdivision of the ultrapotassic rocks with the most extreme enrichment in LILE, very high Mg# and characteristically low CaO, Al_2O_3 and Na_2O contents. Consequently,

K_2O/Al_2O_3 and K_2O/Na_2O are very high (see reviews by Bergman 1987, Foley et al. 1987).

The high-SiO_2, high-MgO rocks considered here are mostly lamproites which show less extreme enrichment in Ba, Nb and Ti, leading to negative spikes for these elements in element variation diagrams (Fig. 3.2). These negative spikes are characteristic of island-arc ultrapotassic rocks, and island-arc rocks in general, relative to basalts from oceanic settings (Gill 1981). TiO_2 contents of the 'boninitic' ultrapotassic rocks (1–2%), although low relative to other ultrapotassic rocks, are nevertheless high relative to true boninites, which have 0.1 to 0.5% TiO_2 (Hickey & Frey 1982). The occurrence of

Table 3.3 Representative analyses of high-SiO_2, high-MgO ultrapotassic rocks from the Mediterranean region.

	F	CX	CM	CL	CH	PA	OR	KA
SiO_2	57.5	55.8	57.3	54.6	56.0	56.0	55.3	56.6
TiO_2	1.45	1.52	1.85	1.75	1.40	1.24	1.80	1.45
Al_2O_3	11.8	10.0	9.52	9.52	8.90	11.0	12.0	13.2
Fe_2O_3	5.89[+]	5.37[+]	5.58[+]	5.66[+]	5.66[+]	5.15[+]	3.42	1.70
FeO							2.60	2.87
MnO	0.07	0.07	0.07	0.07	0.11	0.09	0.08	0.08
MgO	9.15	12.3	10.2	13.3	9.40	9.27	8.35	8.24
CaO	2.93	3.41	2.76	2.87	4.20	4.11	4.28	3.42
Na_2O	1.14	1.18	1.85	0.69	1.90	1.29	1.25	1.21
K_2O	6.84	8.78	8.98	8.86	9.60	9.07	8.11	8.33
P_2O_5	0.79	0.98	0.91	0.95	1.20	1.09	0.62	0.32
LOI	2.28	0.28	0.91	1.60	1.24	1.51	2.46	2.34
Cr	624	816	643	638	586	600		
Ni	416	600	501	657	315	396		
Rb	495	539	557	541	443	569		
Sr	514	828	668	526	1242	544		
Ba	1450	1695	2110	2050				
Zr	620	700	1045	680	730	592		
Nb	34	39	53	32	40	50		
Y	30	28	34	24	55	47		
La	69	109	97	91	164			
Ce	204	301	257	262	357	258		
Nd	129	184	177	169	227	175		
Sm	24.1	33.3	33.6	31.1				
Eu	3.89	5.40	5.49	5.04	7.65	5.33		
Yb	1.82	1.84	2.01	1.61	2.34	1.70		
Ce/Yb	112	164	128	163	153	152		

F = Cabecitos Negros, Fortuna; CX = Sierra de las Cabras, Cancarix; CM = Cerro del Monagrillo, Las Minas de Hellin; CL = Cerro Negro (quarry), Calasparra (Venturelli et al. 1984a, Nixon et al. 1984); CH = Champoluc, Upper Mascognaz Valley; PA = Plan d'Albard, close to Bard, Aosta Valley (Venturelli et al. 1984b); OR = Orciatico, Tuscany (Barberi & Innocenti 1967); KA = Koudiat el Anzazza, NE Algeria (Vila et al. 1974).
LOI = loss on ignition. + = Total Fe as Fe_2O_3.

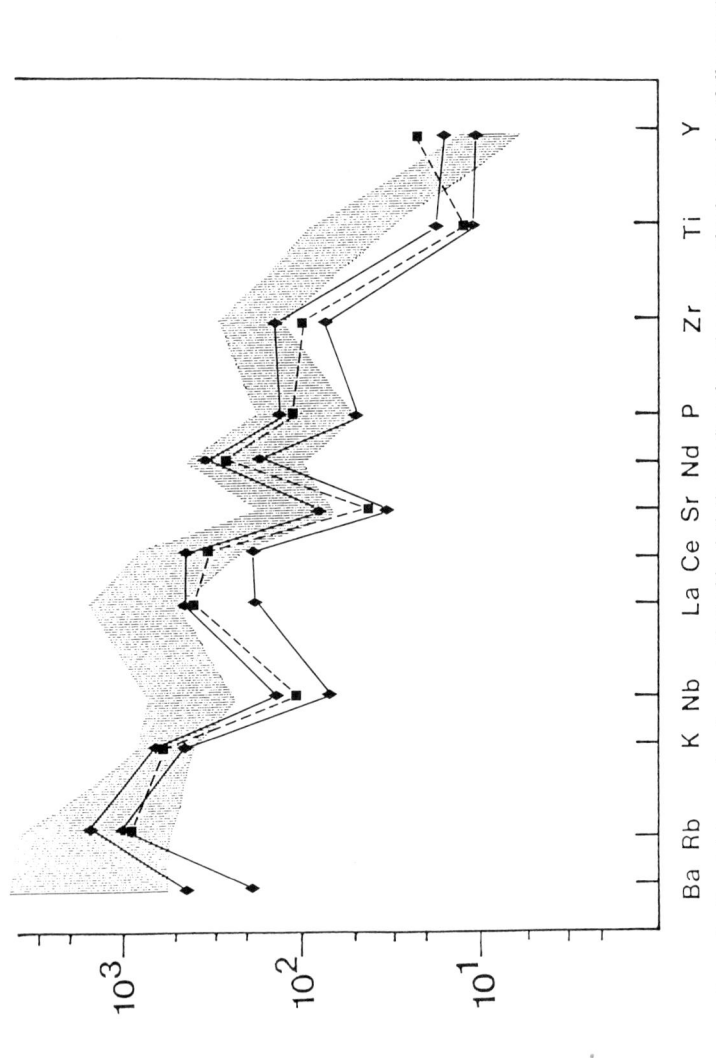

Figure 3.2 Spidergram patterns for Spanish lamproites (range indicated by full diamond) and northwestern Alps lamproites (full square) compared to the range for the West Kimberley lamproites (vertical shading). Sample value are chondritic mantle normalized using the values of Thompson (1982). Data sources: Nixon *et al.* (1984) and Venturelli *et al.* (1984a).

high-SiO_2, high-MgO potassic rocks in post-collisional tectonic settings, therefore, appears to correlate with some degree of arc-like geochemistry.

In addition to relatively low Ti, Nb and Ba, the Spanish and Italian lamproites exhibit REE patterns that are convex upwards in the LREE (Fig.3.3). This is a peculiar characteristic of the W Mediterranean rocks, but is seen to a much less marked degree in lamproites from Priestly Peak, East Antarctica (Foley *et al.* 1987), and in some from the Leucite Hills (Kay & Gast 1973). Unfortunately, no REE analyses are available for the olivine orendites from the latter locality.

The Mediterranean lamproites also have rather lower Sr and Zr contents than most other lamproites. The negative Sr spike is not a characteristic of

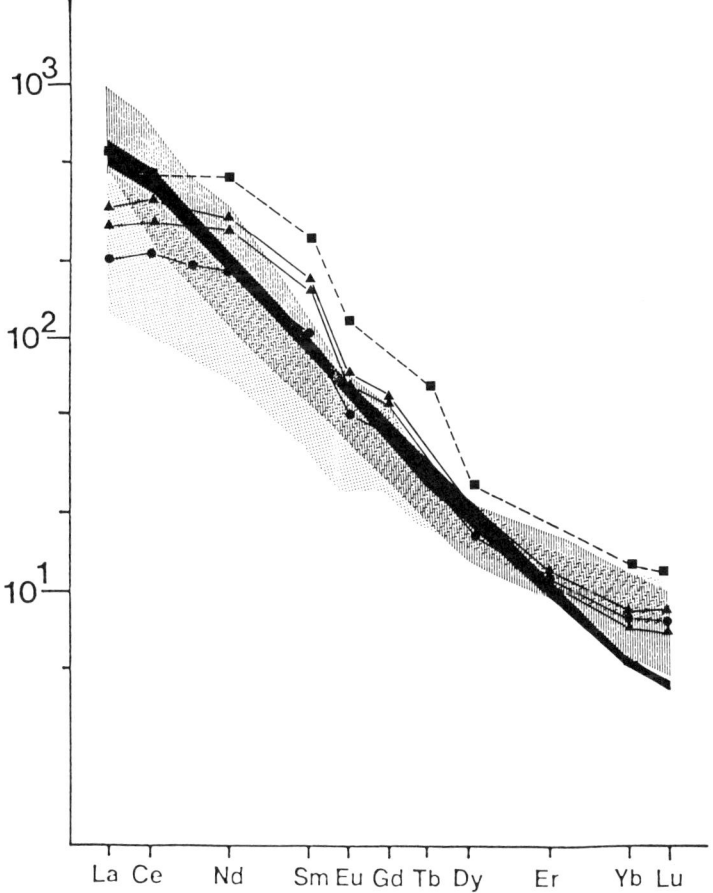

Figure 3.3 Rare-earth-element (REE) patterns for Spanish cancalites (full triangle) and fortunites (full circle) and northwestern Alps lamproites (full square) (Nixon *et al.* 1984, Venturelli *et al.* 1984a, b) compared to patterns for the ultrapotassic rock groups (Foley *et al.* 1987): vertical shading = Group I; solid = Group II; dot shading = Group III.

island-arc ultrapotassic rocks, but occurs in a number of lamproite localities. These include Western Australia, but not the Leucite Hills.

Isotope data for Nd, Sr and Pb on the rocks discussed here are reported only by Powell & Bell (1970) and Nelson *et al.* (1986) for the Spanish rocks, and by Venturelli *et al.* (1984b) for the Alpine lamproites, and a Sr isotope composition from Afyon was reported by Keller (1983). Initial $^{87}Sr/^{86}Sr$ for the Mediterranean rocks are high (mostly > 0.717) and ε_{Nd} values are strongly negative (-11 for the Spanish lamproites), suggesting involvement of an ancient, enriched sedimentary component in the petrogenesis of these rocks. In other areas, data are available only for related, non-ultrapotassic rock types (e.g. Leucite Hills: Vollmer, *et al.* 1984; Navajo: Alibert *et al.* 1986). In most areas, data are too sparse to permit rigorous evaluation of isotopic signatures, since local characteristics and variations are likely to be important.

3.4 Petrogenesis

Here, we concentrate on the W Mediterranean high-MgO, high-SiO$_2$ ultrapotassic rocks because the majority of rocks in these suites have high MgO and high SiO$_2$ contents and are probably derived from high-SiO$_2$ parental melts. High-MgO, high-SiO$_2$ rocks from other areas (e.g. Navajo and central Bohemian regions) are outlying compositions for their suites, and are most likely derived from mafic alkaline lamprophyric parental melts not closely related to boninitic melts in either composition or origin (Roden 1981, Rock 1984, Esperanca & Holloway 1986a,b). The origin of the Mediterranean ultrapotassic rocks is assessed in the light of their affinities with both ultrapotassic rocks and boninitic rocks by considering the viewpoints of hypotheses for the origin of both of these rock groups.

From the viewpoint of boninitic rocks, the high-MgO, high-SiO$_2$ ultrapotassic rocks, besides having similarly high MgO and SiO$_2$ contents, share the low CaO, Al$_2$O$_3$ and Na$_2$O contents characteristic of a strongly depleted mantle source composition. The major difference is in the extreme enrichment in K and other incompatible elements; the enhanced TiO$_2$ contents of these rocks relative to boninites derived from similarly refractory sources suggest that TiO$_2$ was a significant component of the enriching fluids. For characterization of this enrichment, it is best to examine the ultrapotassic rocks in which several types of enrichment signature can be identified, and are developed to their extremes.

From the ultrapotassic rock viewpoint, the high-MgO, high-SiO$_2$ rocks are mostly lamproitic or have lamproitic affinities, but their LILE signatures are transitional between those of lamproites from stable continental regions and island-arc ultrapotassic rocks. They possess intermediate *relative* depletions in HFS elements such as Ti and Nb (1.2–2.1% TiO$_2$ compared to 0.5–1.2% in Italian and Indonesian island-arc ultrapotassic rocks, and 2–7% in continental lamproites; summarized in Fig. 7 of Foley *et al.* (1987). However, examination of the patterns on multi-element variation diagrams shows that the differences between the W Mediterranean lamproites and arc ultrapotassic rocks are of degree rather than kind, with the Spanish lamproites showing the greatest level of enrichment. The enriching agent, therefore, probably has a similar origin in both groups of rocks, which appears to be in some way related to subduction.

The origin of the LILE-enriching agents in island-arc rocks, particularly potassic rocks, is complex and controversial (e.g. Ringwood 1974, Green 1980, Foden & Varne 1980, Gill 1981, Varne 1985, Arculus & Powell 1986), but the relatively simple forearc setting of boninites, together with experimental evidence for a shallow depth of origin and the involvement of H_2O in their genesis (Green 1976, Tatsumi 1982), favours fluids derived by dehydration of the subducted oceanic crust (Dobson & O'Neil 1987). The greatly enhanced K_2O contents in the high-Mg, high-Si lamproites relative to boninites, and the evidence for TiO_2 enrichment of the depleted sources of the former, suggest that hydrous fluids responsible for the LILE and Ti enrichments in the lamproite sources were significantly different in composition from the boninite-generating metasomatic fluids. Experimental studies of K solubility in high-temperature aqueous mantle fluids (Ryabchikov & Boettcher 1980) showed that K is increasingly soluble at pressures from 20 to 30 kbar at constant temperature; in contrast, Na seems more soluble at increasingly low pressures (Ryabchikov *et al.* 1982). Thus, above the subducted slab of oceanic crust, Na may be mobilized and enriched at shallow depths (< 15 kbar), whereas K may be retained in the slab to greater depths, eventually to be transported up into the mantle wedge when the pressure exceeds perhaps 15–20 kbar. Phlogopite forms from reactions between peridotite minerals in the refractory mantle wedge and K-rich hydrous fluids released from the slab.

The ultrapotassic rocks in the W Mediterranean were emplaced in a more complex post-collisional tectonic environment than forearc boninites, so that the full range of suggested enriching agent origins must be considered. Isotopic studies suggest that the high contents of K_2O and other LILE are not due to crustal contamination during ascent, but that involvement of a crust-derived component in the source region appears important (e.g. Nelson *et al.* 1986, Alibert *et al.* 1986). The low Al_2O_3 contents of lamproites also preclude significant crustal incorporation during ascent, because Al_2O_3 contents would rapidly rise due to reaction with aluminous sialic material (Jaques & Foley 1985). The elevated LILE contents thus appear to be a mantle-derived feature.

Experimental studies of both boninitic compositions and *silicic* lamproites have stressed the importance of relatively low-pressure melting of a H_2O-bearing mantle composition with strongly depleted geochemical character. Studies of boninitic rocks have suggested that these rocks are formed either by relatively high-temperature ($\sim 1300°C$) melting at 10–15 km depth (3–5 kbar) with low H_2O contents (1–3%), or by lower-temperature melting with higher H_2O content at 30–40 km depth (Green 1976, Tatsumi 1982,

Cawthorn & Davies 1983, Duncan & Green 1987). The silica content of the melts is enhanced by incongruent melting of enstatite at these low pressures (Kushiro 1974). Liquidus studies of silicic lamproites have sought to contrast these with other alkaline rock types in stressing that the presence of H_2O and *virtual absence* of CO_2 is required to produce lamproites with $> 50\%$ SiO_2 (Barton & Hamilton 1982, Foley 1986). Any CO_2 in the source region during melting would cause depression of the SiO_2 content of the melt. Foley (1986) suggested that the Gaussberg lamproite (51% SiO_2) originated by melting of mica-harzburgite at less than 20 kbar, but noted that the degree of silica saturation increases at lower pressures, so that more silicic lamproites may be produced at shallower depths akin to those proposed for boninite genesis (< 30 km) The silica content in this case is enhanced by a peritectic melting reaction involving the incongruent melting of phlogopite (Foley *et al.* 1986). The predominance of silicic but primitive lamproites in the W Mediterranean, therefore, most likely indicates shallow melting of a mica-bearing mantle in this region.

Boninites with low CaO/Al_2O_3 are believed to be characteristic of forearc settings, and other highly refractory melts transitional to boninites but with higher CaO/Al_2O_3 (> 0.7) occur in backarc basins and mid-ocean ridge settings (Duncan & Green 1987). The post-collisional, Andean-type setting in SE Spain cannot be easily related to any of these, although the details of plate movements and direction of subduction of oceanic crust during the time they were produced is not well understood (Torres-Roldan 1979). The occurrence of potassic rocks in Spain to the north of the slightly older calc-alkaline and shoshonitic rocks (Lopez Ruiz & Rodriguez Badiola 1980) was used by Arana & Vegas (1974) to suggest a direct relationship of K_2O with depth to a north-dipping subducting slab. However, the relationship of K_2O content of volcanics with depth to the Benioff zone does not hold in all modern arc settings (Arculus & Johnson 1978, Foden & Varne 1980). Other assessments of regional plate dynamics in the W Mediterranean at this time consider subduction to have been south-directed, and to have ceased several million years before the ultrapotassic magmatic activity (Torres-Roldan 1979, Puga 1980) so that, as in Central Italy (Cundari 1980) and the NW Alps (Venturelli *et al.* 1984b), no direct relationship exists between potassic magmatism and active subduction, despite the strong subduction-zone signature of the magmas erupted in these regions. A possible explanation for this is that the suprasubduction zone mantle wedge, enriched in LILE by millions of years of contributions from the slab, is reactivated at some stage following cessation of subduction, and development of a tensional regime leads to diapiric ascent and 'delayed' partial melting of enriched mantle (Johnson *et al.* 1978).

In summary, the primitive, high-Si, high-Mg Spanish lamproites are considered to originate by moderate- to shallow-level (< 25 kbar, perhaps 8–15 kbar) partial melting of a strongly depleted mantle which, prior to generation of the ultrapotassic rocks, was affected by an enrichment event that resulted in crystallization of hydrous minerals, probably mica. The lamproitic characteristics are derived from the depleted mantle composition, whereas the intermediate continental-arc nature of the trace-element signature is a result of subduction-related enrichment at an Andean-type plate margin. Mixed continental and arc source regions have been proposed before for Indonesian arc lavas (Varne 1985, Varne & Foden 1986), where decoupled subcontinental upper mantle is considered to have been overridden by the arc; the Andean-type margin is dominantly the converse case, in which arc-related components are introduced into depleted subcontinental mantle still coupled to the overlying crust. The mixed geochemical signatures of the W Mediterranean potassic rocks result from an anomalous enriched mantle (Puga 1980) which is regional in extent, encompassing the NW Alps, Algeria and SE Spain. This anomalous mantle results from complex patterns of subduction and collision related to movements of microplates in the W Mediterranean during the Mesozoic and Tertiary (e.g. Dewey *et al.* 1973), so that potassic rocks probably relate to a collection of past subduction events rather than to active subduction (Cundari 1980, Venturelli *et al.* 1984b). The Italian potassic rocks which lack the depleted source signature (Barton 1979, Foley *et al.* 1987) may be derived from deeper levels of this anomalous mantle mixture, in which previously depleted mantle is only a minor component.

Acknowledgements

We are grateful to A. L. Jaques and R. Varne for comments which have greatly improved the manuscript.

References

Alibert, C., A. Michard & F. Albarede 1986. Isotope and trace element geochemistry of Colorado Plateau volcanics. *Geochim. Cosmochim. Acta* 50, 2735–50.

Arai, S. 1986. 'Iron meteorite paragenesis', a new group of mineral inclusions in diamond. *Neues Jahrb. Min. Monats.* 463–6.

Arana, V. & R. Vegas 1974. Plate tectonics and volcanism in the Gibraltar arc. *Tectonophysics* 24, 197–212.

Arculus, R. J. & R. W. Johnson 1978. Criticism of generalised models for the magmatic evolution of arc–trench systems. *Earth Planet. Sci. Lett.* 36, 118–26.

Arculus, R. J. & R. Powell 1986. Source component mixing in the regions of arc magma generation. *J. Geophys. Res.* 91, 5913–26.

Barberi, F. & F. Innocenti 1967. Le rocce selagitiche di Orciatico e Monticatini in Val di Cecina. *Atti Soc. Toscani Sc. Nat. Ser. A* 74, 139–86.

Barton, M. 1979. A comparative study of some minerals occurring in the potassium-rich alkaline rocks of the Leucite Hills, Wyoming, the Vico volcano, western Italy, and the Toro–Ankole region, Uganda. *Neues Jahrb. Mineral. Abh.* 137, 113–34.

Barton, M. & D. L. Hamilton 1982. Water-undersaturated melting experiments bearing upon the origin of potassium-rich magmas. *Mineral. Mag.* 45, 267–78.

Bellon, H., P. Bordet & C. Montenat 1983. Chronologie du magmatisme Néogene des Cordilleres Betiques (Espagne Meridionale). *Bull. Soc. Geol. Fr.* **25**, 205–17.

Bellon, H., C. Lepvrier, J. Magne & F. Raymond 1975. L'activité eruptive dans l'Algérois: nouvelles données géochronologiques. *Rev. Geol. Medit.–Ann. Univ. Provence* **4**, 291–9.

Bergman, S. C. 1987. Lamproites and other potassium-rich igneous rocks: a review of their occurrence, mineralogy and geochemistry. Geol. Soc. Lond. Spec. Pap. no. 30, 103–90.

Carmichael, I. S. E. 1967. The mineralogy and petrology of the volcanic rocks from the Leucite Hills, Wyoming. *Contrib. Mineral. Petrol.* **15**, 24–66.

Cawthorn, R. G. & G. Davies 1983. Experimental data at 3 kbars pressure on parental magma to the Bushveld Complex. *Contrib. Mineral. Petrol.* **83**, 128–35.

Cundari, A. 1980. Role of subduction in the genesis of leucite-bearing rocks: facts or fashion? *Contrib. Mineral. Petrol.* **73**, 432–4.

Dal Piaz, G. V., J. C. Hunziker & G. Martinotti 1973. Excursion to the Sesia Zone of the Schweiz. Mineralogische und Petrographische Gesellschaft, September 30th to October 3rd 1973. *Schweiz. Min. Petr. Mitt.* **53**, 477–90.

Dal Piaz, G. V., G. Venturelli & A. Scolari 1979. Calc-alkaline to ultrapotassic postcollisional volcanic activity in the internal northwestern Alps. *Mem. Sci. Geol.* **32**, 1–16.

De Marco, L. 1958. *Su alcuni filoni radioattivi del complesso Sesia-Lanzo.* Studi i Ric. Div. Geomineraria I.

Dewey, J. F., W. C. Pitman, W. B. F. Ryan & J. Bonin 1973. Plate tectonics and the evolution of the Alpine system. *Bull. Geol. Soc. Am.* **84**, 3137–80.

Dobson, P. F. & J. R. O'Neil 1987. Stable isotope compositions and water contents of boninite series volcanic rocks from Chichi-jima, Bonin Islands, Japan. *Earth Planet. Sci. Lett.* **82**, 75–86.

Duncan, R. A. & D. H. Green 1987. The genesis of refractory melts in the formation of oceanic crust. *Contrib. Mineral. Petrol.* **96**, 326–42.

Ehrenberg, S. N. 1982. Rare earth element geochemistry of garnet lherzolite and megacrystalline nodules from minette of the Colorado Plateau province. *Earth Planet. Sci. Lett.* **57**, 191–210.

Esperanca, S. & J. R. Holloway 1986a. The origin of high-K latites from Camp Creek, Arizona: constraints from experiments with variable f_{O_2} and a_{H_2O} *Contrib. Mineral. Petrol.* **93**, 504–12.

Esperanca, S. & J. R. Holloway 1986b. On the origin of some mica lamprophyres: experimental evidence from a mafic minette. *Contrib. Mineral. Petrol.* **95**, 207–16.

Foden, J. D. & R. Varne 1980. The petrology and tectonic setting of Quaternary–Recent volcanic centres of Lombok and Sumbawa, Sunda arc. *Chem. Geol.* **30**, 201–26.

Foley, S. F. 1985. The oxidation state of lamproitic magmas. *Tschermaks. Min. Petr. Mitt.* **34**, 217–38.

Foley, S. F. 1986. The genesis of lamproitic magmas in a reduced, fluorine-rich mantle. In *Fourth Int. Kimberlite Conf.* (Abstr.), 16, 173–5.

Foley, S. F., W. R. Taylor & D. H. Green 1986. The role of fluorine and oxygen fugacity in the genesis of the ultrapotassic rocks. *Contrib. Mineral. Petrol.* **94**, 183–92.

Foley, S. F., G. Venturelli, D. H. Green & L. Toscani 1987. The ultrapotassic rocks: characteristics, classification and constraints for petrogenetic models. *Earth Sci. Rev.* **24**, 81–134.

Fuster, J. M., P. Gastesi, J. Sagredo & M. L. Fermoso 1967. Las rocas lamproiticas del SE de Espana. *Estud. Geol.* **9**, 35–69.

Gill, J. B. 1981. *Orogenic andesites and plate tectonics.* Berlin: Springer.

Green, D. H. 1976. Experimental testing of 'equilibrium' partial melting of peridotite under water-saturated, high pressure conditions. *Can. Mineral.* **14**, 255–68.

Green, T. H. 1980. Island arc and continent-building magmatism – a review of petrogenetic models based on experimental petrology and geochemistry. *Tectonophysics* **63**, 367–85.

Gupta, A. K. & K. Yagi 1980. *Petrology and genesis of the leucite-bearing rocks.* Berlin: Springer.

Hickey, R. L. & F. A. Frey 1982. Geochemical characteristics of boninite series volcanics: implications for their source. *Geochim. Cosmochim. Acta* **46**, 2099–115.

Holub, F. V. 1977. Petrology of inclusions as a key to petrogenesis of the durbachitic rocks from Czechoslovakia. *Tschermaks. Min. Petr. Mitt.* **24**, 133–50.

Jaques, A. L. & Foley, S. F. 1985. The origin of Al-rich spinel inclusions in leucite from the leucite lamproites of Western Australia. *Am. Mineral.* **70**, 1143–50.

Jaques, A. L., J. D. Lewis & C. B. Smith 1986. *The kimberlites and lamproites of Western Australia.* Geol. Surv. W. Austr. Bull. no. 132.

Johnson, R. W., D. E. McKenzie & I. E. Smith 1978. Delayed partial melting of subduction-modified mantle in Papua New Guinea. *Tectonophysics* **46**, 197–216.

Kay, R. W. & P. W. Gast 1973. Rare earth element content and origin of alkali-rich basalts. *J. Geol.* **81**, 653–82.

Keller, J. 1983. Potassic lavas in the orogenic volcanism of the Mediterranean area. *J. Volcanol. Geotherm. Res.* **18**, 321–35.

Keller, J. & L. Villari 1972. Rhyolitic ignimbrites in the region of Afyon (central Anatolia). *Bull. Volcanol.* **36**, 342–58.

Kramer, W. 1976. Genese der lamprophyre im Bereich der Fichtelgebirgisch – Erzgebirgischen Antiklinalzone. *Chem. Erde* **35**, 1–49.

Kuehner, S. M., A. D. Edgar & M. Arima 1981. Petrogenesis of the ultrapotassic rocks from the Leucite Hills, Wyoming. *Am. Mineral.* **66**, 663–77.

Kushiro, I. 1974. Melting of hydrous upper mantle and possible generation of andesite magma: an approach from synthetic systems. *Earth Planet. Sci. Lett.* **22**, 294–9.

Lepvrier, C. & D. Velde 1976. A propos des intrusions tertiares de la marge nord-africaine entre Cherchel et Tenes (Algérie). *Bull. Soc. Geol. Fr.* **18**, 991–8.

Lopez Ruiz, J. & E. Rodriguez Badiola 1980. La region volcanica neogena del sureste de Espana. *Estud. Geol.* **36**, 5–63.

Meyer, H. O. A. & F. R. Boyd 1972. Composition and origin of crystalline inclusions in natural diamonds. *Geochim. Cosmochim. Acta* **36**, 1255–73.

Mitchell, R. H. 1985. A review of the mineralogy of lamproites. *Trans. Geol. Soc. S. Afr.* **88**, 411–38.

Nelson, D. R., M. T. McCulloch & S.-S. Sun 1986. Origins of ultrapotassic rocks as inferred from Sr, Nd and Pb isotopes. *Geochim. Cosmochim. Acta* **50**, 231–45.

Nemec, D. 1974. Petrochemistry of the dyke rocks of the Central Bohemian Pluton. *Neues Jahrb. Min. Monats.* **118**, 193–206.

Nixon, P. H., M. F. Thirlwall, F. Buckley & C. J. Davies 1984. Spanish and Western Australian lamproites: aspects of whole-rock geochemistry. In *Kimberlites I: kimberlites and related rocks*, J. Kornprobst (ed.), 285–96. Amsterdam: Elsevier.

Nobel, F., P. A. M. Andriessen, E. H. Hebeda, H. N. A. Priem and H. E. Rondeel 1981. Isotopic dating of the post-Alpine Neogene volcanism in the Betic Cordillera. *Geol. Mijnbouw* **60**, 209–14.

Powell, J. L. & K. Bell 1970. Strontium isotopic studies of alkalic rocks: localities from Australia, Spain and western United States. *Contrib. Mineral. Petrol.* **17**, 1–10.

Puga, E. 1980. Hypothèse sur la genèse des magmatismes calcoalcalins, intra-orogenique et postorogenique alpins, dans les Cordilleres Betiques. *Bull. Soc. Geol. Fr.* **22**, 243–50.

Raoult, J.-F. & D. Velde 1971. Decouverte de trachytes potassiques à olivine et d'andesites en coulées dans le Miocène continental au Sud du Kef Hahouner (Nord du Constantinois, Algérie). *C. R. Acad. Sci. Paris Ser. D* **272**, 1051–4.

Ringwood, A. E. 1974, Petrological evolution of island arc systems. *J. Geol. Soc. Lond.* **130**, 183–204.

Rock, M. N. S. 1984. Nature and origin of calc-alkaline lamprophyres: minettes, vogesites, kersantites and spessartites. *Trans. R. Soc. Edinburgh Earth Sci.* **74**, 193–227.

Roden, M. F. 1981. Origin of coexisting minette and ultramafic breccia, Navajo Volcanic Field. *Contrib. Mineral. Petrol.* **77**, 195–206.

Roden, M. F. & D. Smith 1979. Field geology, chemistry and petrology of Buell Park minette diatreme, Apache County, Arizona. In *Kimberlites, diatremes and diamonds: their geology, petrology and geochemistry*, F. R. Boyd & H. O. A. Meyer (eds), 364–381. Am. Geophys. Union.

Ryabchikov, I. & A. L. Boettcher 1980. Experimental evidence at high pressure for potassic metasomatism in the mantle of the Earth. *Am. Mineral.* **65**, 915–19.

Ryabchikov, I., W. Schreyer & K. Abraham 1982. Compositions of aqueous fluids in equilibrium with pyroxenes and olivines at mantle temperatures and pressures. *Contrib. Mineral. Petrol.* **79**, 80–4.

Tatsumi, Y. 1982. Origin of high-magnesian andesites in the Setouchi volcanic belt, southwest Japan, II. Melting phase relations at high pressures. *Earth Planet. Sci. Lett.* **60**, 305–17.

Thompson, R. N. 1982. Magmatism of the British Tertiary Volcanic Province. *Scot. J. Geol.* **18**, 49–107.

Torres-Roldan, R. L. 1979. The tectonic subdivision of the Betic Zone (Betic Cordilleras, Southern Spain): its significance and one possible geotectonic scenario for the westernmost Alpine belt. *Am. J. Sci.* **279**, 19–51.

van Breemen, O., M. Aftalion, D. R. Bowes, A. Dudek, Z. Misar, P. Povondra & S. Vrana 1982. Geochronological studies of the Bohemian Massif, Czechoslovakia, and their significance in the evolution of Central Europe. *Trans. R. Soc. Edinburgh Earth Sci.* **73**, 89–108.

Varne, R. 1985. Ancient subcontinental mantle: a source of K-rich orogenic volcanics. *Geology* **13**, 405–8.

Varne, R. & J. D. Foden 1986. Geochemical and isotopic systematics of Eastern Sunda arc volcanics: implications for mantle sources and mantle mixing processes. In *The origin of arcs*, F. C. Wezel (ed.), 364–81. Amsterdam: Elsevier.

Venturelli, G., S. Capedri, G. Di Battistini, A. J. Crawford, L. N. Kogarko & S. Celestini 1984a. The ultrapotassic rocks from southeastern Spain. *Lithos* **17**, 37–54.

Venturelli, G., R. S. Thorpe, G. V. Dal Piaz, A. Del Moro & P. J. Potts 1984b. Petrogenesis of calc-alkaline, shoshonitic and associated ultrapotassic Oligocene volcanic rocks from the northwestern Alps, Italy. *Contrib. Mineral. Petrol.* **86**, 209–20.

Venturelli, G., E. Salvioli Mariani, S. F. Foley, S. Capedri & A. J. Crawford 1988. Petrogenesis and conditions of crystallization of Spanish lamproitic rocks. *Can. Mineral.* **26**, 67–80.

Vila, J. M., J. Hernandez & D. Velde 1974. Sur la présence d'un filon de roche lamproitique (trachyte potassique à olivine) recoupant le flysch de type Guerrouch entre Azzaba (ex-Jemmapes) et Hammam-Meskoutine, dans l'Est du Constantinois (Algérie). *C. R. Acad. Sci. Paris Ser. D.* **278**, 2589–92.

Vollmer, R., P. Ogden, J.-G. Schilling, R. H. Kingsley & D. G. Waggoner 1984. Nd and Sr isotopes in ultrapotassic volcanic rocks from the Leucite Hills, Wyoming. *Contrib. Mineral. Petrol.* **87**, 359–68.

Wagner, C. & D. Velde 1985. Mineralogy of two peralkaline arfvedsonite-bearing minettes. A new occurrence of Zn-rich chromite. *Bull. Mineral.* **108**, 173–87.

Wagner, C. & D. Velde 1986. The mineralogy of K-richterite-bearing lamproites. *Am. Mineral.* **71**, 17–37.

4 Experimental studies on boninite petrogenesis

SUSUMU UMINO & IKUO KUSHIRO

Abstract

Melting experiments were performed in the pressure range 0–20 kbar at 1150–1320°C on five magnesian aphyric boninites: CH414 (the most Mg-rich of Umino's Type II boninite, with olivine and bronzite microphenocrysts); CH1002 (Type II); CH634 (the most Mg-rich Type 1 boninite, with olivine, clinoenstatite and bronzite microphenocrysts); CH21 (the most evolved Type 1 boninite); and CH88 (a Type IV boninite with bronzite, pigeonite and augite microphenocrysts). Liquidus temperatures range from 1290 to 1315°C at 1 atm.

Sample CH414 crystallizes olivine and bronzite just below the liquidus at 17 kbar and 1130°C under H_2O-saturated conditions, and at 8 kbar and 1256°C under H_2O-undersaturated (4.7%) conditions. Phase relationships suggest a minimum H_2O content of 4% for the CH414 magma. Samples CH1002 and CH634 crystallize both olivine and bronzite near the liquidus under H_2O-saturated conditions greater than 6 kbar. Protoenstatite and bronzite coexist near the liquidus only at pressures below 2 kbar for these three samples. At 1 atm CH88 crystallizes bronzite on the liquidus at 1280°C, followed by pigeonite at 1200°C and then augite.

These results indicate that the most primitive Type II boninite (CH414) could coexist with harzburgite above 8 kbar with water contents exceeding 5%. The absence of clinoenstatite in CH414 indicates that crystallization occurred at pressures greater than 3 kbar.

4.1 Introduction

Hydrous melting experiments on synthetic systems have demonstrated that the liquidus volume of olivine expands relative to that of low-Ca pyroxene under hydrous conditions. This has led to the suggestion that primary andesitic magmas may be generated in peridotitic upper mantle (Yoder 1969, Kushiro 1972, 1974). Glasses produced by hydrous melting of peridotites are quartz normative, and some are broadly andesitic (Green 1973, 1976, Kushiro 1974,

Mysen & Boettcher 1975), although these liquid compositions are modified by crystallization of quench crystals (Nicholls & Ringwood 1973, Green 1976, Takahashi & Kushiro 1983). Liquidus phase relations of hydrous natural basaltic compositions indicate that magmas coexisting with mantle peridotite at pressures less than 20 kbar are silica-saturated tholeiite (Nicholls & Ringwood 1973, Nicholls 1974). Thus, although silica-saturated magmas may be produced by hydrous melting of upper mantle, they are compositionally unlike typical orogenic andesites.

Some criteria for a lava to represent a primary magma (which could be in equilibrium with mantle peridotite) are high Mg# (> 0.70), high Ni content, and presence of magnesian, nickeliferous olivine ($Fo_{>88}$). This is based on the following observations (e.g. Sato 1977): (a) Fe–Mg–Ni exchange partitioning between olivine and silicate (especially basaltic) melts is almost independent of pressure and temperature; (b) olivine in upper-mantle peridotite is generally $Fo_{90\pm3}$ and has approximately 0.4% NiO. Since olivine is the major liquidus phase in most basaltic magmas, the first olivine crystallizing from primary magma must be compositionally similar to mantle olivine.

Boninites, also referred to as high-Mg andesites, are known from the circum-Pacific region (e.g. Kuroda *et al.* 1978, Jenner 1981, Hickey & Frey 1982, Tatsumi 1981, 1982, Umino 1986, Bloomer & Hawkins 1987) and some ophiolites (e.g. Coish *et al.* 1982, Cameron 1985, Crawford & Cameron 1985) and are characterized by high Ni and Cr concentrations and low HREE and HFS element contents. Some of these boninites, though not all, have olivine phenocrysts with high Fo (>88) and Ni contents which could be in equilibrium with the groundmass (liquid) composition on the basis of Fe–Mg partitioning (Roeder & Emslie 1970). These may be considered as primary magmas, although they are compositionally distinct from orogenic andesite magmas; they are minor components of the crust but have apparently been erupted from Precambrian to Recent. Their infrequent eruption implies that boninite formation needs extraordinary physical and tectonic conditions. However, only a few cases are known which give direct information on, and clues towards understanding the tectonic setting and mechanism of, boninite production.

Those who favour models of direct partial melting of upper mantle for boninite genesis propose a subduction-related origin, specifically, at the very earliest stage of island-arc development (e.g. Shiraki *et al.* 1978, Cameron *et al.* 1979, Meijer 1980), during the very earliest stages of backarc-basin opening (Crawford *et al.* 1981, Flower & Levine 1987), or during and shortly after subduction of a spreading centre (Crawford & Keays 1987, Saunders *et al.* 1987, Rogers & Saunders 1989, Tatsumi & Maruyama 1989). The LILE and LREE enrichment seen in most boninites is generally interpreted to have been caused by fluids derived from the subducted slab, although an origin involving fluids derived from deep, more fertile mantle has been suggested by Hickey-Vargas & Reagan (1987) and Bloomer (1987).

There are, of course, some boninitic-type (i.e. high-Si, high-Mg) magmas which are not primary magmas. For example, Tertiary high-Mg andesites from Greenland are interpreted to have been produced through assimilation of upper crust by high-temperature picritic magmas (Pedersen 1985), and siliceous high-Mg basalts in some Archaean greenstone belts have been shown convincingly to be crustally contaminated komatiites (Arndt & Jenner 1986). Clearly, there are several different petrogenetic scenarios for the formation of high-Si, high-Mg magmas.

Geochemical and petrographic data favour hydrous partial melting of peridotite for the origin of W Pacific boninites. High Mg#, Ni and Cr concentrations and the presence of nickeliferous olivine (0.4% NiO and $Fo_{>90}$) and Cr-rich spinel (Cr# = 0.8–0.9) suggest derivation from refractory peridotitic upper mantle. Depletions in HREE and HFSE and enrichment in LILE suggest a model involving melting of refractory residual mantle following MORB extraction, with melting induced by introduction of slab-derived LILE-rich fluids (Sun & Nesbitt 1978, Cameron *et al.* 1983). Dobson & O'Neil (1987) have shown that these fluids probably come from subducted oceanic crust, because primary δD values of boninite (-55) are higher than those of normal mantle (-80 to -75).

The observations noted above clarify to some extent the chemical features of boninite sources, including the fluids, but they do not throw light on P–T conditions required for boninite generation. Experimental studies on boninitic rocks ('sanukitoids') with relatively low MgO contents (<10%) from the Setouchi region (Tatsumi 1981, 1982) have shown that conditions for producing these magmas are 1020–1120°C at 10–15 kbar with H_2O contents > 7%. However, these data are not directly applicable to more primitive boninites such as those studied here. Higher concentrations of MgO and FeO, plus the presence of clinoenstatite in the Bonin Is. boninites, suggest higher temperatures and/or lower pressures for their generation. Accordingly, this chapter presents the results of high-P, high-T experiments on boninites from Chichijima (Bonin Islands) and discusses the $P-T-X_{H_2O}$ conditions of boninite petrogenesis.

4.2 Petrography and mineralogy of selected boninites

4.2.1 Introduction

Umino (1986) classified Bonin Is. boninites into five types on the basis of microphenocryst assemblages: (I) olivine + clinoenstatite + bronzite; (II) olivine + bronzite; (III) bronzite; (IV) bronzite + pigeonite + augite; (V) clinoenstatite + bronzite. Among these, Types III, IV and V have relatively evolved bulk compositions and mineralogy. Type IV boninite is distinguished from the others in having lower incompatible element abundances and higher

Table 4.1 Wholerock and modal analyses of the Chichi-jima boninites used in melting studies reported in this chapter.

	CH414 (Type II)		CH1002 (Type II)		CH634 (Type I)		CH21 (Type I)		CH88 (Type IV)	
	1a	1b	2a	2b	3a	3b	4a	4b	5a	5b
SiO_2	54.95	57.57	57.13	58.16	57.30	57.77	54.31	57.89	55.03	56.61
TiO_2	0.25	0.26	0.11	0.11	0.10	0.10	0.19	0.20	0.11	0.11
Al_2O_3	10.55	11.05	10.29	10.48	9.24	9.32	11.46	12.21	10.93	11.24
Fe_2O_3	2.85		9.73		9.87		3.18		9.15	
FeO	5.82	8.78	nd	8.91	nd	8.96	5.27	8.66	nd	8.47
MnO	0.17	0.17	0.14	0.14	0.17	0.17	0.15	0.15	0.13	0.13
MgO	11.74	12.30	12.34	12.56	15.04	15.06	9.37	9.98	11.83	12.17
CaO	7.39	7.74	7.89	8.03	6.91	6.97	7.33	7.81	9.70	9.98
Na_2O	1.57	1.64	1.18	1.20	1.19	1.20	2.21	2.35	0.74	0.76
K_2O	0.41	0.42	0.40	0.41	0.35	0.35	0.63	0.67	0.51	0.52
P_2O_5	0.02	0.02	0.00	0.00	0.00	0.00	0.03	0.03	0.00	0.00
H_2O^-	0.47		nd		nd		1.92		nd	
H_2O^+	3.65		nd		nd		3.89		nd	
total	99.84		99.21		100.17		99.94		98.13	

FeO*/MgO	0.71	0.71	0.59	0.87	0.70
100Mg#	71.4	71.5	75.0	67.3	71.9
Ni	288	212	261	100	67
Cr	995	848	1402	611	952
microphenocrysts					
olivine	7.0	4.9	5.3	0.4	–
clinoenstatite	–	–	0.5	0.7	–
bronzite	11.9	15.5	13.1	10.1	8.8
augite & pigeonite	–	–	–	–	6.6
spinel	trace	0.1	0.1	trace	trace
groundmass					
Cpx-microlites	46.8	32.3	36.7	25.5	26.3
glass	34.3	47.0	44.1	60.3	48.4
vesicles	–	0.3	0.2	2.8	10.0
total points	2400	2000	2000	2000	2000

In columns b of oxide wt% lists, total Fe is recalculated to FeO and all totals are normalized to 100 wt% on an anhydrous basis. nd = Not determined.

Mg# = Mg/(Mg + Fe^{2+}) mol ratio (Fe^{2+} is calculated as total Fe for wholerocks).

Major elements of the samples CH1002, CH634 and CH88 were analysed by x-ray fluorescence (XRF) and samples CH414 and CH21 were analysed by conventional wet method (analyst = H. Haramura).

Ni and Cr were determined by XRF.

CaO contents, and is thought to be derived from a somewhat different parent magma. Melting experiments were carried out on five aphyric or sparsely porphyritic boninites. Sample CH634 (Type I) is one of the most magnesian sparsely porphyritic boninites, and samples CH414 and CH1002 are amongst the most magnesian Type II boninites. Sample CH21 (Type I) is the most evolved Type I aphyric boninite, and CH88 is a Type IV aphyric boninite. The latter boninites have been used in 1 atm runs to examine pyroxene phase relations, which will be reported in detail elsewhere (Umino & Kushiro 1989). Wholerock and modal analyses of the samples studied are listed in Table 4.1.

4.2.2 CH414 (primitive Type II)

CH414 has Mg# = 0.71 and high Ni (290 ppm) and Cr (1000 ppm) contents. Olivine microphenocrysts are mostly fresh and range from Fo_{90-87}, with a maximum NiO content of 0.4%. Olivine is sometimes partially surrounded by bronzite microphenocrysts and/or clinopyroxene microlites, but is mostly free of reaction coronae. Bronzite microphenocrysts are euhedral-to-subhedral prisms with Mg# of 0.89–0.85. Cr-spinel occurs as euhedral octahedra embedded in olivine, bronzite and glass, and has Cr# from 0.8 to 0.85 and $Fe^{3+}/(Cr + Al + Fe^{3+})$ from 0.05 to 0.1. Groundmass is composed of clinopyroxene microlites and fresh or partially palagonitized glass. The Fe–Mg exchange partition coefficient K_D(Fe–Mg) between the most magnesian olivine and bulk composition is 0.28–0.37, assuming that $Fe^{3+}/\Sigma Fe$ is 0–0.25. These values are within the experimental ranges determined at 1 atm for basaltic compositions (0.30–0.33; Roeder & Emslie 1970) and basalt–basaltic andesite compositions (0.27–0.38; Takahashi 1978). Thus the most magnesian olivine microphenocrysts could coexist in equilibrium with a liquid with this wholerock composition, suggesting that this boninite is close to a primary magma composition.

4.2.3 CH1002 (primitive Type II)

The textural relations of microphenocrysts and groundmass minerals in CH1002 are the same as those for CH414; pyroxenes are fresh, but olivine and glass are slightly more altered in CH1002. Microphenocrystic olivine is more restricted in composition than in CH414, ranging from $Fo_{89.5}$ with 0.3% NiO to $Fo_{88.5}$ with 0.2% NiO. Bronzite microphenocrysts have core compositions of $Wo_{2-3.5}$ and Mg# of 0.87–0.85, less magnesian than coexisting olivine. Two types of spinels are present: one consists of large discrete crystals with Cr# = 0.88–0.89 and $Fe^{3+}/(Cr + Al + Fe^{3+}) = 0.05$, and the other is of small inclusions in glass and other mineral phases with Cr# = 0.82–0.84 and $Fe^{3+}/(Cr + Al + Fe^{3+}) = 0.05–0.1$.

CH1002 has a nearly identical bulk composition to that of CH414 but is slightly depleted in Cr (850 ppm), Ni (210 ppm) and modal olivine (4.9

modal% (Table 4.1). K_D(Fe–Mg) between the most magnesian olivine and bulk composition is 0.29–0.38, assuming $Fe^{3+}/\Sigma Fe = 0$–0.25, again suggesting that it could be in equilibrium with liquid of the wholerock composition. Lower Ni and Fo contents in olivine suggest that a small amount of olivine fractionation preceded crystallization of this sample.

4.2.4 CH634 (primitive Type I)

Sample CH634 is one of the most primitive Type I aphyric boninites; olivine microphenocrysts are completely altered, although pyroxenes are fresh. Bronzite has Mg# = 0.89–0.85, and clinoenstatite has Mg# = 0.90–0.88. The Cr# of most spinels exceeds 0.83 and is higher than those of spinels in CH21 and CH414 but lower than those of large discrete spinels in CH1002.

Among the boninites selected for this experimental study, CH634 has the highest Mg# (0.75) and Cr and Ni contents of 1400 ppm and 260 ppm, respectively. The primitive bulk composition and highly magnesian pyroxenes suggest that it is a primary magma. Lack of compositional data of olivine makes it impossible to test whether or not it has fractionated olivine.

4.2.5 CH21 (evolved Type I)

Sample CH21 is the most differentiated Type I aphyric boninite, with fresh microphenocrysts of olivine, clinoenstatite and bronzite. Olivine ($Fo_{88.5-87}$) has NiO = 0.3 ± 0.05% Clinoenstatite and bronzite range in Mg# from 0.90 to 0.88, and 0.89 to 0.86, respectively. Spinel is highly chromian with Cr# = 0.79–0.88, while $Fe^{3+}/(Cr+Al+Fe^{3+})$ is less than 0.1. The K_D(Fe–Mg) between the most Fo-rich olivine and the bulk composition is 0.36 if $Fe^{3+}/\Sigma Fe = 0$, but it reaches 0.43 if $Fe^{3+}/\Sigma Fe = 0.25$. Higher K_D values than experimentally determined equilibrium values and those for CH414 and CH1002 suggest that CH21 has fractionated olivine or had significantly lower Fe^{3+}/Fe^{2+} than CH414 and CH1002.

4.2.6 CH88 (Type IV)

This sample is a rare Type IV aphyric boninite with fresh orthopyroxene and clinopyroxene microphenocrysts. The latter, which always occur as composite crystals rimming bronzite microphenocrysts, change compositionally from pigeonite within narrow zones against bronzite, almost continuously through subcalcic augite to augite. Although this compositional trend is similar to that of groundmass clinopyroxene microlites, miscibility gaps probably exist between bronzite and pigeonite, and between pigeonite and augite.

CH88 is distinct from other boninites in that it has the highest CaO (10%) content, in spite of its relatively high Mg# (0.72), and lowest Na_2O content of 0.76% (Table 4.1) of the Bonin Is. boninites. Type IV boninites cannot be

Table 4.2 Analyses of run products for experiments on Chichi-jima boninites.

Product Sample Run no.	Glass CH414 RAN4-1	RAN4-16	PW-13	PW-09	CH634 RAN4-2	RAN5-3	RAN5-14	CH21 RN-7b	CH88 RN-7d	Olivine core CH414 PW-29	PW-29	PW-03	PW-03	RAN4-10	RAN4-12	RAN4-15
SiO_2	57.96	58.13	55.38	54.88	58.60	54.05	54.36	57.00	55.74	41.03	40.93	40.06	39.89	40.50	39.97	40.34
TiO_2	0.17	0.17	0.14	0.14	0.10	0.12	0.10	0.15	0.10	0.00	0.00	0.00	0.00	0.01	0.00	0.01
Al_2O_3	11.15	11.45	10.18	10.18	9.90	8.69	8.81	11.69	10.46	0.11	0.09	0.04	0.05	0.02	0.05	0.02
FeO	7.69	8.10	6.13	6.01	7.96	6.93	7.78	8.18	8.22	7.95	7.90	10.43	10.57	10.30	9.99	9.62
MnO	0.16	0.01	0.10	0.18	0.16	0.18	0.25	0.12	0.14	0.06	0.13	0.16	0.15	0.13	0.15	0.21
MgO	11.93	12.45	11.54	12.01	14.32	13.47	14.36	10.51	12.01	50.57	50.95	49.17	48.60	49.02	49.78	49.17
CaO	7.31	7.72	7.15	7.13	6.75	6.49	6.78	7.41	9.68	0.08	0.11	0.13	0.11	0.17	0.18	0.17
Na_2O	1.66	1.73	1.46	1.51	0.89	0.93	0.90	2.12	0.75	0.04	0.04	0.00	0.00	0.00	0.02	0.00
K_2O	0.44	0.31	0.32	0.35	0.35	0.28	0.25	0.68	0.45	0.01	0.00	0.00	0.00	0.00	0.00	0.00
Cr_2O_3	0.12	0.21	0.16	0.16	0.19	0.17	0.22	0.04	0.14	0.00	0.00	0.00	0.00	0.08	0.00	0.03
NiO	0.00	0.00	0.05	0.05	0.02	0.04	0.00	0.02	0.03	0.42	0.45	0.40	0.35	0.31	0.24	0.00
total	98.59	100.28	92.61	92.60	99.24	91.35	93.81	97.92	97.72	100.27	100.60	100.39	99.72	100.54	100.38	99.57
100Mg #	73.45	73.34	77.04	78.07	76.23	77.60	76.68	69.61	72.26	91.90	92.00	89.37	89.12	89.46	89.88	90.11
P(kbar)	0	0	10	9	0	7.2	8.5	0	0	17	17	8	8	0	0	0
$T(°C)$	1300	1290	1277	1265	1320	1281	1276	1291	1291	1130	1130	1256	1256	1220	1259	1285
$H_2O(\%)$	0	0	4.7	4.7	0	5.2	5.2	0	0	19.8	19.8	4.7	4.7	0	0	0

Product	Clinopyroxene core						Orthopyroxene core						
Sample	CH414						CH414						
Run no.	RAN4-10	RAN4-12	RAN4-12	RAN4-16	R6-16a	RAN4-10	RAN4-10	RAN4-12	RAN4-12	RAN4-15	RAN4-16	R6-16a
SiO_2	57.58	57.35	57.16	58.08	57.38	56.55	57.46	56.33	56.59	57.35	58.45	56.57
TiO_2	0.02	0.00	0.00	0.00	0.08	0.03	0.03	0.02	0.00	0.09	0.05	0.20
Al_2O_3	1.06	0.64	0.77	0.55	0.88	1.12	0.73	0.86	0.72	0.26	1.51	0.69
FeO	6.01	5.81	6.51	4.72	7.37	7.62	7.60	7.97	7.68	8.25	5.59	7.74
MnO	0.21	0.19	0.19	0.09	0.18	0.17	0.17	0.18	0.20	0.15	0.25	0.22
MgO	34.22	35.13	34.60	35.37	32.82	32.23	32.63	32.54	33.34	32.78	33.78	31.87
CaO	1.19	0.87	0.93	0.42	0.96	1.61	1.49	1.53	1.42	1.26	1.00	1.16
Na_2O	0.00	0.02	0.02	0.12	0.40	0.00	0.02	0.02	0.01	0.25	0.39	0.22
K_2O	0.01	0.01	0.00	0.00	0.02	0.01	0.00	0.01	0.01	0.07	0.00	0.00
Cr_2O_3	0.36	0.35	0.43	0.39	0.33	0.71	0.55	0.62	0.33	0.48	0.39	0.25
NiO	0.15	0.12	0.09	0.07	0.06	0.12	0.06	0.06	0.08	0.11	0.26	0.00
total	100.81	100.49	100.70	99.81	100.48	100.17	100.74	100.14	100.38	100.75	101.67	98.92
100Mg#	91.04	91.51	90.46	93.04	88.81	88.29	88.44	87.92	88.55	87.62	91.50	88.01
Wo	2.22	1.60	1.72	0.79	1.83	3.06	2.82	2.88	2.64	2.36	1.90	2.26
En	89.02	90.05	88.91	92.30	87.19	85.58	85.95	85.39	86.22	85.56	89.76	86.02
Fs	8.76	8.35	9.38	6.91	10.98	11.35	11.23	11.73	11.15	12.08	8.34	11.72
P(kbar)	0	0	0	0	0	0	0	0	0	0	0	0
T(°C)	1220	1259	1259	1290	1290	1220	1220	1259	1259	1285	1290	1290
H_2O(%)	0	0	0	0	0	0	0	0	0	0	0	0

derived from the other Bonin Is. boninitic magmas by crystallization differentiation (Umino 1986). Lack of olivine and low Ni and Cr concentration suggest that CH88 is not primary.

4.3 Experimental methods

Melting experiments at 1 atm were carried out with a siliconate furnace. Oxygen fugacity was controlled to lie within the range of 10^{-8} to 10^{-9} bar at temperatures between 1160 and 1320°C (between QFM and WM) by mixing of CO_2 and H_2. Finely ground rock powders of each sample were pressed to 1–2 mm thick hard disks and then crushed and made into 'cubes' 3 mm long and 5 mm wide, weighing 60–100 mg. These pieces were suspended in a furnace with 0.05 mm thick Pt wire. In some runs on CH414, CH1002 and CH634, the starting materials were melted at superliquidus temperatures for 10–15 min and then held at desired temperatures. For runs on CH21 and CH88, glass beads were prepared by melting the rock powders in a Pt crucible which had been preheated with Fe at 1 atm and above 1300°C for 10–15 min.

High-pressure melting experiments were performed with a solid-media, piston–cylinder apparatus above 5 kbar and with an internally heated, gas-media pressure vessel below 5 kbar. Finely ground rock powders were heated at 1100°C in air for more than 24 h and then stored in an oven at 110°C. In experiments on CH1002, dried powder with graphite capsules was used for anhydrous runs. Both hydrated glass and dried rock powder with H_2O were used for hydrous runs. In experiments on CH414, hydrated glass with 4.7% H_2O, and dried rock powder and H_2O were used. The hydrated glass was made as follows: dried rock powder of CH414 and 4.7% H_2O were sealed in a Pt tube and heated at 1250°C and 5 kbar for 30 min. Pt capsules were used for runs above 1180°C and $Ag_{70}Pd_{30}$ capsules were used for runs at lower temperatures. Oxygen fugacities were not controlled during the runs; however, those for the furnace assembly of the piston–cylinder were probably below the Ni–NiO buffer (Merrill and Wyllie 1975). Graphite capsules sealed in Pt tubes were also used for some higher-temperature experiments in order to check the effect of Fe loss to the capsules and the change in f_{O_2} caused by the different sample containers. In some H_2O-saturated runs, both graphite/Pt and Pt sample containers were used, and yielded identical results. Temperatures were measured with Pt–Pt$_{90}$Rh$_{10}$ thermocouples and were controlled within ±1–5°C. The temperature fluctuation was less than 3°C in most runs.

Starting materials were analysed by conventional wet chemical analysis (analyst: H. Haramura) and x-ray fluorescence spectrometry (Rigaku model 3064, using matrix corrections of Matsumoto & Urabe (1980)). Analyses of the run products were carried out by a JEOL JXA-733 electron probe microanalyser (EPMA) at the Ocean Research Institute of the University of Tokyo. Accelerating voltage was 15 kV and specimen currents were 2.0×10^{-8} to

1.2×10^{-8} A. Corrections were made according to Bence & Albee (1968) and Nakamura & Kushiro (1970). Tables of run conditions for all samples are available on request from the senior author.

The Fe loss from charges to capsules was estimated by analysing with the EPMA glass produced in the runs. Analyses of glass produced above the liquidi are listed in Table 4.2; Fe loss is less than 1% for glass produced in 1 atm experiments. Analyses of some hydrated glasses produced near the liquidi show up to 2% (anhydrous basis) Fe loss to Pt capsules (Table 4.2). The Fe loss results in enrichment of SiO_2 and MgO in the melt, so that orthopyroxene and protopyroxene tend to crystallize instead of olivine and protopyroxene. The stability field of olivine is, therefore, somewhat larger than that shown by the present results. However, CH634 glasses produced in the hydrous runs show less Fe loss (0.7–1.4% on an anhydrous basis: RAN5-3, RAN5-14; Table 4.2), but olivine was not observed in any runs.

Identification of mineral phases was first done by an optical microscope and then confirmed by EPMA. Identification of pyroxenes was made as follows. Clinoenstatite crystals (considered to be inverted from protoenstatite) have characteristic polysynthetic twinning and terminal faces crossing at acute angles. Orthopyroxene crystals have terminal faces making obtuse angles with one another. EPMA analyses have shown that presumed clinoenstatite has Wo content up to 2.2 (Table 4.2), which is rather high for protoenstatite (Huebner 1982). However, repeated analyses have shown that presumed orthopyroxene coexisting with clinoenstatite has Wo contents exceeding 2.8, so that a compositional gap exists between the clinoenstatite and orthopyroxene (e.g. RAN4-10 in Table 4.2). The slightly high Wo content of clinoenstatite may be due to contamination by glass surrounding the tiny crystals, and thus distinction between clinoenstatite and orthopyroxene on the basis of crystal habit seems to be valid. To avoid uncertainties in determining pyroxene species by composition, we analysed 5–10 crystals of each mineral species identified in an individual run product and checked for the presence of compositional gaps between the mineral species identified optically.

4.4 Experimental results

4.4.1 CH414 (Type II)

In 1 atm runs RAN4-16, RAN4-17, RAN4-19 and R7-5, the starting materials were melted at superliquidus temperatures for 15–20 min. In all other runs, samples were heated directly to planned temperatures. Run product mineral and glass analyses are given in Table 4.2. Spinel, olivine, protoenstatite and bronzite are present from 1290 to 1220°C. At 1210°C both olivine and protoenstatite disappear, while spinel and bronzite continue to crystallize until at least 1160°C.

Under H_2O-saturated conditions the liquidus phase is olivine up to 17 kbar and orthopyroxene at higher pressures (Fig. 4.1). Protopyroxene is stable up to about 2 kbar near the liquidus and coexists with olivine and orthopyroxene. At 3 and 17 kbar, both olivine and orthopyroxene crystallize near the liquidus. At 8–10 kbar and 1160–1150°C olivine alone crystallizes as a liquidus phase, joined by orthopyroxene at lower temperatures. The olivine-only field is very small below several kbar and was not detected. The 4.7% H_2O liquidus intersects the H_2O saturation curve between 2 and 3 kbar, indicating that the solubility of H_2O in this boninite magma is about 5% at 2.5 kbar. This solubility is close to that of basalt melt determined by Hamilton *et al.* (1964) over a similar pressure range. Under H_2O-undersaturated conditions (4.7%

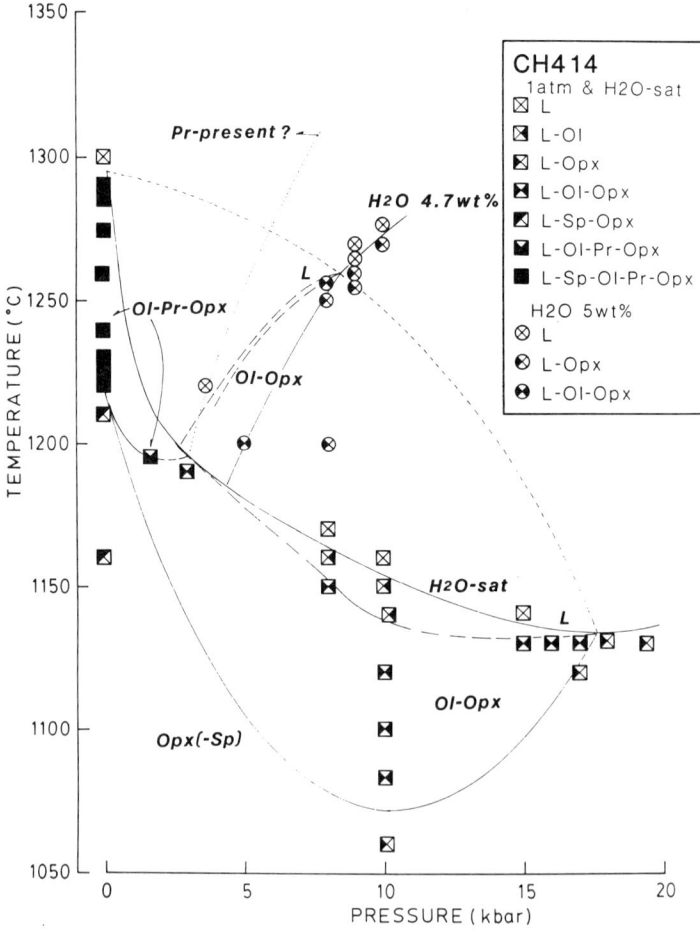

Figure 4.1 Liquidus phase relations of CH414 (Type II aphyric boninite) under anhydrous, H_2O-undersaturated (5%) and H_2O-saturated conditions.

H_2O), olivine and orthopyroxene are on the liquidus at 8 kbar and 1256°C. Orthopyroxene is on the liquidus at higher pressures.

Olivine produced in runs under both H_2O-saturated and H_2O-undersaturated conditions is Fo_{92-89} with NiO contents of 0.45–0.35%; Fo contents higher than those of natural olivine microphenocrysts (Fo_{90}) may be due to Fe loss during the runs. However, these compositions are well within the range for mantle olivines, and support the argument that this boninite was formed by partial melting of upper-mantle peridotite.

4.4.2 CH1002 (Type II)

The 1 atm liquidus phases are orthopyroxene and protopyroxene at 1300°C and both minerals continue to crystallize to 1280°C. At lower temperatures, protopyroxene disappears and orthopyroxene alone crystallizes. In anhydrous runs, both orthopyroxene and protopyroxene appear near the liquidus up to 10 kbar. Protopyroxene is not stable above this pressure and at lower temperatures (Fig. 4.2). Both olivine and orthopyroxene coexist near the liquidus under H_2O-saturated conditions at about 9 kbar and 1150°C. Orthopyroxene is on the liquidus at 2 kbar but olivine doe not crystallize.

4.4.3 CH634 (Type I)

Runs on CH634 were conducted at pressures from 1 atm to 12 kbar; compositions of glasses produced are listed in Table 4.2. In 1 atm experiments, starting materials were heated at superliquidus temperatures (1320°C) for 15–20 min before the runs. Liquidus phases are Cr-spinel, protoenstatite and bronzite, which coexist to 1225°C, below which protoenstatite disappears (Fig. 4.3).

Under H_2O-saturated conditions, olivine and orthopyroxene occur near the liquidus from 6 to 12 kbar (Fig. 4.3). A $P-T$ region where olivine (+ Cr-spinel) alone is stable is expected to be present at higher temperatures. The results of runs at 10 kbar indicate that such a temperature interval is less than 10°C, narrower than that for sample CH414. Multiple saturation of olivine and bronzite occurs at about 5 kbar and 1195°C. Below 5 kbar, protoenstatite and bronzite are the liquidus phases.

Runs with 5% H_2O were performed in the pressure range 6–9 kbar and from 1200 to 1280°C. The liquidus phase is bronzite. Olivine does not appear at any temperature and pressure. The solubility of H_2O is estimated to be 5% at 4 kbar, lower than that for CH414 and the basaltic melt of Hamilton *et al.* (1964).

4.4.4 CH21 (Type I)

Experiments at 1 atm were performed on CH21; a resulting glass composition is given in Table 4.2. Olivine and pyroxene compositions are plotted in Figure

EXPERIMENTAL STUDIES ON BONINITE PETROGENESIS

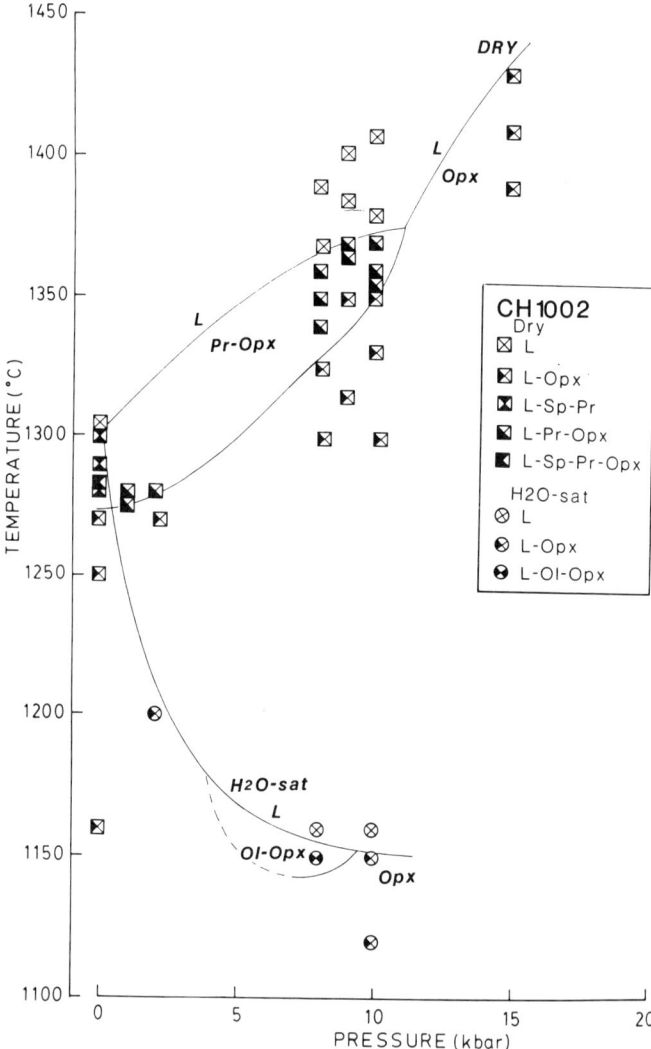

Figure 4.2 Liquidus phase relations of CH1002 (Type II aphyric boninite) under anhydrous and H_2O-saturated conditions.

4.4. Olivine occurs on the liquidus at 1282°C in the run using rock powder. However, a run using a glass bead at 1282°C produced only glass. Both olivine and bronzite crystallize in runs using glass beads and rock powder from 1277 to 1248°C. Protopyroxene crystallizes in the run at 1277°C; olivine disappears and bronzite crystallizes alone below 1235°C.

Olivine produced at 1282°C from rock powder is Fo_{90-88}, slightly higher than Fo contents of natural microphenocrysts. Absence of olivine in the run with a glass bead may be due to Fe loss. In the rus at 1277 and 1248°C, both

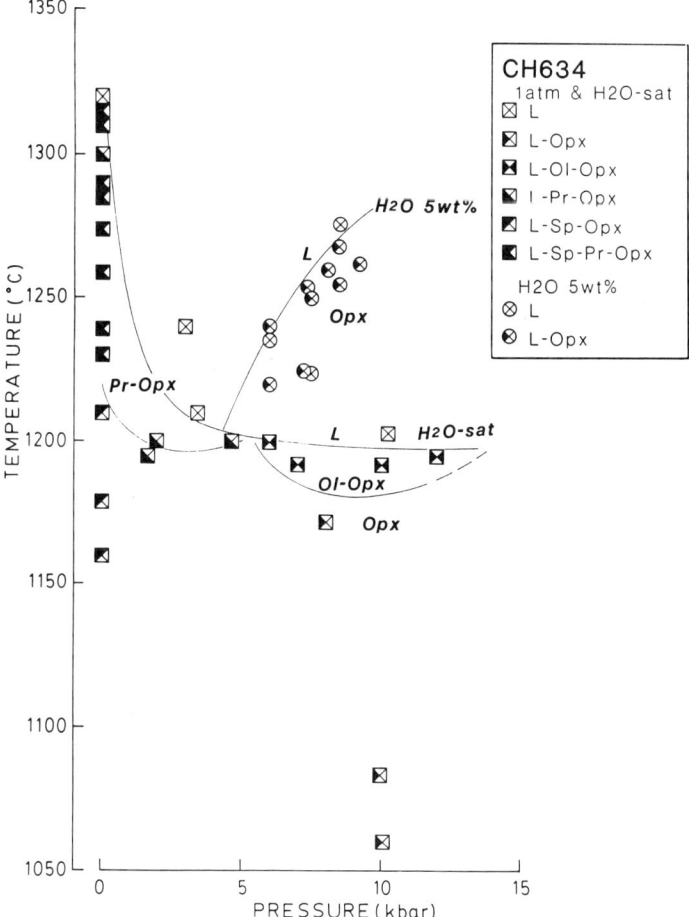

Figure 4.3 Liquidus phase relations of CH634 (Type I aphyric boninite) under anhydrous, H_2O-undersaturated (5%) and H_2O-saturated conditions.

olivine and bronzite crystallized from glass beads are slightly more magnesian than those crystallized from rock powder. From 1235 to 1201°C, orthopyroxene produced in runs with glass beads has similar compositions to those produced in the runs with rock powder; however, the latter have a slightly broader spectrum of compositions.

4.4.5 *CH88 (Type IV)*

Melting experiments on CH88 were conducted at 1 atm. The liquidus temperature is between 1291 and 1282°C. Bronzite is the liquidus phase and it continues to crystallize alone until 1223°C. At 1201°C bronzite is replaced by

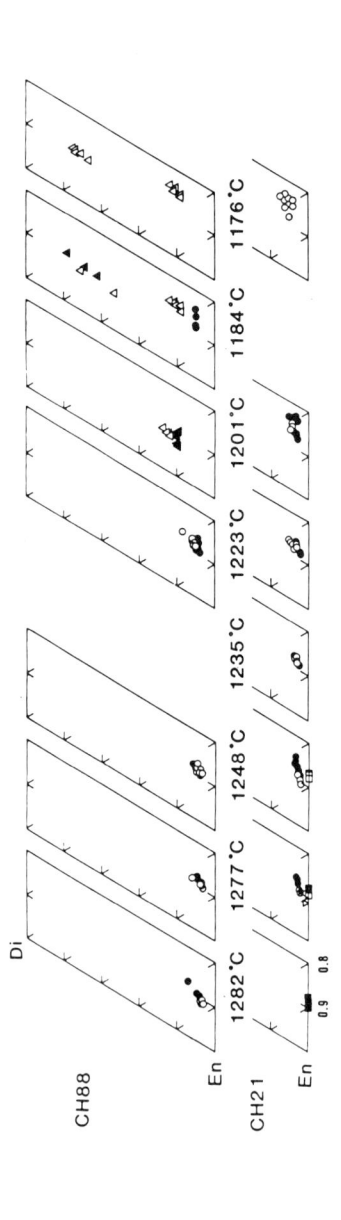

Figure 4.4 Olivine and pyroxene compositions produced in 1 atm runs on CH21 (lower) and CH88 (upper), plotted on pyroxene quadrilateral. Open symbols are products in runs using glass beads and closed symbols are for runs using rock powder.

pigeonite. Runs at 1184°C gave two different results. Run RN-10a with a glass bead produced pigeonite + augite + liquid, while run RN-10b with rock powder gave bronzite + augite + liquid. In RN-10b bronzite may be a relic from the starting material; owing to the low run temperature (1184°C), equilibration may not have been achieved. Only a glass bead was used in runs at 1176°C, in which pigeonite + augite + liquid were produced.

Orthopyroxene changes composition as the temperature falls from 1282 to 1223°C. Bronzite in run RN-9a with a glass bead has Mg# = 0.91–0.90, while bronzite in run RN-9b with rock powder has Mg# = 0.89–0.87, which is similar to that of bronzite microphenocrysts in the original rock. Because orthopyroxene is stable at the run temperature (1282°C), relics of bronzite microphenocrysts remained in the cores in run RN-10b, although starting material clinopyroxene melted. In runs from 1277 to 1223°C, orthopyroxenes from both glass beads and rock powder are compositionally similar. Runs with glass beads produced pigeonite, compositions of which become progressively more Fe-rich from 1201 to 1176°C. Pigeonite crystallized from rock powder (RN-5d) is more magnesian than that from a glass bead (RN-5c), suggesting disequilibrium.

4.5 Conditions of boninite generation and crystallization

High-pressure experiments on the most magnesian Type II aphyric boninite CH414 indicate that this magma could coexist with harzburgite under H_2O-saturated and H_2O-undersaturated conditions in the pressure range from 18 kbar to at least 8 kbar. The composition of olivine produced near the liquidus is within the range for mantle olivine, supporting the possibility of primary origin of this boninite.

A dotted curve in Figure 4.1 connects the maximum $P-T$ stability limit of protopyroxene for CH1002 under H_2O-free conditions and that for CH414 under H_2O-saturated conditions. On the low-pressure side of this line, protopyroxene is stable. Under H_2O-saturated conditions, protopyroxene does not crystallize in runs with CH1002. The protopyroxene field is larger in runs with CH634 than in the runs with CH414. The stability field of protopyroxene was not determined under dry conditions at high pressure on sample CH414, but comparison with H_2O-saturated experiments on samples CH634 and CH1002 suggests that the protopyroxene field would be larger for CH414 than for CH1002. Because CH414 exhibits no evidence of protopyroxene crystallization, CH414 microphenocrysts must have crystallized on the high-pressure side of this line; the minimum pressure is 3 kbar, given by the protopyroxene-free field on the H_2O-saturated liquidus.

The minimum H_2O content of the magma which precludes protopyroxene crystallization was not defined, because liquidi of lower H_2O contents at low pressures and high temperatures were not determined. Considering the results with 3.7% H_2O at 2.1 and 3.2 kbar, the liquidus with 4% H_2O must run through a point below 1240°C at 3 kbar and intersect the H_2O-saturated liquidus below 2 kbar. This indicates that if the H_2O content had been < 4%, protopyroxene would have crystallized from this boninite magma. Therefore, the minimum H_2O content would not be <4%, which is higher than that estimated by Dobson (1986).

Although solubility of H_2O in boninitic melts is close to that in basaltic melts, tholeiitic magmas are considered to have <1% H_2O (Tatsumi et al. 1983, Presnall et al. 1979). Tholeiitic basalts might reach the Earth's surface without losing their primary water, especially when they erupt on the deep sea floor. In contrast, it is clear from Figure 4.1 that boninite magmas with >4% H_2O would reach H_2O saturation early in their ascent path. It is unlikely that they could retain such high H_2O contents as they approach eruption; separation of excess H_2O as hydrothermal fluids must occur during ascent. High H_2O contents are also suggested by evidence of explosive eruptions seen in volcanic breccias and scoriaceous fragments of the Marubewan Formation on Chichi-jima (Umino 1985, Dobson 1986). Therefore, H_2O contents in boninites are unlikely to reflect primary values.

If the composition of CH634 has not been modified late in its history by fractionation of microphenocrysts, $P-T-X_{H_2O}$ conditions can be found to reproduce the natural microphenocryst assemblage olivine + clinoenstatite + bronzite. However, olivine does not crystallize from this composition even in runs at low pressures. Therefore, fractionation of some small amount of olivine microphenocrysts from the magma occurred before eruption. As the $P-T-X_{H_2O}$ conditions of CH634 crystallization were not reproduced, the upper pressure limit of crystallization of boninite magmas was not directly determined.

In the system $MgO-SiO_2$, protoenstatite is stable up to 15 kbar under dry conditions (Chen & Presnall 1975). At higher pressures, protoenstatite transforms to orthoenstatite. Under H_2O-saturated conditions, the stability field of protoenstatite shrinks relative to that under dry conditions and occurs only up to 8 kbar and 1390°C (Kushiro et al. 1968). Because the pressure limit of protopyroxene is reduced in the natural boninite system, the CH634 magma must have crystallized its protoenstatite and bronzite microphenocrysts at pressures < 8 kbar. These observations suggest that the crystallization pressures of boninite magmas are probably between 3 and 8 kbar.

The olivine and bronzite microphenocryst assemblage of sample CH1002 can be reproduced below 8 kbar (and above 2 kbar), which is within the pressure limit noted above, However, olivine crystallizes only above 1150°C (and probably below 1200°C) under H_2O-saturated conditions. The $P-T$ range of the olivine (+ orthopyroxene) field under H_2O-undersaturated conditions has not been determined for this composition, but it would be lower than 1200°C and probably close to the H_2O-saturated liquidus. If this boninite magma was in equilibrium with upper-mantle peridotite at pressures >5 kbar, it must have been formed under almost H_2O-saturated conditions. The magma intersects the H_2O-saturated liquidus soon after segregation, and begins to crystallize olivine and orthopyroxene while still in the mantle (crustal thickness beneath the Bonin Islands is 8–12 km (e.g. Ishihara et al. 1981)). It is unlikely that such a magma could reach the surface without fractionating crystals. Thus CH1002 is not considered to be a primary magma.

Umino (1986) has shown that there are at least two parental magmas for Bonin Is. boninites. One is parental to rare high-Ca Type IV boninites, and the other formed the low-Ca Types I, II, III and V boninites. Compositional variations of the Type V aphyric boninites can be explained by mixing of the most Mg-rich boninite and bronzite andesite magmas (Umino 1986).

The present experiments, however, demonstrate that Mg-rich sample CH634 is not primary, but that less magnesian sample CH414 may be primary. This suggests that at least three primary magmas are required to generate the spectrum of Bonin Is. boninites. Compositions of olivine and Cr-spinel, and wholerock compositions of Bonin Is. boninites, also suggest the necessity of a number of parent magmas (Umino 1986, 1987). The most magnesian Type I boninites have Fo-rich, Ni-poor olivine ($Fo_{<91.4}$ and $NiO < 0.33\%$) and Cr-rich spinel ($Cr\# = 0.85$–0.90), whereas some of the most magnesian Type II boninites are less magnesian than Type I boninites but contain more Ni-rich olivine (Fo_{90} and $NiO < 0.4\%$) and less Cr-rich Cr-spinel ($Cr\# = 0.80$–0.85). However, as shown by Umino (1986), some of the observed chemical variation in this suite may be due to mixing of primitive and more evolved (boninite-derived bronzite andesite) magmas.

REE patterns confirm that Type II boninites could not be derived from magnesian Type I boninites by crystallization differentiation (Masuda and Mori, pers. comm. 1987). These features indicate that local variations in degree of partial melting and mantle source heterogeneity may generate a spectrum of boninitic primary magma compositions. Some Type I boninites must have been derived from a parental magma formed by relatively high-degree partial melting of a mantle source with high LREE/HREE. These highly magnesian magmas fractionated olivine during ascent, so that olivines formed later are relatively Ni-poor. In contrast, some primary magmas (Type II) formed by lower degrees of melting erupted little modified by crystallization differentiation, and thus crystallized more Ni-rich olivine at low pressures.

What geologic situations are required for boninite petrogenesis? The present study supports a petrogenetic model of partial melting of mantle peridotite under hydrous conditions. Dehydration of the subducted slab is the most plausible source of water (e.g. Cameron et al. 1979). However, clinoenstatite-bearing boninites are not known from any recent arc-trench systems, implying that mantle wedges which yielded boninite magmas had different thermal structures and histories than those beneath modern island arcs from which boninites are unrecorded.

CH414 boninite magma separated from mantle harzburgite at pressures and temperatures between 8 kbar/1256°C and 17 kbar/1140°C. This temperature range is 100–200°C lower than that for arc basalts determined by Tatsumi *et al.* (1983). According to Tatsumi (1986), dehydration of subducted oceanic crust occurs at 30 kbar and the mantle above the slab is metasomatized to form a hydrous mantle wedge consisting of amphibole peridotite. This amphibole peridotite is dragged downwards along the subducting slab surface and dehydrates at <35 kbar, corresponding to the depth of the subducting slab beneath the volcanic front in most island arcs. Water released by dehydration of amphibole in peridotite migrates upwards and forms an amphibole peridotite 'curtain'. Since there is a hot region ($T > 1400°C$) above this 'curtain' beneath modern island arcs, melting of the amphibole peridotite 'curtain' occurs, producing island-arc tholeiite magmas. The high pressures (>15 kbar) within the zone of magma generation beneath arc volcanoes prohibit generation of boninite magmas.

There are two ways in which boninites may be produced in this scheme. One involves a low-temperature mantle wedge. If the region above the amphibole peridotite 'curtain' is cooler than about 1300°C, partial melting of rising diapirs will not proceed to the extent required to generate tholeiitic magmas. The other possibility involves melting of the hydrous mantle wedge beneath the forearc, which does not melt under normal thermal conditions in subduction zones (Crawford *et al.* 1981, Tatsumi & Maruyama 1989). These authors have suggested that an unusually high-temperature geotherm in the subforearc mantle wedge may be caused by injection of MORB-source diapirs associated with backarc-basin opening. Petrogenetic models should be constrained not only by experimental data but also by reconstructions of the tectonic setting of the Bonin Islands during the time these type-location boninites were erupted (Eocene). This is discussed in more detail in Tatsumi & Maruyama (1989).

4.6 Summary

Solubility of H_2O in primitive boninite CH414 melt is about 5% at 2.5 kbar, similar to that in basaltic melt as determined by Hamilton *et al.* (1964), whereas in primitive boninite CH634 melt, it is about 5% at 4 kbar.

Of the samples studied, most magnesian aphyric boninite CH634 (Type I) is not a primary magma, although less magnesian aphyric boninite CH414 (Type II) may be a primary magma. CH414 melt can coexist with harzburgite at pressures and temperatures ranging from 17 kbar and 1130°C under H_2O-saturated conditions to 8 kbar and 1256°C under H_2O-undersaturated (4.7% H_2O) conditions. The experimental results, olivine and spinel chemistry and REE patterns of near-aphyric boninites suggest the existence of at least three primary magmas for Bonin Is. boninites.

Because boninite CH414 shows no evidence of protopyroxene crystallization, it must have crystallized its microphenocryst assemblage at pressure >3 kbar. The minimum H_2O content in CH414 melt to preclude protopyroxene crystallization is 4%. Although conditions for crystallization of boninite CH634 were not reproduced experimentally, the maximum pressure for crystallization of protopyroxene cannot be >8 kbar, on the basis of data from high-pressure experiments in synthetic systems. These aphyric boninites, therefore, crystallized their microphenocryst assemblage at pressures between 3 and 8 kbar.

Acknowledgements

The authors thank Drs T. Fujii and K. Ozawa of the University of Tokyo and Y. Tatsumi of the University of Kyoto for their helpful suggestions on experimental technique. Professor A. Masuda of the University of Tokyo kindly permitted us to quote unpublished REE data for boninites from Chichi-jima. Thanks are also due to two reviewers for their constructive comments.

References

Arndt, N. T. & G. A. Jenner 1986. Crustally contaminated komatiites and basalts from Kambalda, Western Australia. *Chem. Geol.* **56**, 229–55.

Bence, A. E. & A. L. Albee 1968. Empirical correction factors for the electron microanalysis of silicates and oxides. *J. Geol.* **76**, 382–403.

Bloomer, S. H. 1987. Geochemical characteristics of boninite- and tholeiite-series volcanic rocks from the Mariana forearc and the role of an incompatible element-enriched fluid in arc petrogenesis. Geol. Soc. Am. Spec. Paper no. 215, 151–64.

Bloomer, S. H. & J. W. Hawkins 1987. Petrology and geochemistry of boninite series volcanic rocks from the Mariana Trench. *Contrib. Mineral Petrol.* **97**, 361–77.

Cameron, W. E. 1985. Petrology and origin of primitive lavas from the Troodos ophiolite, Cyprus. *Contrib. Mineral. Petrol.* **89**, 239–55.

Cameron, W. E., E. G. Nisbet & V. J. Dietrich 1979. Boninites, komatiites and ophiolitic basalts. *Nature* **280** 550–3.

Cameron, W. E., M. T. McCulloch & D. A. Walker 1983. Boninite petrogenesis: chemical and Nd–Sr isotopic constraints. *Earth Planet. Sci. Lett.* **65**, 75–89.

Chen, C.-H. & D. C. Presnall 1975. The system Mg_2SiO_4–SiO_2 at pressures up to 25 kilobars. *Am. Mineral.* **60**, 398–406.

Coish, R., R. Hickey & F. A. Frey 1982. Rare earth element geochemistry of the Betts Cove ophiolite, Newfoundland: complexities in ophiolite formation. *Geochim. Cosmochim. Acta* **46**, 2117–34.

Crawford, A. J. & W. E. Cameron 1985. Petrology and geochemistry of Cambrian boninites and low-Ti andesites from Heathcote, Victoria. *Contrib. Mineral. Petrol.* **91**, 93–104.

Crawford, A. J. & R. R. Keays 1987. Petrogenesis of Victorian Cambrian tholeiites and implications for the origin of associated boninites. *J. Petrol.* **28**, 175–109.

Crawford, A. J., L. Beccaluva & G. Serri 1981. Tectonomagmatic evolution of the West Philippine–Mariana region and the origin of boninites. *Earth Planet. Sci. Lett.* **54**, 346–56.

Dobson, P. F. 1986. The petrogenesis of boninite: a field, petrologic and geochemical study of the volcanic rocks of Chichi-Jima, Bonin Islands, Japan. Ph.D. Thesis, Stanford Univ.

Dobson, P. F. & J. R. O'Neil 1987. Stable isotope compositions and water contents of boninite volcanic rocks from Chichijima, Bonin Islands, Japan. *Earth Planet. Sci. Lett.* **82**, 75–86.

Flower, M. F. J. & H. Levine 1987. Petrogenesis of a tholeiite–boninite sequence from Ayios Mamas, Troodos Ophiolite: evidence for splitting of a volcanic arc? *Contrib. Mineral. Petrol.* **97**, 509–24.

Green, D. H. 1973. Experimental melting studies on a model upper mantle composition at high pressure under water-saturated and water-undersaturated conditions. *Earth Planet. Sci. Lett.* **19**, 37–53.

Green, D. H. 1976. Equilibrium testing of 'equilibrium' partial melting of peridotite under water-saturated, high-pressure conditions. *Can. Mineral.* **14**, 255–68.

Hamilton, D. L., C. W. Burnham & E. F. Osborn 1964. The solubility of water and effect of oxygen fugacity and water content on crystallization in mafic magmas. *J. Petrol.* **5**, 21–39.

Hickey R. L. & F. A. Frey 1982. Geochemical characteristics of boninite series volcanics: implications for their source. *Geochim. Cosmochim. Acta* **46**, 2099–115.

Hickey-Vargas, R. & M. K. Reagan 1987. Temporal variation of isotope and rare earth element abundances in volcanic rocks from Guam: implications for the evolution of the Mariana Arc. *Contrib. Mineral. Petrol.* **97**, 497–508.

Huebner, J. S. 1982. Pyroxene phase equilibria at low pressures. In *Reviews in mineralogy*, vol. 7 C. T. Prewitt (ed.), 254–99. Mineral. Soc. Am.

Ishihara, T., F. Murakami, T. Miyazaki & K. Nishimura 1981. Gravity survey. In *Geological investigation of the Ogasawara (Bonin) and Northern Mariana Arcs*, E. Honza, E. Inoue and T. Ishihara (eds), 45–78. Cruise Report Geol. Surv. Japan no. 14.

Jenner, G. A. 1981. Geochemistry of high-Mg andesites from Cape Vogel, Papua New Guinea. *Chem. Geol.* **33**, 307–32.

Kuroda, N., K. Shiraki & H. Urano 1978. Boninite as a possible calc-alkaline primary magma. *Bull. Volcanol.* **41**, 563–75.

Kushiro, I. 1972. Effect of water on the composition of magmas formed at high pressures. *J. Petrol.* **13**, 311–34.

Kushiro, I. 1974. Melting of hydrous upper mantle and possible generation of andesitic magma: an approach from synthetic systems. *Earth Planet. Sci. Lett.* **22**, 294–9.

Kushiro, I., H. S. Yoder & M. Nishikawa 1968. Effect of water on the melting of enstatite. *Geol. Soc. Am. Bull.* **79**, 1685–92.

Matsumoto, R. & T. Urabe 1980. An automatic analysis of major elements in silicate rocks with x-ray fluorescence spectrometer using fused disc samples. *Ganseki Kobutsu Kosho Gakkaishi* **75**, 272–8 (in Japanese).

Meijer, A. 1980. Primitive arc volcanism and a boninite series: examples from western Pacific island arcs. In *Tectonic and geologic evolution of southeast Asian seas and islands*, D. E. Hayes (ed.) 269–82. Am. Geophys. Union Monogr. no. 23.

Merrill, R. B. & P. J. Wyllie 1975. Kaersutite and kaersutite eclogite from Kakanui, New Zealand – water-excess and water-deficient melting to 30 kilobars. *Geol. Soc. Am. Bull.* **86**, 555–70.

Mysen, B. O. & A. L. Boettcher 1975. Melting of a hydrous upper mantle: II. Geochemistry of crystals and liquids formed by anatexis of mantle peridotite at high pressures and high temperatures as a function of controlled activities of water, hydrogen and carbon dioxide. *J. Petrol.* **16**, 549–93.

Nakamura, Y. & I. Kushiro 1970. Compositional relations of coexisting orthopyroxene, pigeonite and augite in a tholeiitic andesite from Hakone Volcano. *Contrib. Mineral. Petrol.* **26**, 265–75.

Nicholls, I. A. 1974. Liquids in equilibrium with peridotitic mineral assemblages at high water pressures. *Contrib. Mineral. Petrol.* **45**, 289–316.

Nicholls, I. A. & A. E. Ringwood 1973. Effect of water on olivine stability in tholeiites and the production of silica-saturated magmas in the island-arc environment. *J. Geol.* **81**, 285–300.

Pedersen, A. K. 1985. *Reaction between picrite magma and continental crust: early Tertiary silicic basalts and magnesian andesites from Disko, West Greenland.* Bull. Geol. Surv. Greenland no. 152.

Presnall, D. C., J. R. Dixon, T. H. O'Donnell & S. A. Dixon 1979. Generation of mid-ocean ridge tholeiites. *J. Petrol.* **20**, 3–35.

Roeder, P. L. & R. F. Emslie 1970. Olivine–liquid equilibrium. *Contrib. Mineral. Petrol.* **29**, 275–89.

Rogers, G. & A. D. Saunders 1989. Magnesian andesites from Mexico, Chile and the Aleutian Islands: implications for magmatism associated with ridge–trench collision (this volume).

Sato, H. 1977. Nickel content of basaltic magmas: identification of primary magmas and a measure of the degree of olivine fractionation. *Lithos* **10**, 113–20.

Saunders, A. D., G. Rogers, G. F. Marriner, D. J. Terrell & S. P. Verma 1987. Geochemistry of Cenozoic rocks, Baja California, Mexico: implications for the petrogenesis of post-subduction magmas. *J. Volcanol. Geotherm. Res.* **32**, 223–45.

Shiraki, K., N. Kuroda, S. Maruyama & H. Urano 1978. Evolution of the Tertiary volcanic rocks in the Izu Mariana Arc. *Bull. Volcanol.* **41**, 548–62.

Sun, S.-S. & R. W. Nesbitt 1978. Geochemical regularities and genetic significance of ophiolitic basalts. *Geology* **6**, 689–93.

Takahashi, E. 1978. Partitioning of Ni^{2+}, Co^{2+}, Fe^{2+}, Mn^{2+} and Mg^{2+} between olivine and silicate melts: compositional dependence of partition coefficient. *Geochim. Cosmochim. Acta* **42**, 1829–44.

Takahashi, E. & I. Kushiro 1983. Melting of a dry peridotite at high pressures and basalt magma genesis. *Am. Mineral.* **68**, 859–79.

Tatsumi, Y. 1981. Melting experiments on a high-magnesian andesite. *Earth Planet. Sci. Lett.* **54**, 357–65.

Tatsumi, Y. 1982. Origin of high magnesian andesites in the Setouchi volcanic belt, southwest Japan, II. Melting phase relations at high pressures. *Earth Planet. Sci. Lett.* **60**, 305–17.

Tatsumi, Y. 1986. Origin of subduction zone magmas. *Bull. Volcanol. Soc. Japan* **30**, 153–72 (Special number; in Japanese).

Tatsumi, Y. & S. Maruyama 1989. Boninites and High-Mg andesites: tectonics and petrogenesis (this volume).

Tatsumi, Y., M. Sakuyama, H. Fukuyama & I. Kushiro 1983. Generation of arc basalt magmas and thermal structure of the mantle wedge in subduction zones. *J. Geotherm. Res.* **88**, 5815–25.

Umino, S. 1985. Volcanic geology of Chichijima, the Bonin Islands (Ogasawara Islands). *J. Geol. Soc. Japan* **91**, 505–23.

Umino, S. 1986. Magma mixing in boninite sequence of Chichijima, Bonin Islands. *J. Volcanol. Geotherm. Res.* **29**, 125–57.

Umino, S. 1987. Parent magmas of boninite suggested by olivine and spinel chemistry. *Bull. Volcanol. Soc. Japan* **32**, 145 (Abstract; in Japanese).

Umino, S. & I. Kushiro 1989. In preparation.

Yoder, H. S. 1969. Calc-alkalic andesites: experimental data bearing on the origin of their assumed characteristics. In *Proc. Andesite Conf.*, A. R. McBirney (ed.), 77–89. Dept. Geol. Mineral Indust., Oregon, Bull. no. 65.

5 Experimental evidence for the origin of boninites: near-liquidus phase relations to 7.5 kbar

SIEGER R. VAN DER LAAN,
MARTIN F. J. FLOWER &
A. F. KOSTER VAN GROOS

Abstract

Near-liquidus phase relations from 1 atm to 7.5 kbar of boninite containing 0–8% H_2O were determined from crystallization experiments on three compositions: a high-Ca, low-Si (Troodos), an intermediate-Ca intermediate-Si (Bonin), and a low-Ca, high-Si (Cape Vogel) boninite. Low-Ca pyroxene (orthopyroxene–clinoenstatite) is the dominant crystalline phase, and clinoenstatite stability is enhanced by low pressure and low Ca content of melts. The olivine crystallization field is located near the H_2O-saturated liquidus, extending to a broad range of undersaturated conditions for Troodos, and restricted to a narrow range of relatively high H_2O contents for Cape Vogel. The derived olivine compositions (Fo_{88-91}) imply a slightly evolved character for the studied compositions. The experiments restrict boninite genesis to MORB-type or slightly hotter geotherms at low pressures but in excess of 3 kbar. Mantle–melt mass-balance calculations suggest that high-Ca boninite (Troodos) forms through eutectic-type melting processes while intermediate-Ca boninite (Bonin) and low-Ca boninite (Cape Vogel) are peritectic-type melts.

Two tectonic settings seem to satisfy the petrogenetic conditions demanded for boninite production: interference of a subducting slab with a pre-existing MORB-type geotherm at the initiation of subduction, or a tensional regime inducing deeply rooted, adiabatic diapirism of metasomatized refractory mantle in an established volcanic arc.

5.1 Introduction

The combination of high MgO and SiO_2 contents in primitive mafic magma is unusual and distinguishes boninite from other primitive magma types. Boninite seems to be restricted to supra-subduction zone settings (Cameron et al. 1979) associated with volcanic arcs (Reagan & Meijer 1984, Wood et al. 1982, Bloomer 1983, Kuroda et al. 1978) and ophiolites of arc and backarc affinity (Cameron 1985, Coish & Church 1979, Sameshima & Paris 1983, Jenner 1981). Petrographic definitions prescribe a phenocryst mineralogy dominated by orthopyroxene and/or clinoenstatite, commonly accompanied by olivine (Fo_{88-94}) and high-Cr spinel. Plagioclase is characteristically absent, and boninite groundmasses consist of glass and spinifex pyroxene (Kuroda et al. 1978, Shiraki et al. 1980, Umino 1986). The presence of mantle-equivalent olivine (Fo_{88-90}) is commonly cited as evidence for their primitive nature.

There is, nonetheless, confusion concerning the petrologic definition of boninite and the significance of its genetic conditions in the evolution of volcanic arcs and backarc basins. Chemically, the term has been extended to include all primitive high-MgO, high-SiO_2, low-TiO_2 magmas (Meijer 1980), which range from high-Ca to low-Ca types (Cameron et al. 1983), and appear to reflect a primary melt spectrum distinct from the relatively narrow range of compositions for primary arc tholeiite, MORB and komatiite. Enrichment of LILE and LREE in boninite (Hickey & Frey 1982, Cameron et al. 1983) provides evidence for hydrous contamination of the source, while depleted HFS elements appear to reflect a refractory (magma-depleted) source character. Boninite H_2O contents, estimated from stable isotope studies, lie between 1 and 2% (Kyser et al. 1986), e.g. 1.6–1.7% for parental magmas from Chichi-jima, Bonin Is. (Dobson & O'Neil 1987). These general characteristics of boninite reflect genesis in a variably metasomatized, refractory peridotite.

In this chapter we report results of a near-liquidus experimental study of three representatives of the boninite compositional spectrum, from Cyprus, Cape Vogel and the Bonin Is. Phase relations were investigated in the pressure range from 1 atm to 7.5 kbar, focusing on the extent of olivine crystallization as a function of pressure, temperature, bulk composition and H_2O content. Olivine saturation in these melts provides valuable information about the possible conditions of magma genesis from an olivine-bearing source. A simple system analogue model and experimental results presented here provide a qualitative understanding of boninite genesis, and suggest that differences in melt composition and the observed fractional crystallization trends in boninite suites may be attributed to eutectic- and peritectic-type melting regimes. The model is supported by petrographic evidence and by least-squares mass-balance calculations of melting and fractionation behaviour.

5.2 The simple system analogue for boninite

In order to understand crystal–liquid phase equilibria of complex multicomponent systems, simple system analogues are useful if reliable phase equilibrium data exist such that phase relations can be accurately depicted with respect to all compositional variables simultaneously. Boninites comprise over 98 wt% of CaO, MgO, FeO, Al_2O_3 and SiO_2 (dry weight, and assuming all Fe to be Fe^{2+}). Although primary boninites range from relatively high-Ca, high-Al variants with low Si and low Mg, to low-Ca, low-Al lavas with high Si and high Mg, they all represent melts that were equilibrated with residual mantle phases and can be represented to a close approximation in the system olivine–anorthite–clinopyroxene–silica (Ol–An–Cpx–Si), which is part of the system $CaO–MgO–FeO–Al_2O_3–SiO_2$ (CMFAS). Unfortunately, only the Fe-free system has been studied and therefore a simplifying approach must be taken. Assuming that FeO behaves similarly to MgO, the relevant phase relations can be presented in the quaternary system $CaO–(MgO+FeO)–Al_2O_3–SiO_2$, which may parallel the system $CaO–MgO–Al_2O_3–SiO_2$ (CMAS).

The solid–liquid phase relations in the forsterite–anorthite–diopside–silica (Fo–An–Di–Si) subsystem of CMAS at 1 atm and 7 kbar are presented in Figure 5.1 (Presnall *et al.* 1979). Refractory peridotites (Stosch & Seck 1980, Jackson & Wright 1970) comprise dominant olivine and orthopyroxene and minor amounts of clinopyroxene and plagioclase and/or spinel. Initial melts of such assemblages in CMAS at 7 kbar would have the isobaric invariant

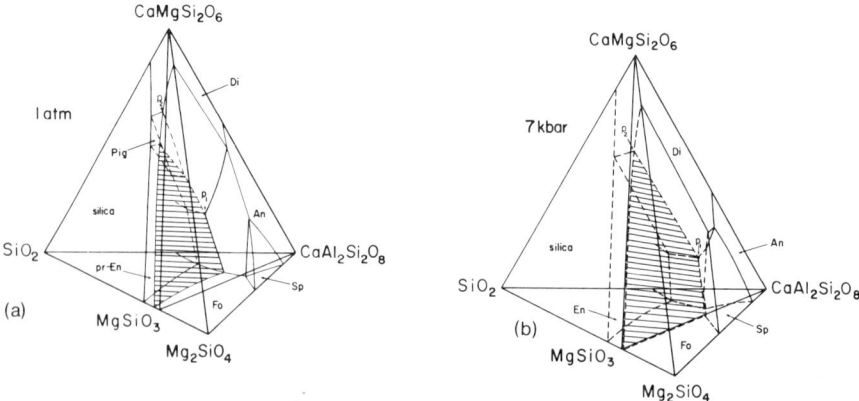

Figure 5.1 Liquidus phase relations in CMAS at 1 atm and 7 kbar (Presnall *et al.* 1979). (a) Initial melts of refractory peridotite have composition p_1 in equilibrium with forsterite (Fo), anorthite (An), diopside (Di) and enstatite (En). (b) Progressive melting will exhaust Di or An and melt compositions will subsequently migrate towards p_2 (An absent) or p_3 (Di absent). After exhaustion of a second phase, melts will lie on the Fo–En surface (shaded) and evolve along paths like m and n respectively. Note that silica contents of the melt in CMAS increase along m and n. At 1 atm the Fo + En + L surface is displaced towards more silica-rich compositions.

composition of p_1 (Fig. 5.1b) in equilibrium with Fo–An–Di–En. Progressive equilibrium melting exhausts clinopyroxene or anorthite, depending on the An/Di ratio in the source, and melt compositions will migrate towards p_3 (Di absent) or p_2 (An absent). After exhaustion of a second phase, An or Di, both melt compositions lie on the Fo + En + L surface (shaded), and will evolve along paths such as m and n, respectively, increasing in silica content towards the Fo–Si join. Comparison of phase equilibria at 1 atm and 7 kbar (or higher pressures) reveals that silica contents at the Fo + En + L surface increase with lower pressures (Presnall *et al.* 1979) owing to expansion of the olivine primary phase volume. Incongruent melting of orthopyroxene occurs at $P < 5$ kbar, but persists to at least 20 kbar (in Fo–Di–Si) under H_2O-saturated conditions (Kushiro 1969). A shift towards higher silica content for melts in equilibrium with olivine and orthopyroxene is also expected at low pressures in the presence of H_2O and/or alkalis (Kushiro 1975).

Based on the assumption that the boninite chemistry can be represented in the system CaO–$(MgO + FeO)$–Al_2O_3–SiO_2 and that this system behaves similarly to CMAS, phase equilibria can explain the following important factors in boninite genesis.

(a) High-Si, high-Mg contents in boninite are enhanced by advanced partial melting in equilibrium with olivine and orthopyroxene, and boninite segregation leaves a harzburgite residuum.
(b) Ca/Al ratios of boninites reflect clinopyroxene and plagioclase contents of the source prior to melting.
(c) The correlation of low Ca, low Al with high Si, high Mg is a result of dilution of the initial melt during continued melting.
(d) Melting at low pressures gives rise to high silica contents due to peritectic melting of orthopyroxene and the expansion of the olivine phase volume.
(e) Small amounts of H_2O promote silica enrichment in the melt relative to anhydrous conditions, through the expansion of the olivine phase volume and the pressure range of incongruent orthopyroxene melting.

5.3 Previous experimental work relevant to boninite genesis

Few experimental studies have been conducted on actual boninite compositions. Because boninites have been suggested to be parental to andesite (Kuroda *et al.* 1978) and the term high-magnesian andesite (HMA) has often been used in association with boninites, some experimental studies of relevance are concerned with the genesis of andesites. Studies can be divided into three groups: (a) crystallization experiments on boninite and related compositions; (b) anhydrous melting of refractory mantle compositions and crystallization experiments on derived liquids; (c) hydrous melting of fertile mantle and crystallization experiments on derived liquids.

Kushiro (1982) conducted H_2O-saturated and H_2O-undersaturated crystallization experiments on a high-Ca, high-Al boninite from DSDP Site 458 at pressures of 8–14.5 kbar with 10–21% H_2O. He concluded that this composition was not primitive, but had evolved through olivine–orthopyroxene fractionation at shallow depths. Tatsumi (1981) experimentally determined the stability of olivine and orthopyroxene for sanukitoid (TGI), an alkaline boninite, at pressures of 11–17 kbar and temperatures between 1080 and 1120°C in the presence of 8–22% H_2O, and concluded these to be realistic genetic conditions.

Experiments by Sen (1982) and Duncan & Green (1980, 1987) showed that quartz-tholeiitic melts can be derived through second-stage melting of a refractory source at pressures of 9 kbar at 1210–1275°C, and of 7–8 kbar at 1360°C respectively. Sen's experiments on refractory plagioclase-spinel lherzolite showed that progressive melting in equilibrium with olivine and orthopyroxene results in effectively constant SiO_2 contents (51.2%), increasing MgO and FeO, and decreasing Al_2O_3 and CaO contents. Melts from Sen's experiments at 1350°C, and from Duncan & Green's experiments at 1360°C are very similar in composition but the latter are slightly richer (1.2%) in SiO_2 presumably due to their lower pressure of melting. Duncan & Green (1980, 1987) interpreted that such melts would be parental, via olivine fractionation, to the high-Ca, high-Al boninite series from Troodos. They also inferred, based on unpublished experimental studies by Jenner, that low-Ca, low-Al boninites from Cape Vogel could have formed at similar pressures (7–8 kbar) with 2–3% H_2O, or at shallow depth (2–3 kbar) with less than 1% H_2O.

Melting studies of hydrous and anhydrous mantle compositions conducted in the early 1970s encountered many difficulties (summarized by Takahashi & Kushiro 1983). Alteration of partial melt compositions during quenching was generally observed, such that equilibrium melt compositions were derived from mass balances of the bulk composition and crystalline phase proportions. Nicholls (1974) used an iterative approach to reconcile melt compositions derived in melting experiments and multiple-saturation experiments. In these experiments, H_2O-saturated andesitic melt compositions in equilibrium with olivine and orthopyroxene were found. They contained approximately 59–60% H_2O at 5 and 10 kbar and 1000°C, but were less silicic at higher pressures and had markedly higher Na and Al and lower Mg and Fe contents than boninites.

Despite the tangential application to boninite, results of these experiments are consistent with predictions from CMAS, viz. that factors of lower pressures, refractory source, and H_2O content all favour generation of high-Mg, high-Si melts. As a consequence, our study was designed to establish experimentally the validity of these factors in genesis of boninites.

5.4 Experiments

5.4.1 Experimental rationale

High Mg#, Cr and Ni contents in boninite suggest that even the highly silicic variants were not derived through mafic phase fractionation, but represent primary melts. This was pointed out by Cameron et al. (1980) for the Troodos boninites, and seems to apply also to Cape Vogel and Bonin types (Jenner 1981, Umino 1986). Primary melts that are richer in silica than the compositional span defined by the phases contributing in the melting process necessarily involve incongruent melting of orthopyroxene. Experimental crystallization of olivine on the liquidus of such a peritectic olivine–orthopyroxene saturated melt will not occur at its conditions of origin (Mysen & Kushiro 1974). The phase equilibrium model based on CMAS and experimental data on natural systems suggests that olivine stability is enhanced in a melt at lower pressure and with increased H_2O content. Therefore, to attain supersaturation in order to crystallize olivine, not only lower temperatures but, in addition, lower pressures or higher H_2O contents than those during melt genesis are required. For a primary boninite, olivine–orthopyroxene saturation on the liquidus related to the last equilibration of that melt with harzburgite will thus give an estimate of the minimum pressure of melting or of the maximum H_2O content. Maximum olivine stability is reached for H_2O-saturated conditions at low pressure. In consequence, primary boninites are predisposed to crystallize olivine during their ascent. H_2O saturation requires unrealistically large H_2O contents at high pressure, and the combination of low pressure and small amounts of H_2O seems more realistic for boninite genesis. H_2O contents between 1 and 3%, similar to values determined by Kyser et al. (1986) and Dobson and O'Neil (1987), were assumed to be realistic for the starting compositions.

5.4.2 Sample selection

Three primitive boninite compositions were selected for polybaric phase equilibrium studies. They span the chemical spectrum of boninites and include a high-Ca boninite from Troodos (Cyprus), an intermediate-Ca boninite from Chichi-jima (Bonin Islands), and a low-Ca boninite from Cape Vogel (Papua New Guinea). The selected boninites were sparsely phyric, to assure their correspondence to melts unaffected by olivine or orthopyroxene accumulation, and have among the highest Mg# and Cr and Ni contents. Ideally, samples should have Mg# that match the Fo contents of the most refractory mantle olivine found. Whereas recent petrological data suggest that more primitive compositions could have been selected (Umino 1986), abundant information on the three locations (Cameron 1985, Flower & Levine 1987, Jenner 1981) confirms the primitive character of our selected samples. Nevertheless, because of our selection criteria, in which the aphyric character was emphasized, our samples approach, but may not represent, primary compositions. This is based on calculated equilibrium olivine compositions for these samples, assuming that up to 30% of the total Fe content of the sample is Fe^{3+}. All compositions contained olivine phenocrysts that were close to, or could have been in equilibrium with, bulk rock based on apparent Fe–Mg partitioning for which a K_D value of 0.3 was used (Roeder & Emslie 1970, this work). The presence of olivine guarantees that under certain conditions olivine supersaturation would occur allowing olivine crystallization from peritectic olivine–orthopyroxene saturated melts in the experiments.

For candidate compositions, normative Fo–Cpx–An–Si components were calculated based only on CMFAS contents (Elthon 1983). Other components were ignored. The compositions with highest silica were adopted for the experiments and are shown in Figure 5.2 in the Ol–Cpx–An–Si tetrahedron.

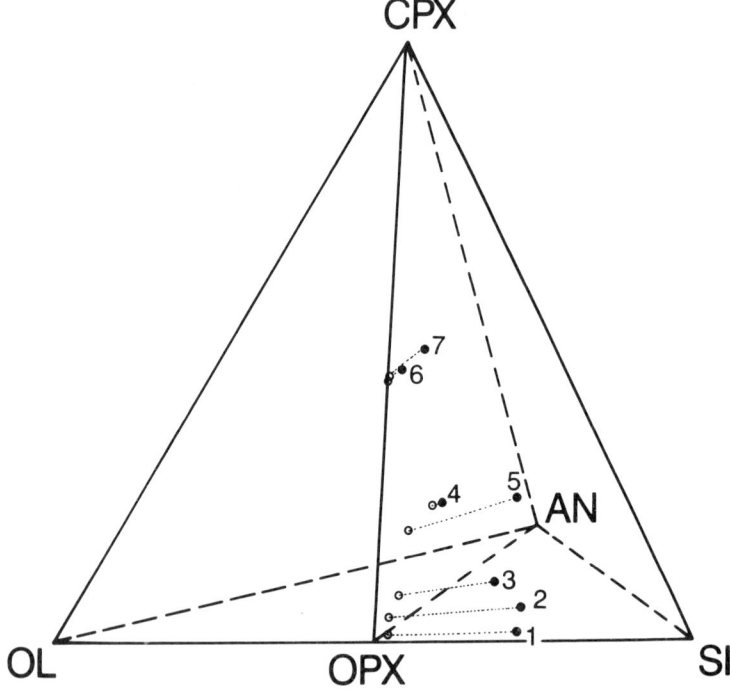

Figure 5.2 Molar projection of Elthon (1983); norms calculated for CMFAS components of the starting compositions. (1) CV-46; (2) Bonin; (3) Margi; (4) 7 kbar Fo + En + Di + An + L CMAS-invariant liquid composition (p_1 in Fig. 5.1b); (5) 1 atm CMAS-invariant liquid composition (p_1 in Fig. 5.1a); (6) 7 kbar Fo + En + Di + L CMS-invariant liquid composition (p_2 in Fig. 5.1b); (7) 1 atm CMS-invariant liquid composition. Compositions are from Presnall *et al.* (1979). Open circles represent corresponding compositions projected from Ol onto the Opx–Di–An plane. Points (1), (2) and (3) suggest an olivine stability volume for boninite of similar extent to the 1 atm CMAS forsterite phase volume (point (5)) (cf. Fig. 5.1).

Comparing the plotted positions to CMAS invariant liquid compositions, the extent of the low-pressure olivine stability volume for boninite is similar to that of forsterite at 1 atm. Mantle–melt mass-balance calculations, presented in a later section, position the Troodos boninite very close to the critical plane of silica saturation, while the Bonin and Cape Vogel compositions are silica saturated and require incongruent melting. Compositions of the selected samples are listed in Table 5.1, and a brief description follows below.

MARGI 38: NORTHERN RIM OF THE TOODOS OPHIOLITE, CYPRUS

The sample is from black glassy pillows near the village of Margi, and contains 8 modal% olivine phenocrysts (< 0.9 mm, Fo_{88}) in a groundmass of quench pyroxene and fresh glass. Compared to other Troodos boninites (Cameron 1985, Flower & Levine 1987) it has an exceptionally high Mg# and is silica-rich. Because the most magnesian olivines from Troodos have $Fo_{90.6}$ (Duncan & Green 1987), and based on calculated equilibrium olivine compositions, this composition is probably slightly evolved.

Table 5.1 Gel compositions, Mg# and CaO/Al_2O_3 of the starting materials, and Cr, Ni contents of the natural samples. Calculated equilibrium olivine compositions (Roeder & Emslie 1970) for various mol% Fe^{3+} are also listed for each sample.

Sample	Margi[a]	Bonin[b]	CV-46[c]
SiO_2	54.93	58.82	59.34
TiO_2	0.51	0.14	0.51
Al_2O_3	13.15	11.06	10.62
FeO	8.20	8.43	9.27
MgO	11.52	11.71	12.57
CaO	10.21	7.79	5.74
Na_2O	1.17	1.61	1.37
K_2O	0.26	0.42	0.49
P_2O_5	0.04	0.02	0.10
total	99.99	100.00	100.01
100Mg#	71.5	71.2	70.7
CaO/Al_2O_3	0.77	0.70	0.54
Cr	807	695	910
Ni	202	194	227
calculated forsterite content of equilibrium olivine ($K_D = 0.30$)			
0% Fe^{3+}	88.0	87.9	87.6
10% Fe^{3+}	89.2	89.1	88.9
20% Fe^{3+}	90.4	90.3	90.1
30% Fe^{3+}	91.6	91.5	91.3

[a] Margi 38, Troodos, Cyprus (unpubl.).
[b] Bonin 106, Chichi-jima, Bonin Islands (Cameron et al. 1983).
[c] CV-46.2, Cape Vogel, Papua New Guinea (Walker & Cameron 1983).

BONIN 106: CHICHI-JIMA, BONIN ISLANDS

Compositional data for this sample were published by Cameron et al. (1983), and additional petrographical data were made available upon request. The sample contains very few olivine phenocrysts, altered to smectite, and a trace of clinoenstatite ($En_{89.8}$) overgrown by orthopyroxene (En_{87-88}). Olivine phenocrysts of related samples are Fo_{88}. The groundmass contains skeletal olivine ($Fo_{87.7-88.3}$) and orthopyroxene (En_{87-88}). This sample conforms to the slightly more calcic variety, Type II boninites, described by Umino (1986). Umino reports $Fo_{91.5}$ olivine microphenocrysts and $Fo_{92.2}$ phenocrysts associated with wholerock Mg# = 0.75 for the most primitive aphyric compositions.

CV-46.2: CAPE VOGEL, PAPUA NEW GUINEA

This sample is from a pillow lava and was described by Walker & Cameron (1983). It is the least magnesian aphyric boninite in the Cape Vogel sequence and contains 2 modal% clinoenstatite and orthopyroxene (~ 1 mm size), and a trace of olivine (0.5 mm, $Fo_{92.7}$). The groundmass consists of quench clinoenstatite and orthopyroxene–pigeonite crystals in unaltered glass. The presence of vesicles may reflect H_2O saturation of the magma. Samples from Cape Vogel are commonly porphyritic, and may contain olivine up to Fo_{94} (Jenner 1981). CV-46.2 is not the most primitive composition according to Walker & Cameron (1983), but their more primitive samples were porphyritic or olivine-free and could therefore not satisfy our selection criteria. In addition CV-46.2 had the highest normative Si component in the Ol–Di–An–Si projection (Elthon 1983).

Gel starting materials were preferred to glasses or crystalline mixtures because of their greater reactivity. Each composition was therefore prepared as a Na–Mg–Al–Si–P–K–Ca gel (Hamilton & Henderson 1968), and mechanically mixed with TiO_2, Fe and Fe_2O_3 (to yield FeO), all below 200 mesh. The mixtures were stored over P_2O_5 to prevent adsorption of H_2O.

Because we were primarily interested in genetic conditions of boninite rather than its fractionation products, the experiments were restricted to near-liquidus conditions, with the aim of determining liquidus phase variation as a function of P, T and H_2O contents.

5.4.3 Experimental techniques

Runs at 1 atm were made in a gas mixing furnace of vertical quench type. The f_{O_2} was buffered with a CO_2–H_2 gas mixture, and experiments were conducted in the NNO–WM range. Three types of experiments were performed differing only in the treatment of Fe loss to platinum: (a) very thin Pt wire (0.08 mm) was used to minimize Fe loss; (b) the starting compositions were adjusted with 10% extra FeO to compensate Fe loss; or (c) presoaked Pt hooks were used to eliminate Fe loss. Run durations were from 12 to 48 h. Details of the techniques used are to be published elsewhere (Van der Laan *et al*. 1989).

Experiments between 1 and 7.5 kbar were carried out using an internally heated pressure vessel (IHPV) (Holloway 1971). A double-wound tungsten furnace allowed very rapid heating rates and run temperatures to 1300°C were reached within 10 min. The pressure medium was commercial-grade argon, and Mg metal at 200–500°C provided an ultra-reducing atmosphere inside the vessel to prevent oxidation of furnace windings during experiments. Temperature was controlled within 3°C but a gradient of about 25°C was usually observed with respect to a second thermocouple at 6.5 mm vertical distance. Reported temperatures are averages of the two readings but are believed to be precise to within 5°C in view of the identical capsule–thermocouple geometry maintained throughout the runs, and the reproducibility of the experiments. Pressure was determined using calibrated high-precision Heise gauges and a Harwood manganin cell, and is believed to be accurate within 1%.

To prevent Fe loss to Pt in the high-pressure experiments, starting materials were contained in a foil envelope of Pt–Fe alloy placed inside a 2 mm diameter Pt capsule. The Pt–Fe alloys were prepared by electroplating Fe on 0.04 mm thick Pt foil, followed by annealing at 1400°C. Because the foils buffer the activity of Fe metal in the charge, the associated f_{O_2} and Fe^{2+}/Fe^{3+} were controlled, and determined by the alloy composition (Van der Laan and Koster van Groos 1989). Run durations were limited to 25–30 min because of H_2O loss in oxidation reactions with Fe and the related H diffusion from the capsules. The change in Fe^{2+}/Fe^{3+} was, however, buffered by the foil composition. Run durations, though short, were adequate considering that run products showed good evidence that equilibrium was closely approached. This is probably because the starting materials, being gels and containing H_2O, were very reactive. Also, recrystallization during the heating stage was minimized with very rapid heating rates of up to 300°C min^{-1} approaching the desired run temperature. The runs were quenched within seconds by turning the vessel to a vertical position, allowing the capsules to drop to the cooler part of the furnace. Run charges were between 12 and 15 mg, and charge and H_2O contents were weighed to within 0.01 mg using a 1712 MP 8 Satorius balance. After the run the capsules were inspected and weighed, then sectioned using a 0.3 mm diamond saw, polished and examined microscopically.

5.5 Results

5.5.1 Phase identification

Experimental conditions and results are given in Table 5.2. Mineral phases observed in runs include olivine (Ol), orthopyroxene (Opx), clinoenstatite (Cen) and magnetite (Mgt), identified from the following criteria. Ol has a

Table 5.2 Experimental conditions and results for three boninite compositions (L, glass; Q, quench crystals; Opx, orthopyroxene; Cen, clinoenstatite; Mgt, magnetite; Px, Opx or Cen; Ol, olivine; r., rounded, resorbing; tr., trace; (Cen), some Cen); V, vapour.

(a) Margi.

Run	Pressure (kbar)	Temp. (°C)	Initial H$_2$O (wt%)	Fe in Pt (wt%)	Phases	100Mg# in Px (%)	K_d for Px	100Mg# in Ol (%)	K_d for Ol	Fe gain (wt%)	Fe^{2+}/Fe total	Final H$_2$O (wt%)
31.5	1.0	1215	2.8	6.0	L + r. Ol	—	—	88.9	0.22	1.8	0.73	2.0
31.4	1.0	1215	3.0	6.1	L + Ol + tr. Opx	85.7–86.3	0.30	88.3–88.8	0.24	1.5	0.80	2.4
31.6	1.0	1215	3.9	5.8	L	—	—	—	—	2.2	—	3.3
25.2	1.0	1220	1.6	7.6	L + Ol + Opx	85.8–87.1	0.24	87.2	0.24	0.8	0.80	1.2
25.4	1.0	1220	4.1	7.1	L	—	—	—	—	—	—	—
32.1	1.0	1240	0.5	6.7	L + Opx	—	—	—	—	—	—	—
26.1	1.0	1240	1.0	6.2	L + Opx	86.3	0.30	—	—	0.4	0.99	0.9
26.2	1.0	1240	1.1	6.4	L + Ol + Opx	86.0–87.8	0.26	86.4	0.29	0.4	0.96	1.0
26.3	1.0	1240	1.6	6.6	L + Ol + Opx	84.6–86.7	0.30	87.7	0.27	1.0	0.89	1.2
36.5	3.0	1237	2.5	5.3	L	—	—	—	—	—	—	—
45.3	3.0	1240	2.0	4.3	L	—	—	—	—	0.0	—	—
45.4	3.0	1240	2.1	3.6	L	—	—	—	—	−0.2	—	—
33.1	3.0	1261	1.0	6.2	L + Opx + r. Ol	85.7–86.4	0.30	79.5	0.50	1.2	1.00	0.7
42.2	4.0	1200	3.1	5.5	L + Opx + tr. r. Ol	—	—	—	—	—	—	—
42.1	4.0	1200	4.0	5.7	L(Q) + Ol + tr. Opx	88.6	0.24	88.8	0.25	1.3	0.82	3.5
41.2	4.0	1225	3.0	5.4	L	—	—	—	—	—	—	—
43.6	5.5	1195	4.5	3.8	L + Opx	88.3–89.0	0.25	—	—	0.5	0.83	4.2
44.3	6.0	1170	5.5	4.8	L(Q) + Opx	89.4–89.8	0.26	—	—	0.4	0.86	5.3
44.4	6.0	1170	6.2	3.4	L(Q)	—	—	—	—	0.4	—	6.0
37.5	7.0	1209	5.3	5.0	L	—	—	—	—	—	—	—
35.1	7.0	1241	3.0	5.1	L + Opx	—	—	—	—	—	—	—
35.2	7.0	1241	3.2	4.7	L + Opx	89.5	0.27	—	—	−0.3	0.90	3.0
34.1	7.0	1270	1.0	6.2	L + Opx	—	—	—	—	—	—	—
34.2	7.0	1270	1.5	5.9	L + Opx	—	—	—	—	—	—	—

(b) Bonn.

Run	Pressure (kbar)	Temp. (°C)	Initial H$_2$O (wt%)	Fe in Pt (wt%)	Phases	100Mg# in Px (%)	K_d for Px	100Mg# in Ol (%)	K_d for Ol	Fe gain (wt%)	$\frac{Fe^{2+}}{Fe\ total}$	Final H$_2$O (wt%)
32.4	1.0	1240	3.0	5.3	L + V + r. Ol + Px		—		—	—	—	—
32.5	1.0	1240	3.5	5.1	L + V + r. Ol + Px		—		—	—	—	—
19.2	1.0	1246	4.0	10.0	L(Q) + V		—		—	2.3	—	3.5
20.2	1.0	1257	4.0	7.0	L		—		—	2.0	—	3.5
27.5	1.0	1270	1.6	5.3	L(Q)		—		—	0.3	—	1.5
27.6	1.0	1270	1.6	6.0	L(Q)		—		—	0.1	—	1.6
30.4	1.0	1278	1.0	5.2	L + Ol + Px	91.0–91.6	0.24		—	−0.4	0.95	1.0
29.4	1.0	1290	0.5	5.5	L + Px	88.4	0.30		—	0.3	1.00	0.4
46.1	1.8	1210	4.2	3.2	L + Ol + Px	91.6–92.3	0.20	90.2–91.1	0.23	0.3	0.78	4.0
46.2	1.8	1210	4.7	3.2	L + Px + Ol	91.2–91.7	0.21	90.7	0.24	0.4	0.79	4.5
40.1	2.0	1181	5.0	5.1	L + r. Ol + Px	88.6	0.21		—	1.8	0.66	4.1
39.3	2.0	1216	4.0	5.1	L + r. Ol + tr. Px	89.0	0.23		—	1.8	0.73	3.2
39.2	2.0	1216	5.0	5.1	L(Q)		—		—	1.5	—	4.7
38.1	2.0	1228	4.0	5.5	L(Q)		—		—	—	—	—
38.2	2.0	1228	4.4	5.3	L + Ol + Px	89.7	0.21		—	1.6	0.70	3.6
24.2	2.0	1250	3.3	6.5	L + Px	88.3	0.24		—	0.9	0.81	2.7
24.3	2.0	1250	5.2	6.8	L		—		—	1.8	—	4.7
47.2	3.0	1180	6.1	3.8	L(Q) + Ol		—	90.3–91.2	0.24	0.5	0.80	5.9
47.3	3.0	1180	6.6	3.9	L(Q) + V + Ol		—	90.4–91.0	0.22	0.0	0.73	6.3
36.1	3.0	1237	4.0	4.8	L		—		—	—	—	—
45.5	3.0	1240	3.5	3.3	L		—		—	−0.1	—	—
16.3	3.0	1250	2.0	8.9	L + Px		—		—	1.1	—	1.7
16.4	3.0	1250	2.7	8.8	L + Px		—		—	2.5	—	2.1
21.1	3.0	1260	4.0	6.9	L		—		—	1.9	—	3.5
33.6	3.0	1261	3.5	4.4	L		—		—	—	—	—
42.3	4.0	1200	5.5	5.7	L(Q)		—		—	—	—	—
41.3	4.0	1225	3.8	5.4	L + Px	86.3	0.28		—	1.2	0.93	3.4
43.2	5.5	1195	5.5	3.3	L + Px	86.4	0.21		—	1.4	0.70	4.8
44.5	6.0	1170	6.9	3.5	L(Q) + Px	88.2–89.8	0.19		—	1.1	0.66	6.2
37.1	7.0	1209	6.0	4.9	L(Q)		—		—	—	—	—
35.6	7.0	1241	3.5	4.5	L + Px	89.4	0.25		—	−0.6	0.83	3.3
34.5	7.0	1270	2.5	4.6	L + Px		—		—	—	—	—
34.6	7.0	1270	3.5	2.4	L		—		—	—	—	—
49.2	7.5	1107	10.9	4.1	L + Px + Mgt	87.2–87.8	—		—	—	—	—

(Continued)

Table 5.2 (*Continued*)

(c) CV-46.2.

Run	Pressure (kbar)	Temp. (°C)	Initial H$_2$O (wt%)	Fe in Pt (wt%)	Phases	100Mg# in Px (%)	K_d for Px	100Mg# in Ol (%)	K_d for Ol	Fe gain (wt%)	$\frac{Fe^{2+}}{Fe\ total}$	Final H$_2$O (wt%)
32.6	1.0	1240	4.0	5.0	L + V + Px(Cen)		—		—	—	—	—
19.1	1.0	1246	4.0	9.9	L(Q) + V + Cen	90.0	0.21		—	1.5	0.70	3.2
20.3	1.0	1257	2.0	7.0	L + Cen	87.0–88.1	0.28		—	0.6	0.95	1.8
27.3	1.0	1270	2.1	5.5	L + Cen	87.7	0.28		—	1.1	0.92	1.7
30.1	1.0	1278	2.0	5.2	L + tr. Cen	90.0–91.0	0.25		—	0.2	0.83	1.8
30.2	1.0	1278	2.5	5.5	L + tr. Cen	90.0	0.26		—	0.2	0.87	2.3
30.3	1.0	1278	3.0	5.3	L		—		—	0.2	—	2.9
29.1	1.0	1290	1.7	5.5	L + Cen	88.4–89.0	0.25		—	1.5	0.83	1.1
29.3	1.0	1290	3.0	5.4	L		—		—	1.9	—	2.5
46.3	1.8	1210	4.5	3.2	L + Px	91.6	0.21		—	0.2	0.70	4.1
46.4	1.8	1210	4.9	3.2	L + Px + Ol	91.4–91.7	0.20	89.7–90.0	0.25	0.4	0.83	4.7
40.2	2.0	1181	6.0	5.3	L(Q) + V + Ol + Cen	88.7–89.4	0.19	87.2–89.1	0.20	2.2	0.66	5.0
40.3	2.0	1181	7.1	5.6	L + V + Ol + Cen	87.4–89.7	0.20	87.1–89.4	0.21	1.5	0.70	6.4
39.1	2.0	1216	6.1	5.4	L(Q) + V		—		—	1.6	—	5.7
38.3	2.0	1228	5.0	5.3	L + Cen	87.4–89.0	0.21		—	1.4	0.71	4.2
24.6	2.0	1250	6.0	7.0	L(Q) + V		—		—	—	—	—
47.1	3.0	1180	6.1	3.8	L(Q) + Px	88.1–89.1	0.25		—	0.2	0.85	5.9
36.3	3.0	1237	5.0	4.5	L(Q)		—		—	—	—	—
36.4	3.0	1237	5.9	4.4	L(Q)		—		—	—	—	—
16.1	3.0	1250	4.0	9.6	L(Q) + Cen	85.5–86.0	0.28		—	2.5	0.93	3.3
21.3	3.0	1260	4.0	6.9	L		—		—	—	—	—
42.5	4.0	1200	6.0	5.9	L + Px(Cen)		—		—	—	—	—
43.3	5.5	1195	6.5	3.6	L(Q) + Px	87.3–88.5	0.19		—	1.1	0.63	5.8
43.4	5.5	1195	7.1	3.5	L(Q) + Px	87.6	0.21		—	1.5	0.70	6.3
44.6	6.0	1170	8.0	3.3	L + Px	89.8–90.7	0.24		—	0.2	0.80	7.7
37.3	7.0	1209	6.0	5.0	L + Px	88.2–89.8	0.27		—	0.0	0.90	5.9
49.1	7.5	1107	9.5	4.2	L(Q) + Px + Mgt	83.9–86.3	—		—	—	—	—

prismatic habit and high birefringence. Opx is prismatic, columnar or acicular and lacks fracturing and lamellar twinning, whereas Cen, also of prismatic, columnar and acicular habit, is invariably fractured when exposed at polished surfaces and in twinned specimens. Cen commonly develops irregular lamellar twins and only such specimens show the characteristic oblique low-angle extinction (Longhi & Boudreau 1980, Carlson 1986). In Table 5.2, clinoenstatite is listed when optically recognized in the run products. Mgt appeared as tiny black specks throughout the glass.

With increasing pressure, characteristic optical features of clinoenstatite vanished and pyroxenes had to be distinguished by other means. The large amount of glass in the run products prohibited determination of pyroxene structures by conventional x-ray techniques. We therefore relied on the continuity of compositional trends to assess the occurrence of one or two kinds of pyroxene for a given starting composition (cf. Biggar 1985). Quench crystallization products (Q) appeared as V-shaped outgrowths on crystal extremities when minor, and when major, as wavy bundles of acicular crystals. However, equilibrium crystallization products were easily distinguished in quench groundmass because of their pronounced birefringence. Runs were analysed using a JEOL 35CF scanning electron microscope (SEM) with Tracor Northern EDS using ZAF data correction for Pt–Fe and Bence & Albee (1969) correction factors for silicate analyses. Although large crystals were analysed whenever possible, care had to be taken during microanalysis because quench overgrowth of pyroxene appearing as light-coloured rims on SEM backscatter images was commonly present.

5.5.2 The effect of bulk compositional changes on phase compositions

Loss of H_2O and changes in iron content and Fe^{2+}/Fe^{3+} speciation of the experimental charge are expected to bear critically on olivine stability. Therefore careful monitoring of Fe content and Fe^{2+}/Fe^{3+} of runs was maintained.

In the 1 atm experiments, crystal–liquid equilibria were interpreted with respect to the measured Fe content of the glass phase. The Fe content of the melt was affected by both crystallization and by Fe loss to platinum determined by the $Pt-Fe-f_{O_2}$ equilibrium. However, using different experimental techniques the observed range of Fe contents caused by these techniques encompassed the original starting composition, assuring that the observed 1 atm phase equilibria were not biased by compositional changes.

In the experiments at elevated pressure, bulk compositional changes involved loss of H_2O through oxidation reactions with Fe, subsequent H loss from the capsule, and Fe gain from the foil which buffered the Fe speciation. The amount of H_2O loss was estimated from the amount of Fe gain and the change in Fe^{2+}/Fe^{3+}, using the stoichiometry of the oxidation reactions. The bulk Fe content was estimated from the glass composition in runs with very little crystallization, or in runs with quench crystallization, from scans of several large surfaces areas. The Fe^{2+}/Fe^{3+} in the melt was estimated from the observed K_D for Fe–Mg partitioning between olivine and melt or orthopyroxene and melt, assuming $K_D = 0.3$ for Fe^{2+}–Mg partitioning (Roeder & Emslie 1970), and Fe^{3+}-free crystalline phases, were $K'_D = (\Sigma Fe/Mg)^{xl}/(\Sigma Fe/Mg)^{melt}$ and $K_D = (Fe^{2+}/Mg)^{xl}/(Fe^{2+}/Mg)^{melt}$. Listed K'_D values in Table 5.2 apply to crystal rim–liquid equilibria. K'_D values for Fe-poor crystals are very sensitive to small compositional changes and become susceptible to variations associated with analytical error. For equal Mg#, olivine K_D values are less susceptible to the analytical error than pyroxene because of the larger Fe content of olivine. A difference in analysis resulting in a change in Mg# of orthopyroxene or olivine from 0.91 to 0.92 results in a 12% higher K_D value for the crystal–liquid equilibrium. Considering this sensitivity to slight compositional changes, K'_D values for olivine and orthopyroxene matched in the hydrous experiments. Therefore, the same calculations for pyroxene–melt equilibria were performed (using $K_D = 0.30$) to estimate Fe^{3+} in the melt in the absence of olivine. The initial H_2O content was then corrected in proportion to the Fe gain ($PtFe + H_2O = Pt + FeO + H_2$) and the amount of Fe^{2+} oxidation ($Fe^{2+} + \frac{1}{2}H_2O = Fe^{3+} + \frac{1}{2}H_2 + \frac{1}{2}O^{2-}$). The very limited zoning of mineral phases indicated that Fe^{2+} contents stayed approximately constant or decreased only slightly during a run. On the basis of these calculations it was concluded that the bulk compositional changes in the experiments effectively were Fe^{3+} addition of up to 2.5% (but usually less than 1%), and H_2O loss between 0 and 1.0% (but usually less than 0.5%). Fe^{3+} percentages (Table 5.2) of between 10 and 25% lie, at 1 atm, in the WM–NNO f_{O_2} range (cf. Sack et al. 1980, Van der Laan & Koster van Groos 1989). Olivine and pyroxene compositions were not influenced by this gain in Fe^{3+} of the starting composition, as is evident from the absence of correlations between melt Fe^{3+} content and mineral compositions.

5.5.3 Olivine

Olivine occurs in runs of all three starting compositions but is more abundant in Ca-rich than in Ca-poor types. Comparing Fo contents and stability of olivine, the high-Ca boninite (Margi) shows the largest olivine stability field and is the only composition that crystallizes olivine at 1 atm. It shows the lowest olivine Fo content ($Fo_{86.6-89.4}$ at 1 atm and $Fo_{86.4-88.9}$ at elevated pressure). The intermediate-Ca boninite (Bonin) has the highest Fo contents ranging from $Fo_{90.0-91.2}$, but a more restricted stability field, and runs on the low-Ca boninite CV-46.2 show a very limited olivine stability field and compositions from $Fo_{87.1-90.0}$

Occasional opaque inclusions in olivine have been noted. This local disequilibrium causes olivine to crystallize outside its primary phase field (e.g.

Margi run 33.1), and Fe–Mg partitioning values (K'_D) are much higher than 0.3 for olivine-glass pairs in such cases. There is no indication that an increase in mafic components through Fe gain has any effect on olivine stability in the experiments. Zonation in olivine is always minor, rarely reaching 2% in Fo content. It invariably reflects progressive oxidation of the charge through increasing Fo towards the rim. Instability of olivine in some runs might be inferred from rounded edges but resorption features are often limited to some edges only whereas faces developed elsewhere on the crystal. A summary of the olivine compositions from the IHPV experiments is given in Table 5.2.

5.5.4 Pyroxenes

Owing to the relative complexity of pyroxene phase relationships, the following discussion is confined to pyroxenes from the high-pressure experiments. The 1 atm results are discussed elsewhere (Van der Laan *et al.* 1989). For each starting composition the high-pressure pyroxene compositions are plotted in the pyroxene quadrilateral (Fig. 5.3). Correlation of pyroxene Mg# and wollastonite (Wo) content with the variables P, T, H_2O content and Fe gain, and correlations between Mg# and wollastonite, were examined using statistical techniques.

In the high-Ca boninite (Margi), clinoenstatite was found only in 1 atm runs within 3°C of the liquidus. It usually coexisted with orthopyroxene with a 1–1.6% higher Wo content and lower Mg#, around 0.90. At higher pressures only orthopyroxene was present. Wo content in orthopyroxenes was always more than 3% (Figs 5.3a & 5.4c) in the IHPV experiments but lower at 1 atm (above 2% Wo). There was an increase in Mg# with pressure (Fig. 5.4a) and a slight positive correlation between Mg# and Wo content (Fig. 5.3a). A slight negative correlation was found between Wo content and pressure (Fig. 5.4c). The 1 atm orthopyroxenes had Mg# similar to those of 7 kbar runs, and thus do not conform to the observed pressure trend.

Clinoenstatite with distinct optical characteristics was invariably present at or below 2 kbar in low-Ca boninite (CV-46.2), less frequently at 3 and 4 kbar, and was absent at higher pressures. Wo contents, however, showed a simple continuous increase with pressure (Fig. 5.4d), implying the absence of distinct pyroxene compositions in this sample. No correlation other than Wo–pressure was present. Mg# at 1 atm and at elevated pressures (Fig. 5.4b) showed a similar range.

In the intermediate-Ca composition (Bonin), clinoenstatite was the liquidus phase at 1 atm. For coexisting clinoenstatite and orthopyroxene at 1 atm, clinoenstatite has a Mg# ~ 0.90, compared with values around 0.885 for orthopyroxene, and a 1–1.1% lower Wo content. At higher pressures clinoenstatite was not recognized optically, but runs below 2 kbar crystallized pyroxenes compositionally similar to those of CV-46.2 but slightly higher in Wo content, which did not appear at higher pressures (Fig. 5.4d). Pyroxenes

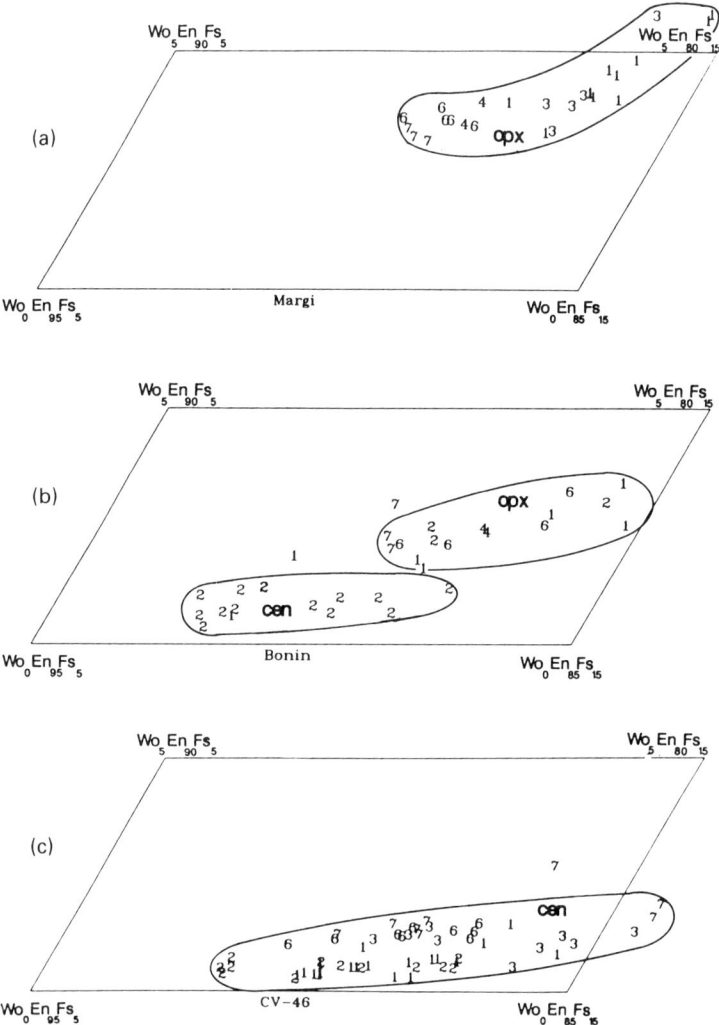

Figure 5.3 Experimental pyroxene compositions plotted in a portion of the pyroxene quadrilateral according to Wo, En and Fs content. The Wo content of pyroxene reflects the Ca content of the boninite. Symbol is pressure of the experiment in kbar rounded to nearest integer value. (a) Margi; (b) Bonin; (c) CV-46.2. Fs and Wo contents decrease with pressure in orthopyroxenes of Margi and Bonin. Wo content increases with pressure in clinoenstatites of CV-46.2.

below 2 kbar defined two separate compositional groups, with a 1% difference in Wo content. Analogous to those at 1 atm, the low-Wo pyroxene was designated clinoenstatite and the high-Wo, orthopyroxene. Orthopyroxenes exhibited the same positive correlation of Mg# with pressure as observed in the Margi composition (Fig. 5.4a). Mg# of orthopyroxenes from 1 atm experiments also deviated from the pressure trend.

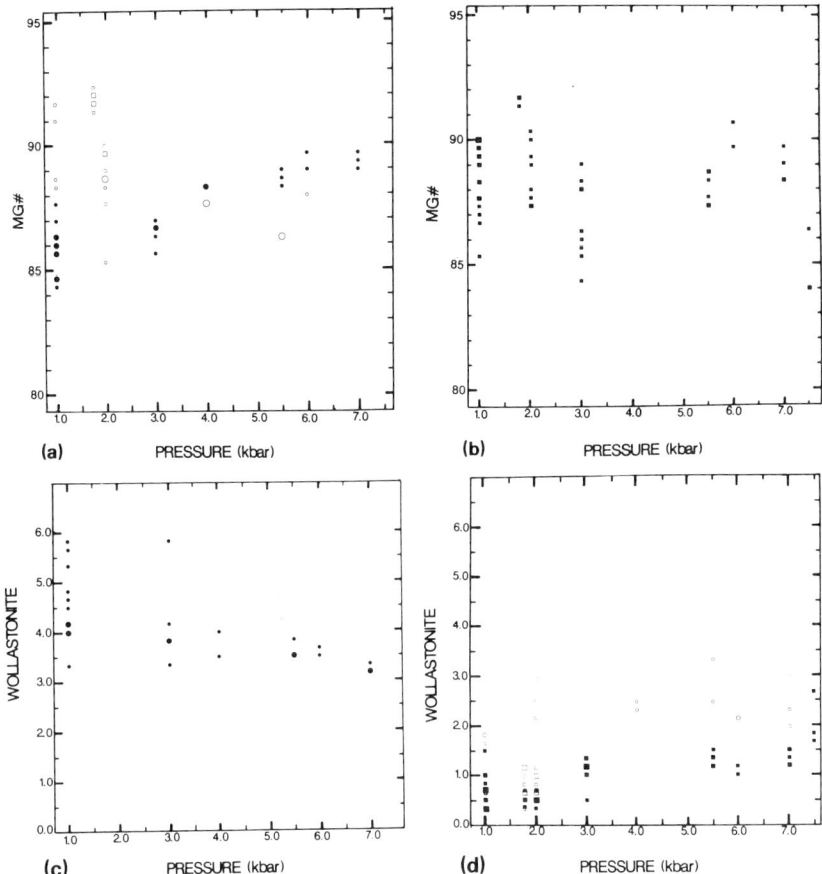

Figure 5.4 Compositional trends with pressure for clinoenstatite (squares) and orthopyroxene (circles). Open symbols are the Bonin, closed circles are for Margi, and closed squares are for CV-46.2. Small symbols indicate a single analysis, intermediate symbols 2–3, and large symbols 4–7 analyses. (a) Bonin and Margi orthopyroxenes (circles) increase in Mg# with pressure (b) Absence of Mg# trends with pressure for clinoenstatite in CV-46.2. (c) A weak trend for Wo content with pressure for orthopyroxene in Margi. (d) Increase in wollastonite content of clinoenstatite (squares) with pressure for Bonin and CV-46.2, and the absence of any trend for Bonin orthopyroxene (circles).

In summary, the pyroxene variation shows a critical dependence on bulk composition, and also pressure and temperature conditions of the experiments.

5.5.5 Near-liquidus phase relations

Near-liquidus phase relations of the three compositions are presented in both P–T projections and isobaric T–X_{H_2O} sections (Fig. 5.5). The P–T relations represent projections of the phase assemblage of a run along the H_2O axis onto

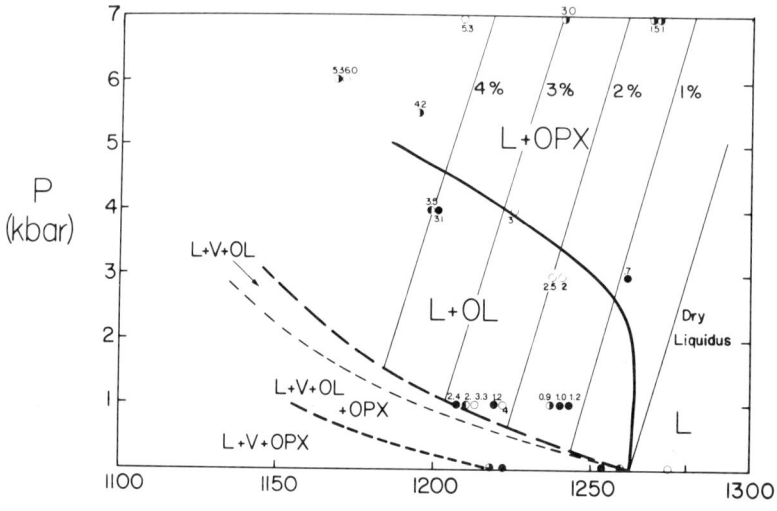
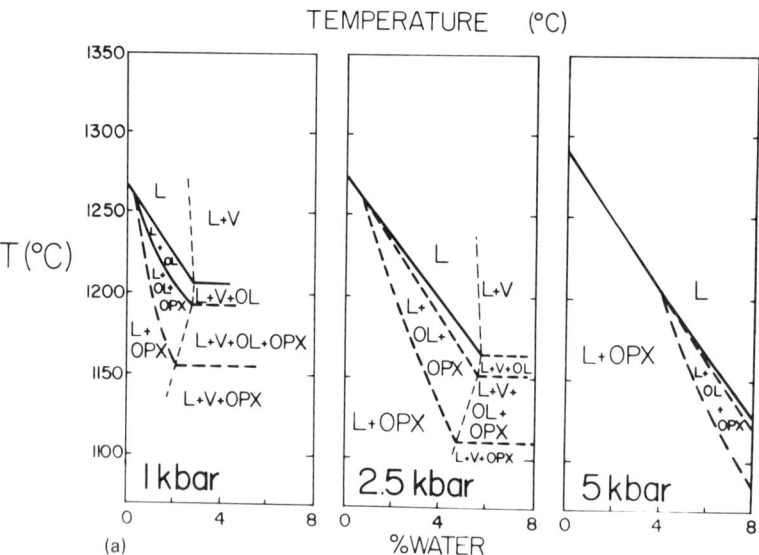

Figure 5.5 P–T projections and T–X_{H_2O} diagrams for boninite. Stability fields of olivine (Ol) and orthopyroxene (Opx)–clinoenstatite (Cen) on the liquidus are marked. Isopleths are labelled for their respective weight percentages of H_2O. Run symbols: open, all liquid (L); closed, L + Ol + Opx/Cen; left side closed, L + Ol; right side closed, L + Opx/Cen. Corrected per cent H_2O marked above run symbol, uncorrected marked underneath (see text). (a) P–T–liquidus and T–X_{H_2O} diagram for Margi. The large liquidus stability field for olivine extends also relatively far below the liquidus in T–X_{H_2O} diagrams. (b) P–T–liquidus and T–X_{H_2O} diagram for Bonin. Olivine crystallization is confined to the H_2O-saturated liquidus. This is also apparent from the T–X_{H_2O} diagrams. Cen stability field is omitted. (c) P–T–liquidus and T–X_{H_2O} diagram for CV-46. Note the narrow range of near to H_2O-saturated conditions for which olivine crystallizes.

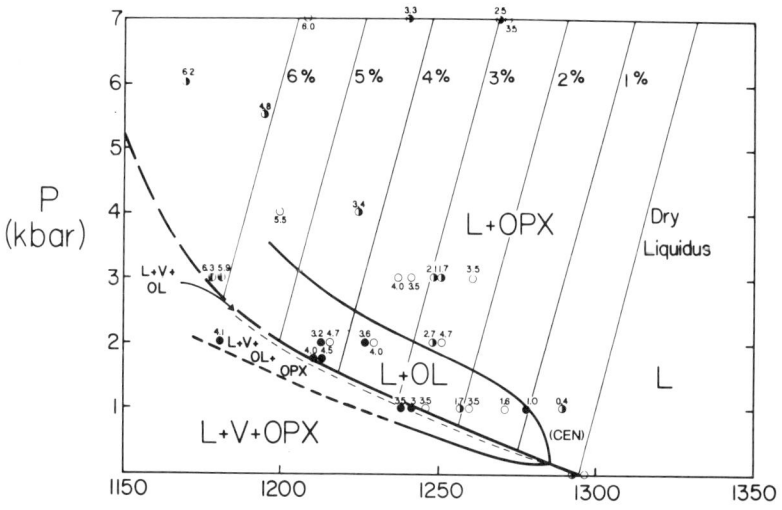

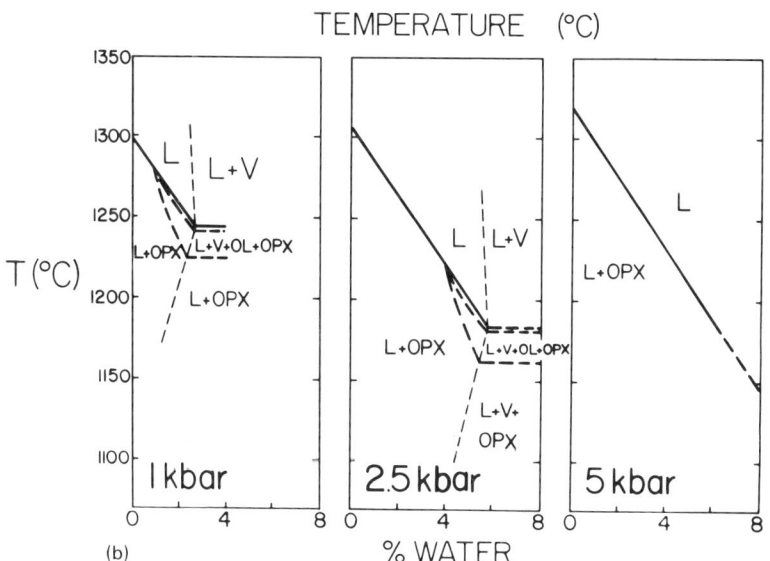

Figure 5.5 *Continued*

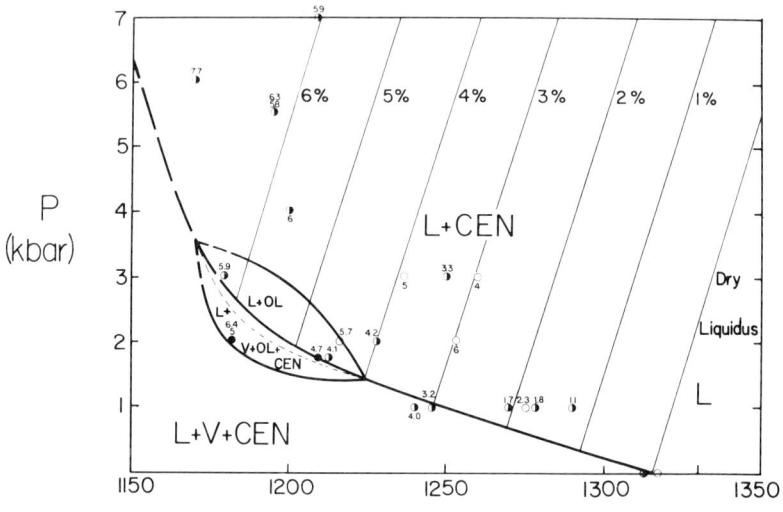

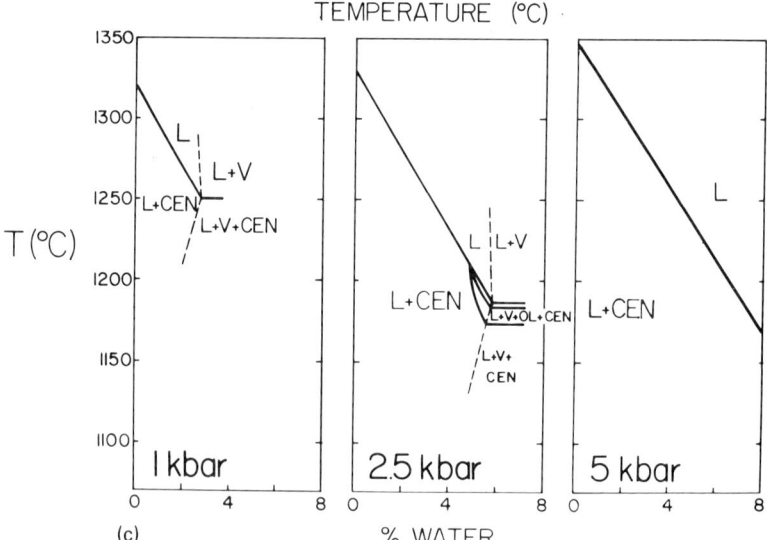

Figure 5.5 *Continued*

the liquidus surface. This approach has the advantage that the position of a run becomes independent of its H$_2$O content. Accurate assessment of the H$_2$O contents is crucial, however, for the positioning of H$_2$O isopleths on the liquidus surface.

The outlined olivine–orthopyroxene–clinoenstatite stability fields in the P–T diagrams are valid for the liquidus surface only, and are inferred from near-liquidus experiments. The relationship between the liquidus surface and the T–X_{H_2O} sections is schematically illustrated in Figure 5.6. Most of the H$_2$O contents of runs were adjusted for losses through oxidation of Fe and these corrected values are listed above run symbols. Uncorrected values appear below symbols. Corrected H$_2$O contents are believed to be accurate within about 0.2%.

In our first-order interpretation, the liquidus surface for H$_2$O-undersaturated conditions is a plane which is reflected in parallel, equally spaced isopleths; this surface is theoretically curved, but the relatively small ranges

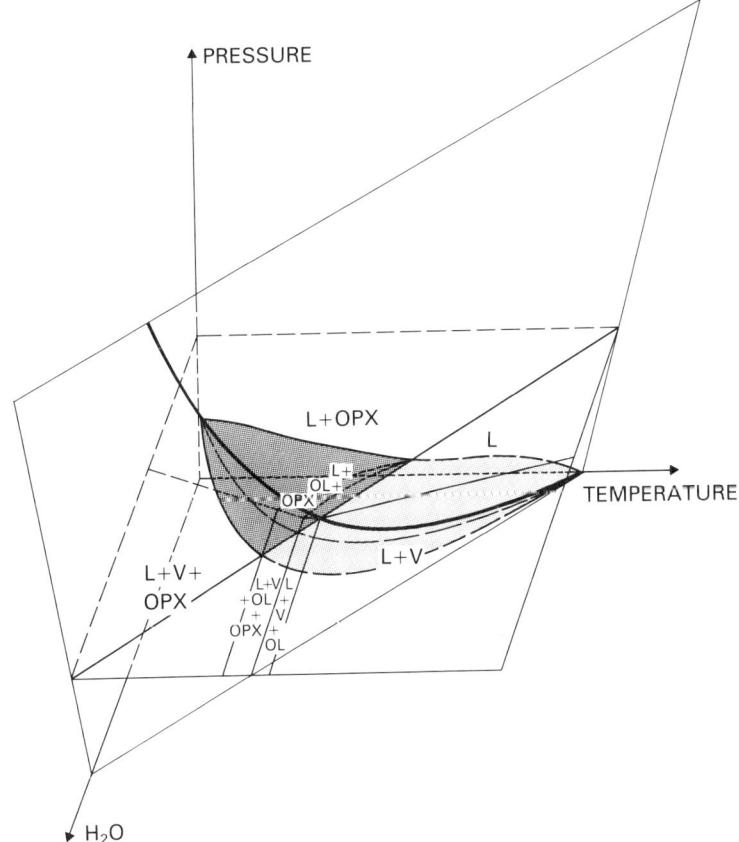

Figure 5.6 Near-liquidus P–T–X_{H_2O} phase relations for boninites. Forwardly tilted plane is the liquidus surface, and contains the shaded olivine stability field. Outside this field Opx(Cen) is the liquidus phase. The horizontal plane is an isobaric T–X_{H_2O} cross section. The extent of olivine stability below the liquidus is outlined. Field boundaries are dotted for Ol + L and Ol + Opx + L.

of P, T and H_2O content considered in this study preclude the need for higher-order interpretation. The position of the H_2O-saturated liquidus was determined from the phase relations in the $T-X_{H_2O}$ sections. At 1 kbar the solubility of H_2O is close to 3% in all three compositions, at 2 kbar it is 5% and at 3 kbar it is estimated to be 6.5%. The resulting liquidus depression is $25 \pm 1°C$ per 1% H_2O for CV-46.2 and $21 \pm 1°C$ per 1% for Bonin and Margi.

The H_2O solubility in boninites in our experiments is between that for basalt and andesite melt at $1100°C$, as determined by Hamilton *et al.* (1964). The liquidus depression per 1% H_2O in the pressure range of this study is nearly linear. It is lower than the approximate $30°C$ per 1% H_2O observed by Eggler & Burnham (1973) for Mt Greenwood andesite, but similar to values of about $20°C$ per 1% H_2O determined by Baker & Eggler (1987) on some Aleutian andesites. The liquidus depression, however, is much lower than commonly observed values for basalts of around $50°C$ per 1% H_2O (Baker & Eggler 1987).

The effect of H_2O expanding the olivine phase volume towards silica-rich compositions is generally ascribed to depolymerization of the silica network of the melt by dissolved dissociated H_2O. The olivine stability limit with increasing silica content of a melt is expected to be reached for the optimum combination of minimum pressure and maximum dissociated H_2O content. Stolper (1982) showed for a variety of glass compositions that, for H_2O contents above 5%, molecularly dissolved H_2O contents increase rapidly whereas the amount of dissociated bond-breaking H_2O (hydroxyl groups) stabilizes. Therefore it is interesting to note that olivine crystallization in CV-46.2 coincides with the minimum pressure necessary to dissolve 5–6% H_2O in this melt.

Olivine as the sole liquidus phase is rare in our runs. More commonly it coexists with pyroxene. Olivine crystals dominate the run products in the centre of the Ol-liquidus stability field, but decrease in abundance towards the margin. Conditions well below the liquidus yielded only pyroxene. This very limited olivine stability confined to near-liquidus conditions is illustrated in the isobaric $T-X_{H_2O}$ sections. The presence of the assemblage L + Ol + Opx in the Ol-liquidus field results from the close proximity of the bulk composition to the L + Ol + Opx phase field. That is, because the experiments were, by necessity, conducted slightly below the liquidus, the Opx-in reaction had already occurred.

At 1 atm, the high-Ca composition (Margi) had olivine on the liquidus at $1268°C$ for f_{O_2} conditions between WM and NNO. Olivine was not formed at $> 40°C$ below the liquidus, and orthopyroxene was the sole crystalline phase. In Margi runs, traces of clinoenstatite were found only at the 1 atm liquidus. With small amounts of H_2O (1.0–4.0%), olivine stability increased to pressures of 3 kbar at $1250°C$, and to 4.5 kbar at $1200°C$.

The intermediate-Ca composition (Bonin) had, at 1 atm, clinoenstatite on the liquidus at 1295°C, and coexisting orthopyroxene and clinoenstatite between 30 and 70°C below the liquidus. At lower temperatures orthopyroxene is the only crystalline phase. At higher pressures, and in the presence of H_2O, the Bonin sample showed a narrow range of conditions, confined to the H_2O-saturated melt curve, where olivine was stable on the liquidus. The upper pressure range of olivine stability extended from 1 kbar/1280°C and 1.0% H_2O to 2.5 kbar/1220°C and 4.0% H_2O. Clinoenstatite was found in the 1 and 2 kbar runs close to the liquidus, always coexisting with orthopyroxene. At higher pressure, however, only orthopyroxene was present. Magnetite was formed in one run with very high H_2O content (10.9%) at 7.5 kbar and 1107°C, possibly as a result of excessive oxidation and Fe gain.

At 1 atm, the low-Ca boninite (CV-46.2) had clinoenstatite on the liquidus at 1315°C, and this was the only crystallizing phase for at least 100°C below the liquidus. In the IHPV experiments, olivine was found in three runs. It appears, therefore, that olivine was marginally stable between 1.5 kbar at 1220°C and 3.5 kbar at 1170°C with H_2O contents of 5–6%. Magnetite was noted in one run with very high H_2O content (9.5%) at 7.5 kbar/1107°C.

5.6 Discussion

5.6.1 Comparison of natural and experimental mineral compositions

A summary of some experimentally derived mineral compositions in the near-liquidus range, and, for comparison, the natural compositions, are given in Table 5.3. Calculated (Table 5.1) and experimental olivine compositions are in excellent agreement and, for Margi, correspond with the natural composition. Experimental olivines have higher Fo contents than natural olivine phenocrysts for Bonin and lower for CV-46.2. Calculated olivine compositions for various wholerock Fe^{3+} contents (Table 5.1) indicate that natural olivines in the Bonin sample probably crystallized under more reducing conditions than our experiments. The discrepancy for CV-46.2 could reflect minor olivine accumulation in this sample. All olivines from the experiments are less forsteritic than the most refractory olivine from their sample localities. A slightly evolved character for the selected compositions is therefore likely.

Experimental and natural pyroxenes in Bonin and CV-46.2 exhibit compositional deviations similar to the olivines. Because orthopyroxenes equilibrated only after 200 h in the experiments of Longhi & Boudreau (1980), it is possible that in our short runs orthopyroxene compositions are not equilibrium compositions (Fig. 5.4). The absence of distinctive polysynthetic twinning in clinoenstatite in high-pressure runs is attributed to the pressure at which the quench occurred. These experiments were quenched at elevated pressures whereas natural lavas quenched at low pressure. The observed compositional range in pyroxene Mg# from 0.84 to 0.92 (Figs 5.3 & 5.4) is a composite effect

Table 5.3 Summary of best-matching natural and experimentally (IHPV) derived mineral phases (phX, phenocryst; mphX, microphenocryst).

	Olivine Fo	Clinoenstatite			Orthopyroxene		
		Wo	En	Fs	Wo	En	Fs
Margi natural phX	88.0	–			–		
run 31.5	88.9	–			–		
run 31.4	88.8	–			4.0–4.1	82.3–82.9	13.1–13.7
run 42.1	88.8	–			3.5–3.9	84.9–85.4	11.1–11.2
Bonin natural phX	89.5	0.8	89.8	9.4	1.8–2.3	87.0–88.0	10.0–11.0
natural mphX	87.7–88.3						
run 47.2–47.3	90.3–91.2	–			–		
run 46.1–46.2	90.7	0.4–1.1	91.4–90.6	7.6–8.3	–		
run 30.4	90.0	0.6	91.0	8.4	1.9	90.2	8.9
CV-46.2 natural phX	92.7	n.d.	93.3	6.7	1.5	84.1	14.4
run 46.4	89.7–90.0	0.5	91.0–91.2	8.3–8.5	–		
run 40.3	87.1–89.4	0.5–0.6	87.1–89.3	10.1–12.4	–		

of many factors. All experimental pyroxenes are, however, entirely within the compositional range of their natural counterparts found in boninites.

5.6.2 *Interpretation of experimental phase equilibria*

If the investigated compositions represent primary melts, each must have equilibrated with a dunitic or harzburgitic assemblage, at $P-T$ conditions close to the respective upper limits of olivine stability for the appropriate H_2O contents in these melts. For Margi and Bonin, the upper pressure limits of olivine stability are undetermined, whereas for Cape Vogel it is confined to a narrow pressure interval approximately between 1.5 and 3.5 kbar (Fig. 5.5). Since LILE enrichment in boninites is associated with hydrous metasomatism, the relative importance of a metasomatic component for each composition can be used as an estimate of H_2O content (Cameron *et al.* 1983, Flower & Van der Laan 1989). Inferred conditions of magma segregation are: for Margi, 3 kbar, 1240°C and 1–2% H_2O; for Bonin, 3 kbar, 1240°C and 3% H_2O; and for Cape Vogel, 3 kbar, 1200°C and 5–6% H_2O. Contents of H_2O needed for olivine crystallization in the experiments were expected to be higher (or pressure estimates lower) than for the original melting conditions. Comparing these results to the magmatic H_2O contents for boninites as determined by Kyser *et al.* (1986) and Dobson & O'Neil (1987), our values are higher by about 1% for Bonin and by 2.5% for CV-46.2. If the compositions are primary, the magma segregation temperature estimates are in the same range as commonly observed MORB liquidi (Walker *et al.* 1979, Fisk *et al.* 1980, Fuji & Bougault 1983) implying a mid-oceanic-ridge-type geotherm for boninite, and very shallow magma segregation (~ 10 km) from a dunitic or harzburgitic residuum.

Our results obtained for the Margi boninite differ significantly from those of Duncan & Green (1980, 1987) on an inferred parental magma composition for the Troodos Upper Pillow Lavas (see Section 5.3). Their parent magma is less siliceous and more magnesian and was in equilibrium with a harzburgite residue at 7–8 kbar and 1360°C. Duncan & Green (1987) inferred the presence of 0.5–1% H_2O in the primary Troodos magmas, lowering the melting temperature by about 50°C. The presence of H_2O seems essential to generate this melt, because its anhydrous liquidus temperature exceeds that for primary MORB at 9 kbar by 70–80°C (Presnall et al. 1979, Fuji & Scarfe 1985). We suggest that a less magnesian composition with a lower liquidus temperature would be a more likely candidate for parental magma of the Upper Pillow Lavas. Cameron et al. (1980) likewise disagreed with the premises of Duncan & Green (1980) that a high-Mg# Upper Pillow Lava parent would be related to the erupted magmas through substantial olivine fractionation. They pointed out that less magnesian, more silicic lavas had been reported with Mg# similar to Duncan & Green's inferred parental magma. Less magnesian, more silicic parents would be in equilibrium with a harzburgite residue at lower pressures or at higher H_2O contents than Duncan & Green's composition. As a consequence, Duncan & Green's experiments only provide an upper $P-T$ limit for Troodos boninite generation. Because our results give a minimum pressure estimate for the Troodos Upper Pillow Lavas, parental magmas segregated between 3 and 7 kbar. Our results imply a MORB-type geotherm based on the liquidus temperature of the erupted lava. Because the Margi composition is closer to a plausible parental magma candidate, and the $P-T$ conditions demanded are reasonable, we prefer conditions for Margi primary magma generation to be 4–5 kbar and 1250–1260°C, with 1% H_2O in the primary magma. These conditions are not very different from those preferred by Duncan & Green (1987) (8–9 kbar, 1310°C with 0.5–1% H_2O in the magma) and are consistent with the differences in Fo content of olivine (Fo_{88} in our composition, and $Fo_{91.7}$ in that of Duncan & Green).

Depending on pressure and H_2O content, melting in equilibrium with Opx–Ol may be of eutectic (Ol + Opx → L) or peritectic (Opx → Ol + L) type. If Margi, Bonin and CV-46.2 melts represent primary magmas, all must have formed through incongruent melting of mantle pyroxene because their compositions lie outside the Alkemade tetrahedron defined by the four solid phases contributing during melting (Fig. 5.2).

The compositions studied may not represent primary melts but may have undergone (slight) fractional crystallization. In this case the experiments yield the last $P-T$ condition of Ol + Opx fractionation, rather than a precise $P-T$ condition at magma segregation. The size of the olivine stability field indicates the tendency for each composition to fractionate olivine. For Margi and Bonin, the occurrence of olivine phenocrysts in the natural samples is reasonable for realistic magmatic H_2O contents of $< 2.5\%$, because these compositions could pass through an olivine phase volume before eruption. However, CV-46.2 is not capable of olivine crystallization under such conditions, and pyroxene (clinoenstatite) must be the fractionating phase. Outside the olivine stability field, these magmas will evolve through pyroxene fractionation. For Margi, orthopyroxene fractionation will cause a decrease in SiO_2 content and direct the melt composition towards the Ol + Opx phase volume boundary. The absence of pyroxene in eutectic, sparsely phyric boninites is a prerequisite to be a primary magma candidate. However, olivine–phyric boninites like Margi still need not be primary because olivine fractionation cannot be ruled out.

In the experiments, H_2O in excess of juvenile contents would nonetheless stabilize olivine in slightly evolved compositions. The discrepancy between natural and experimental olivine compositions and H_2O content for CV-46.2 probably reflects such stabilization, and juvenile H_2O contents were, by implication, lower than 4%.

If, as is likely, some of the studied compositions reflect small degrees of fractionation, initial eutectic primary melts could have evolved to compositions outside the mantle assemblage tetrahedron. The distinctions between eutectic- and peritectic-type melt would be emphasized rather than obscured due to fundamental differences in their isobaric and polybaric fractionation regimes. This is illustrated schematically in Figure 5.7, which shows the topology of the Ol + Opx boundary with respect to the direction of compositional change through fractional crystallization of Opx. Compositions of eutectic (E) and peritectic (P) melts of an assemblage of Ol + Opx are indicated respectively for 'low' and 'high' H_2O contents at high pressure, on the Ol–Opx reaction lines. E- and P-type melts will both fractionate Opx (with a trace of

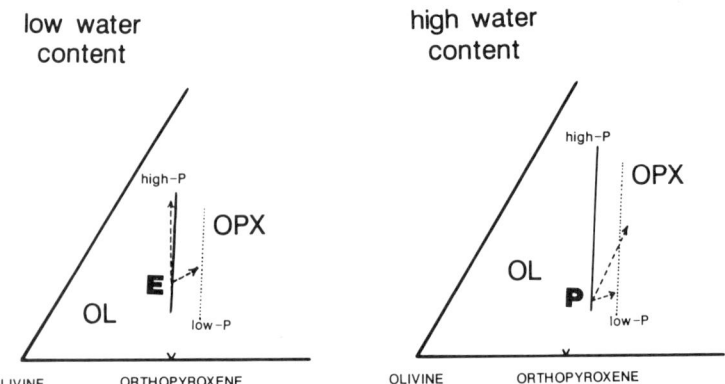

Figure 5.7 Distinct fractionation behaviour of eutectic (E) and peritectic (P) boninite. Direction of compositional evolution through fractional crystallization is indicated by a vector for olivine and for orthopyroxene. After high-pressure fractionation, E-type boninite still crystallizes olivine at low pressure, while P-type boninite crystallizes orthopyroxene–clinoenstatite.

olivine for E) at high pressure. However, for minor fractionation, E-type melts will still lie in the Ol field at lower pressures while P-type melts will lie in the Opx field. Polybaric fractionation of Ol will result in both compositions, owing to the shift of the Ol–Opx phase field boundary towards silica upon pressure release. The 'maximum expansion' of the Ol stability field is determined by the pressure for which H_2O saturation of the melt is attained, and all compositions evolve through this condition before eruption. However, P-type boninites that underwent isobaric Opx fractionation are unable to crystallize olivine at the 'maximum expansion' condition, whereas E-type boninites always crystallize Ol at its 'maximum expansion' stability. Thus, olivine is expected to be absent from P-type boninite, but present in primitive E-type melts.

Because of their sparsely phyric character, the selected samples represent liquid compositions. Therefore, if the experimental compositions are evolved, their liquidus $P-T$ array for realistic H_2O content still lies on or even above petrogenetic conditions for MORB and implies a MORB-type adiabat for their genesis.

5.6.3 Melting and fractionation mass balances

In order to test this model further, least-squares mass-balance calculations were made to reconcile phase equilibria, petrographic characteristics and melt–residue relationships for the boninites studied. These calculations were based on premises that (a) boninite melts are buffered by a restricted range of assemblages, i.e. dunite, harzburgite, Cpx-harzburgite, or (less likely) lherzolite; (b) the boninite source is a (variably?) refractory peridotite; and (c) the variation in melt composition results from a combination of eutectic- versus peritectic-type melting, and differing degrees of partial melting. Compositional mass balances involved in boninite genesis are clearly more complex than those pertaining to MORB, assuming the latter to be generated in a 'fertile', essentially homogeneous, mantle.

For our purposes, a range of plausible source compositions was selected, including a computed residue from N-MORB generation, and lherzolite compositions of variably refractory character (Table 5.4). Primary boninite melt candidates were selected from the experimentally studied compositions, together with others from the literature. The N-MORB melt residue R-9 (Table 5.4; Viereck *et al.* 1989) was computed assuming that a high-Mg # N-MORB tholeiite (ARP-74; Fuji & Bougault 1983, Fuji & Scarfe 1985) was generated through ~ 15% partial melting of a fertile plagioclase- or spinel-lherzolite (Ib/8; Stosch & Seck 1980) at the lherzolite solidus 'cusp' (Presnall *et al.* 1979, Presnall & Hoover 1981). Other potential (basalt-depleted) sources for boninite were selected from xenolith suites from Dreiser Weiher (Ib/3, Ib/24, Ia/110, Ia/412; Stosch & Seck 1980) and Oahu (68SAL3; Jackson & Wright

Table 5.4 Major-element compositions of fertile (Ib/8) and refractory source compositions (others) and calculated mode options used in mass-balance calculations (see text).

	Ib/8	R-9	Ib/3	68SAL3	Ia/412	Ib/24	Ia/110
SiO_2	46.71	45.96	44.58	44.79	43.88	44.17	44.28
Al_2O_3	4.35	2.58	2.18	2.15	1.10	1.17	1.61
FeO^*	7.62	7.65	8.22	7.51	7.87	7.77	7.89
MgO	36.66	40.85	43.65	43.58	46.06	46.17	44.51
CaO	4.10	2.79	1.26	1.60	0.96	0.63	1.53
Na_2O	0.36	0.09	0.08	0.29	0.11	0.08	0.16
K_2O	0.02	0.01	0.02	0.02	0.03	0.02	0.03
TiO_2	0.23	0.07	0.06	0.08	0.04	0.03	0.04
P_2O_5	nd	nd	nd	0.03	nd	nd	nd
CaO/Al_2O_3	0.943	1.081	0.578	0.744	0.872	0.538	0.950
100Mg #	90.5	91.3	91.3	92.0	92.0	92.1	91.8
calc. modes							
Ol	56.71	64.14	74.13	75.09	81.56	81.87	77.89
Opx	22.14	23.24	19.80	15.58	13.65	14.09	14.39
Cpx	8.96	6.22	0.32	2.87	1.87	0.96	3.32
Pl	12.22	6.42	5.77	6.48	2.98	3.08	4.43
Ol	38.75	54.33	64.41	63.70	74.95	73.38	69.44
Opx	38.06	31.93	28.54	28.75	21.96	25.02	23.93
Cpx	16.86	10.38	4.05	4.80	1.99	6.40	4.72
Sp	6.34	3.38	3.02	2.77	1.11	1.22	1.93

nd = not determined.

1970), assuming these to be residua from tholeiite magma generation (discussed in Viereck *et al.* 1989). Of the residual compositions, R-9 is the least refractory (Mg# = 0.913, CaO/Al_2O_3 = 1.081) while the natural xenolith compositions range from Ib/3 (Mg# = 0.914, CaO/Al_2O_3 = 0.578) to Ib/24 (Mg# = 0.921, CaO/Al_2O_3 = 0.538) (Table 5.4). Model calculations were therefore set up for each candidate melt in relation to each of the selected residues.

The mass-balance solutions obtained are summarized in Table 5.5, and were judged according to best-fit criteria (oxide residuals preferably < 0.1) (Wright 1974, Flower & Robinson 1981), and petrologic validity (e.g. computed olivine compositions preferably in the range Fo_{88-92}, melt fraction $\geq$ 5%), plausible residue assemblage (see above), and positive contributions of solid phases to the melt, unless incongruent melting of Opx is suspected.

The best numerical solutions are consistent with sources Ia/110 and Ia/412 (for Margi, Bonin and Cape Vogel 'melt' compositions), 68SAL3 (Margi, Bonin) and Ib/24 (Cape Vogel), which indicate harzburgite ($\pm$ accessory Cpx) or (Opx) dunite residues, and degrees of partial melting in the range 7–13%. Perhaps the most significant result is the distinctively higher proportion of Opx relative to Ol in the Margi residue compared to Bonin and Cape Vogel

Table 5.5 Summary of (a) calculated melt residues and weight fraction of (F) melt produced, and (b) solid fractions entering melt, assuming plagioclase-'lherzolite' and spinel-'lherzolite' mantle assemblages. Assumed: all Pl and Sp consumed during partial melting (from Table 5.4).

(a)

	Margi				Bonin				Cape Vogel			
	Cpx	Opx	Ol	F	Cpx	Opx	Ol	F	Cpx	Opx	Ol	F
Source												
R-9	4.3	11.8	67.1	16.8	5.0	2.5	71.9	20.6	6.9	–	72.4	20.7
1b/3	–	9.6	76.7	13.7	–	1.2	81.1	17.7	0.6	–	81.1	18.3
68SAL3	–	9.0	75.6	15.4	0.5	0.3	80.1	19.1	2.4	–	79.6	18.0
1a/412	0.7	10.3	81.9	7.1	0.7	5.9	84.2	9.1	1.8	5.3	84.1	8.8
1b/24	–	13.0	80.5	6.5	–	8.5	82.9	8.6	–	7.3	83.0	9.7
1a/110	1.6	9.0	78.7	10.8	1.9	2.8	81.9	13.4	3.3	1.2	82.2	13.4

(b)

	Margi			Bonin			Cape Vogel		
	Cpx	Opx	Ol	Cpx	Opx	Ol	Cpx	Opx	Ol
Pl-Lh									
R-9	0.31	0.49	– 0.05	0.20	0.89	– 0.12	– 0.11	1.0	– 0.13
1b/3	1.0	0.40	– 0.03	1.0	0.94	– 0.09	– 0.02	1.0	– 0.09
68SAL3	1.0	0.25	– 0.01	0.84	0.98	– 0.07	0.15	1.0	– 0.06
1a/412	1.0	0.15	0.00	0.60	0.56	– 0.03	0.02	0.61	– 0.03
1b/24	1.0	0.01	0.02	1.0	0.40	– 0.01	1.0	0.48	– 0.01
1a/110	0.91	0.25	– 0.01	0.42	0.81	– 0.05	0.02	0.92	– 0.06
Sp-Lh									
R-9	0.59	0.63	– 0.23	0.52	0.92	– 0.3	0.33	1.0	– 0.33
1b/3	1.0	0.66	– 0.19	1.0	0.96	– 0.2	0.84	1.0	– 0.26
68SAL3	1.0	0.69	– 0.19	0.91	0.99	– 0.2	0.49	1.0	– 0.25
1a/412	0.64	0.53	– 0.09	0.63	0.73	– 0.1	0.10	0.7	– 0.12
1b/24	1.0	0.48	– 0.10	1.0	0.66	– 0.1	1.0	0.7	– 0.13
1a/110	0.66	0.63	– 0.13	0.59	0.88	– 0.1	0.31	0.9	0.18

solutions, irrespective of source compositions used in the calculation. Estimates of melt fraction contributed by the solid phases were calculated assuming two alternative mineral modes for each source: one with spinel ('Sp-Lh') and one with plagioclase ('Pl-Lh') accompanying Cpx, Opx and Ol (Table 5.4). These determine the respective proportions of pyroxene and olivine contributed to the melt, from which computed mass balances should give an indication of whether olivine contributes negative ('peritectic') or positive ('eutectic') mass to the melt. While the Sp-Lh models indicate negative Ol contributions in all solutions, the numerically preferred Pl-Lh models (see above) all differentiate between Margi (small positive or negative Ol contribution) on the one hand, and Bonin and Cape Vogel (negative Ol contributions) on the other, consistent with a eutectic model to produce high-Ca boninite, and a peritectic model to produce low-Ca boninite. Owing to the uncertainties concerning both bulk source and primary boninite compositions, our objective was not to attempt unique mantle–melt mass-balance solutions for each case, but to constrain the relative importance of factors indicated by the experimental and theoretical phase equilibria.

Phase equilibria in CMAS and field evidence for ophiolite residue lithologies suggest that Cpx-rich (e.g. wehrlite) residues, of the type indicated for Cape Vogel and Bonin melts in relation to the R-9 source (Table 5.5), are not expected. In all other cases, the computed residues consist of harzburgite or dunite, with accessory amounts of Cpx.

Solutions which do not involve Cpx or Opx indicate that a Cpx- or Opx-free source could yield the same partial melt. Phase equilibria in CMAS below 20 kbar (Presnall et al. 1979) preclude melting in the presence of Cpx, Opx, Ol, Pl without Pl, Opx and Cpx contributing to the melt. This means that in order to melt substantial amounts of Opx, the small amount of Cpx and Pl in the source needs to have entered the melt (value of 1 for Cpx in Table 5.5b). On this basis R-9, Ib/3, 68SAL3 and Ia/412 seem plausible sources for Margi, Ib/3 and Ib/24 for Bonin, and Ib/3 (ignoring the solution for Cpx because this source is almost Cpx-free) and Ib/24 for CV-46.2. From the discussion above, the numerically acceptable mass-balance solutions, we suggest two models:

(a) Boninite variants can be derived from a single refractory source, through similar degrees of partial melting (e.g. Ib/3; 13.7–18.3%), but for different H_2O contents, which control the extent of peritectic melting. This is supported by the sharp differences in Ol/Opx residue solutions for Ca-rich vs. Ca-poor boninite melts.

(b) More extreme peritectic melting with increasing H_2O is accompanied by a decrease in CaO/Al_2O_3, CaO and Al_2O_3 contents, and an increase in Mg# in respective bulk sources (Table 5.4), such that less refractory sources Ia/110 and 68SAL3 yield high-Ca boninites (with relatively low H_2O content), and more refractory sources such as Ib/3 and Ib/24 yield low-Ca boninite (with relatively high H_2O content) for variable degrees of partial melting.

Mass balances were also computed for representative fractional crystallization models for Margi, Bonin and Cape Vogel boninite series, using appropriate lava and mineral phase compositions from the literature. The solutions reflect the dominance of Opx (+ Sp + Cpx) in low-Ca boninite suites (Bonin, Cape Vogel), and of Ol + Opx (+ Cpx) in high-Ca suites, consistent with our predictions for P- and E-type melts. Similar observations concerning the predominance of fractionating phases for these suites were made by Dallwitz (1968), Walker & Cameron (1983), Cameron (1985), Umino (1986) and Flower & Levine (1987).

5.7 Tectonic implications

Boninite magma generation unquestionably reflects a unique combination of source composition (refractory peridotite, significant H_2O) and physical conditions (low pressure, high temperature), implying adiabatic melting due to diapirism in a supra-subduction environment. The imprint of subduction-related metasomatism (e.g. Hickey & Frey 1982, Cameron *et al.* 1983, Flower & Levine 1987) suggests that diapirism is deeply rooted and genetically associated with slab dehydration. Experimental indications that boninites can form at temperatures around $1240°C$ and shallow depths (but > 3 kbar) appear to reflect even more inflected geotherms than those inferred for mid-ocean-ridge axes (Presnall *et al.* 1979, Presnall & Hoover 1981, Fuji & Scarfe 1985). Any model attempting to reconcile magma variation and tectonic evolution at colliding plate boundaries must recognize these thermal and petrologic constraints.

Hawkins *et al.* (1984) proposed that partial melting of serpentinized harzburgite during the initiation of subduction along a transform fracture zone would yield boninite. Although satisfying some petrologic requirements, this model fails to satisfy thermal criteria, since it implies reheating of cool, lower-lithosphere material to MORB type, or hotter, temperatures in the thermal environment of a subducted slab. The proposal for second-stage melting of a single mantle diapir at 7–8 kbar (Duncan & Green 1980, 1987) also encounters a thermal problem unless access of water to flux melting is also a factor. The thermal problem of second-stage melting of a single diapir is also resolved if the first-stage melting is at ~ 20 kbar/$1450°C$ yielding picritic magmas which are parental to MORB (Duncan & Green 1987, Fig. 5.7). Although accelerated diapirism at shallow depths might yield second-stage melts from a single MORB-source diapir, an even more critical obstacle derives from the LILE-enriched and hydrous character of boninite melts, presumably an effect of slab-derived metasomatism. As first-stage melting would inevitably strip a mobile diapir of magmaphile elements (including H_2O), it is difficult to conceive of a mechanism whereby refractory diapirs, relieved at shallow levels of a basaltic magma fraction, may be further enriched in H_2O and LILE-rich fluid.

Meijer (1980) proposed boninite genesis in a static non-convecting mantle wedge. Continuous fluid supply would first give rise to arc-tholeiite magmas from the hottest interior of the mantle wedge followed by boninitic magma compositions while the source became progressively more refractory. Increasing amounts of H_2O involved in melting would satisfy the thermal requirements. Our experimental results, however, call for MORB-type or hotter geotherms indicating that arc-tholeiite thermal conditions are probably too cold for boninite genesis.

In contrast to Hawkins *et al.* (1984) and Meijer (1980), we concur with Duncan & Green (1980, 1987) to the extent that diapirism is the principal factor promoting partial melting of refractory mantle at low pressures. We propose instead that different primary magmas (boninite, IAT, MORB) derive from intrinsically different diapirs, and that mantle differentiation must pre-date the initiation of diapirism. We also propose that H_2O serves a pivotal role in the melting behaviour (especially) of orthopyroxene, hence generation of P- and E-type boninite, and in a more remote sense, the initiation of diapiric activity in the immediate supra-subduction region.

The rarity of boninite may reflect the comparative difficulty of refractory diapirs in reaching upper-mantle regions, and the temporal–spatial relations of boninite to MORB and IAT therefore have significant tectonic implications. Boninite precedes arc volcanics in the southern Mariana forearc (Reagan & Meijer 1984, Hawkins *et al.* 1984, Crawford *et al.* 1981) and Bonin Islands (Umino 1986), but also occurs following eruption of arc volcanics further north, at DSDP Hole 458 (Meijer 1980) and elsewhere (e.g. Cameron 1985, Flower & Levine 1987) and preceding MORB lavas (Coish & Church 1979, Crawford & Keays 1987) in ophiolite sequences.

We suggest two petrogenetic scenarios for boninite which appear to satisfy petrologic and thermal constraints. The first model is subduction under an active or dying oceanic ridge system (Meijer 1980), for instance at the initiation of subduction along a transform fault. Boninites might form when slab-derived fluids invade the former spreading region still occupied by a MORB-type geotherm, and shallow melting of a refractory source could occur. This model explains both the local presence of boninite in forearcs and that boninite magmas can precede arc volcanism.

The second model may explain boninite eruption following a 'normal' arc-volcanic stage. In this model, slab-induced convection in the mantle wedge would transport pockets of depleted mantle to deeper levels and allow hydration near the subducted slab. The hydrated (Cpx-) harzburgite would become buoyant in the garnet lherzolite mantle facies. Evidence for garnet-buffered metasomatism of the boninite source is furnished by HFSE contents. Increasing abundance of some HFSE with decreasing Al and Ca contents in boninites implies a selective enrichment of the boninite source and HREE retention with respect to elements such as Ti and P (Flower & Van der Laan 1989). Increasing propensity for diapirism might be created by tensional forces giving rise to a hotter geothermal path for the ascending diapir. Initial melting would then produce sufficient buoyancy to propagate ascent.

This interplay of tension and change in thermal regime occurs at the initiation of backarc spreading or arc sundering. This tectonic regime for boninite genesis is similar to that proposed by Crawford *et al.* (1981) and explains boninite magmatism following an arc-volcanic stage. Sufficient amounts of partial melt for extraction will not be generated until very shallow levels (10–15 kbar). Fractional crystallization will therefore be minor, and P-type boninites will evolve through orthopyroxene–clinoenstatite fractionation, while E-type boninites will evolve through olivine + orthopyroxene fractionation.

Acknowledgements

We acknowledge the support from the US National Science Foundation, Grant EAR 82-12738, and from the UIC Research Board. S.R.L. gratefully acknowledges the receipt of a University of Illinois fellowship, thanks W. E. Cameron for providing analyses on one of the samples, and G. A. Jenner for sharing his views in a stimulating discussion in the early stages of this work. D. C. Presnall is thanked for his comments on our CMAS model and reviews by Professor D. H. Green and Dr I. A. Nicholls considerably improved our original manuscript.

References

Baker, D. R. & D. H. Eggler 1987. Compositions of anhydrous melts coexisting with plagioclase, augite, and olivine or low-Ca pyroxene from 1 atm to 8 kbar: application to the Aleutian volcanic center of Atka. *Am. Mineral.* **72**, 12–28.

Bence, A. E. & A. L. Albee 1969. Empirical correction factors for the electron microanalysis of silicates and oxides. *J. Geol.* **76**, 382–403.

Biggar, G. M. 1985. Calcium-poor pyroxenes: phase relations in the system $CaO-MgO-Al_2O_3-SiO_2$. *Mineral. Mag.* **49**, 49–58.

Bloomer, S. H. 1983. Distribution and origin of igneous rocks from the landward slope of the Mariana Trench: implications for its structure and evolution. *J. Geophys. Res.* **88**, 7411–28.

Cameron, W. E. 1985. Petrology and origin of primitive lavas from the Troodos ophiolite, Cyprus. *Contrib. Mineral. Petrol.* **89**, 239–55.

Cameron, W. E., R. A. Duncan & D. H. Green 1980. Comment and reply on 'Role of multistage melting in the formation of oceanic crust'. *Geology* **8**, 562–3.

Cameron, W. E., M. T. McCulloch & D. A. Walker 1983. Boninite petrogenesis: chemical and Nd–Sr isotopic constraints. *Earth Planet. Sci. Lett.* **65**, 75–89.

Cameron, W. E., E. G. Nisbet & V. J. Dietrich 1979. Boninites, komatiites and ophiolitic basalts. *Nature* **280**, 550–3.

Carlson, W. D. 1986. Reversed pyroxene phase equilibria in $CaO-MgO-SiO_2$ from 925 to 1175°C at one atmosphere pressure. *Contrib. Mineral. Petrol.* **92**, 218–24.

Coish, R. A. & W. R. Church 1979. Igneous geochemistry of mafic rocks in the Betts Cove Ophiolite, Newfoundland. *Contrib. Mineral. Petrol.* **70**, 29–39.

Crawford, A. J. & R. R. Keays 1987. Petrogenesis of Victorian Cambrian tholeiites and the origin of related boninites. *J. Petrol.* **28**, 175–209.

Crawford, A. J., L. Beccaluva & G. Serri 1981. Tectono-magmatic evolution of the West Philippine–Mariana region and the origin of boninites. *Earth Planet. Sci. Lett.* **54**, 346–56.

Dallwitz, W. B. 1968. Chemical composition and genesis of clinoenstatite-bearing volcanic rocks from the Cape Vogel area, Papua: a discussion. *23rd Int. Geol. Congr.* **2**, 229–42.

Duncan, R. A. & D. H. Green 1980. Role of multistage melting in the formation of oceanic crust. *Geology* **8**, 22–6.

Duncan, R. A. & D. H. Green 1987. The genesis of refractory melts in the formation of oceanic crust. *Contrib. Mineral. Petrol.* **96**, 326–42.

Dobson, P. F. & J. R. O'Neil 1987. Stable isotope compositions and H_2O contents of boninite

series volcanic rocks from Chichi-jima, Bonin Islands, Japan. *Earth Planet. Sci. Lett.* **82**, 75–86.
Eggler, D. H. & C. W. Burnham 1973. Crystallization and fractionation trends in the system andesite–H_2O–CO_2–O_2 at pressures to 10 kb. *Geol. Soc. Am. Bull.* **84**, 2517–32.
Elthon, D. 1983. Isomolar and isostructural pseudo-liquidus phase diagrams for oceanic basalts. *Am. Mineral.* **68**, 506–11.
Fisk, M. R., J.-G. Schilling & H. Sigurdsson 1980. An experimental investigation of Iceland and Reykjanes ridge tholeiites: 1. Phase relations. *Contrib. Mineral. Petrol.* **74**, 361–74.
Flower, M. F. J. & H. Levine 1987. Petrogenesis of a tholeiite–boninite sequence from Ayios Mamas, Troodos Ophiolite: evidence for splitting of a volcanic arc. *Contrib. Mineral. Petrol.* **97**, 509–24.
Flower, M. F. J. & P. T. Robinson 1981. Basement drilling in the Western Atlantic Ocean. 1. Magma fractionation and its relation to eruptive chronology. *J. Geophys. Res.* **86**, 6273–98.
Flower, M. F. J. & S. R. Van der Laan 1989. In preparation.
Fuji, T. & H. Bougault 1983. Melting relations of a magnesian abyssal tholeiite and the origin of MORB. *Earth Planet. Sci. Lett.* **62**, 283–95.
Fuji, T. & C. M. Scarfe 1985. Composition of liquids coexisting with spinel lherzolite at 10 kbar and the genesis of MORB. *Contrib. Mineral. Petrol.* **90**, 18–28.
Hamilton D. L. & M. Henderson 1968. The preparation of silicate compositions by a gelling method. *Mineral. Mag.* **36**, 832–8.
Hamilton, D. L., C. W. Burnham & E. F. Osborn 1964. The solubility of H_2O and effects of oxygen fugacity and H_2O content on crystallization in mafic magmas. *J. Petrol.* **5**, 21–39.
Hawkins, J. W., S. H. Bloomer, C. A. Evans & J. T. Melchior 1984. Evolution of intra-oceanic arc–trench systems. *Tectonophysics* **102**, 175–205.
Hickey, R. & F. A. Frey 1982. Geochemical characteristics of boninite series volcanics: implications for their source. *Geochim. Cosmochim. Acta* **46**, 2099–115.
Holloway, J. R. 1971. Internally-heated pressure vessels. In *Research techniques for high pressure and high temperature*, G. C. Ulmer (ed.), 217–58. New York: Springer.
Jackson, E. D. & T. L. Wright 1970. Xenoliths in the Honolulu volcanic series. *J. Petrol.* **11**, 405–30.
Jenner, G. A. 1981. Geochemistry of high-Mg andesite from Cape Vogel, Papua New Guinea. *Chem. Geol.* **33**, 307–32.
Kuroda, N., K. Shiraki & H. Urano 1978. Boninite as a possible calc-alkaline primary magma. *Bull. Volcanol.* **41**, 563–75.
Kushiro, I. 1969. The system forsterite–diopside–silica with and without H_2O at high pressures. *Am. J. Sci.* **267A**, 269–94.
Kushiro, I. 1975. On the nature of silicate melt and its significance in magma genesis: regularities in the shift of the liquidus boundaries involving olivine, pyroxene, and silica minerals. *Am. J. Sci.* **275**, 411–31.
Kushiro, I. 1982. Petrology of high-MgO bronzite andesite resembling boninite from Site 458 near the Mariana trench. Init. Rep. DSDP Leg 60, 731–4.
Kyser, T. K., W. E. Cameron & E. G. Nisbet 1986. Boninite petrogenisis and alteration history: constraints from stable isotope compositions of boninites from Cape Vogel, New Caledonia and Cyprus. *Contrib. Mineral. Petrol.* **93**, 222–6.
Longhi, J. & A. Boudreau 1980. The orthoenstatite liquidus field in the system forsterite–diopside–silica at one atmosphere. *Am. Mineral.* **65**, 563–73.
Meijer, A. 1980. Primitive arc volcanism and a boninite series: examples from western Pacific island arc. In *Tectonic and geologic evolution of southeast Asian seas and islands*, D. E. Hayes (ed.), 269–82. Am. Geophys. Union Monogr. no. 23.
Mysen, B. O. & I. Kushiro 1974. A possible origin for andesitic magmas: discussion of a paper by Nicholls and Ringwood. *Earth Planet. Sci. Lett.* **21**, 221–9.

Nicholls, I. A. 1974. Liquids in equilibrium with peridotitic mineral assemblages at high H_2O pressures. *Contrib. Mineral. Petrol.* **45**, 289–316.

Presnall, D. C. & J. D. Hoover 1981. Analytical uncertainties and the generation of picritic versus tholeiitic primary magmas at mid-ocean ridges (abstract). *Geol. Soc. Am. Abstr. Prog.* **13**, 533.

Presnall, D. C., J. R. Dixon, T. H. O'Donnell & S. A. Dixon 1979. Generation of mid-ocean ridge tholeiite. *J. Petrol.* **20**, 3–35.

Reagan, M. K. & A. Meijer 1984. Geology and geochemistry of early arc-volcanic rocks from Guam. *Geol. Soc. Am. Bull.* **95**, 701–13.

Roeder, P. L. & R. F. Emslie 1970. Olivine–liquid equilibrium. *Contrib. Mineral. Petrol.* **29**, 275–89.

Sack, R. O., I. S. E. Carmichael, M. Rivers & M. S. Ghiorso 1980. Ferric–ferrous equilibria in natural silicate liquids at 1 bar. *Contrib. Mineral. Petrol.* **75**, 369–76.

Sameshima, T. & J.-P. Paris 1983. Clinoenstatite-bearing lava from Nepoui, New Caledonia *Am. Mineral.* **68**, 1076–82.

Sen, G. 1982. Composition of basaltic liquids generated from a partially depleted lherzolite at 9 kbar pressure. *Nature* **299**, 336–8.

Shiraki, K., N. Kuroda, H. Urano & S. Maruyama 1980. Clinoenstatite in boninites from the Bonin Islands, Japan. *Nature* **285**, 31–2.

Stolper, E. 1982. The speciation of H_2O in silicate melts. *Geochim. Cosmochim. Acta* **46**, 2609–20.

Stosch, H.-G. & H. A. Seck 1980. Geochemistry and mineralogy of two spinel peridotite suites from Dreiser Weiher, West Germany. *Geochim. Cosmochim. Acta* **44**, 457–70.

Takahashi, E. & I. Kushiro 1983. Melting of a dry peridotite at high pressures and basaltic magma genesis. *Am. Mineral.* **68**, 859–79.

Tatsumi, Y., 1981. Melting experiments on a high-magnesian andesite. *Earth Planet. Sci. Lett.* **54**, 357–64.

Umino, S. 1986. Magma mixing in boninite sequence of Chichijima, Bonin Islands. *J. Volcanol. Geotherm. Res.* **29**, 125–57.

Van der Laan, S. R. & A. F. Koster van Groos 1988. Platinum–iron alloys applied to high pressure research on hydrous iron-bearing systems. *Am. Mineral.* (under review).

Van der Laan, S. R., et al. 1989. In preparation.

Viereck, et al. 1989. In preparation.

Walker, D. A. & W. E. Cameron 1983. Boninite primary magmas: evidence from the Cape Vogel Peninsula, PNG, *Contrib. Mineral. Petrol.* **83**, 150–8.

Walker, D., T. Shibata & S. E. Delong 1979. Abyssal tholeiites from the Oceanographer Fracture Zone. II: Phase equilibria and mixing. *Contrib. Mineral. Petrol.* **70**, 111–25.

Wood, D. A., N. G. Marsh, J. Tarney, J.-L. Joron, P. Fryer & M. Treuil 1982. Geochemistry of igneous rocks recovered from a transect across the Mariana trough, arc, fore-arc, and trench; Sites 453 through 461, Deep Sea Drilling Project Leg 60. *Int. Rep. DSDP Leg 60*, 611–45.

Wright, T. L. 1974. Presentation and interpretation of chemical data for igneous rocks. *Contrib. Mineral. Petrol.* **48**, 233–48.

6 Geochemistry and petrogenesis of Archaean and early Proterozoic siliceous high-magnesian basalts

SHEN-SU SUN, ROBERT W. NESBITT & MALCOLM T. McCULLOCH

Abstract

Chemical and isotope systematics of Archaean and Proterozoic siliceous high-Mg basalts (SHMB) are evaluated in terms of different mechanisms of magma generation, i.e. whether they are (1) Archaean equivalents of modern boninites; (2) derived from komatiite through crustal assimilation–fractional crystallization (AFC) processes; or (3) derived from 'subduction' of modified refractory continental lithospheric mantle. Close examination indicates that the chemical and mineralogical similarities of these SHMB to modern boninites are superficial. It is shown that chemical and isotope data for practically all Archaean spinifex-textured SHMB can be interpreted in terms of AFC processes starting with a komatiitic parent magma. Modelling indicates that it is possible to derive a SHMB with 10% MgO from a komatiite parent with 24% MgO, by assimilation of ~ 14% felsic crust and 35 wt% crystal fractionation. Archaean SHMB from Kambalda (Yilgarn Block) and Negri Volcanics (West Pilbara), Western Australia, have a wide range of MgO but constant initial ε_{Nd} within each suite (-2 and -3 respectively). These data support the prediction of fluid-dynamic models that assimilation of crust by komatiite magma takes place initially at depth and is then followed by crystal fractionation accompanied by minor assimilation at shallow depths in layered flows/sills. Initial ε_{Nd} of SHMB combined with chemical data can offer information on age and rock types (e.g. tonalite versus granite) of the contaminant in the basement rocks. Although crustal contamination also seems important for generation of early Proterozoic SHMB dykes and layered complexes, some of these clearly were not derived from komatiite via AFC processes. Instead, they appear to be generated by second-stage melting of subduction-modified refractory harzburgitic Archaean lithosphere and mixing with asthenosphere-derived melts (picrite±komatiite), followed by AFC processes. These dykes are LREE-enriched but poor in CaO, Sr, P_2O_5, TiO_2, Nb and Ta, reflecting a source history involving sediment subduction.

Large layered complexes (e.g. Stillwater and Bushveld) with associated magmatic PGE mineralization commonly have SHMB as parent magmas. PGE mineralization in these bodies is related to the sulphide-undersaturated character of the parent magma and magma chamber mixing processes including crustal contamination, which eventually result in sulphide saturation and concentration of the PGE. The mineralization does not appear to be due to a mantle source enriched in PGE, as all Archaean komatiites appear to have a uniform Ti/Pd ($\sim 3 \times 10^5$). Pd/S and Pd/Cu estimated for parent magmas of these layered complexes provide a useful criterion to distinguish between different origins for the parent magmas. Parent magmas derived from komatiite through AFC processes have Pd/S (2 to 4×10^{-5}) and Pd/Cu (1.5 to 2.0×10^{-4}) close to the estimates for the primitive mantle, whereas parent magmas mainly derived from refractory harzburgitic continental lithosphere have much higher Pd/S and Pd/Cu ratios (e.g. 1×10^{-4} and 5×10^{-4} respectively).

Since crustal contamination can greatly affect trace-element and isotope compositions of mantle-derived magmas, and sulphides precipitated from these magmas, interpretation of the trace-element (e.g. Th/U, Nb/Th) and isotope data of these rocks in terms of mantle evolution must take these effects into account.

6.1 Introduction

The genesis of siliceous high-Mg basalts (SHMB) in Archaean greenstone belts has been a controversial issue. These basalts commonly have 51–55% SiO_2 and 10–16% MgO. In contrast, normal komatiitic basalts have $\leqslant 50\%$ SiO_2 with the same range of MgO. The common association of SHMB with komatiites suggests a possible genetic link through magma differentiation. However, it has been shown that SHMB have LREE-enriched patterns, in contrast to komatiites, which are generally LREE-depleted (e.g. Sun & Nesbitt 1978a). Thus, SHMB cannot be related to LREE-depleted komatiites simply by crystal fractionation of olivine and pyroxenes, as these minerals do not fractionate REE efficiently enough to generate the observed LREE-enriched patterns. REE and other trace-element and isotopic data suggest that SHMB were derived from different mantle sources to komatiites (e.g. Sun & Nesbitt 1978a) or that their genetic link to komatiites has been camouflaged by crustal contamination (Longhi et al. 1983, Barley 1986, Arndt & Jenner 1986).

Since SHMB are characterized by high MgO and SiO$_2$ and are variably enriched in LREE, K, Rb, Zr and PGE and depleted in Nb, Ta and Ti, they have been regarded as Archaean analogues of Tertiary boninites (Redman & Keays 1985). Chemical and isotopic studies of boninites have been interpreted in terms of second-stage melting of a refractory harzburgitic mantle, induced either by passage of a high-temperature mantle diapir (Crawford et al. 1981) or by introduction of fluid or hydrous melt derived from the subducted oceanic crust (e.g. Sun & Nesbitt 1978b). However, if SHMB are not Precambrian analogues of boninite, then their siliceous, high-Mg characteristics may be ascribed to crustal contamination, or melting of metasomatized asthenosphere–continental lithosphere with primary melt separation at shallow depth, perhaps < 12 kbar (Weaver & Tarney 1981, Sun & Nesbitt 1978b, Fisk 1986, McCallum 1987, Jaques & Green 1980, Green 1981). The depth of melt separation can be slightly deeper if these SHMB are not primary magma (i.e. olivine fractionation was involved).

The crustal contamination model (i.e. SHMB are genetically related to the komatiites through AFC processes) has recently gained strong support from considerations of thermal budgets and fluid dynamics relevant to injection of high-temperature komatiite in dykes and magma chamber processes (Huppert & Sparks 1985). It has also received support from chemical and isotope modelling (Longhi et al. 1983, Arndt & Jenner 1986). However, it is crucial to realize that, even if crustal contamination is indeed important in the petrogenesis of SHMB, the composition of the primary magma prior to contamination might not necessarily have been komatiite.

SHMB also occur in early Proterozoic terrains, often as dykes and layered complexes as well as lava flows; some of these occurrences have komatiite–SHMB association (e.g. Arndt et al. 1987). On the basis of a detailed petrological and geochemical study of Scourie dykes in Scotland, Weaver & Tarney (1981) suggested an origin from modified Archaean lithospheric mantle, and for the Bushveld complex, a boninite connection has been postulated (e.g. Sharpe & Hulbert 1985).

In this chapter, we present new chemical and Nd isotope data for (1) late Archaean SHMB from Kambalda and Mt Hunt in the Eastern Goldfield of Western Australia, (2) Negri Volcanics from the western part of the Pilbara Block, Western Australia, and (3) early Proterozoic dykes from Cowarna Rocks (Yilgarn Block) and eastern Antarctica. These data are integrated with additional information from the literature to evaluate the three different hypotheses mentioned above.

We are concerned with the following issues:

(a) How to evaluate different hypotheses quantitatively as well as qualitatively using chemical and isotope criteria? For example, how to distinguish the effect of crustal contamination from modification of source mantle by sediment subduction.

(b) How to reconstruct the characteristics of the primary magmas by filtering out the effect of crustal contamination.
(c) Is there a secular variation in relative importance of different magma generation processes? For example, does melting of subduction-modified continental lithosphere become more important in the Proterozoic than in the Archaean?
(d) What are the implications of this study for the chemical and isotopic composition of the Archaean and early Proterozoic convecting mantle and the continental lithosphere?
(e) What is the genetic significance of Archaean and early Proterozoic SHMB to large layered complexes with chromite and PGE mineralization?

6.2 Brief description of sample occurrences

The presence of 2.7 Ga SHMB in Kambalda and nearby Mt Hunt in the Eastern Goldfield, Yilgarn Block of Western Australia has been recently documented (Redman & Keays 1985, Arndt & Jenner 1986). These rocks occur at the same stratigraphic position, towards the base of the sequences in the Norseman–Wiluna belt, and are metamorphosed to upper greenschist or lower amphibolite facies. The term spinifex-textured basalt was used by Sun & Nesbitt (1978a) to describe those SHMB with skeletal clinopyroxene ± olivines.

The Negri Volcanics (~2.8 Ga) occur in the Whim Creek belt, in the western part of the Pilbara Block, Western Australia. They unconformably overlie the ~3.0 Ga old Whim Creek Group sedimentary and felsic volcanic sequence (Barley 1987). The Negri Volcanics contain a sequence (200 m–2000 m) of SHMB and differentiates with an MgO range of 16% to 5%. Samples used in this study have been described in Sun & Nesbitt (1978a). They contain spinifex-textured clinopyroxene and skeletal olivine (now chlorite–serpentine) in a quenched, devitrified matrix.

A SHMB sample from the Cowarna Rocks comes from the 2.4 Ga Celebration Dyke system in the Eastern Goldfield (Hallberg 1987). This sample has been described by Purvis & Moeskops (1981) as a porphyritic orthopyroxene–clinopyroxene basalt with quench texture and devitrified groundmass. It contains magmatic sulphide droplets. Chemical data reported in Table 6.1 for this sample were supplied by W. Cameron (pers. comm. 1987).

Two samples of high-Mg basaltic dykes from the Rayner Glacier, Enderby Land and Vestfold Hills, and one sample of a tholeiitic dyke from Priestley Peak, Enderby Land (E Antarctica) used in this study came from the 2.4 Ga dyke system in that area (Sheraton & Black 1981). These dykes were emplaced at deep crustal levels (up to 8–10 kbar) after an amphibolite–granulite facies metamorphic event at 2.5 Ga. The petrology and geochemistry of these dykes have been described by Sheraton & Black (1981) and Kuehner (1989).

Two sediment samples from the Black Flag Beds (shale–greywacke) (Eastern Goldfields of the Yilgarn Block, Western Australia) have been analysed for Nd isotope composition. The Black Flag Beds overlie SHMB in the Kalgoorlie/Mt Hunt region. They have been described by Nance & Taylor (1977). Nd isotope data for these two samples can be combined with those for sediments underlying the Hangingwall Basalts of Kambalda (Arndt & Jenner 1986) and granites exposed in this region (McCulloch et al. 1983) to constrain the effect of contamination of komatiites by the upper crust.

6.3 Analytical methods and results

Major and trace elements other than REE were analysed by x-ray fluorescence (University of Adelaide) following the method described by Nesbitt et al. (1976). REE concentrations were determined by the isotope dilution method described by Sun & Nesbitt (1978a). Nd isotope compositions and Sm and Nd concentrations were determined at the Research School of Earth Sciences, Australian National University. Experimental procedures used are those described by McCulloch & Chappell (1982) with uncertainty in $^{147}Sm/^{144}Nd$ of $\pm\ 0.2\%$ and in ε_{Nd} of $\pm\ 0.4$ unit (2σ). During the period of data collection the $^{143}Nd/^{144}Nd$ value for BCR-1 was 0.511833 ± 10 $(n = 7)$ and for the Ndα it was 0.511101 ± 8 $(n = 8)$. Some discrepancy (up to 20%) in concentration of Sm and Nd (but not Sm/Nd) between isotope dilution data obtained at ANU and the University of Adelaide has been observed. The difference is most likely due to different methods of sample decomposition, rather than sample heterogeneity: the open beaker method was used at Adelaide whereas Teflon digestion bombs were used at ANU to ensure complete equilibration between the decomposed sample and spikes. The REE data reported in Table 6.1 have been normalized to be consistent with the Nd and Sm concentrations as determined at ANU.

Major- and trace-element data for samples analysed in this study are presented in Table 6.1. $^{143}Nd/^{144}Nd$ isotope composition, ε_{Nd} and Sm and Nd concentration data are presented in Table 6.2. Since alkali and alkaline-earth elements (K, Rb, Ba, Sr, Ca) can be mobile during alteration and metamorphism, they are not treated in detail in the following discussion. However, internal consistency of CaO and Sr abundances in SHMB within each suite of rocks is usually observed. These data are presented in Table 6.3 along with data for immobile elements (Ti, Zr, REE), and are used to support an AFC model involving contamination by felsic crustal materials depleted in CaO and Sr.

Table 6.1 Major- and trace-element data of some late Archaean and early Proterozoic SHMB from Western Australia and eastern Antarctica.

	Western Australia								Eastern Antarctica		
	Negri Volcanics, Pilbara				Kambalda	Mt Hunt		Cowarna Rocks	Rayner Glacier	Vestfold Hills	Priestley Peak
	331/337	331/338	331/339	331/646	331/477	331/626	331/628	4329	77284758	65280007	77284828
SiO_2	51.48	55.39	54.73	56.76	53.08	52.95	51.66	54.67	55.0	55.7	50.80
TiO_2	0.39	0.43	0.39	1.01	0.63	0.63	0.59	0.55	0.62	0.67	0.93
Al_2O_3	9.93	11.69	11.92	12.08	12.88	12.92	11.94	11.18	14.09	11.07	13.53
Fe_2O_3	10.63	10.26	10.63	11.59	10.95	11.22	11.49	9.60	10.43	11.36	12.80
MnO	0.23	0.22	0.19	0.18	0.18	0.20	0.17	0.14	0.17	0.16	0.20
MgO	16.50	12.26	11.71	6.06	9.54	10.56	13.22	11.04	7.60	10.67	7.54
CaO	9.06	7.86	7.51	7.44	10.67	9.10	7.91	7.63	8.68	7.05	11.83
Na_2O	0.98	1.92	2.72	3.65	1.88	2.11	1.73	2.12	1.69	1.48	2.2
K_2O	0.34	0.48	0.07	0.98	0.36	0.41	0.94	0.76	1.14	1.14	0.40
P_2O_5	0.04	0.04	0.04	0.13	0.06	0.08	0.07	0.07	0.09	0.09	0.11
LOI	4.06	1.76	2.47	1.20	1.52	1.95	2.65	2.4	–	–	–
Rb	8	13	4	30	7.5	13	41	23	57	56	11
Sr	30	111	140	252	149	118	92	154	109	111	186
Ba	135	222	46	505	155	134	145	235	237	253	165
Ti	2340	2580	2340	6120	3696	3860	3480	3300	3720	4020	5580
Nb	2.8	3.0	3.2	7.0	3.5	2.8	3.5	3.0	5	6	3
Zr	41	49	50	130	63	63	57	74	93	106	60
Y	11	14	13	20	17	16	15	12	17	18	21
Sc	28	33	32	27	39	43	39	27	–	34	–
V	147	169	177	196	200	208	193	157	197	182	250
Cr	2535	1287	1474	561	836	910	1460	1150	462	993	84
Ni	471	172	191	198	177	211	341	286	139	236	51
Co	70	50	58	74	(58)	68	72	–	–	–	–
La	7.12	8.68	8.28	21.4	7.86	9.66	8.34	8	18	18	9
Ce	15.1	17.5	15.9	41.3	17.0	19.6	16.6	20	30	33	16
Nd	6.02	6.43	6.50	18.3	8.32	8.49	8.20	9.84	13.86	15.05	10.39
Sm	1.44	1.50	1.52	3.98	2.02	2.04	1.83	2.16	2.92	3.22	2.73
Eu	0.392	0.474	0.416	1.17	0.662	0.663	0.507	–	–	–	–
Gd	1.65	1.65	1.66	3.93	–	2.26	2.08	–	–	–	–
Dy	1.93	1.90	1.81	3.49	2.64	2.45	2.26	–	–	–	–
Er	1.24	1.24	1.15	1.96	1.69	1.55	1.44	–	–	–	–
Yb	1.24	1.20	1.10	1.75	1.66	1.51	1.41	–	–	–	–
Lu	–	0.189	–	0.257	0.248	0.233	–	–	–	–	–

Table 6.2 Nd isotope and Sm, Nd concentration data of some Archaean and early Proterozoic SHMB, two late Archaean sediments from Western Australia and three early Proterozoic dykes from eastern Antarctica.

	Sample no.	$^{143}Nd/^{144}Nd$ ($\pm 2\sigma$)	Sm (ppm)	Nd (ppm)	$^{147}Sm/^{144}Nd$	Initial ε_{Nd}
Mt Hunt	331/626	0.510808 ± 16	2.043	8.49	0.1456	−2.5
Negri Volcanics	331/338	0.510644 ± 16	1.503	6.43	0.1414	−3.2
Negri Volcanics	331/339	0.510650 ± 16	1.524	6.50	0.1417	−3.3
Negri Volcanics	331/646	0.510470 ± 14	3.978	18.25	0.1318	−3.2
Cowarna Rocks	4329	0.510714 ± 20	2.156	9.84	0.1326	−2.3
Rayner Glacier	4758	0.510656 ± 14	2.923	13.86	0.1276	−2.2
Vestfold Hills	8007	0.510695 ± 16	3.216	15.05	0.1293	−1.9
Priestley Peak	828	0.511273 ± 12	2.731	10.39	0.1589	+0.7
Black Flag	KH44	0.510462 ± 14	3.361	16.56	0.1227	−1.3
Black Flag	85-4	0.510919 ± 14	5.485	23.29	0.1425	+0.8

6.4 Geochemical distinction between Archaean and early Proterozoic SHMB and modern boninites

Although both boninite and Archaean–early Proterozoic SHMB (Fig. 6.1) are siliceous, high-Mg and have low abundances of the HFSE, Ti, Nb, Ta, Zr and P, there are major differences. For example, boninite magmas contain about 2% primary magmatic water (e.g. Cameron *et al.* 1983) whereas SHMB are basically dry and contain only minor hydrous minerals (e.g. Redman & Keays 1985). Total FeO content in boninites (with 10–12% MgO) is 6–7% (reflecting a refractory source character) in contrast to about 10% for SHMB. Other differences include the U-shaped REE patterns that are commonly observed in boninites (e.g. Sun & Nesbitt 1978b, Hickey & Frey 1982, Cameron *et al.* 1983), whereas SHMB have either very flat to slightly sloping HREE patterns, or less commonly show continuous depletion from MREE to HREE. Cattell (1987) showed that in the Abitibi Belt (Canada) SHMB with the latter pattern are associated with Barberton-type komatiites which also have HREE depletion. An exception is one 'Type 3' komatiite (No. 5092) from Barberton (South Africa) which has a U-shaped REE pattern (Jahn *et al.* 1982).

Sun & Nesbitt (1978b), Hickey & Frey (1982) and Cameron *et al.* (1983) attributed the U-shaped REE patterns of boninites, along with other chemical characteristics, to be a product of mixing of two components in a subduction-zone environment. Component A is the refractory harzburgitic mantle located above the subducted slab, and is considered to be residual following previous melting event(s). It is generally severely LREE-depleted with normalized REE abundances decreasing continuously from Lu. Component B is a water-rich fluid or melt derived from the subducted oceanic lithosphere (MORB ± seamounts ± sediment).

Boninites share certain chemical and isotope similarities with island-arc basalts. This interpretation has been supported by more recent chemical and isotope studies of boninites (e.g. Jenner 1981, McCulloch & Cameron 1983). Hickey & Frey (1982) also pointed out that, if amphibole or sphene is stable in the subducted slab, the MREE could be retained in the residue relative to Zr, causing a Zr spike relative to Sm (Fig. 6.1a). Nb (and Ta) depletion relative to La is a general feature of boninites except for Cyprus boninites with U-shaped REE patterns (Fig. 6.1a; Cameron *et al.* 1983). The latter case might be a result of an enrichment process involving intraplate melt migration; alternatively, our preferred view is that there was a lack of a suitable mineral phase (sphene or rutile) to hold back Nb and Ta during the second-stage melt addition process, owing to the high temperature ($\geqslant 1000°C?$) postulated to exist in the subduction zone during boninite genesis (Crawford *et al.* 1981). Sphene and rutile are not stable under such conditions (e.g. Ryerson & Watson 1987).

Although there are some similarities in geochemistry of boninites and Archaean SHMB (Fig. 6.1), the contrast is clear from Gd to Al. SHMB show a continuous decrease from Gd to Lu and a U-shaped pattern from Lu to Al, whereas the majority of boninites show two different types of patterns. One type, represented by Cyprus sample No. 9 of Cameron *et al.* (1983), has a continuous increase from Gd to Al, and the second type, represented by more LREE-enriched Cape Vogel sample No. 3 of Cameron *et al.* (1983), has a minimum at the HREE (Fig. 6.1a). These fundamentally different patterns lead to quite different Al_2O_3/TiO_2 (mostly 30–60 versus 20–27) and Sc/Y (5–8 versus 2–3) for boninites and Archaean SHMB. In comparison, these ratios

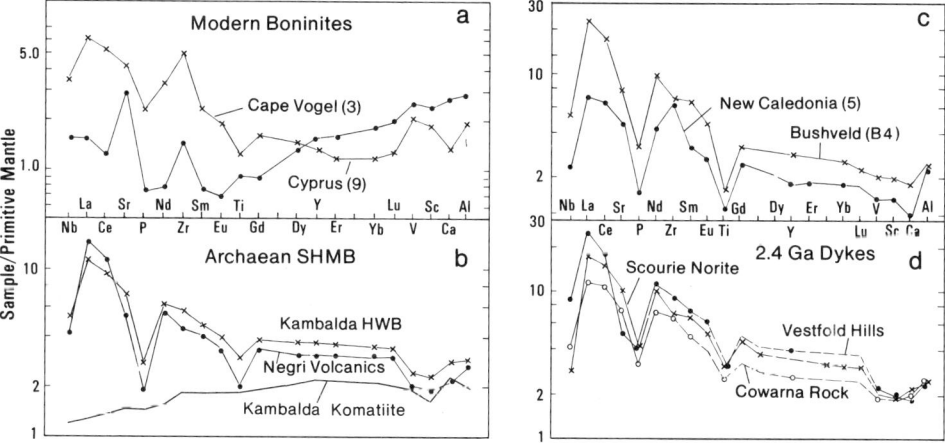

Figure 6.1 Comparison of trace-element abundance patterns of Archaean and early Proterozoic SHMB with modern boninites; data sources given in text. Sr concentration in the New Caledonian sample (no. 5) (c) was inferred from other fresh New Caledonian boninites (Cameron 1989). A pattern for a Kambalda komatiite is also shown along with the Archaean SHMB. Early Precambrian SHMB have similar patterns and show a relative depletion of Nb, P and Ti, suggesting a connection with felsic crust. Consistent but different magnitudes of Sr depletion are observed in samples from a number of areas, suggesting these are primary magmatic rather than alteration features. All data have been normalized to values of an estimated primitive mantle composition (Sun & McDonough 1988).

for the primitive mantle are 21 and 3.8, respectively. An exception to this generalization is the boninite samples from New Caledonia (Cameron 1989), which have basic patterns similar to Cape Vogel sample No. 3 but which show lower V, Sc and Ca (but not Al) normalized abundances than the HREE (see Fig. 6.1c). It is very likely that element abundances in the New Caledonian sample reflect the added second component, which is enriched in LREE but poor in V, Sc and Ca.

Boninites similar to Cape Vogel sample No. 3 have comparable LREE abundances to Archaean SHMB and are also depleted in P and Ti. However, such boninites show a clear enrichment of Zr relative to Sm and Nd and have Sr/Nd greater than chondrites (e.g. Hickey & Frey 1982). In contrast, Archaean SHMB do not show Zr spikes and always have Sr/Nd lower than chondrite (Figs 6.1a & b). As a result, Archaean SHMB have significantly higher Ti/Zr (50–75) than boninites (commonly 20–40). Archaean SHMB have Ti/Sc and Ti/V equal to or greater than primitive mantle values (74 and 15.5, respectively) whereas the ratios for modern boninites are commonly only about half of these values. Furthermore, boninites commonly have very high Al_2O_3/TiO_2 and CaO/TiO_2 (30 to 60 and occasionally more) reflecting the refractory nature of their source (e.g. Sun & Nesbitt 1978b), whereas Archaean SHMB (Fig. 6.1b) have ratios only slightly higher than primitive mantle ratios (about 21 and 18, respectively).

The only Archaean sample we know of that has a trace-element pattern similar to modern boninite is a komatiite (No. 5092 with Nb $\leq$ 2 ppm) from Barberton reported by Jahn et al. (1982). If the REE pattern is not a result of alteration, this sample may represent mixing of two components similar to that found in modern boninites, but not necessarily having the same tectonic connotation.

Figure 6.1d shows the element abundance patterns of early Proterozoic (2.4 Ga) SHMB dykes. They have patterns generally similar to the Archaean SHMB (Fig. 6.1b) but quite distinctive from boninites (Fig. 6.1a). On the basis of the above discussion, it is concluded that the chemical similarity between modern boninites and SHMB of Archaean and early Proterozoic age is only superficial, and they are not genetically related.

6.5 SHMB: komatiitic parent magmas and the role of crustal contamination

6.5.1 *Evidence supporting AFC processes*

The idea that Archaean SHMB were derived from komatiitic magma through assimilation–fractional crystallization (AFC) processes is attractive for the following reasons:

(a) Komatiites and Archaean SHMB often occur together in a continuous stratigraphic sequence. Al-undepleted (i.e. Al/Ti ~ chondritic) komatiites with flat HREE patterns are associated with SHMB having flat HREE patterns. Likewise, SHMB associated with Al-depleted and HREE-depleted komatiites also have depletion in HREE (e.g. Cattell 1987).
(b) On the basis of thermal and fluid-dynamic considerations, it is highly likely that high-temperature komatiites (liquidus temperature of up to $1650°C$ at 1 atm) will be subject to extensive (up to 50%) crustal contamination during dyke injection and differentiation in a magma chamber (e.g. Campbell 1985, Huppert & Sparks 1985).
(c) Old zircon (up to 3.45 Ga) xenocrysts are present in the 2.7 Ga old SHMB from the Hangingwall Basalts in Kambalda, suggesting sialic contamination (Compston et al. 1986).
(d) Model calculations of AFC using komatiite as starting material and felsic crust as contaminant can in many cases generate the observed element abundance patterns and match the Nd isotope data of the Kambalda Archaean SHMB (Chauvel et al. 1985, Arndt & Jenner 1986).

In this section we will present further chemical and isotopic data for Archaean and early Proterozoic SHMB to support the AFC model. However, it is first necessary to point out some complications involved in this process.

6.5.2 Complications involved in AFC processes

On the basis of thermal budget, AFC processes with komatiite starting material could assimilate more crustal material in the early stages of differentiation at greater depths, when the temperature of the melt and wall rocks are high; at a later stage (lower temperature and high crustal level) crystal fractionation in layered flows and sills accompanied by limited crustal assimilation may be the rule. Fluid-dynamic modelling suggests that the ascending komatiitic magma is likely to become contaminated by continental crust as the magma flows turbulently through the crust in a dyke. But this does not lead to the AFC processes, because if crystallization does accompany contamination, there is no method of separating crystals from melt while the melt is rising through the crust (Campbell 1985, pers. comm. 1988). Campbell (1985) further suggested that crystal fractionation by growth on dyke walls becomes possible when the process changes from assimilation of the walls to crystallization at the walls. It may occur at a lower level than the change-over from turbulent to laminar flow. The AFC processes require magma to pond in a crustal magma chamber where melting of roof rocks can accompany crystal fractionation in the magma chamber. Consequently, although the two processes (assimilation and crystal fractionation) are thermally linked, we are dealing with two distinct processes: assimilation occurs initially, followed by fractional crystallization accompanied by minor assimilation. Potential variables involved in the AFC process include: the composition of the parent magma and the contaminant (which could be a partial melt instead of wall rock); changing mass ratio (r) between the contaminant and crystal fractionate (very large at the early stage then diminishes); extent of crystal fractionation involving combination of liquidus minerals at different stages; the possibility of open system magma chamber processes; and variable mineral/melt distribution coefficient (D) for a specific element. Quality of the analytical data is an additional variable. At present, a unique numerical solution based on geochemical data cannot be determined. Nevertheless, chemical and isotopic data of SHMB can be used to test possible solutions.

6.5.3 Enrichment/depletion of elements associated with the AFC processes

Owing to the complications involved in the AFC processes mentioned above, two different approaches are taken in the following treatment to evaluate the changes of elemental concentration in the fractionated magma. The first approach assumes that contamination occurs early, involving assimilation of χ per cent of crustal material by the komatiite parent magma followed by simple crystal fractionation; the second approach follows the conventional AFC calculation (DePaolo 1981) assuming that the mass ratio (r) between the contaminant and crystal fractionate is constant during the AFC process.

The equations governing the two approaches are:

$$C_m = [C_m^\circ (1 - \chi) + C_a \chi] F^{D-1} \tag{6.1}$$

and

$$C_m = C_m^\circ \{F^{-z} + C_a [r/(r-1)](1 - F^{-z})/z C_m^\circ\} \tag{6.2}$$

where

$$z = (r + D - 1)/(r - 1)$$

and C_m°, C_m and C_a are concentrations of an element in the parent komatiite, fractionated magma and the contaminant, respectively; F is (mass of fractionated magma)/(mass of initial magma). For Equation 6.1, the initial magma constitutes a mixture of the komatiite and contaminant. D is the partition coefficient of the specific element between the fractionated minerals and the coexisting melt. When $D \ll 1$ (e.g. for Sc, Y, Ti, LREE, etc., when olivine is the fractionating mineral) and $C_a/C_m^\circ < 1$ (e.g. Sc $\leqslant$ 10 ppm in felsic contaminant as compared to 27 ppm in the komatiite), AFC processes will result in C_m lower than would be produced by simple crystal fractionation. However, when $C_a > C_m$ (e.g. Y ~ 20 ppm in the felsic contaminant as compared to 10 ppm in the komatiite), the situation is reversed. These points are illustrated in Fig. 6.1b and Table 6.3, in which abundances of some selected elements in

Table 6.3 Enrichment factors of some trace elements in Kambalda 331/477 relative to an idealized komatiite with 24% MgO. Data of other samples except for the estimated composition of the contaminant were normalized by 331/477 to show that they have general consistent trends.

	Kambalda komatiite (ppm or %)	Kambalda (331/477) / Komatiite	Estimated contaminant for 331/477	Kambalda			Eastern Goldfield, W Australia					Negri Volcanics, Pilbara		Vestfold Hills, E Antarctica 65280007	Bushveld B4
				HWB3	HWB10	21	Mt Hunt 626+628	Hays Hill 23	Kilgoorlie 25	Ravensthorpe 28	Cowarna 4329	338	646		
MgO	9.5	24	—	15.5	7.0	9.8	12	8.8	7.7	10.0	11	12.3	6.1	10.7	8.9
Sc	27	1.30	(≤10)	0.91	1.17	—	1.05	1.02	0.91	0.95	0.77	0.85	0.69	0.97	—
V	160	1.25	(≤100)	1.00	1.15	0.99	1.00	1.04	0.97	1.04	0.79	0.85	0.98	0.91	0.84
CaO	8.6	1.23	(4?)	0.82	1.02	0.97	0.80	0.78	0.81	1.05	0.72	0.74	0.70	0.66	0.63
TiO_2	0.38	1.66	(0.7)	0.87	1.14	0.94	1.00	1.13	1.08	0.95	0.87	0.68	1.60	1.06	0.54
Al_2O_3	8.6	1.51	(16?)	0.85	1.16	0.96	0.97	1.01	1.09	0.95	0.87	0.91	0.94	0.86	0.89
Yb	0.99	1.68	1.9	0.85	1.12	1.03	0.88	—	—	0.95	—	0.72	1.05	—	0.69
Y	10	1.70	20	0.88	1.23	1.06	0.94	—	1.17	0.95	0.70	0.82	1.18	0.87	0.76
P_2O_5	0.03	2.0	0.10	0.7	1.3	1.0	1.3	1.2	1.3	1.0	1.2	0.7	2.2	1.5	1.2
Sm	0.78	2.59	4.7	0.83	1.25	1.11	0.96	—	—	0.76	1.07	0.74	1.97	1.59	1.31
Zr	20	3.15	172	0.90	1.23	1.00	0.95	1.24	1.03	0.67	1.17	0.78	2.06	1.68	1.22
Nb	0.8	4.4	11	0.9	1.3	1.0	0.9	—	—	—	0.9	0.9	2.0	1.7	1.0
Sr	30	5.0	510	(0.36)	0.79	1.53	0.70	0.93	1.00	0.57	1.03	0.74	1.69	0.74	1.06
Nd	1.98	4.2	27	0.88	1.27	1.12	1.00	—	—	0.57	1.18	0.77	2.19	1.81	1.44
La	0.82	9.6	30	0.96	—	1.13	1.15	—	—	0.39	1.18	1.10	2.72	2.29	1.91

Data sources: Arndt & Jenner (1986), Redman & Keays (1985), Purvis & Moeskops (1981), Sheraton & Black (1981), Cameron (pers. comm. 1988) and this study.

Kambalda komatiites (with 24% MgO) are given, based on our own data and data from Arndt & Jenner (1986). Table 6.3 also includes element abundances of a SHMB from Kambalda Hangingwall Basalts HWB 331/477 (Table 6.1) normalized to abundances in the komatiite to show the enrichment factor for each element. This sample appears to be compositionally representative of Kambalda SHMB. Additional data for other SHMB (Table 6.1) and data from the literature (Redman & Keays 1985, Arndt & Jenner 1986) are also represented. They have been normalized to HWB 331/477 to emphasize the strong similarity among them. Elements in Table 6.3 are arranged in order of increasing enrichment factors for sample 331/477. This order reflects a combination of partitioning of an element between fractionating mineral and liquid and the character of the contaminant, which is most likely the felsic crust. Normalized patterns for SHMB from the Eastern Goldfield relative to 331/477 show generally consistent (except for Ca and Sr) values due to different extents of crystal fractionation and minor increases (more contaminant) or decreases (less contaminant) towards La, except for the sample from Ravensthorpe, which obviously is much less contaminated. An inconsistency in Sc and V normalized values may reflect laboratory differences. It is possible that Sc data determined at the University of Adelaide are about 10% too high relative to other laboratories. To make them consistent with data of Redman & Keays (1985) and Arndt & Jenner (1986), a Sc value of 35 ppm (instead of 39 ppm in Table 6.1) for 331/477 will be used in Table 6.3 and later calculations.

Variation of Ca and Sr in other SHMB relative to the Kambalda sample 331/477 could be due either to low-temperature post-eruption mobility of these elements (Arndt & Jenner 1986) or to different abundances of Ca and Sr in their contaminants. In connection with the latter factor, it is important to bear in mind that a more fractionated felsic contaminant would be expected to have lower Ca, Sr, Ti, Nb and P abundances. Element systematics and large negative Eu anomalies (20%) of Negri Volcanics (Tables 6.1 & 6.3) are consistent with a contaminant of more felsic composition than that inferred for the Kambalda SHMB, since they have more depletion of Ca, Sr, P, Nb and Ti on the normalized plot (Fig. 6.1b).

6.5.4 Estimation of the extent of crystal fractionation for Kambalda SHMB 331/477

As outlined earlier, the enrichment factor of a specific element in an AFC-affected fractionated magma depends on the extent of fractional crystallization, the mass ratio of the contaminant to crystals fractionated, the K_D of that element between the bulk fractionate and melt, and the element abundance ratio of the contaminant relative to the starting material. To generate SHMB from komatiite, the most appropriate contaminant is likely to be

Archaean felsic igneous rocks and their sedimentary derivatives, which have low Sc (< 10 ppm), V ($\leqslant 100$ ppm) and CaO ($\leqslant 4\%$) contents relative to komatiites. Archaean tonalites and granodiorites have higher TiO_2 ($\geqslant 0.4\%$), P_2O_5 ($\geqslant 0.06\%$) and generally higher Y (> 10 ppm) than komatiites. Adamellites and granites have lower TiO_2 ($< 0.4\%$) and P_2O_5 ($\leqslant 0.03\%$). A komatiite undergoing AFC processes will initially crystallize olivine, followed in more fractionated derivatives by clinopyroxene, low-Ca pyroxene and plagioclase (e.g. Redman & Keays 1985). Since the K_D values for Sc, V, Y and Yb between an olivine-dominated bulk fractionate and a high-temperature mafic–ultramafic melt are small ($\leqslant 0.1$; Sun et al. 1979, Nielsen 1988), the extent of crystal fractionation can be constrained by the enrichment factors of Sc (1.30) and Y (1.70), which correspond to 23% (Sc) to 41% (Y) fractional crystallization. More accurate estimates may be obtained when the effect of contamination is taken into account. For the first approach (i.e. early crustal contamination followed by dominantly olivine fractionation (Eqn 6.1)), if we assume $D = 0$ and the contaminant is an Archaean granodiorite (with Sc $\leqslant 10$, Y = 10–20, Zr = 150–190 and La = 30 ppm), with a komatiite with 24% MgO (Table 6.3) as the starting material and Kambalda Hangingwall Basalts 331/477 (Table 6.1) as the end-product, we obtain $F \sim 0.68$ (i.e. 32% of crystal fractionation) and $x = 15\%$ contamination. For $D \ll 1$ but $D \neq 0$, the F value should be slightly lower than 0.68. The weight ratio (r) between the contaminant and crystal fractionate is about 0.50. Results of this approach also match major-element data reasonably well.

For the second approach, with a constant r factor, when $D = 0$, and $C_a/C_m^\circ \ll 1$ (e.g. for Sc), Equation 6.2 becomes $C_m = C_m^\circ/F$, where $F = 1 - C + A$, and C and A are weight percentages of crystal fractionate and assimilate respectively. With $r = A/C = 0.4$ (following Arndt & Jenner 1986) and Sc = 35 ppm for Kambalda sample 331/477 and 27 ppm for the komatiite parent, we obtain $F = 27/35 = 1 - C + 0.40C$. A value of 0.38 is obtained for crystal fractionation; this is consistent with the estimate obtained using the Zr versus Mg# plot in Fig. 9 of Arndt & Jenner (1986).

6.5.5 Estimation of the chemical and isotopic composition of the contaminant for SHMB 331/477

Using Equations 6.1 and 6.2, the enrichment factor C_m/C_m° for each element in Table 6.3 can be used along with $F = 0.68$ and $x = 0.15$ (Eqn 6.1) and $F = 0.77$ and $r = 0.4$ (Eqn 6.2) to estimate the concentration of each element in the contaminant. Some change of r value (e.g. 0.3 to 0.5) will not significantly affect the result. Estimates are more accurate for elements with high enrichment factors. Owing to the effect of clinopyroxene and plagioclase, CaO and Al_2O_3 cannot be estimated. The estimated composition of the contaminant using Equation 6.1 is listed in Table 6.3. The results obtained using Equation 6.2 are basically the same as those obtained using Equation 6.1. Although this estimated contaminant composition is not entirely unique, its element abundances and ratios (e.g. La/Sm, La/Nb, Sr/Nd) are all consistent with data for Archaean felsic igneous rocks (tonalite to granodiorite).

The ε_{Nd} value of the contaminant for Kambalda SHMB with $\varepsilon_{Nd} = -2.0$ (Chauvel *et al.* 1985) can be estimated by mass-balance calculation. If the parental komatiite magma had $\varepsilon_{Nd} = +4$ (Chauvel *et al.* 1985) a value of $\varepsilon_{Nd} = -4.5$ is obtained for the contaminant. This value agrees well with $\varepsilon_{Nd} = -4$ suggested by Chauvel *et al.* (1985). Data for Mt Hunt (Tables 6.2 & 6.3) give the same result, whereas $\varepsilon_{Nd} = -6.4$ is calculated for the contaminant of the Negri Volcanics. These ε_{Nd} values suggest that the contaminants are early Archaean felsic rocks. This conclusion is consistent with ion probe age dating of the xenocryst zircons from the Eastern Goldfield (Compston *et al.* 1986) and the geology of the western Pilbara Block.

It is important to note that, among both the Kambalda Upper Hangingwall Basalts (Arndt & Jenner 1986) and Negri Volcanics (this study) with variable MgO content (24–6% and 16–6%, respectively), there is no change with increasing fractionation in their calculated initial ε_{Nd} values (-2.0 and -3.2 respectively) and there is only a minor change in their La/Sm (3.9–4.4 and 4.9–5.4 respectively). These data support the idea that assimilation took place initially in the deep crust and simple crystal fractionation (and olivine accumulation for Kambalda samples with very high MgO) took over at lower temperature and shallow depth to produce layered flows/sills. This interpretation is consistent with ε_{Nd} data of Archaean sediments from the Eastern Goldfields. Archaean granite and sediments both below and above SHMB from the Eastern Goldfields have ε_{Nd} values close to zero at 2.7 Ga (Arndt & Jenner 1986, McCulloch *et al.* 1983, Table 6.2), and are therefore not suitable contaminants to generate $\varepsilon_{Nd} = -2.0$ observed in SHMB. Further support comes from data for PGE and Cu. Redman & Keays (1985) emphasized that Archaean SHMB from the Eastern Goldfields are sulphide-undersaturated and have the same PGE (Pd, Ir) abundances as sulphide-undersaturated high-Mg basalts unambiguously derived from fractionation of komatiites. Thus PGE data require that the contaminant be sulphide-poor (e.g. felsic igneous rocks), but not near-surface sulphide-rich sediments.

A genetic connection of Eastern Goldfield SHMB to komatiites is further supported by the fact that the enrichment factors of Pd and Cu (both incompatible elements in the absence of sulphide) in SHMB are about 10% lower than Y and Ti (Table 6.4). This dilution effect is caused by the felsic contaminants (tonalite–granodiorite) having lower Pd and Cu but similar Y and Ti concentrations as the komatiites.

ε_{Nd} values similar to those of the Kambalda SHMB have been reported for late Archaean and early Proterozoic SHMB dykes in Eastern Goldfields, W Australia: $\varepsilon_{Nd} = -2.5$ for Mt Hunt, -2.3 for Cowarna Rocks (Table 6.2), -1.1 ± 0.3 for the 2.7 Ga Ora Banda sill (Chauvel *et al.* 1985) and -2.1 ± 0.6 for 2.4 Ga Jimberlana dyke (Fletcher *et al.* 1987). Furthermore,

Table 6.4 Correlation of Pd, Cu, Ti and Y abundances in the primitive mantle, average modern boninites, Archaean komatiites and SHMB from the Eastern Goldfield, Western Australia. Data for average Bushveld SHMB type B4 dykes and estimates for parent magma of the Stillwater Complex based on sulphide-undersaturated orthocumulates below the JM Reef (Hamlyn pers. comm. 1988) are also shown.

	MgO (%)	Pd (ppb)	Cu (ppm)	Ti (ppm)	Y (ppm)	Pd/Y ($\times 10^{-3}$)	Pd/Cu ($\times 10^{-3}$)	Cu/Y	Ti/Y
primitive mantle	38	4	28	1330	4.6	0.9	0.14	6.1	283
komatiites	24	9	55	2280	10	0.9	0.16	5.5	228
Kambalda SHMB (7)	10.3	15.6	94	3660	18	0.9	0.17	5.2	203
Mt Hunt SHMB (4)	12.3	13.1	70	3480	16	0.8	0.19	4.4	217
boninites	12	15	20	1200	6	2.5	0.75	3.7	200
Bushveld SHMB	12	12	56	2160	13	0.9	0.21	4.3	166
Stillwater[a]	–	16	49	–	–	–	0.33	–	–

Data sources: Keays (1982), Redman & Keays (1985), Sun (1982), Hamlyn *et al.* (1985), Ross & Keays (1979), Sharpe & Hulbert (1985), Davies & Tredoux (1985), Brügmann *et al.* (1987) and this study.

[a]Stillwater parent magma has Pd/S = 9.5×10^{-5} as compared to about 30×10^{-5} for average boninites and $(2-4) \times 10^{-5}$ for komatiite magma (Hamlyn pers. comm. 1988).

two SHMB dykes (2.4 Ga) from Rayner Glacier and Vestfold Hills, eastern Antarctica have $\varepsilon_{Nd} = -2.2$ and -1.9 respectively (Table 6.2). These initial ε_{Nd} values are similar and all negative, in contrast to estimated positive ε_{Nd} values ($\sim +4$) for the depleted convecting mantle in the late Archaean and early Proterozoic. In terms of the AFC model, these Nd isotope data suggest that similar quantities of LREE-enriched felsic rocks of early Archaean age are the most common contaminant. In contrast, interpretation of ε_{Nd} values from other early Archaean SHMB samples (e.g. Barberton komatiite samples 5067 and 5088 of Jahn *et al.* (1982), and tholeiites of North Star Basalts, Pilbara Block, of Gruau *et al.* (1987)) remains conjectural due to a lack of reliable age determinations.

6.5.6 Komatiite as starting material for AFC?

It is emphasized that, unless there is strong geological and petrological evidence for extensive crustal contamination, the AFC process discussed above is not a unique solution for generation of SHMB with element abundance patterns of felsic rocks. This character could also be produced via melting of the continental lithospheric mantle which has been modified by subduction processes. Furthermore, it is quite possible that, after filtering away the effect of crustal contamination on SHMB, the restored composition of the parent magma may not be komatiite.

Weaver & Tarney (1981) and Tarney & Weaver (1987) emphasized that crustal contamination cannot explain the geochemical and isotopic character of 2.4 Ga Scourie dykes of NW Scotland. For example, La/Nb of the noritic dykes (4–6) are much higher, and those of olivine gabbro dykes (~1) are much lower than Archaean felsic rocks (~2). Furthermore, the olivine gabbros are rich in TiO_2 (2.0% at 12% MgO). Weaver & Tarney (1981) suggested that bronzite-bearing picrite and norite dykes were derived from refractory harzburgitic continental lithosphere which was modified by subduction processes whereas quartz tholeiite and olivine gabbro dykes were derived from more fertile mantle sources.

Involvement of subduction-modified refractory harzburgitic continental lithosphere in generation of the SHMB parent magma may also explain why the Negri Volcanics have considerably lower Ti, Sc and V contents (<2400, 32 and 170 ppm respectively at 12% MgO) than Archaean SHMB of the Eastern Goldfields. An alternative (but less likely?) explanation is that the komatiite parents for the Negri Volcanics had lower contents of these elements (~1800, 24 and 120 ppm respectively) than komatiite parents for the Hangingwall Basalts of Kambalda (~2400, 27 and 160 ppm respectively at 24% MgO). Many early Proterozoic SHMB (Table 6.3 & Fig. 6.1), for example, Cowarna Rocks, Scourie noritic dykes, eastern Antarctic and southern West Greenland (Hall & Hughes 1987) noritic dykes and especially Bushveld B4 dykes, have similar character to the Negri Volcanics (i.e. low CaO, Ti, Nb, Sc and T). The same general genetic model is also applicable to these occurrences.

6.5.7 Low initial $^{87}Sr/^{86}Sr$ and low $\delta^{18}O$: inconsistent with crustal contamination?

Late Archaean and early Proterozoic layered complexes and dykes with SHMB affinities commonly have initial $^{87}Sr/^{86}Sr$ around 0.7020. Depleted convecting mantle at 2.4–2.7 Ga had $^{87}Sr/^{86}Sr = 0.7006$–0.7010 (e.g. Jahn et al. 1982). It has been argued by some authors that such low initial $^{87}Sr/^{86}Sr$ (0.7020) is not consistent with extensive contamination by an early Archaean felsic crust. On the other hand, Weaver & Tarney (1981) pointed out that the 2.4 Ga Scourie dykes have initial $^{87}Sr/^{86}Sr$ (0.7022 ± 1) too high to have been contaminated by underlying Lewisian granulites with very low initial $^{87}Sr/^{86}Sr$.

To evaluate the validity of these arguments, a mass-balance calculation for the Kambalda SHMB using a hypothetical $^{87}Sr/^{86}Sr = 0.7007$ for the komatiite magma, an estimated $^{87}Rb/^{86}Sr = 0.19$ (Rb = 35 ppm, Sr = 510 ppm) and a $^{87}Sr/^{86}Sr = 0.7027$ for the crustal contaminant at the time of contamination (2.7 Ga) gives an initial $^{87}Sr/^{86}Sr = 0.7020$ for the SHMB. Interestingly, the reported initial $^{87}Sr/^{86}Sr$ (Roddick 1974) for Kambalda SHMB (Hangingwall Basalts) is 0.70129 ± 15, considerably lower than the modelled value (0.7020), but similar to the initial $^{87}Sr/^{86}Sr$ (0.70130 ± 15) of the Footwall Basalt, which, based on Nd isotope data, is only slightly contaminated by crustal material (Chauvel et al. 1985). Since these basalts were erupted below sea water in a submerged continental rift, it is quite possible that the low initial

$^{87}Sr/^{86}Sr$ for the Hangingwall Basalts was a result of reaction with sea water. It is well established (Veizer 1985) that the Sr isotope ratio of Archaean sea water was buffered by the mantle values (0.7006–0.7010).

On the basis of an integrated Pb, Sr, Nd and Ce isotopic study (and also some experimental petrology investigations) of the effect of crustal contamination on early Tertiary basalts erupted in NW Scotland, Dickin et al. (1987) concluded that these basalts were contaminated by the low-temperature melting fraction of granitic sheets in the Lewisian granulites. According to Morrison et al. (1985), present-day $^{87}Sr/^{86}Sr$ values of the Lewisian granulites range from 0.702 to 0.708. It is therefore possible that high-temperature Scourie norite dykes may also have been contaminated by a low-temperature melt fraction at 2.4 Ga ago, which raised their initial $^{87}Sr/^{86}Sr$ to 0.7022. However, this exercise only indicates that the AFC process is a possible, but not a unique, explanation (see also Section 6.5.6).

A low estimated $\delta^{18}O_{SMOW}$ value of $+6.2$ (with the NBS-28 quartz standard = $+9.6$) reported for the SHMB parent magma of the Stillwater Complex has been used as an argument against significant crustal contamination (McCallum 1987). The Stillwater Complex has low initial ε_{Nd} (-2), a high initial $^{87}Sr/^{86}Sr$ (~ 0.7020) and high μ character (~ 9.4) from the Pb isotope studies. These 'crustal' isotopic characteristics have been considered to be derived from partial melting of a harzburgitic or pyroxenitic mantle source which was modified by the addition of subducted sediments (McCallum 1987, Wooden & Mueller 1988). However, because some fresh olivines from Archaean komatiites appear to have $\delta^{18}O$ values (Kyser 1987) of about $+5.0$ (magmatic values?) and Archaean felsic volcanics and high-grade metamorphics commonly have low $\delta^{18}O = +6$ to $+9$, the low $\delta^{18}O$ values estimated for the Stillwater parent magma may not necessarily mean a lack of crustal contamination. Under favourable circumstances, it is therefore still possible to generate the SHMB parent magma of the Stillwater Complex via a large amount ($\geqslant 15\%$) of crustal contamination of a komatiitic magma with AFC processes (Longhi et al. 1983).

6.6 Continental lithosphere as a magma source for SHMB

In addition to the crustal contamination AFC process, there are other mechanisms capable of generating Archaean and early Proterozoic SHMB. These include the following:

(a) Subduction of sediments into the mantle followed by relatively large degrees of partial melting of the contaminated refractory mantle wedge at shallow depth ($\leqslant 12$ kbar?). This mechanism has been suggested to be responsible for generation of Mesozoic flood basalts, especially the

Tasmanian dolerites associated with splitting up of Gondwanaland (e.g. Hergt 1987, Sun & McDonough 1988).

(b) Intrusion or underplating of a high-temperature asthenospheric diapir into the refractory harzburgitic continental lithosphere. Melting of refractory lithospheric mantle (olivine + orthopyroxene) at low pressure will generate basaltic melt relatively enriched in silica (Jaques & Green 1980, Fisk 1986). Mixing of asthenosphere-derived and lithosphere-derived melts could generate chemical and isotope characteristics of SHMB, if the lithosphere was previously modified by incorporation of melts derived from subducted sediments. The modified lithosphere-derived melt would control the incompatible element contents and Pb, Sr and Nd isotopic composition of the combined melt. However, contamination following intrusion of this high-temperature magma into a crustal environment is highly probable and often supported by chemical and isotope data and geological observations.

It is well known that early Proterozoic mafic dykes and layered complexes commonly have bimodal, tholeiitic and noritic (high-Si, high-Mg) compositions (e.g. Longhi et al. 1983). The latter have lower abundances of Ti, Zr and HREE than tholeiites with similar MgO contents, but generally have much higher LREE, K, Rb, Th, U and Ba contents. Tholeiites generally have flat REE patterns similar to MORB, although they often show some limited crustal character in trace-element abundances and ratios. Our data for 2.4 Ga dykes from E Antarctica (Tables 6.1, 6.2 & 6.3) indicate that the noritic (SHMB) dykes have initial $\varepsilon_{Nd} = -2$ and chemical compositions similar to Archaean SHMB (see Fig. 6.1), whereas the tholeiite dyke from Priestley Peak has higher ε_{Nd} (+0.7) and therefore significantly less crustal component. In terms of the AFC model, the lower-temperature tholeiitic magma has lower capability of assimilation/fractional melting than the higher-temperature high-Mg noritic magma. However, in terms of the asthenosphere–lithosphere interaction model, the noritic magma may have been generated at higher temperature and involved extensive melting of modified lithosphere. This explanation is consistent with the model of Hatton & Sharpe (1989).

6.7 PGE mineralization in Archaean and early Proterozoic layered complexes

A boninite connection is often suggested for parent magmas of Archaean and early Proterozoic layered complexes containing PGE mineralization and having an affinity with SHMB (e.g. Sharpe & Hulbert 1985, Hamlyn et al. 1985). A close examination of the factors which may affect PGE mineralization, however, indicate that a boninite connection is a possible but not a necessary condition for PGE mineralization. These factors include: (i) heterogeneous distribution of PGE in the mantle sources; (ii) magma generation processes which concentrate some PGE (Pd, Pt, Re) and Au into the melt; (iii) concentration of PGE in the magma chamber (layered complex) by sulphide saturation and precipitation as thin layered cumulates; and (iv) possible later hydrothermal mobilization and enrichment (e.g. Stumpfl & Ballhaus 1986). A key factor for the development of PGE mineralization seems to be a strongly sulphide-undersaturated parent magma, thereby delaying sulphide saturation in the magma chamber (Keays 1982). Final sulphide saturation through magma mixing (e.g. input of a new pulse of sulphide-saturated magma into the sulphide-undersaturated magma in the chamber), and/or differentiation, will result in collection of PGE from a large volume of melt in the magma chamber into thin sulphide cumulate layers (e.g. Campbell *et al.* 1983). The effect of an enriched mantle source on PGE mineralization seems to be minor. Studies of PGE abundances in ultramafic xenoliths and komatiites of different ages (e.g. Chou 1978, Sun 1982, Brügmann *et al.* 1987) show that PGE distribution in fertile mantle is quite homogeneous, with chondritic relative abundances, and Pd ~ 4 ppb and Ti/Pd ~ 3×10^5 throughout geological time. Mantle sources for boninites have comparable or lower PGE abundances to komatiite sources, accompanied by lower Cu and much lower sulphur (Hamlyn *et al.* 1985) than the komatiite sources (with Pd ~ 4 ppb, Cu ~ 30 ppm and S ⩽ 250 ppm; e.g. Sun 1982). This depletion is probably due to extraction by a silicate melt (sulphide-saturated) during first-stage melting.

In Archaean and early Proterozoic terrains, komatiites and their differentiates, including SHMB with MgO contents down to 7%, are abundant and sulphide-undersaturated (e.g. Redman & Keays 1985). Furthermore, these SHMB have comparable Pd, Pt and Au concentration to boninites (Table 6.4). Consequently, there is no obvious reason why a boninite parent magma should be more favourable than komatiite for developing PGE mineralization. However, during the Phanerozoic, komatiite melts formed by high-temperature, large degrees of melting are rare (e.g. the Gorgona Island early Tertiary komatiites with 18% MgO). Consequently, small degrees of melting of refractory harzburgitic continental lithosphere or mantle wedge above the subduction zone poor in sulphur (⪡ 50 ppm?) may be important for igneous complexes and dykes with PGE and Au mineralization. Stable cratons also offer an environment to establish large magma chambers.

6.8 Development of a continental lithosphere modified by subduction processes

If crustal contamination is considered important for the generation of Archaean and early Proterozoic SHMB and their associated sulphide mineralization, then isotope data for these SHMB have also to be interpreted according to this contamination model. Figure 6.2 shows Pb isotope data of

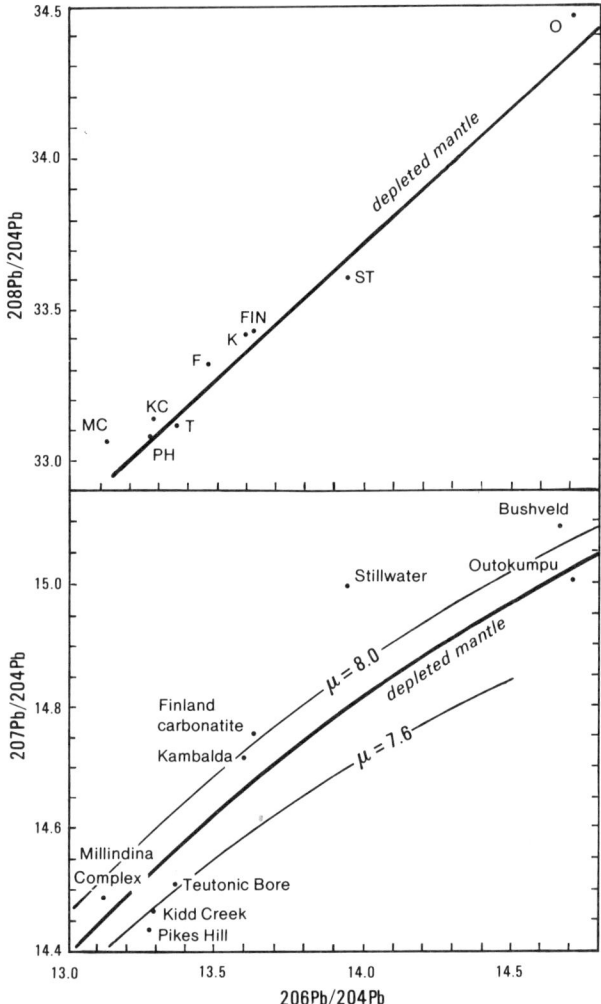

Figure 6.2 Pb isotope ratio plots for some late Archaean and early Proterozoic basalts and sulphides from Teutonic Bore and Kambalda (Eastern Goldfield) and the Millindina Complex (Pilbara), W Australia; Kidd Creek and Pikes Hill (Abitibi Belt), Canada; Stillwater Complex, Wyoming, USA; Bushveld Complex, S Africa; and 1.9 Ga Outokumpu 'ophiolite', Finland. Samples with affinities to SHMB have high μ values. This is consistent with the trace-element data, which suggest some involvement of the felsic crust.

some late Archaean and early Proterozoic mafic rocks and magmatic sulphides and their relationship to the inferred evolution curves of depleted mantle. Sulphides from Kidd Creek and Pikes Hill are from the 2.7 Ga Abitibi Belt, Canada (Tilton 1983), massive sulphide from Teutonic Bore is associated with calc-alkaline volcanics of the Eastern Goldfields (Gulson 1986), and Outokumpu galena is from the 1.9 Ga Finnish ophiolite (Vaasjoki 1981). These samples have μ values lower than the reference curve for depleted mantle. In contrast, samples with SHMB connection shown in Figure 6.2 all have higher μ values relative to the reference curve. These include the 2.84 Ga Millindina Complex, near the Negri Volcanics in the western Pilbara, W Australia (Korsch & Gulson 1986), 2.7 Ga nickel sulphide ore associated with Kambalda komatiites (Roddick 1984), plagioclases of the 2.7 Ga Stillwater Complex (Czamanske *et al.* 1986), and an estimated initial ratio of the Bushveld Complex (Manhes *et al.* 1980). These contrasting results can be interpreted in terms of the age of underlying basement in the area. Preservation of mantle Pb isotope compositions in Kidd Creek, Pikes Hill and Teutonic Bore samples reflect a lack of early Archaean basement (e.g. Chauvel *et al.* 1985, R. Hill 1988 pers. comm.), whereas SHMB with high μ values occur in areas with early Archaean basement.

The same consideration applies to Sr and Nd isotope studies of komatiites, their differentiates including SHMB and clinopyroxene separates (e.g. Chauvel *et al.* 1985), as well as to elemental ratios (e.g. Th/U, Nb/Th). In addition, the effect of alteration and metamorphism may in some cases further complicate isotopic studies of these rocks.

The alternative interpretation, that SHMB were derived from modified lithospheric or a subduction-zone environment, requires involvement of melts derived from subduction of early Archaean sediments and implies recycling of a considerable amount of continental crust back into the mantle at this early stage of Earth history. Evidence for the existence of ancient LILE-enriched continental lithosphere comes from studies of modern continental basalts and ultrapotassic rocks (e.g. Carlson 1984, Nelson *et al.* 1986). It is not clear, however, how much of this modified Archaean lithosphere has been consumed in formation of younger basalts. A refractory harzburgitic continental lithosphere enriched in incompatible elements may be an important mantle reservoir for incompatible elements and needs to be considered in geochemical modelling of the mantle–crust evolution (e.g. Carlson 1984).

6.9 Concluding remarks

Archaean and early Proterozoic SHMB have similar geochemical characteristics, which suggest involvement of felsic crust in their petrogenesis. However, more than one mechanism is capable of generating these features and includes crustal contamination of ascending komatiite, sediment subduction into the upper mantle, and melting of modified harzburgitic continental lithosphere. Their 'uncontaminated', parent magmas may have compositions ranging from komatiite to 'boninite' generated by melting of refractory lithospheric mantle.

On the basis of integrated geological and petrological observations, fluid-dynamic and thermal-budget considerations together with trace-element and isotope modelling, we concluded that SHMB in Archaean greenstone belts were probably derived from komatiites through AFC processes. Contamination by different types of felsic crust can be detected in the chemistry of the SHMB. Contamination by tonalite causes relative increases in Ti, P and Sr and a positive Eu anomaly but low SiO_2 content ($\leqslant 52\%$) in the SHMB, whereas contamination by granite results in low Ti, Nb, Ca, P and Sr, large negative Eu anomaly and high SiO_2 content (54–55%) with a crystallization sequence of olivine followed by orthopyroxene.

Concentration ratios of Pd/Cu, Pd/S, Pd/Ti and Pd/Y can be used as effective criteria to distinguish komatiite from boninite parent magmas. The felsic crustal contaminants are generally Cu- and Pd-poor but have Y and Ti concentrations comparable to or lower than komatiites. According to Hamlyn et al. (1985), mantle sources of boninites are depleted in S, Cu, Ti and Y relative to Pd (with the order of depletion $S > Cu \geqslant Ti > Y \gg Pd$). As a result, sulphide-undersaturated boninites have considerably higher Pd/S, Pd/Cu, Pd/Y and Pd/Ti than komatiites (see Table 6.4). Using these ratios, it can be shown that Archaean SHMB in the Eastern Goldfields of Western Australia are related to komatiites. It would be desirable to apply this test to major layered complexes, when high-quality trace-element data (including PGE) for their parent magmas become available.

Acknowledgements

We are grateful to Nick Arndt, Warrington Cameron, Ian Campbell, Tony Crawford, Andrew Glikson, Paul Hamlyn and John Sheraton for their constructive comments and suggestions. Shen-su Sun publishes with the permission of the Director of Bureau of Mineral Resources, Geology and Geophysics, Australia.

References

Arndt, N. T. & G. A. Jenner 1986. Crustally contaminated komatiites and basalts from Kambalda, Western Australia. *Chem. Geol.* **56**, 229–55.

Arndt, N. T., G. E. Brügmann, K. Lenhert, B. W. Chappell & C. Chauvel 1987. Geochemistry, petrogenesis and tectonic environment of circum-Superior Belt basalts, Canada. In *Geochemistry and mineralization of proterozoic volcanic suites*, T. C. Pharaoh, R. D. Beckinsdale & D. Rickard (eds), 133–46. London: Blackwell Scientific.

Barley, M. E. 1986. Incompatible-element enrichment in Archaean basalts: a consequence of contamination by older sialic crust rather than mantle heterogeneity. *Geology* **14**, 947–50.

Barley, M. E. 1987. The Archaean Whim Creek Belt, an ensialic fault-bounded basin in the Pilbara Block, Australia. *Precambrian Res.* **37**, 199–215.

Brügmann, G. E., N. T. Arndt, A. W. Hoffmann & H. J. Tobschall 1987. Noble metal abundances in komatiite suites from Alexo, Ontario and Gorgona Island, Columbia. *Geochim. Cosmochim. Acta* **51**, 2159–69.

Cameron, W. E. 1989. Contrasting boninite–tholeiite associations from New Caledonia (this volume).

Cameron, W. E., M. T. McCulloch & D. A. Walker 1983. Boninite petrogenesis: chemical and Nd–Sr isotopic constraints. *Earth Planet. Sci. Lett.* **65**, 75–89.

Campbell, I. H. 1985. The difference between oceanic and continental tholeiites: a fluid dynamic explanation. *Contrib. Mineral. Petrol.* **91**, 37–43.

Campbell, I. H., A. J. Naldrett & S. J. Barnes 1983. A model for the origin of the platinum-rich sulfide horizons in the Bushveld and Stillwater complexes. *J. Petrol.* **24**, 133–65.

Carlson, R. W. 1984. Isotopic constraints on Columbia River flood basalt genesis and the nature of the subcontinental mantle. *Geochim. Cosmochim. Acta* **48**, 2357–72.

Cattell, A. 1987. Enriched komatiitic basalts from Newton Township, Ontario: their genesis by crustal contamination of depleted komatiite magma. *Geol. Mag.* **124**, 303–9.

Chauvel, C., B. Dupré & G. A. Jenner 1985. The Sm–Nd age of Kambalda volcanics is 500 Ma too old! *Earth Planet. Sci. Lett.* **74**, 315–24.

Chou, C. L. 1978. Fractionation of siderophile elements in the Earth's upper mantle. *Proc. Lunar Sci. Conf.* **9**, 219–30.

Compston, W., I. S. Williams, I. H. Campbell & J. J. Gresham 1986. Zircon xenocrysts from the Kambalda volcanics: age constraints and direct evidence for old continental crust below the Kambalda–Norseman Greenstones. *Earth Planet. Sci. Lett.* **76**, 299–311.

Crawford, A. J., L. Beccaluva & G. Serri 1981. Tectono-magmatic evolution of the West Philippine–Mariana region and the origin of boninites. *Earth Planet. Sci. Lett.* **54**, 346–56.

Czamanske, G. K., J. L. Wooden & M. L. Zientak 1986. Pb isotopic data for plagioclase from the Stillwater complex. *EOS, Trans. Am. Geophys. Union* **44**, 1251.

Davies, G. & M. Tredoux 1985. The platinum-group element and gold contents of the marginal rocks and sills of the Bushveld Complex. *Econ. Geol.* **80**, 838–48.

DePaolo, D. J. 1981. Trace element and isotopic effects of combined wallrock assimilation and fractional crystallization. *Earth Planet. Sci. Lett.* **53**, 189–202.

Dickin, A. P., N. W. Jones, M. F. Thirlwall & R. N. Thompson 1987. A Ce/Nd isotope study of crustal contamination processes affecting Paleocene magmas in Skye, Northwest Scotland. *Contrib. Mineral. Petrol.* **96**, 455–64.

Fisk, M. R. 1986. Basalt magma interaction with harzburgite and the formation of high-magnesian andesites. *Geophys. Res. Lett.* **13**, 467–70.

Fletcher, I. R., W. G. Libby & K. J. R. Rosman 1987. Sm–Nd dating of the 2411 Ma Jimberlana dyke, Yilgarn Block, Western Australia. *Austr. J. Earth Sci.* **34**, 523–5.

Green, D. H. 1981. Petrogenesis of Archaean ultramafic magmas and implications for Archaean tectonics. In *Precambrian plate tectonics*, A. Kroner (ed.), 469–88. Amsterdam: Elsevier.

Gruau, G., B. M. Jahn, A. Y Glikson, R. Davy, A. H. Hickman & C. Chauvel 1987. Age of the Archean Talga-Talga Subgroup, Pilbara Block, Western Australia, and early evolution of the mantle: new Sm–Nd isotopic evidence. *Earth Planet. Sci. Lett.* **85**, 105–16.

Gulson, B. L. 1986. *Lead isotopes in mineral exploration*. Amsterdam: Elsevier.

Hall, R. P. & D. J. Hughes 1987. Noritic dykes of southern West Greenland: early Proterozoic boninitic magmatism. *Contrib. Mineral. Petrol.* **97**, 169–82.

Hallberg, J. A. 1987. Post cratonization mafic and ultramafic dykes of the Yilgarn Block. *Austr. J. Earth Sci.* **34**, 135–49.

Hamlyn, P. R., R. R. Keays, W. E. Cameron, A. J. Crawford & H. M. Waldron 1985. Precious metals in magnesian low-Ti lavas: implications for metallogenesis and sulfur saturation in primary magmas. *Geochim. Cosmochim. Acta* **49**, 1797–811.

Hatton, C. J. & M. R. Sharpe 1989. Significance and origin of boninite-like rocks associated with the Bushveld Complex (this volume).

Hergt, J. M. 1987. The origin and evolution of the Tasmanian dolerites. Ph.D. Thesis, Australian National University.

Hickey, R. L. & F. A. Frey 1982. Geochemical characteristics of boninite series volcanics: implications for their source. *Geochim. Cosmochim. Acta* **46**, 2099–115.

Huppert, H. E. & R. S. J. Sparks 1985. Cooling and contamination of mafic and ultramafic lavas during ascent through continental crust. *Earth Planet. Sci. Lett.* **74**, 371–86.

Jahn, B.-M., G. Gruau & A. Y. Glikson 1982. Komatiites of the Onverwacht Group, S. Africa: REE geochemistry, Sm/Nd age and mantle evolution. *Contrib. Mineral. Petrol.* **80**, 25–40.

Jaques, A. L. & D. H. Green 1980. Anhydrous melting of peridotite at 0–15 kb and the genesis of tholeiitic basalts. *Contrib. Mineral. Petrol.* **73**, 287–310.

Jenner, G. A. 1981. Geochemistry of high-Mg andesites from Cape Vogel, Papua New Guinea. *Chem. Geol.* **33**, 307–32.

Keays, R. R. 1982. Palladium and iridium in komatiites and associated rocks: application to petrogenetic problems. In *Komatiites*, N. T. Arndt & E. G Nisbet (eds), 435–57. London: Allen & Unwin.

Korsch, M. J. & B. L. Gulson 1986. Nd and Pb isotopic studies of an Archaean layered mafic–ultramafic complex, Western Australia, and implications for mantle heterogeneity. *Geochim. Cosmochim. Acta* **50**, 1–10.

Kuehner, S. M. 1989. Petrology and geochemistry of early Proterozoic high-Mg dykes from the Vestfold Hills, Antarctica (this volume).

Kyser, T. K. 1987. Stable isotope variations in the mantle. In *Reviews in mineralogy*, Vol. 16: *Stable isotopes in high temperature geological processes*, 141–64. Mineral. Soc. America.

Longhi, J., J. L. Wooden & K. D. Coppinger 1983. The petrology of high-Mg dikes from the Beartooth Mountains, Montana: a search for the parent magma of the Stillwater complex. *J. Geophys. Res.* **88**, B53–69.

McCallum, I. S. 1987. Evidence for crustal recycling during the Archean: the parent magmas of the Stillwater complex (Abstract). *Workshop on the Growth of Continental Crust*, 13–16 July, 56–8. Oxford.

McCulloch, M. T. & W. E. Cameron 1983. Nd–Sr isotopic study of primitive lavas from the Troodos ophiolite, Cyprus: evidence for a subduction-related setting. *Geology* **11**, 727–31.

McCulloch, M. T. & B. W. Chappell 1982. Nd isotopic characteristics of S- and I-type granites. *Earth Planet. Sci. Lett.* **58**, 51–64.

McCulloch, M. T., W. Compston & D. Froude 1983. Sm–Nd and Rb–Sr dating of Archaean gneisses, eastern Yilgarn Block, Western Australia. *J. Geol. Soc. Austr.* **30**, 149–53.

Manhes, G., C. J. Allégre, B. Dupré & B. Hamelin 1980. Lead isotopic study of basic–ultrabasic layered complexes: speculations about the age of the earth and primitive mantle characteristics. *Earth Planet. Sci. Lett.* **47**, 370–82.

Morrison, M. A., R. N. Thompson & A. P. Dickin 1985. Geochemical evidence for complex magmatic plumbing during development of a continental volcanic center. *Geology* **13**, 581–4.

Nance, W. B. & S. R. Taylor 1977. Rare earth element patterns and crustal evolution – II. Archaean sedimentary rocks from Kalgoorlie, Australia. *Geochim. Cosmochim. Acta* **41**, 225–31.

Nelson, D. R., M. T. McCulloch & S.-S. Sun 1986. The origins of ultrapotassic rocks as inferred from Sr, Nd and Pb isotopes. *Geochim. Cosmochim. Acta* **50**, 231–45.

Nesbitt, R. W., H. Mastins, G. W. Stolz & D. R. Bruce 1976. Matrix corrections in trace element analysis by x-ray fluorescence: an extension of the Compton scattering technique to long wavelengths. *Chem. Geol.* **18**, 203–13.

Nielsen, R. L. 1988. A model for the simulation of combined major and trace element liquid line of descent. *Geochim. Cosmochim. Acta* **52**, 27–38.

Purvis, A. C. & P. G. Moeskops 1981. Nickel–copper sulfide-rich Proterozoic dikes at Cowarna Rocks, Western Australia. *Econ. Geol.* **76**, 1597–605.

Redman, B. A. & R. R. Keays 1985. Archaean basic volcanism in the Eastern Goldfields Province, Yilgarn Block, Western Australia. *Precambrian Res.* **30**, 113–52.

Ryerson, F. J. & E. B. Watson 1987. Rutile saturation in magmas: implications for Ti–Nb–Ta depletion in island-arc basalts. *Earth Planet. Sci. Lett.* **86**, 225–39.

Roddick, J. C. 1974. Responses of strontium isotopes to some crustal processes. Ph.D. Thesis, Australian National University.

Roddick, J. C. 1984. Emplacement and metamorphism of Archaean mafic volcanics at Kambalda, Western Australia – geochemical and isotopic constraints. *Geochim. Cosmochim. Acta* **48**, 1305–18.

Ross, J. R. & R. R. Keays 1979. Precious metals in volcanic-type sulfide deposits in Western Australia: I. Relationship with the composition of the ores and their host rocks. *Can. Mineral.* **17**, 417–35.

Sharpe, M. R. & L. J. Hulbert 1985. Ultramafic sills beneath the Eastern Bushveld Complex: mobilized suspensions of early Lower Zone cumulates in a parental magma with boninitic affinities. *Econ. Geol.* **80**, 849–71.

Sheraton, J. W. & L. P. Black 1981. Geochemistry and geochronology of Proterozoic tholeiite dykes of East Antarctica: evidence for mantle metasomatism. *Contrib. Mineral. Petrol.* **76**, 305–17.

Stumpfl, E. F. & C. G. Ballhaus 1986. Stratiform platinum deposits: new data and concepts. *Fortschr. Mineral.* **64**, 205–14.

Sun. S.-S. 1982. Chemical composition and origin of the Earth's primitive mantle. *Geochim. Cosmochim. Acta* **46**, 179–92.

Sun, S.-S. & W. F. McDonough 1988. Chemical and isotopic systematics of oceanic basalts: implications for mantle composition and processes. In *Magmatism in ocean basins*, A. D. Saunders & M. J. Norry (eds), Spec. Publ. 42, 313–45. London: Geological Society.

Sun, S.-S. & R. W. Nesbitt 1978a. Petrogenesis of Archaean ultrabasic and basic volcanics: evidence from rare earth elements. *Contrib. Mineral. Petrol.* **65**, 301–25.

Sun, S.-S. & R. W. Nesbitt 1978b. Geochemical regularities and genetic significance of ophiolitic basalts. *Geology* **6**, 689–93.

Sun, S.-S. R. W. Nesbitt & A. Y. Sharaskin 1979. Geochemical characteristics of mid-ocean ridge basalts. *Earth Planet. Sci. Lett.* **44**, 119–38.

Tarney, J. & B. L. Weaver 1987. Geochemistry and petrogenesis of early Proterozoic dyke swarms. In *Mafic dyke swarms*, H. C. Halls & W. F. Fahrig (eds), 81–94. Geol. Assoc. Can., Spec. Paper no. 34.

Tilton, G. R. 1983. Evolution of depleted mantle: the lead perspective. *Geochim. Cosmochim. Acta* **47**, 1191–7.

Vaasjoki, M. 1981. *The lead isotopic composition of some Finnish galenas.* Geol. Surv. Finland Bull. no. 316.

Veizer, J. 1985. Carbonates and ancient oceans: isotopic and chemical record on time scales of 107–109 years. In *The carbon cycle and atmospheric CO_2: natural variations Archean to Present*, E. T. Sundquist & W. S. Broecker (eds), 595–601. Am. Geophys. Union, Geophys. Monogr. no. 32.

Weaver, B. L. & J. Tarney 1981. The Scourie Dyke Suite: petrogenesis and geochemical nature of the Proterozoic sub-continental mantle. *Contrib. Mineral. Petrol.* **78**, 175–88.

Wooden, J. L. & P. A. Mueller 1988. Pb, Sr and Nd isotopic compositions of a suite of late Archean igneous rocks, eastern Beartooth Mountains: implications for crust–mantle evolution. *Earth Planet. Sci. Lett.* **87**, 59–72.

7 Significance and origin of boninite-like rocks associated with the Bushveld Complex

CHRISTOPHER J. HATTON &
MARTIN R. SHARPE

Abstract

The marginal rocks of the Bushveld Complex are divided into pre- and syn-Bushveld sills. The syn-Bushveld sills provide the best available estimate of the magma compositions parental to the layered series. Among these, the quench-textured micro-orthopyroxenites are representative of the parental magma of the lower zone, and are boninitic in composition. This boninitic magma played an important role in the genesis of the chromitite and PGE-enriched layers of the Bushveld Complex.

The geochemistry of the boninitic rocks indicates derivation from a depleted source, since TiO_2 contents are low, but high K_2O, LREE and Zr contents suggest an enriched source. The possibility that these magmas originated from a depleted source, then attained their enriched signature by upper crustal contamination during ascent, is considered unlikely. The preferred hypothesis is that the Bushveld boninite was generated from mantle which had been previously depleted by basalt extraction, then enriched by mantle metasomatism. The metasomatic component is thought to be ultimately derived from subducted sediment.

7.1 Introduction

The parental magmas of the Bushveld Complex were intruded 2050 Ma into metasedimentary and volcanic rocks of the 2100–2300 Ma Transvaal Sequence. Evidence has been accumulating that the first major magma inputs into the Bushveld Complex, which gave rise to up to 2000 m of orthopyroxenitic cumulates, were not basaltic in composition as had historically been assumed (cf. Wager & Brown 1968) but were unusually rich in SiO_2 and MgO (i.e. boninitic) (Sun & Nesbitt 1978, Irvine & Sharpe 1982). Later magmas that were added to the Bushveld magma chamber were basaltic.

Evidence of the nature of these parental magmas comes from (a) suites of sills, dykes and marginal rocks that are spatially associated with the Complex, and (b) the geochemistry of the cumulates themselves. The sills that intrude the Transvaal Sequence (Sharpe 1984) are of two main types. Pre-Bushveld sills pre-date, and were cut and thermally metamorphosed by, the marginal zone of the Bushveld Complex. Syn-Bushveld sills have widely ranging metamorphic grades and degrees of alteration, but can nevertheless be clearly related to the various zones of the Complex. The compositions of the marginal rocks change in a similar manner, with pyroxenitic rocks abutting the lower zone and lower critical zone, and suites of basaltic rocks the upper critical zone and main zone.

The quench-textured micro-orthopyroxenites that occur within the syn-Bushveld sill group and in the pyroxenitic marginal rock suite are the prime subject of this chapter. We describe the petrography and geochemistry of these boninitic rocks, examine their relations with the various zones of the Bushveld Complex and discuss their properties as liquids. The intrusion and subsequent behaviour of the liquids in the magma chamber is reviewed and a model for the genesis of boninites from an enriched mantle source is developed.

7.2 Geological setting and stratigraphy of the Eastern Bushveld Complex

7.2.1 The Transvaal Sequence

Details of the stratigraphy of the Transvaal Sequence and Bushveld Complex are given in Figures 7.1 and 7.2, and by Button (1976). The Witwatersrand Sequence (2.7 Ga) and the Ventersdorp lavas (2.65 Ga; Armstrong *et al.* 1986) were the first major groups of rocks to cover the Kaapvaal craton. Rocks of the 2.3–2.1 Ga old Transvaal Sequence host the Bushveld Complex. The Rooiberg Felsites were considered to be separate from the Transvaal Sequence (SACS 1980) but recent work by Schweitzer (1987) has shown that the Rooiberg Felsites are an upward continuation of the Dullstroom Volcanics. Since the Dullstroom Volcanics are part of the Transvaal Sequence (SACS 1980) the Rooiberg Felsites are also included. The Transvaal Sequence may then be divided into five Groups:

(a) the Wolkberg Group, consisting of quartzites, shales and lava;
(b) the Black Reef Group, which is mostly quartzitic;

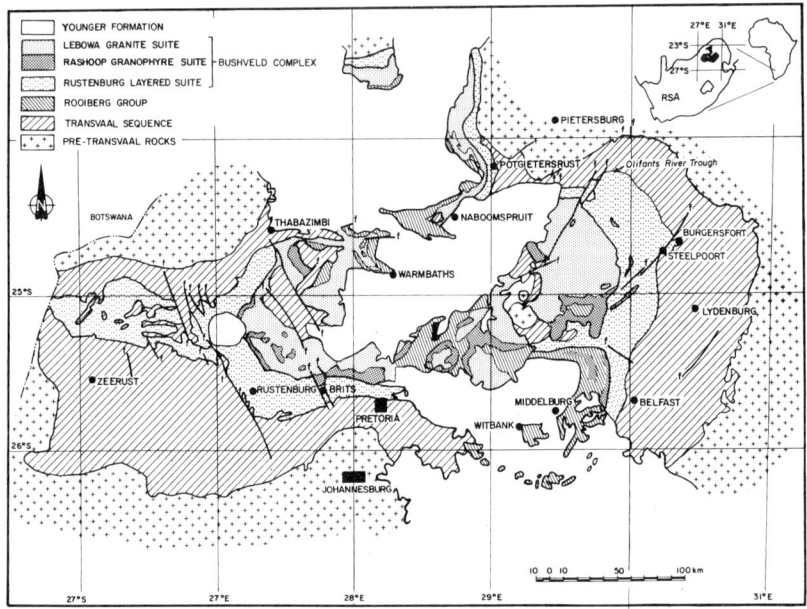

Figure 7.1 Locality map for the Bushveld Complex.

(c) the Chuniespoort Group, in which the Malmani and Duitschland Dolomite Formations sandwich the Penge Ironstone Formation:
(d) the Pretoria Group, which is an alternating succession of shallow-water quartzites and shales with some volcanics and is topped by the Dullstroom Volcanics;
(e) the Rooiberg Group, which is made up of felsites with minor sedimentary components (Twist & French 1983).

The emplacement of the Complex was preceded by the sub-aerial extrusion of 5 km of Dullstroom Volcanics and Rooiberg Felsites, which blanketed the Transvaal rocks (Twist 1985) and intrusion of 2 km of mafic sills (Sharpe & Snyman 1980, Sharpe 1984).

7.2.2 The Bushveld Complex

The Bushveld mafic rocks (formally known as the Rustenburg Layered Suite) were emplaced into the Transvaal Sequence 2050 Ma ago (Hamilton 1977, Sharpe 1985). They form horseshoe-shaped belts that surround and dip inwards beneath an acid central mass, which consists of pre-Bushveld felsite, intrusive granophyres and post-Bushveld granite (Fig.7.1). Geophysical work indicates that the layered rocks are generally 6–9 km thick. The layered rocks of the Complex are, in general, minimally deformed and essentially

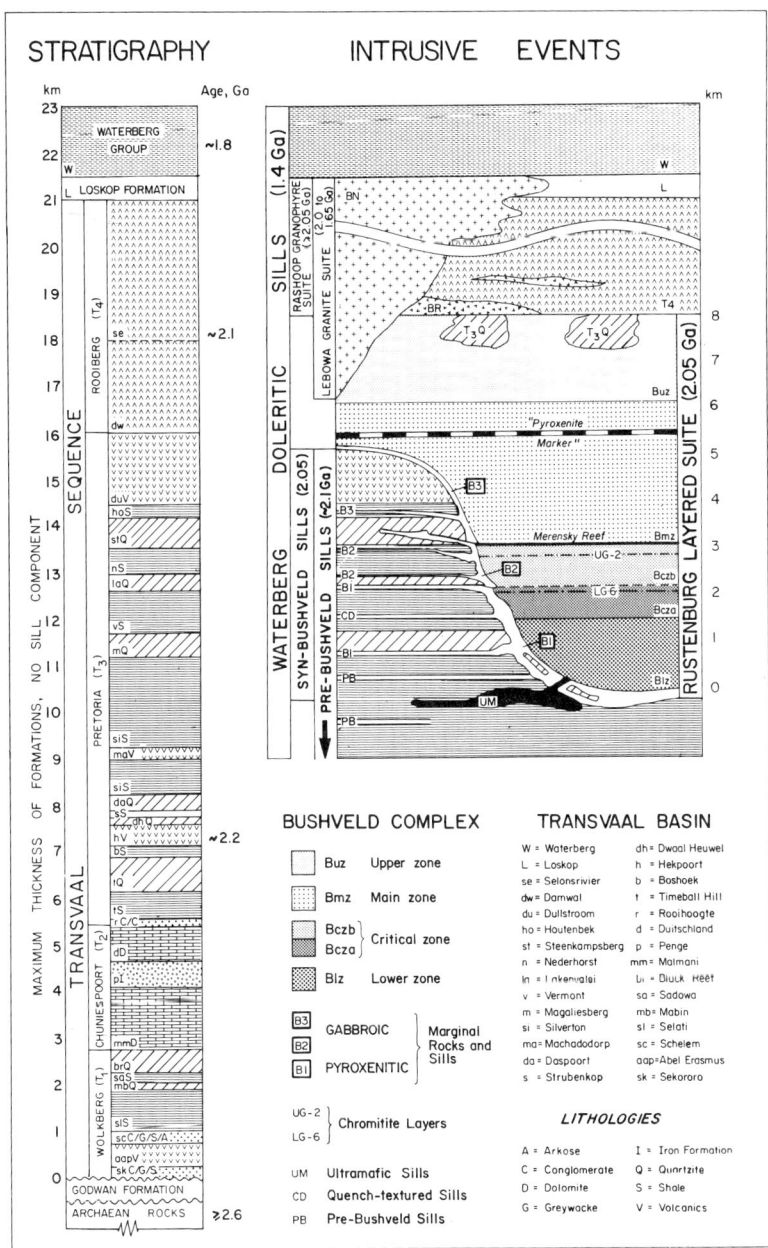

Figure 7.2 Lithostratigraphic column for the Bushveld Complex and Transvaal Sequence in the eastern Transvaal. Thicknesses of formations are maxima, and total thickness of the Transvaal Sequence at any one locality is generally less than 10 km. From Von Gruenewaldt et al. (1985).

unmetamorphosed. Younger Waterberg and Karoo rocks cover parts of the mafic sequences.

The stratigraphy of the layered rocks derived from magmas supplied from the various feeders is broadly similar throughout the Complex. For example, the correlation of chromitite layers (Hatton & Von Gruenewaldt 1987) and the continuity of the Merensky Reef and the upper-zone lithostratigraphy (Von Gruenewaldt 1973, Molyneux 1974) are remarkable. Emplacement of an early boninitic magma into the shallow, cone-like feeder zones was followed by introduction of larger amounts of basaltic magma. With addition of basaltic magma, the originally separate bodies coalesced into the continuous magma chambers of the eastern and western Bushveld Complex.

In eastern Transvaal, the relation between the base of the layered suite and the marginal rocks and sills is particularly well exposed. At Burgersfort, the Bushveld incises deeply into the floor rocks, some 700 m down into shales of the Silverton Formation (Fig.7.3). Southwards, as the altitude of the country increases, so the contact between the layered rocks and the underlying floor climbs through the stratigraphy of the Transvaal Sequence to its highest level of exposure at the Dullstroom Volcanics, at which level the adjoining layered rocks are well within the main zone. A section that follows the contact between the Bushveld Complex and Transvaal Sequence through 5–6 km of stratigraphy is thus available for inspection (Fig.7.2)

THE LOWER ZONE

The lower zone is best exposed in the Olifants River trough of the NE Bushveld (Cameron 1978). It consists of four units with an aggregate thickness of 1700 m. The lowermost unit (subzone A) consists of feldspathic bronzitites which become compositionally more primitive upwards, similar to the basal parts of the Stillwater (Raedeke & McCallum 1984, Czamanske & Zientek 1985), Duluth (Grant & Molling 1981) and Insizwa (Lightfoot et al. 1984) Complexes. Subzone B is made up of monomineralic bronzitites of constant composition, as is subzone D. Subzone C, on the other hand, consists of a series of cyclic units of dunite–harzburgite–bronzitite. Elsewhere in the Complex, the lower-zone successions are broadly similar, and are generally characterized by the absence of chromitite layers and the low chromite content of the component rock types.

THE CRITICAL ZONE

The base of the critical zone is defined as the level at which cumulus chromite first becomes abundant (to form chromitite layers) and the amount of intercumulus plagioclase exceeds a few modal per cent. However, it is essentially an artificial, unmappable boundary, both because of poor exposure and the fact that bronzitites of subzone D of the lower zone and subzone A of the critical zone are compositionally and petrologically more similar than many lithologies from along-strike facies variations within the lower zone

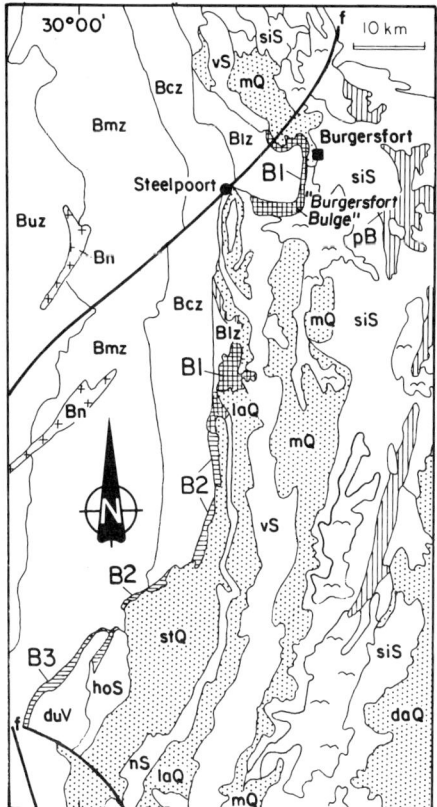

Figure 7.3 Generalized geology of the eastern Bushveld Complex, south of Steelpoort. Symbols as in Figure 7.2.

itself (Lee & Sharpe, 1989). For a comprehensive account of the lower- and critical-zone lithologies of the E Bushveld Complex and the chromitite layers within them, see papers by Cameron (1980, 1982) and Hatton & Von Gruenewaldt (1987).

The upper critical zone hosts the economically significant PGE deposits of the UG-2 chromitite layer and the Merensky Reef. Their geological setting, distribution of the platinum-group minerals within them and the mineralization in the Platreef (in the Potgietersrus compartment) are described in special issues of *Economic Geology* (**77, 80** and **81**). The geological setting of the Merensky Reef, which caps the critical zone, has been described in great detail in extensive literature. Some of the more recent ideas are summarized by Lee (1983) and Naldrett *et al.* (1987). There are two main schools of thought on its genesis, both of which are essentially controlled by the boninitic parental magma of the Complex. The first group holds that the Reef represents the final stage of a process of precious-metal enrichment in the magma and that its precipitation was the natural, unstimulated results of extended magma fractionation (Vermaak 1977). The second hypothesis, which has recently gained favour, is that the Reef precipitated in response to the intrusion of new magma of different composition into a chamber dominated by derivatives of boninitic magmas (Sharpe 1985, Naldrett *et al.* 1987).

THE MAIN ZONE

The main zone is comparatively homogeneous. Subzone A consists of gabbronorites with rare anorthosite and pyroxenite layers; subzone B comprises gabbros with inverted pigeonite and in subzone C the gabbronorites with cumulus orthopyroxene reappear (Von Gruenewaldt 1973, Molyneux 1974). Subzones B and C are separated by the 'pyroxenite marker', across which many compositional changes occur. These have generally been accepted as reflecting a volumetrically large addition of magma at this level in the intrusion (Von Gruenewaldt 1973). Sharpe (1985) considers that the reversal in orthopyroxene Mg# back to high values marks the interface between the magmas that formed the main zone, and the packet of magma mixtures, mostly derived from the boninitic parental magma, that was raised up when these magmas were intruded at the base of the magma chamber to form the Merensky Reef.

THE UPPER ZONE

The 2.5 km thick upper zone is best exposed in the Steelpoort–Roossenekal–Magnet Heights region of the eastern Transvaal. Upper-zone rocks contain the widest variety of cumulus minerals in the Complex. The lower part of the zone commences with the appearance of abundant magnetite in two-pyroxene gabbros. The magnetite is concentrated into up to 24 magnetitite layers through the 2.5 km thick zone. The magnetite layers are V-rich and Ti-poor at the base, becoming poorer in V and richer in Ti upwards. Olivine (fayalite-rich) reappears in the upper diorite cumulates, with apatite and, later, hornblende as additional cumulus phases.

POST-BUSHVELD ROCKS

Pre-Bushveld Rooiberg Felsite and Transvaal rocks formed the roof of the magma chamber. They were intruded by Bushveld Granite after the formation of the layered rocks. For a complete description see Twist (1985) and Twist & Harmer (1987).

7.3 The parental magmas to the Bushveld Complex

In their study of the Stillwater Complex, Irvine *et al.* (1983) concluded that two very different types of magma are required to account for the harzburgite–pyroxenite–norite and anorthosite–norite–gabbro portions of this body. One would have had olivine followed by orthopyroxene on the liquidus (U_0), and the other plagioclase, followed by olivine or pyroxene (A_0). To a first approximation, a similar approach may be adopted for the Bushveld Complex, since many workers believe that at least two major magma types were involved in its evolution (Cawthorn et al. 1981, Irvine & Sharpe 1982, Harmer & Sharpe 1985, Sharpe 1985). Evidence of the nature of magmas responsible for the various zones of the Complex was obtained from the field relations and geochemistry of rocks along the basal margin of the Complex and from sills in the floor. Sharpe (1981) identified three groups of marginal rocks which might be representative of parental magmas to the layered intrusion. These are the boninitic B1 group and the tholeiitic B2 and B3 groups. The boninites of the B1 group would fall into the U category of Irvine et al. (1983), whereas tholeiites of the B2 and B3 groups fall into the A category.

The B1 marginal rocks occur next to the lower zone and lower critical zone, and are rich in SiO_2 (52–56%), MgO (10–16%), Cr (600–1500 ppm) and incompatible elements (e.g. Rb = 20–50 ppm, Zr = 60–110 ppm, Y = 10–16 ppm, Sm ~ 14 × chondritic), low in Sr (150–200 ppm), LREE-enriched ($(Ce/Yb)_N$ = 10–14) and have low TiO_2 contents and low Ti/Zr. In contrast, the B2 and B3 groups are fairly typical of tholeiitic basalt (SiO_2 = 49–52%, MgO = 6–10%). The gabbronorites of the B2 suite (Sharpe 1981) occur adjacent to the upper critical zone. These rocks have $(Ce/Yb)_N$ ~ 6, but have low Rb (1–3 ppm) and Zr (10–70 ppm), with relatively high Y (10–30 ppm) and Sr (340–370 ppm). Gabbronorites of the B3 suite, abutting the main zone, are relatively depleted in incompatible elements (Zr = 20–35 ppm, Rb = 1–4 ppm, Y = 11–14 ppm, Sm ~ 4 × chondritic), rich in Sr (300–380 ppm) and have flat REE patterns with $(Ce/Yb)_N$ ~ 3. All tholeiitic marginal rocks are Cr-poor and V-rich.

Sr_I (i.e. $^{87}Sr/^{86}Sr$ at 2050 Ma) of the boninitic marginal rocks lie in the range 0.703–0.706, whereas the tholeiitic marginal rocks have Sr_I = 0.706–0.7075 (Harmer & Sharpe 1985, Sharpe 1985). The parental liquids that gave rise to the main zone had SR_I in excess of 0.708 (Sharpe 1985). Because no marginal rocks so enriched in radiogenic Sr have yet been found, it is necessary to propose the existence of a further magma type. For consistency with the earlier definitions of Sharpe (1981) this is designated B4. Considerations of the composition of the main-zone cumulates indicate that this magma must have been relatively Al- and Fe-rich and may be termed an aluminous tholeiite to distinguish it from the high-REE B2 tholeiite and the low-REE B3 tholeiite.

The evolution of the layered suite of the Bushveld Complex may be viewed in terms of the successive addition of the tholeiitic magmas to the boninitic B1 magma, which was the first to be emplaced into the chamber. A feature of the

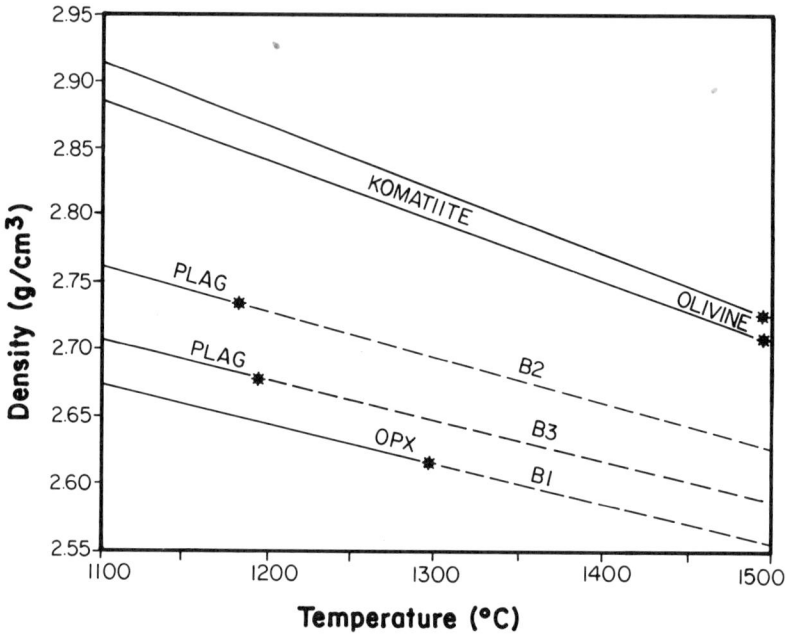

Figure 7.4 Densities of Bushveld magmas of the marginal groups, and densities of komatiites. Stars indicate liquidus temperatures and the mineral on the liquidus is indicated to the left of the stars.

boninitic magma is that its density is less than that of tholeiitic magma (Fig. 7.4) so that the B1 magma was progressively displaced upwards by denser tholeiitic magma (Sharp 1985). Furthermore, the liquidus temperature of the boninite is higher than that of tholeiite (Sharpe & Irvine 1983). These features are central to the interpretation of the economically important chromitite layers and the Merensky Reef (Hatton 1988). The chromitite layers may have formed in response to the emplacement of dense tholeiitic magma below hotter, less dense boninitic B1 magma in the chamber. Because the bulk density of B1 magma plus about 5% orthopyroxene crystals is greater than that of tholeiitic magma, crystallization of orthopyroxene in the boninitic magma would increase its density to the point that slurries of orthopyroxene crystals and boninitic magma would collapse into the tholeiitic magma on the floor. Mixing of the boninite, with orthopyroxene on its liquidus, and tholeiite, with plagioclase on its liquidus, would result in the crystallization of chromite, as proposed by Irvine et al. (1983) and experimentally demonstrated by Sharpe and Irvine (1983). The platiniferous sulphides associated with the chromitite layers may have formed in response to cooling of boninitic magma by tholeiitic magma (Hatton 1988). The low density of boninite is unusual for an ultramafic magma. For example, komatiites have densities which are substantially greater than tholeiitic magma (Fig. 7.4).

7.4 Field relations of boninites

7.4.1 Introduction

Boninites occur in two main settings: as marginal rocks in the floor of the Bushveld Complex and as sills intruding the underlying Transvaal Sequence for up to 120 km laterally away from the basal margin of the Complex. In areas of good exposure, it is seen that some of the sills are extensions of the pyroxenitic marginal zone; in other cases they either crosscut the packet or are cut by it. It is clear that multiple generations of quench-textured sills were intruded from the basal part of the Complex, as shown by their geochemistry and isotopic compositions. The sills thus provide a record of the liquids that existed in the magma chamber during the evolution of the Complex. Whereas most rocks of these families are fresh, some are variably altered to tremolite and mica. They all have orthopyroxene (Mg# = 0.75–0.90), An_{65-75} plagioclase and 1–2 modal% phlogopite. Olivine is rare (0–2 modal%).

7.4.2 Boninitic marginal rocks

The B1 group of marginal rocks consists mostly of orthopyroxenites, norites and subordinate peridotites. It forms a unit of variable thickness between the floor rocks and the lower, pyroxenitic portion of the Complex. North of Steelpoort (Fig. 7.3), where the contact between layering in the lower zone and the bedding in the floor is concordant, these marginal rocks form a 50–200 m thick zone. A 300–500 m thick stratified packet of crosscutting orthopyroxenite, norite, quartz–norite and quench-textured micropyroxenite sheets surrounds the periphery of the 'Burgersfort Bulge' (Sharpe & Hulbert 1985) and intrudes into the floor as sills. The pyroxenitic marginal zone south of Steelpoort occupies the space between the irregular surface of the floor and the overlying layered bronzitites. Thin offshoots extend from these masses. Quench-textured micropyroxenites are found within all these environments.

Quench-textured micropyroxenites are made up of homogeneous rock with randomly orientated, macroscopic needles and sprays of orthopyroxene up to 40 mm long and 0.5 mm wide. The quench cells themselves range in size from 20 mm to 0.4 m across. They are generally hexagonal in section, interfering with their neighbours along sharply defined, straight (planar) contacts. Within the cells, orthopyroxene grains with aspect ratios of up to 40:1 define the ubiquitous striae that give rise to the radiate texture of the cells. Interstices are filled with sheaves of orthopyroxene–plagioclase intergrowths, comb-textured plagioclase and scattered clinopyroxene, chromite and olivine crystals. In the marginal rocks, the quench-textured masses are mostly irregular and are generally found as selvedges around peridotite bodies immediately beneath the Complex. Sharpe & Hulbert (1985) postulated that these micropyroxenites were solidified liquids that had been expelled from the olivine-bearing suspensions that were ultimately to form the peridotite sills. They also form wedge-shaped, structureless masses that penetrate for short distances into the Transvaal rocks along bedding planes, and along the contacts between various sills and their hosts.

7.4.3 Boninitic quench-textured sills

All sills in the lower part of the Transvaal Sequence are metagabbroic amphibolites (pre-Bushveld sills of Sharpe (1984)). Quench-textured, micropyroxenitic B1 group sills are present only within the Pretoria Group, just below the level of deepest penetration of the Bushveld Complex. They cover over a > 6000 m stratigraphic interval between the Machadodorp and Dullstroom Volcanics, but are most common in the Magaliesberg and Vermont Formations (Fig.7.2).

Sills with quench-textured orthopyroxene grains set in a devitrified glass matrix were first documented from the Lydenburg area by Frick (1967). A similar sill, continuous for 100 km along strike just above the Magaliesberg Quartzite, was noted by Sharpe (1978). Sills of comparable morphology and composition, in identical stratigraphic positions, have also been found in the western Bushveld (Cawthorn et al.1981).

The quench-textured sills are 10–100 m thick. It is uncommon for the whole sill to display macroscopic quench textures, which tend to be limited to discontinuous pods and lenses close to the margins. The morphology of the quenched cells in the sills is essentially the same as in the marginal rocks. Therefore, the quench-textured micropyroxenite sills and marginal rocks must be considered as having an identical origin but different intrusion histories. Their petrography and geochemistry will thus be described jointly.

7.4.4 Petrography of the quench-textured micropyroxenites

The quench-textured micropyroxenites can be divided into two broad types: those in which orthopyroxene crystals are evidently xenocrystic because of lack of equilibrium with their matrix compositions, and those in which the orthopyroxenes grew in place.

Xenocryst-bearing micropyroxenites have bimodal grain size distribution. The majority contain orthopyroxenes of 10–40 mm length, the subordinate types 0.3–2 mm long grains. The interstices between grains are filled with variable proportions of sheaf-like orthopyroxene (Mg# $= 0.66$)–plagioclase (An_{65-70}) intergrowths, biotite and devitrified glass, with rare clinopyroxene ($Wo_{35}En_{49}Fs_{16}$), olivine (Fo_{80-85}), magnetite, chromite and sulphides (Fig. 7.5a). All orthopyroxene megacrysts are reverse-zoned (Fig. 7.5b), strongly (Mg# $= 0.82$ centre to Mg# 0.91 edge) where the rock is choked with euhedral grains, and mildly (Mg# $= 0.82-0.86$) where sparse and irregularly distri-

Figure 7.5 Textures in rocks from the B1 group. (a) Subhedral orthopyroxene phenocrysts set in a groundmass of quenched sheaves of orthopyroxene and plagioclase. (b) Zoned orthopyroxene phenocrysts in quenched groundmass. (c) Coarse quench crystals of orthopyroxene. (d) Skeletal and hopper orthopyroxenes in devitrified glass. Width of all photomicrographs is 4.5 mm.

buted. The boundaries of the pyroxene megacrysts are generally sharp, but are commonly embayed by terminations of plagioclase combs. Some grains (Mg# = 0.88–0.90) have skeletal outlines and castellated terminations. In some varieties (Fig.7.5c), coarse, bladed quench crystals of orthopyroxene (Mg# = 0.65) are present. Plagioclase grains (An_{65}) in the coarser-grained rocks are comb-textured and have devitrified glass cores, indicative of quenching (Luhr & Carmichael 1980).

Orthopyroxenes in other quenched rocks appear to have crystallized in place. In typical examples, abundant 0.2–1 mm long orthopyroxene grains of variable morphology and rare subhedral olivine are set in devitrified glass. The orthopyroxene grains and crystallites, which are commonly hollow, have wispy swallowtail terminations that overgrow the matrix as delicate fingers, optically continuous with the bulk of the crystal (Fig. 7.5d). Such orthopyroxenes have the same composition (Mg# = 0.62) as rod-like quenched blades in the coarser micropyroxenites. Tiny (0.01 mm) idiomorphic chromite octahedra are distributed throughout. In these rocks, olivine and orthopyroxene compositions are in equilibrium with each other and with the wholerock composition (Sharpe & Hulbert 1985).

Quench textures are also found within the poorly exposed chilled basal margin of an ultramafic sill (Sharpe & Hulbert 1985). These ultramafic sills are considered to be solidified suspensions of olivine and orthopyroxene grains in parental (boninite-like) liquids from which the Bushveld layered rocks were crystallizing, which were injected into the floor rocks at different times during the evolution of the Complex. In this rock, anhedral grains of Fo_{90} olivine are surrounded by poikilitic orthopyroxene (Mg# = 0.88–0.91) in a quench-textured matrix of plagioclase, biotite and idiomorphic chromite grains with 1–2 mm needles of orthopyroxene.

7.5 Wholerock geochemistry

Analyses of quench-textured micropyroxenites are listed in Table 7.1. It is important to know which of the rocks represent liquid compositions. Most are not quenched melts because they have distinct cumulate textures. The coarse-grained quench-textured micropyroxenites contain orthopyroxene megacrysts and cumulus chromite, which renders them inappropriate as liquid compositions. However, a few have geochemical characteristics consistent with those required to be the quench parental liquids to sections of the lower zone (CO-017, CO-114, DI-225, Table 7.1). Such evidence comes from their uniformly microcrystalline textures, mineral compositions and REE distributions (Harmer & Sharpe 1985).

Most samples have MgO contents of 10–16% and high SiO_2 contents, averaging about 53%. Total alkalis mostly exceed 2.5%, with K_2O values generally around 1%. All have high values of Ni, Cr and incompatible

Table 7.1 Compositions of B1 quench-textured micropyroxenites, ultramafic sills derived from B1 magma, B2 fine-grained marginal rock and B3 fine-grained marginal rock. Major elements (wt%) and trace elements (ppm). Co to Rb were determined by XRF. Sc to Lu were determined by INAA. Analytical techniques are described in Harmer & Sharpe (1985).

	B1					Ultramafic sills		B2	B3
	CO-114	DI-225	CO-230	CO-255	CO-211	CO-113	CO-175	CO-250	CO-045
SiO_2	56.7	56.5	56.4	56.3	55.3	48.6	41.4	51.6	51.7
TiO_2	0.40	0.39	0.27	0.28	0.28	0.13	0.09	0.59	0.19
Al_2O_3	12.7	13.0	10.6	10.6	10.5	5.07	3.16	15.8	16.8
Fe_2O_3	1.83	1.89	1.77	1.78	1.78	1.63	1.59	2.09	1.69
FeO	7.60	7.36	7.55	8.06	7.62	8.14	7.85	8.85	5.56
MnO	0.17	0.16	0.17	0.19	0.17	0.17	0.15	0.20	0.15
MgO	10.2	10.0	14.9	14.4	15.4	30.0	35.2	7.74	7.94
CaO	6.68	7.12	6.18	5.93	6.08	2.83	2.11	11.2	11.8
Na_2O	1.74	2.08	1.61	1.48	1.73	0.63	0.15	2.41	2.31
K_2O	1.22	0.89	0.73	0.77	0.75	0.30	0.26	0.19	0.24
P_2O_5	0.09	0.09	0.06	0.07	0.05	0.03	0.02	0.11	0.02
Cr_2O_3	0.11	0.09	0.19	0.22	0.22	0.72	0.75	0.05	0.08
NiO	0.05	0.05	0.05	0.07	0.06	0.23	0.26	0.03	0.02
LOI	0.25	0.38	0.10	−0.27	−0.05	0.67	6.96	−0.58	−0.03
H_2O^-	0.00	0.16	0.08	0.03	0.05	0.00	0.05	0.12	0.09
total	99.7	100.2	100.7	99.9	99.9	99.2	100.0	100.4	98.6
Co	113	77	96	95	85	244	155	99	81
Cr	766	631	1316	1485	1349	9160	4665	317	547
V	166	175	155	163	163	151	88	198	130
Zn	91	78	73	88	72	79	79	109	56
Cu	60	55	47	65	60	23	63	121	11
Ni	218	208	319	349	403	1487	2044	122	126
Nb	6	3	5	3	16	4	3	3	
Zr	105	97	64	70	236	31	39	28	26
Y	16	15	12	12	32	7	9	18	11
Sr	201	230	164	151	167	63	72	324	329
Rb	46	31	29	29	30	11	24	3	4
Sc	27	33	31	33	30	19	17	36	31
Hf	1.79	1.92	1.5	1.39	1.44	0.45	0.65	0.63	0.41
La	19.9	16.2	12.0	15.1	13.0	7.2	5.9	9.9	3.8
Ce	39.0	32.8	24.2	28.3	24.6	13.1	12.3	18.8	8.3
Nd	17.5	15.1	7.70	11.3	—	5.4	6.0	11.9	—
Sm	3.22	2.92	1.82	2.16	1.96	1.08	1.21	2.77	0.86
Eu	0.82	0.70	0.54	0.58	0.66	0.25	0.33	1.00	0.52
Yb	1.1	1.29	0.92	0.79	0.85	0.29	0.62	1.73	0.64
Lu	0.14	0.21	0.08	0.11	0.09	0.06	0.08	0.230	0.07

elements and have high Rb/Sr (> 0.1). Of the recognized quenched melts, SiO_2 is also 52–55%, but MgO values cluster around 11–12%. The K_2O and Rb contents (0.7–1.0% and 30–50 ppm, respectively) are characteristic. On an AFM diagram (Fig. 7.6) the quench-textured micropyroxenites follow the trend defined by most of the pyroxenitic sills and cumulates. This tholeiitic trend is distinct from that of other suites of marginal rocks and sills from this area, in that it continues the path defined by the ultramafic sills. These rocks plot close to the olivine–orthopyroxene boundary on the Ol–Plag–(Opx + 4Qtz) diagram of Irvine (1970) (Harmer & Sharpe 1985). The ultramafic (olivine-rich) sills may be related to the quench-textured micropyroxenites simply as a result of olivine addition.

The quench-textured rocks are characterized by steep chondrite-normalized REE patterns with $(Ce/Yb)_N$ of 10–14, and high absolute values of LREE with Ce = 25–40 ppm (Table 7.1). These data are an important key to the recognition of parental liquids. Bronzitites and dunites from the lower zone of the eastern Bushveld Complex have flat REE patterns with 1–3× chondritic

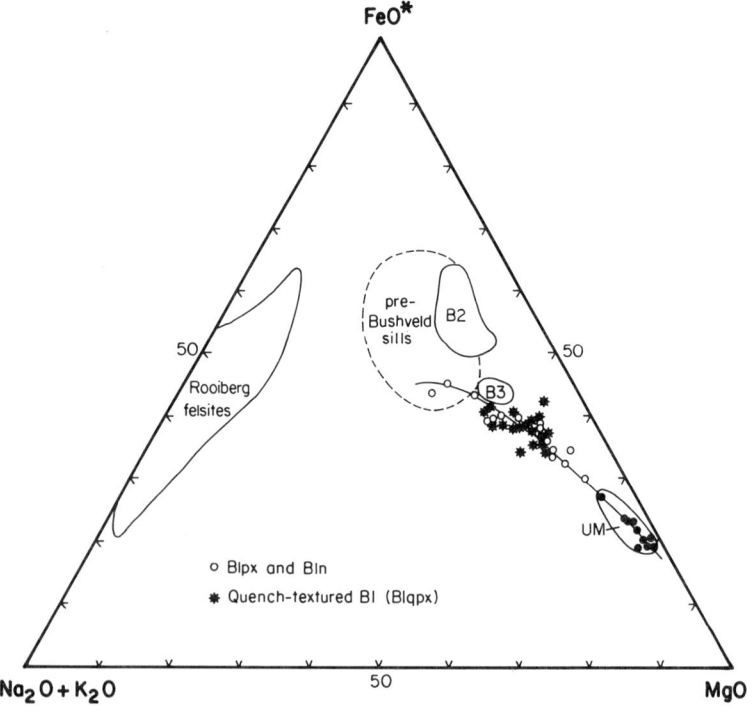

Figure 7.6 Ternary AFM diagram showing the marginal groups, the pre-Bushveld sills, the Rooiberg Felsites and the relation between quench-textured micropyroxenites (possible B1 primary magma) and other members of the B1 group (cumulate-enriched rocks have higher MgO while residual liquids have lower MgO than primary magma). Ultramafic sills (UM) represent olivine-enriched cumulates derived from the B1 magma.

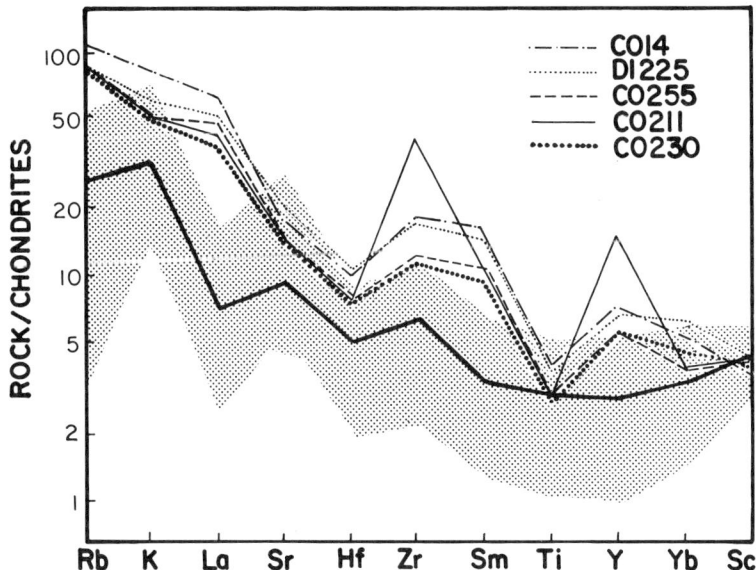

Figure 7.7 Trace-element concentrations (normalized to values used by Hickey & Frey (1982)) in the B1 group and in boninites from the western Pacific. The average for the western Pacific boninites is shown as a thick full line, and the range is given by the dotted field. Data from Jenner (1981), Hickey & Frey (1982), Cameron *et al.* (1983) and Bloomer & Hawkins (1987).

abundances (Harmer & Sharpe 1985). Because the patterns are not LREE-depleted, and orthopyroxene has a strong preference for HREE, it is possible that the bronzitites were derived from liquids with the highly fractionated patterns displayed by the B1 group. This highly fractionated character of the B1 group is shown in Figure 7.7.

Quench-textured micropyroxenites and other members of the pyroxenitic group, when plotted on a conventional Sr isochron diagram (Fig. 7.8), define a linear array with an 'age' of 2015 ± 211 Ma and an Sr_I of 0.7049 ± 13. The large degree of scatter in the data (Harmer & Sharpe 1985) implies that the Sr systematics of these samples are complex, and that a multicomponent source is involved in the genesis of these rocks. All micropyroxenite samples are enriched relative to the bulk-Earth composition of 0.7022 (calculated at 2050 Ma using a Rb/Sr value of 0.03 and a present-day $^{87}Sr/^{86}Sr$ of 0.7047 for bulk Earth (O'Nions *et al.* 1977).

7.6 Discussion

7.6.1 Comparison with other boninites

In terms of major-element concentrations, the most striking feature of boninites is the combination of high MgO and high SiO_2 contents. Figure 7.9

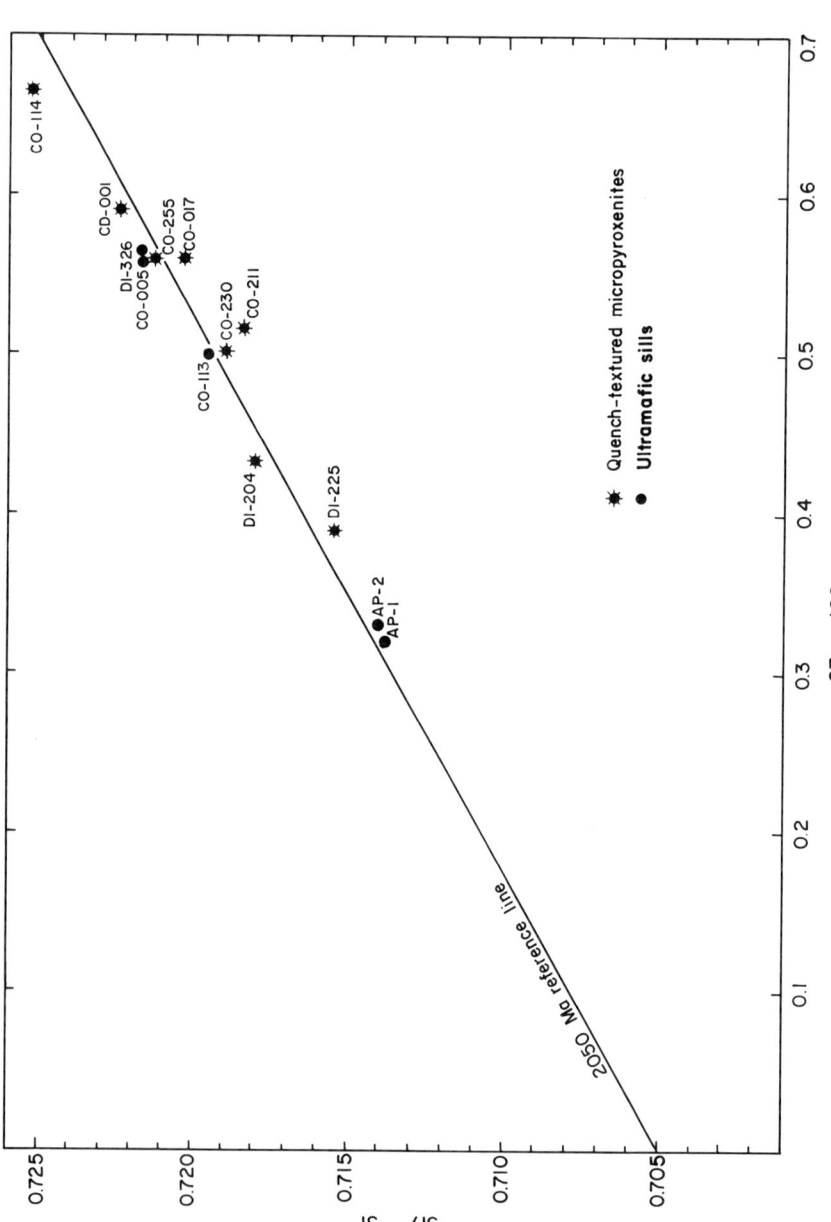

Figure 7.8 Isochron plot for samples of the B1 group. From Harmer & Sharpe (1985).

shows clearly that in this regard the B1 quench-textured micropyroxenites compare closely with boninites from the W Pacific. The CaO and TiO_2 contents of boninites are generally low (Cameron *et al.* 1983) and in this aspect the quench-textured micropyroxenites again fall within the range of W Pacific boninites (Figs 7.7 & 7.9). The quench-textured micropyroxenites are, however, more strongly enriched in Rb and the LREE and MREE than W Pacific boninites. Consequently, the notable enrichment of Zr relative to Sm in W Pacific boninites (Hickey & Frey 1982) is only weakly shown by these rocks (Fig. 7.7).

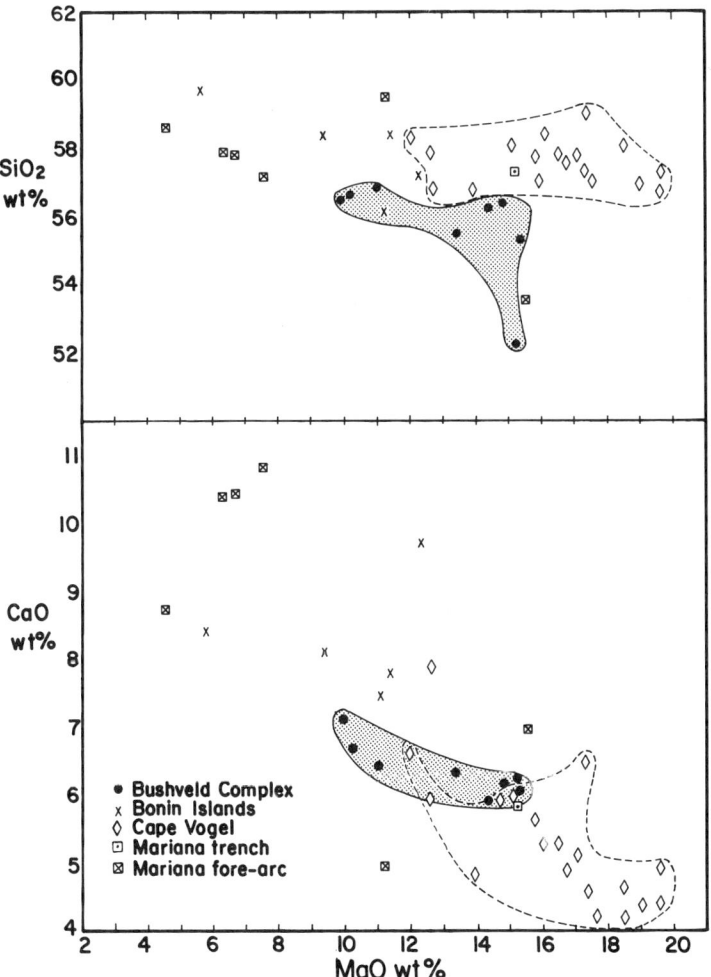

Figure 7.9 Binary plots of SiO_2 and CaO against MgO, comparing the B1 group with boninites from the western Pacific. Data sources as in Figure 7.8.

The presence of an enriched component in the source of the Bushveld quench-textured micropyroxenites is recorded in the high Sr_I values and high concentrations of K_2O, LREE and Zr. On the other hand the low CaO and TiO_2 contents and high Mg# (Table 7.1) are suggestive of a depleted source. This combination of depleted and enriched components is characteristic of boninitic magmas world-wide (Jenner 1981, Hickey & Frey 1982, Cameron et al. 1983, Nelson et al. 1984, Bloomer 1987). It is generally held that the boninite source consists of depleted mantle to which an enriched component has been added. In relation to depleted mantle, the enriched component has low Ti/Zr, high LREE contents and low to very low ε_{Nd}, reflecting recent-to ancient LREE enrichment (Hickey & Frey 1982, Nelson et al. 1984). On the basis of these features the enriched component could be (a) continental crust, (b) subducted sediment, or (c) ancient enriched mantle (Hickey & Frey 1982, Nelson et al. 1984). In an attempt to differentiate between these sources we will concentrate on Ti/Zr.

7.6.2 Nature of the enriched component

On the basis of Ti and Zr contents it can be argued that the Earth's mantle initially formed with a constant Ti/Zr throughout. The reason for this is that carbonaceous chondrites, which are representative of the low-temperature component from which the earth formed (Ringwood 1979), and Ca–Al chondrules in the Allende meteorite, representative of the high-temperature component, have very similar Ti/Zr (100 in carbonaceous chondrite, 98 in Ca–Al chondrules; Wanke et al. 1974) (Fig. 7.10a). Primitive mantle should have Ti/Zr close to 100, and this is indeed the case for primitive mantle compositions estimated by Taylor & McLennan (1985) (T and A in Fig.7.10a).

During extraction of MORB from primitive mantle, the Ti/Zr is little changed. The ratios for the primitive MORB glasses 3-18-17-1 and 3-14-10-1 are 97 and 95, respectively (Frey et al. 1974; full squares in Fig. 7.10a). For MORB analysed by Sun et al. (1979), the ratio ranges between 84 and 123 (dotted field in Fig. 7.10a). The small change in Ti/Zr during extraction of MORB from the mantle and during fractionation en route to the surface is an indication that the K_D for Ti and Zr are similar for all the minerals involved. LeRoex (1987) gives a bulk K_D of 0.03 for both Ti and Zr during melting of mantle containing 70% olivine, 25% orthopyroxene and 5% clinopyroxene. For plagioclase, Zr K_D is low, and for the limited data available it is not clear that there is a significant difference in the values for Ti and Zr (Fig. 2 of Pearce & Norry 1979).

Because Ti and Zr are not significantly fractionated during formation of the Earth or during MORB genesis, primordial or deep undepleted mantle (Kurz et al. 1982, Hart et al. 1983) cannot be invoked as the enriched component involved in boninite genesis. Fractionation of Ti from Zr during formation of continental crust must be considered. Average compositions of lower and

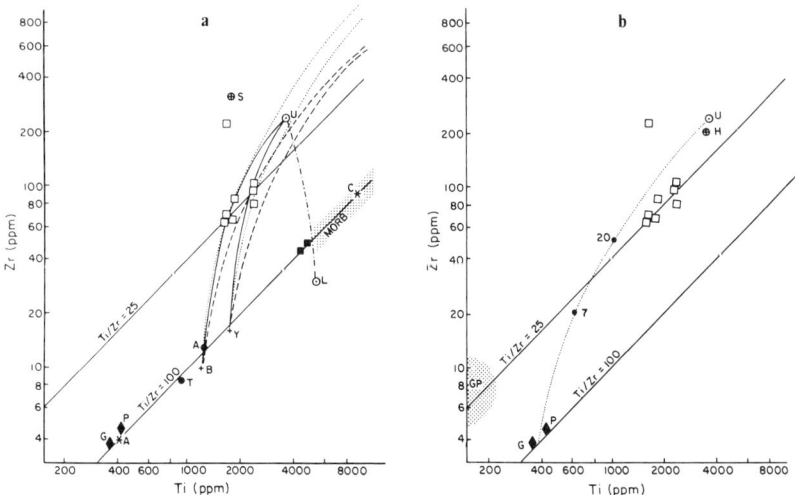

Figure 7.10 Zr and Ti concentrations in a range of rocks relevant to melting in the mantle and in the crust. The quench-textured micropyroxenites of the B1 group are depicted as open squares. With the exception of the anomalous sample CO211 the quench-textured micropyroxenites form a tight cluster with Ti/Zr near to 25. (a) Rocks related to mantle processes have Ti/Zr near to 100. Extraction of MORB (full squares and stippled field) from primitive mantle (full circles A and T) leaves depleted mantle (full diamonds G and P). (Star A is Ca–Al chondrule; star C is carbonaceous chondrite. Crosses B and Y are komatiites from Barberton and Yilgarn. Circled cross S is Swaziland gneiss. Circled points U and L are upper and lower crust. Chain curve is mixing curve between upper and lower crust. Thin full curves are mixing curves between komatiites and upper crust. Broken curves are AFC curves for assimilation of upper crust by komatiites. Dotted curves are AFC curves for assimilation of Swaziland gneiss by komatiites.) (b) Rocks related to processes in the lithosphere below the Kaapvaal craton have Ti/Zr of approximately 25. Simple mixing between depleted mantle (full diamonds) and upper crust (circled point U) is shown as a dotted curve with points labelled 7 and 20 indicating the percentage of upper crustal component which must be added to depleted mantle to obtain the range in Ti/Zr observed in the quench-textured micropyroxenites. In the case of mixing between depleted mantle and a partial melt of upper crust, mixing follows a similar path to that illustrated but the percentage of partial melt that must be added to depleted mantle is lower than for the case of simple mixing. Circled cross H is high-magnesian felsite. Extraction of quench-textured micropyroxenites from depleted mantle contaminated by upper crust leaves granular garnet peridotite (stippled field; Shimizu & Allegre 1978) as a residue. For further explanation and additional references, see text.

upper crust (Taylor 1979) are presented in Fig. 7.10a (open circles L and U). It is of interest that the mixing line joining lower and upper crust passes close to primitive MORB 3-14-10-1; it is possible to model the upper crust as a 10% melt of primitive MORB containing 5000 ppm Ti and 51 ppm Zr, leaving the Ti-enriched lower crust as a residue. Although the upper crust certainly did not form in so simple a fashion, the important point is that Ti appears to have behaved as a compatible element during generation of the upper crust, in contrast to its behaviour as an incompatible element in the mantle.

Considering the formation of continental crust in a little more detail, the behaviour of Ti as a compatible element can be attributed, at least in part, to the presence of an oxide phase in the relevant assemblages.

Taylor (1967) considered that continents initially grow by the addition of andesites and calc-alkaline rocks associated with island arcs. In the calc-alkaline series, Miyashiro & Shido (1975) ascribed the regular decrease of Ti and V with increasing FeO^*/MgO to magnetite fractionation; thus an oxide phase with a high K_D for Ti is involved during this first phase of continental growth. The next phase of continental growth is the generation of granites by partial melting of tonalites and other calc-alkaline rocks formed in the first phase (Anheusser & Robb 1981). Condie (1981) modelled this step with magnetite or ilmenite as a residual phase, and again Ti behaved as a compatible element. Zr, on the other hand, behaves as an incompatible element during the two stages mentioned above. Crustal melting therefore results in lowering of the Ti/Zr, in contrast to mantle processes where the ratio remains approximately constant.

Since the enriched component associated with boninites has low Ti/Zr values, continental crust was very likely involved in boninite genesis. Accepting this, debate then shifts to the question of whether the crustal component was incorporated during magma ascent through continental crust (Sparks 1986) or whether the crustal component was incorporated in the boninite source by subduction of sediments.

7.6.3 Boninites as contaminated komatiites

Sparks (1986) proposed that the boninitic magmas associated with the Bushveld and Stillwater intrusions were formed by assimilation of sediments into an ascending komatiitic magma. In Figure 7.10a, mixing curves between upper crust (U) and a typical komatiite from the Yilgarn block, Western Australia (Y) and komatiite from Barberton (B) are shown as thin full curves. A mix of approximately 70% komatiite with 30% crust can reproduce the Ti and Zr contents in most of the quench-textured pyroxenites. However, assimilation cannot be modelled as a simple mixing process, since the heat required to dissolve the assimilated material must be supplied by crystallization of olivine in the contaminated komatiite. This process, termed assimilation–fractional crystallization (AFC), can be modelled with equations developed by DePaolo (1981). The AFC paths are shown as broken curves in Figure 7.10a. For these paths the mass of assimilated material was assumed to be only slightly smaller than the mass of crystals formed to provide the necessary heat ($r = 0.98$). Taylor (1980) estimated an upper limit of r (mass assimilated/mass crystallized) as 0.3, but DePaolo (1981) suggested that in favourable circumstances, as in the lower crust in regions of high heat flow, r could approach 1. Even with the high value chosen, the AFC path for Yilgarn komatiite–upper crust does not intersect the field of B1 quench-textured micropyroxenites, although the Barberton komatiite–upper crust path intersects the field at its upper end (Fig.7.10a) This divergence of the AFC paths away from the simple mixing lines is a consequence of the fact that Ti and Zr are incompatible in olivine, so that, as assimilation proceeds, olivine crystallization drives the contaminated magma to higher Ti and Zr contents.

The Ti and Zr contents of the quench-textured micropyroxenites can better be reproduced by contamination with gneiss of the type exposed in Swaziland (Condic & Hunter 1975; S in Fig. 7.10a). Zr contents of Swaziland gneiss were not determined but Sharpe *et al.* (1983) considered 300 ppm to be a reasonable estimate. Although a match can be obtained for $r = 0.98$ (not shown), the very rapid enrichment of incompatible elements in the AFC process limits the possible amount of contaminant to less than 1% and this would not significantly change the major-element compositions. To effect a change in SiO_2 content from that of a komatiite ($\sim 45\%$ SiO_2) to that of a boninite ($\sim 53\%$ SiO_2) a value of r equal to 0.6 is required. This results in the addition of 9% contaminant and loss of 15% olivine to yield the required Ti, Zr and SiO_2 contents. AFC can thus yield a boninitic magma by contamination of komatiite with gneiss. The choice of parameters is poorly constrained, and it must be accepted that the AFC model is so flexible that demonstration that AFC can reproduce the geochemistry of a particular rock cannot be regarded as conclusive proof that AFC did in fact take place. We will therefore review and evaluate the reasoning behind the position that komatiites are likely to assimilate rocks within the upper crust.

The concept that komatiite dykes are strongly prone to contamination stems from the hypothesis, advanced by Huppert *et al.* (1984), that komatiite lavas extruded at the surface are easily contaminated. The nickel sulphides associated with komatiites at Kambalda were proposed to have formed in response to incorporation of sulphidic sediments during turbulent flow of komatiite lava over the bedrock (Huppert *et al.* 1984). Claoue-Long & Nesbitt (1985) disputed this, pointing to evidence that the komatiites followed pre-existing channels, rather than cutting their own by thermal erosion. In Zimbabwe, nickel deposits of a similar age and style to the Kambalda ores can be observed in the Trojan nickel mine. Here, field evidence favours the existence of a pre-existing topographic low, because the banded ironstone below the massive sulphide is of limited extent and pinches out away from the ore body (Chimimba pers. comm. 1988). Both the ironstone and the massive sulphides appear to have been deposited in a pre-existing channel. Furthermore, sulphur in the Trojan ores, in contrast to Kambalda, is isotopically distinct from the sulphur in the sediments (Chimimba pers. comm. 1988, Coleman *et al.* 1982). This evidence suggests that thermal erosion was not as extreme as Huppert *et al.* (1984) propose, and reinforces the argument of Claoue-Long & Nesbitt (1985) that 'the real lavas were more restrained and less voracious than their mathematical counterparts'. This restraint might be attributable to formation of a protective margin which cooled sufficiently that the viscosity became too low to permit turbulent flow and erosion of the bedrock.

In their evaluation of contamination during ascent of a komatiitic dyke, Huppert & Sparks (1985a) explicitly omit consideration of the effects of variation of viscosity with temperature of the magma close to the wall rock. Delaney & Pollard (1982) considered that cooling of a basaltic magma to a temperature of 0.9 to 0.95 times the liquidus temperature would result in the effective solidification of the magma. During laminar flow the magma near the walls would thus cool to form a viscous margin which would protect the hotter centre from wall rock contamination (Delaney & Pollard 1982). In komatiitic magmas, flow is calculated to be turbulent over a wide range of conditions (Huppert & Sparks 1985b), so a viscous margin may not develop. However, as contamination by cold and silica-rich country rock proceeds, the magma becomes more viscous and flow may eventually become laminar. At this point, a viscous margin would form and contamination would be halted. The boninitic magmas of the Bushveld Complex have relatively high silica contents. It is not clear that contamination of komatiite could ever proceed far enough to produce such siliceous magmas, since the increase in viscosity during contamination could result in flow becoming laminar and less amenable to further contamination.

An additional feature that is not entirely consistent with the general concept that the boninites of the Complex formed by contamination of komatiite is the existence of reverse-zoned orthopyroxene phenocrysts, as mentioned in Section 7.4.4. Reversed zoning indicates that in the final stages of emplacement the Mg# of the magma was increasing, whereas the contamination process should result in a decline in Mg# (Huppert & Sparks 1985b).

In view of the uncertainties associated with the hypothesis that Bushveld boninitic magmas formed by contamination of komatiitic magma, alternative hypotheses can be considered.

7.6.4 Enrichment of the boninite source by subducted sediments

If, during ascent, the boninitic magmas of the Bushveld Complex were protected from crustal contamination by formation of a viscous margin, then incorporation of material derived from the continental upper crust must have taken place in the source area. Subduction of continental crust in the form of sediment and subsequent mixing with subcontinental upper mantle is then the hypothesis to be evaluated.

Hamlyn and coworkers (Hamlyn *et al.* 1985, Hamlyn & Keays 1986) have stressed the role that depleted mantle, residual after the extraction of MORB, may have played in boninite genesis. Their arguments can briefly be outlined as follows. During extraction of MORB magma from primitive mantle, the system is sulphur-saturated, so immiscible sulphide globules formed remain in the depleted mantle residual after MORB extraction. Because PGE are very strongly partitioned from a silicate melt into a sulphide melt (Campbell & Barnes 1984), these sulphides will retain most of the PGE in the system. In consequence, depleted mantle has lower sulphur contents than primitive mantle, but the sulphides in depleted mantle have very high PGE/sulphur ratios. Hamlyn & Keays (1986) proposed that boninitic magmas in the Bushveld and Stillwater Complexes and Great Dyke were derived from depleted mantle, and that these magmas gave rise to the PGE deposits in these instrusions. Depleted mantle is therefore considered as the dominant component in the boninite source.

As representative of depleted mantle, peridotite xenoliths from the Lashaine volcano in East Africa are chosen. These are very fresh and have undergone minimal reaction with their host ankaramite (Rhodes & Dawson 1975). Metasomatism is limited and has resulted only in a slight enrichment of the alkalis and LREE. The Ti and Zr contents of the average garnet peridotite (G) and average garnet-free peridotite (P) from Lashaine are plotted in Figures 7.10a and b. These samples lie very close to the Ti/Zr = 100 line.

The mixing line between depleted mantle and upper crust is shown as a dotted curve in Figure 7.10b. The important feature of this curve is that it illustrates that, once more than 7% crust has mixed with depleted mantle, the Ti/Zr of the mixture will be equal to that in the boninitic rocks of the Bushveld Complex. When the proportion of crust exceeds 20%, the ratio will exceed that of the boninites, but because this portion of the curve is subparallel to lines of constant Ti/Zr, extensive mixing leads to only small changes in Ti/Zr. For mixing between a partial melt of continental crust and depleted mantle, the shape of the curve is the same, if it is assumed that the partial melt has a similar Ti/Zr to the source. However, the proportion of melt required to effect a significant change in Ti/Zr is much lower; for a 10% partial melt, mixtures of depleted crust with more than 1% of the 10% partial melt will have similar Ti/Zr to the boninitic rocks. The proportion of the curve that is subparallel to lines of constant Ti/Zr is therefore higher than for bulk crust–depleted mantle mixtures, since addition of only 1% melt will produce low Ti/Zr in the contaminated mantle.

The exact mechanism by which subducted sediment mixes with depleted mantle is not well understood at this stage, but it is clear that all the Ti contaminated in the sediment cannot have been incorporated in the boninite source. To change the Ti/Zr from that of depleted mantle to that of the Bushveld boninites results in a mixture which has higher Ti contents than primitive mantle. If a boninite were derived directly from this source, the Ti concentration would probably be higher than that in magma derived from primitive mantle (i.e. higher than primitive MORB). Since an important characteristic of boninites is their low Ti contents relative to MORB (Hamlyn

& Keays 1986), the boninite source cannot have originated directly by bulk assimilation of sediment. The most likely process for the creation of the boninite source is one of metasomatism of depleted mantle by a melt derived from subducted sediment. This metasomatism would be most intense at the interface between subducted sediment and subcratonic lithosphere (Fig. 7.11).

Melting and metasomatism are viewed in the context of hot subducted sediment underlying relatively cooler subcratonic lithosphere. In the initial stages small volumes of sediment-derived melt are envisaged to have risen and reacted with overlying lithosphere (Fig. 7.11b). On cooling, refractory components of the melt would crystallize and react with surrounding lithosphere, and less refractory components would concentrate in the residual melt. This residual melt would have high concentrations of K_2O and TiO_2, since these elements are incompatible in mantle minerals, and would be of broadly alkaline character (Fig. 7.11c). The alkaline rocks of the Phalaborwa Complex are contemporaneous with the Bushveld Complex (Eriksson 1982) and these may possibly be related to this phase of alkaline magmatism. The mantle

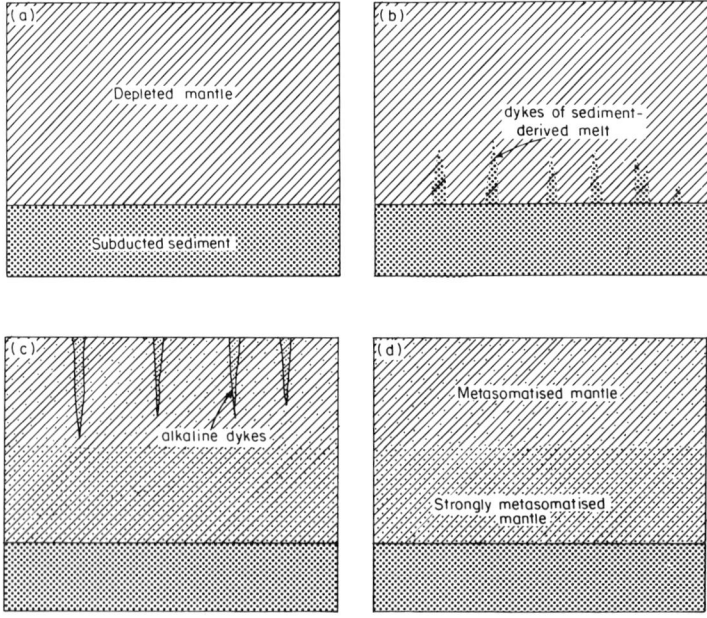

Figure 7.11 Model for the formation of boninite source area. (a) Subducted sediment underlies depleted mantle. (b) Dykes of melt derived from the sediment penetrate the base of the depleted mantle, partially crystallize and strongly metasomatize depleted mantle. (c) The residual melt rises as alkaline dykes which may then further metasomatize overlying mantle during their ascent. (d) Strongly metasomatized mantle (boninite source area) may contain olivine-free zones as a consequence of reaction with sediment-derived melt.

which remained after upward migration of the alkaline dykes would have increased its Ti/Zr by reaction with the sediment-derived melt, but would have relatively low abundances of Ti. This portion of the mantle is designated as 'strongly metasomatised mantle' in Figure 7.11d. This zone lay between silica-rich (coesite-bearing?) subducted sediment and depleted mantle in which olivine was the dominant phase. Interaction between these two bodies would lead to orthopyroxene formation by reaction between olivine and the silica-rich melt derived from the subducted sediment. It may be possible that this reaction could have led to the elimination of olivine from large portions of strongly metasomatised mantle. Such olivine-free portions would be appropriate source areas for the Bushveld boninites. Experimental data show that for Bushveld boninite, olivine only becomes a liquidus phase at pressures < 4 kbar (Cawthorn & Davies 1983), so that if the boninites underwent only limited modification during ascent, they would appear to have originated from an olivine-free source.

If the hypothesis that the boninitic rocks of the Bushveld Complex originated from sediment-contaminated mantle is to be accepted, then the quantities of sediment involved must have been sufficient to account for the large quantities of boninite in the Bushveld Complex. One million cubic kilometres is an upper limit. Subduction of such quantities is not considered to be an insurmountable problem because: (a) some 180 km^3 are presently being subducted down each kilometre line length of the Aleutian Trench (Kvenvolden & von Huene 1986) and at these rates a subduction zone located north of the Bushveld Complex could supply the necessary material in 14 million years; and (b) Nd and Hf isotopic results suggest that, from about 2 Ga, large quantities of continental material were recycled back into the mantle (Allegre & Ben Othman 1980, DePaolo 1983), probably by subduction of sediments. (The fact that the Bushveld Complex formed at the same time as the onset of large-scale crustal recycling may not be coincidental.)

7.6.5 *Implications of the subducted-sediment model for other Bushveld magmatism*

If immense quantities of subducted sediment were involved in Bushveld magmatism, then its signature might be apparent in rocks other than the boninites. Pre-Bushveld volcanism is recorded by the Dullstroom Volcanics and the Rooiberg Felsites. A certain rock type, common to both these successions, is referred to as high-Mg felsite (Twist 1985) in the Rooiberg Felsites, whereas in the Dullstroom Volcanics it is referred to as low-Ti dacite (Schweitzer 1987). For simplicity, the term high-Mg felsite will be used irrespective of the succession in which the rock occurs. High-Mg felsite has major-element concentrations which are very similar to late Archaean to early Proterozoic clastic sediments (Fig. 7.12). REE patterns are also similar to

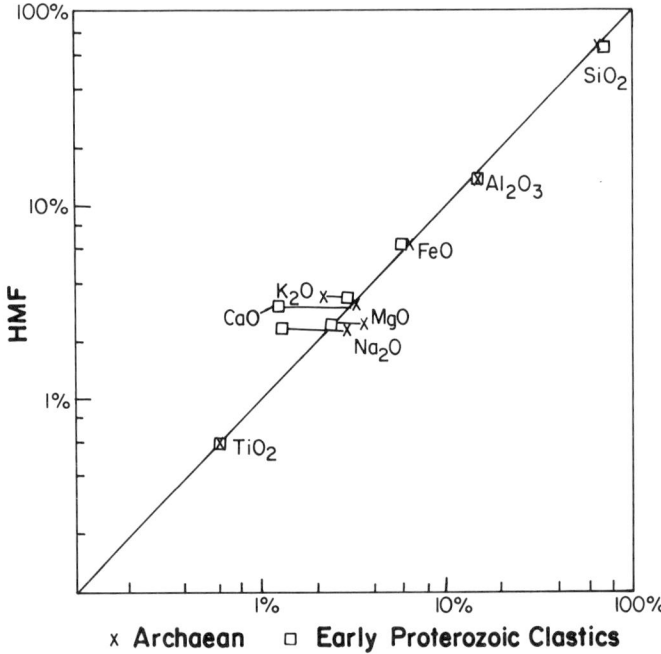

Figure 7.12 Comparison between major-element composition of high-Mg felsites and Archaean–early Proterozoic clastics.

those observed in sediments (Fig. 7.13) and the Ti and Zr concentrations are very similar to upper continental crust (compare high-Mg felsite (H) with upper crust (U) in Fig. 7.10b). Their dissimilarity from other acid lavas and similarity to sediments led Elsthon & Twist (1987) to suggest that these rocks might be impact melts. The alternative is that they represent very high degrees of melting of subducted sediment. If so, then melting of subducted sediment was a major process in the Bushveld magmatic cycle.

The possibility that the B1 magma was derived from depleted mantle which had been contaminated by the passage of high-Mg felsite through it is supported by concentrations of a number of trace elements (Fig. 7.14). Concentrations of Rb, K, La, Zr, Sm, Ti, Y and Yb in the B1 quench-textured micropyroxenite rocks are systematically lower than in the high-Mg felsites, and the trace-element patterns are subparallel. This suggests that one of the components in the B1 source may have been derived from high-Mg felsite. Depleted mantle of the type found in Lashaine has very low Y and Yb contents and in this regard may not be a suitable source component for the B1 rocks. LREE-depleted mantle, as represented by certain spinel lherzolite inclusions in alkali basalts, is more appropriate (Fig. 7.14). Contamination of LREE-depleted mantle by high-Mg felsite yields a rock that is a suitable source for the B1 magmas, at least with regard to the elements mentioned above. Concentra-

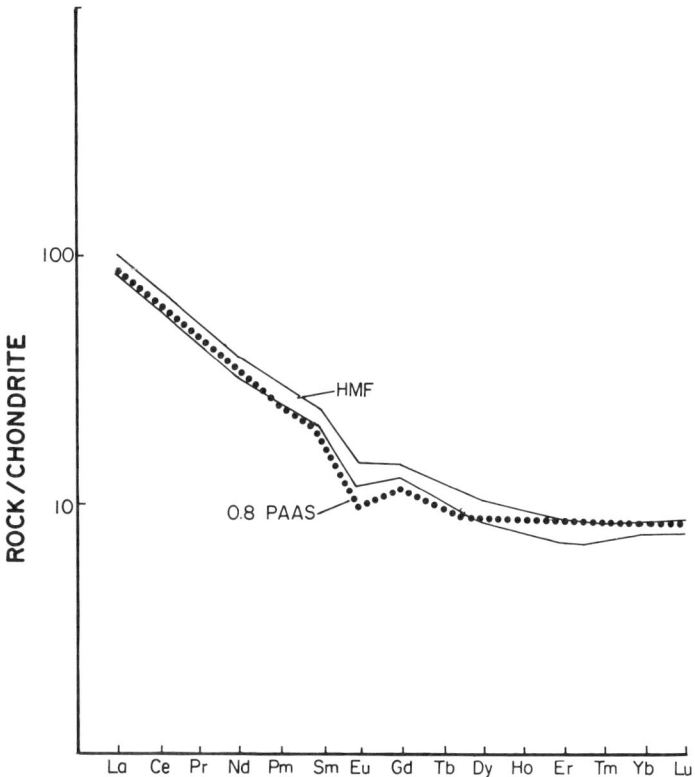

Figure 7.13 Rare-earth-element patterns of high-magnesian felsite (HMF) and 0.8 post-Archaean Australian shale (PAAS), equivalent to REE pattern of post-Archaean clastic sediments.

tions of Sr and Sc in the B1 quench-textured micropyroxenites are higher than in the high-Mg felsite. The tholeiites associated with the Bushveld Complex have relatively high concentrations of these elements, and a tholeiitic component may have been involved in the genesis of the B1 magmas. As an example of this tholeiitic component, the pattern for the B2 marginal rock (Table 7.1) is shown in Figure 7.14. Although involvement of a component of this nature could account for the Sr and Sc contents of the B1 quench-textured micropyroxenites, the evidence for a tholeiitic component in the B1 source is less convincing than the evidence for a component derived from high-Mg felsite.

The suggestion that sediment subduction was important requires that the Bushveld Complex was located above a subduction zone. The tectonic setting of the Bushveld Complex cannot be treated in detail here, but an observed northward increase in the model amount of olivine in the lower and critical zones led Hatton & Von Gruenewaldt (1987) to suggest that the lithosphere thinned to the north, where the plate margin may have been located.

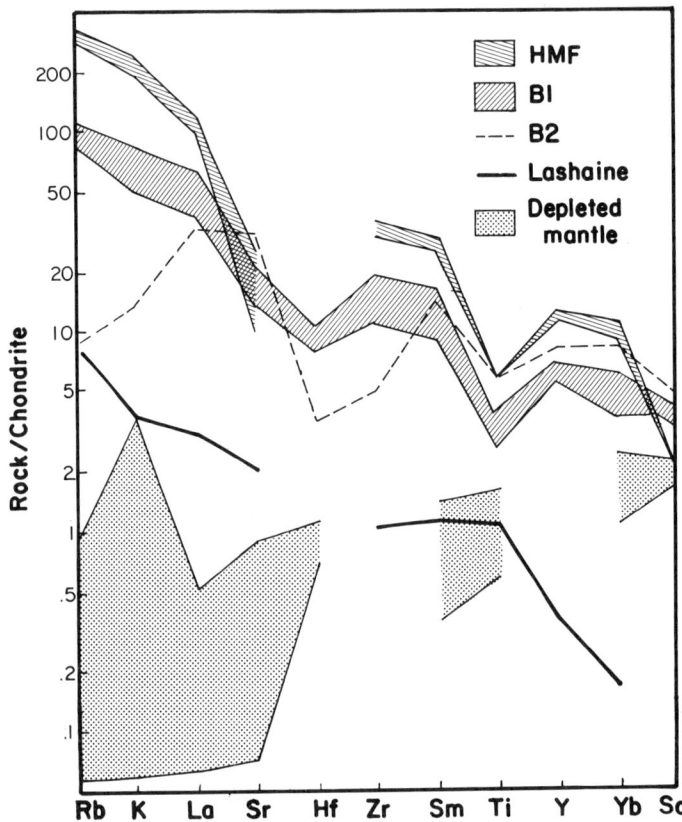

Figure 7.14 Normalized concentrations of trace elements in high-magnesian felsite (HMF), B1 quench-textured micropyroxenite (B1), fine-grained marginal rock (B2), depleted mantle from Lashaine (enriched in LREE), and depleted mantle as represented by spinel lherzolite xenoliths in alkali basalts (depleted in LREE). Normalization is to values given by Hickey & Frey (1982). Gaps in the patterns indicate that no data are available for certain elements. Data from Twist & Harmer (1987), Ridley & Dawson (1975), Rhodes & Dawson (1975), Philpotts *et al.* (1972), Jagoutz *et al.* (1979) and BVSP (1981).

7.7 Summary

The B1 group of marginal rocks and sills is the best available representative of the magma that gave rise to the lower zone and other pyroxenitic rocks of the Bushveld Complex. Economically important layers in the layered suite are invariably associated with pyroxenite and the B1 magma type is regarded as the source for most of the PGE and Cr mineralization in these layers. Quench-textured micropyroxenites are least affected by near-surface fractionation and are closest to the primary B1 magma composition. Their composition is very similar to that of boninites and consideration of their possible origins suggests that they formed in a broadly similar (i.e. subduction-related) tectonic setting and by similar processes to modern W Pacific boninites.

The quench-textured micropyroxenites have low Ti/Zr and consideration of the fractionation of these elements during cosmochemical processes, MORB extraction and formation of continental crust suggests that the low Ti/Zr probably reflects incorporation of a component derived from continental upper crust. Crustal incorporation during ascent of a komatiite through the upper crust is considered to be an unlikely petrogenetic model for these rocks. Instead, a model of incorporation of upper crust into the boninite source via sediment subduction and subsequent metasomatism of subcontinental lithosphere is favoured. Sediment subduction may have played a major role in the formation of the Bushveld Complex.

References

Allegre, C. J. & D. Ben Othman 1980. Nd–Sr isotopic relationship in granitoid rocks and continental crust development: a chemical approach to orogenesis. *Nature* **286**, 335–41.

Anheusser, C. R. & L. J. Robb 1981. Magmatic cycles and the evolution of the Archaean granitic crust in the eastern Transvaal and Swaziland. Spec. Publ. Geol. Soc. Aust. no. 7, 457–67.

Armstrong, R. A., W. Compston, E. A. Retief & H. J. Welke 1986. Ages and isotopic evolution of the Ventersdorp volcanics. *Extended Abstracts: Geocongress '86*, 89–92. Johannesburg: Geol. Soc. S. Afr.

Bloomer, S. H. 1987. Geochemical characteristics of boninite and tholeiite-series volcanic rocks from the Mariana forearc and the role of an incompatible element-enriched fluid in arc petrogenesis. Geol. Soc. Am. Pap. no. 215, 151–63.

Bloomer, S. H. & J. W. Hawkins 1987. Petrology and geochemistry of boninite series volcanic rocks from the Mariana trench. *Contrib. Mineral. Petrol.* **97**, 361–77.

Button, A. 1976. Stratigraphy and relations of the Bushveld floor in the eastern Transvaal. *Trans. Geol. Soc. S. Afr.* **79**, 3–12.

BVSP (Basaltic Volcanism Study Project) 1981. *Basaltic volcanism on the terrestrial planets*. New York: Pergamon.

Cameron, E. N. 1978. The lower zone of the eastern Bushveld Complex in the Olifants River trough. *J. Petrol.* **19**, 437–62.

Cameron, E. N. 1980. Evolution of the lower critical zone, central sector, eastern Bushveld Complex, and its chromite deposits. *Econ. Geol.* **75**, 845–71.

Cameron, E. N. 1982. The upper critical zone of the eastern Bushveld Complex – precursor to the Merensky Reef. *Econ. Geol.* **77**, 1307–27.

Cameron, W. E., M. T. McCulloch & D. A. Walker 1983. Boninite petrogenesis: chemical and Nd–Sr isotopic constraints. *Earth Planet. Sci. Lett.* **65**, 75–89.

Campbell, I. H. & S. J. Barnes 1984. A model for the geochemistry of the platinum-group elements in magmatic sulfide deposits. *Can. Mineral.* **22**, 151–60.

Cawthorn, R. G. & G. Davies 1983. Experimental data at 3 kbars pressure on parental magma to the Bushveld Complex. *Contrib. Mineral. Petrol.* **83**, 128–35.

Cawthorn, R. G., G. Davies, A. R. Clubley-Armstrong & T. S. McCarthy 1981. Sills associated with the Bushveld Complex, South Africa: an estimate of the parental magma composition. *Lithos* **14**, 1–15.

Claoue-Long, J. C. & Nesbitt, R. W. 1985. Contaminated komatiites. *Nature* **313**, 247.

Coleman, M., S. Taylor, L. Chimimba & J. Kramers 1982. *The origin of sulphur in mineralisation at the Trojan nickel mine, Zimbabwe*. Inst. Geol. Sci., Stable Isotope Rep. no. 81.

Condie, K. C. 1981. Geochemical and isotopic constraints on the origin and source of Archaean granites. Spec. Publ. Geol. Soc. Aust. no. 7, 469–79.

Condie, K. C. & D. R. Hunter 1976. Trace element geochemistry of Archaean volcanic rocks from the Barberton region, South Africa. *Earth Planet. Sci. Lett.* **29**, 389–400.

Czamanske, G. K. & M. L. Zientek 1985. *The Stillwater Complex, Montana: geology and guide*. Montana Bur. Mines and Geology, Spec. Publ. no. 92.

Delaney, P. T. & D. D. Pollard 1982. Solidification of basaltic magma during flow in a dyke. *Am. J. Sci.* **282**, 856–85.

DePaolo, D. J. 1981. Trace element and isotopic effects of combined wallrock assimilation and fractional crystallization. *Earth Planet. Sci. Lett.* **53**, 189–202.

DePaolo, D. J. 1983. The mean life of continents: estimates of continent recycling rates from Nd and Hf isotopic data and implications for mantle structure. *Geophys. Res. Lett.* **10**, 705–8.

Elsthon, W. E. & Twist, D. 1987. Is the Rooiberg Felsite (Bushveld Complex) an impact melt, *Contributions to International Workshop on Cryptoexplosions and Catastrophes in the Geological Record*, Parys, South Africa, Section E-2.

Eriksson, S. C. 1982. Aspects of the petrochemistry of the Phalaborwa Complex, northeastern Transvaal, South Africa. Ph.D.Thesis, University of the Witwatersrand, Johannesburg.

Frey, F. A., W. B. Bryan & G. Thompson 1974. Atlantic ocean floor: geochemistry and petrology of basalts from Legs 2 and 3 of the Deep-Sea Drilling Project. *J. Geophys. Res.* **79**, 5507–27.

Frick, C. 1967. The margin of the Bushveld Complex in the vicinity of De Berg, north of Dullstroom. M.Sc. Thesis, University of Pretoria.

Grant, N. K. & P. A. Molling 1981. A strontium isotope and trace element profile through the Partridge River troctolite, Duluth Complex, Minnesota. *Contrib. Mineral. Petrol.* **77**, 296–305.

Hamilton, P. J. 1977. Sr isotope and trace element studies of the Great Dyke and Bushveld mafic phase and their relationship to early Proterozoic magma genesis in southern Africa. *J. Petrol.* **18**, 24–52.

Hamlyn, P. R. & R. R. Keays 1986. Sulfur saturation and second-stage melts: application to the Bushveld platinum metal deposits. *Econ. Geol.* **81**, 1431–45.

Hamlyn, P. R., R. R. Keays, W. E. Cameron, A. J. Crawford & H. M. Waldron 1985. Precious metals in magnesian low-Ti lavas: implications for metallogenesis and sulfur saturation in primary magmas. *Geochim. Cosmochim. Acta* **49**, 1797–811.

Harmer, R. E. & M. R. Sharpe 1985. Field relations and strontium isotope systematics of the marginal rocks of the eastern Bushveld Complex. *Econ. Geol.* **80**, 813–37.

Hart, R., J. Dymond, L. Hogan & J.-G. Schilling 1983. Mantle plume noble gas component in glassy basalts from the Reykjanes Ridge. *Nature* **305**, 403–7.

Hatton, C. J. 1988. Densities and liquidus temperatures of Bushveld parental magmas as constraints on the formation of the Merensky Reef. In *Proc. Fifth Magmatic Sulphide Conf.*, M. D. Prendergast (ed.). Institution of Mining and Metallurgy.

Hatton, C. J. & G. Von Gruenewaldt 1987. The geological setting and petrogenesis of the Bushveld chromitite layers. In *Evolution of chromium ore fields*, C. W. Stowe (ed.), 109–43. New York: Van Nostrand Reinhold.

Hickey, R. L. & F. A. Frey 1982. Geochemical characteristics of boninite series volcanics: implications for their source. *Geochim. Cosmochim. Acta* **46**, 2099–115.

Huppert, H. E. & R. S. J. Sparks 1985a. Cooling and contamination of mafic and ultramafic magmas during ascent through continental crust. *Earth Planet. Sci. Lett.* **74**, 371–86.

Huppert, H. E. & R. S. J. Sparks 1985b. Komatiites. I: Eruption and flow. *J. Petrol.* **26**, 694–725.

Huppert, H. E., R. S. J. Sparks, J. S. Turner & N. T. Arndt 1984. Emplacement and cooling of komatiitic lavas. *Nature* **309**, 19–22.

Irvine, T. N. 1970. Crystallization sequences in the Muskox intrusion and other layered intrusions: I. Olivine–pyroxene–plagioclase relations. Geol. Soc. S. Afr. Spec. Publ. no. 1, 441–76.

Irvine, T. N. & Sharpe, M. R. 1982. Source-rock compositions and depth of origin of Bushveld and Stillwater magmas. *Carnegie Inst. Washington Yearb.* **81**, 294–303.

Irvine, T. N., D. W. Keith & S. G. Todd 1983. The J-M platinum–palladium reef of the Stillwater Complex Montana. II. Origin by double-diffusive convective magma mixing and implications for the Bushveld Complex. *Econ. Geol.* **78**, 1287–334.

Jagoutz, E., H. Palme, H. Baddenhausen, K. Blum, M. Cendales, G. Dreibus, B. Spettel, V. Lorentz & H. Wanke 1979. The abundances of major, minor, and trace elements in the Earth's mantle as derived from primitive ultramafic nodules. *Proc. 10th Lunar Planet. Sci. Conf.*, 2031–50.

Jenner, G. A. 1981. Geochemistry of high-Mg andesites from Cape Vogel. *Chem. Geol.* **33**, 307–32.

Kurz, M. D., W. J. Jenkins & S. R. Hart 1982. Helium isotope systematics of oceanic islands and mantle heterogeneity. *Nature* **297**, 43–7.

Kvenvolden, K. A. & R. von Huene 1986. Natural gas generation in sediments of the convergent margin of the eastern Aleutian trench area. In *Tectonostratigraphic terranes of the circum-Pacific region*, D. G. Howell (ed.), 31–49. Circum-Pacific Council for Energy and Mineral Resources Earth Sci. Ser. no. 1.

Lee, C. A. 1983. Trace and platinum-group element geochemistry and the development of the Merensky Unit of the western Bushveld Complex. *Mineral. Dep.* **18**, 173–90.

Lee, C. A. & M. R. Sharpe 1989. In preparation.

LeRoex, A. P. 1987. Source regions of mid-ocean ridge basalts; evidence for enrichment processes. In *Mantle metasomatism*, M. A. Menzies & C. J. Hawkesworth (eds), 389–422. London: Academic Press.

Lightfoot, P. C., A. J. Naldrett & C. J. Hawkesworth 1984. The geology and geochemistry of the Waterfall Gorge Section of the Insizwa Complex with particular reference to the origin of the nickel sulfide deposits. *Econ. Geol.* **79**, 1857–79.

Luhr, J. F. & I. S. E. Carmichael 1980. The Colima volcanic complex, Mexico. I. Post-caldera andesites from Volcan Colima. *Contrib. Mineral. Petrol.* **71**, 343–72.

Miyashiro, A. & F. Shido 1975. Tholeiitic and calc-alkalic series in relation to the behaviours of titanium, vanadium, chromium and nickel. *Am. J. Sci.* **275**, 265–77.

Molyneux, T. G. 1974. A geological investigation of the Bushveld Complex in Sekhukhuneland and part of the Steelpoort Valley. *Trans. Geol. Soc. S. Afr.* **77**, 329–38.

Naldrett, A. J., G. Cameron, G. Von Gruenewaldt & M. R. Sharpe 1987. The formation of stratiform platinum-group element deposits in layered intrusions. In *Origin of igneous layering*, I. Parsons (ed.), 313–97. Dordrecht: Reidel.

Nelson, D. R., A. J. Crawford & M. T. McCulloch 1984, Nd–Sr isotopic and geochemical systematics in Cambrian boninites and tholeiites from Victoria, Australia. *Contrib. Mineral. Petrol.* **88**, 164–72.

O'Nions, R. K., P. J. Hamilton & N. M. Evensen 1977. Variation in $^{143}Nd/^{144}Nd$ and $^{87}Sr/^{86}Sr$ ratios in oceanic basalts. *Earth Planet. Sci. Lett.* **34**, 13–22.

Pearce, J. A. & M. J. Norry 1979. Petrogenetic implications of Ti, Zr, Y and Nb variations in volcanic rocks. *Contrib. Mineral. Petrol.* **69**, 33–47.

Philpotts, J. A., C. C. Schnetzler & H. H. Thomas 1972. Petrogenetic implications of some new geochemical data on eclogitic and ultrabasic inclusions. *Geochim. Cosmochim. Acta* **36**, 1131–66.

Raedeke, L. D. & I. S. McCallum 1984. Investigations in the Stillwater Complex, part II. Petrology and petrogenesis of the ultramafic series. *J. Petrol.* **25**, 395–420.

Rhodes, J. M. & J. B. Dawson 1975. Major and trace element chemistry of peridotite inclusions from the Lashaine volcano, Tanzania. *Phys. Chem. Earth* **9**, 545–57.

Ridley, W. I. & J. B. Dawson 1975. Lithophile trace element data bearing on the origin of peridotite xenoliths, ankaramite and carbonatite from Lashaine volcano, N Tanzania. *Phys. Chem. Earth* **9**, 559–69.

Ringwood, A. E. 1979, Composition and origin of the Earth. In *The Earth: its origin, structure and evolution*, M. W. McElhinny (ed.), 1–58. London: Academic Press.
SACS (South African Committee for Stratigraphy) 1980. *Stratigraphy of South Africa*, Part 1, L. E. Kent (compiler). Geol. Surv. S. Afr. Handb. no. 8.
Schweitzer, J. 1987. The transition from the Dullstroom basalt formation to the Rooiberg felsite group, Transvaal sequence: a volcanological, geochemical, and petrological investigation. Ph.D. Thesis, University of Pretoria.
Sharpe, M. R. 1978. 'Cone-type' diabases from the Eastern Transvaal – representatives of a quenched magma. *Trans. Geol. Soc. S. Afr.* **81**, 373–8.
Sharpe, M. R. 1981. The chronology of magma influxes to the eastern compartment, Bushveld Complex, as exemplified by its marginal border groups. *J. Geol. Soc. Lond.* **138**, 307–26.
Sharpe, M. R. 1984. *Petrography, classification and chronology of mafic sill intrusions beneath the eastern Bushveld Complex*. Geol. Surv. S. Afr. Bull. no. 77.
Sharpe, M. R. 1985. Strontium isotope evidence for preserved density stratification in the main zone of the Bushveld Complex, South Africa. *Nature* **316**, 119–126.
Sharpe, M. R. & L. J. Hulbert 1985. Ultramafic sills beneath the eastern Bushveld Complex: mobilised suspensions of early lower zone cumulates in a parental magma with boninitic affinities. *Econ. Geol.* **80**, 849–71.
Sharpe, M. R. & T. N. Irvine 1983. Melting relations of two Bushveld chilled margin rocks and implications for the origin of chromitites. *Carnegie Inst. Washington Yearb.* **82**, 295–300.
Sharpe, M. R. & J. A. Snyman 1980. A model for the emplacement of the eastern compartment of the Bushveld Complex. *Tectonophysics* **65**, 85–110.
Sharpe, M. R., R. Brits & J. P. Engelbrecht 1983. *Rare earth and trace element evidence pertaining to the petrogenesis of 2.3 Ga old continental andesites and other volcanic rocks from the Transvaal Sequence, South Africa*. Inst. Geol. Res. Bushveld Complex, Univ. Pretoria, Res. Rep. no. 40.
Shimizu, N. & C. J. Allegre 1978. Geochemistry of transition elements in garnet lherzolite nodules in kimberlites. *Contrib. Mineral. Petrol.* **67**, 41–50.
Sparks, R. S. J. 1986. The role of crustal contamination in magma evolution through geological time. *Earth Planet. Sci. Lett.* **78**, 211–23.
Sun, S.-S. & R. W. Nesbitt 1978. Geochemical regularities and genetic significance of ophiolitic basalts. *Geology* **6**, 689–93.
Sun, S.-S., R. W. Nesbitt & A. Y. Sharaskin 1979. Geochemical characteristics of mid-ocean ridge basalts. *Earth Planet. Sci. Lett.* **44**, 119–38.
Taylor, H. P. 1980. The effects of assimilation of country rocks by magmas on $^{18}O/^{16}O$ and $^{87}Sr/^{86}Sr$ systematics in igneous rocks. *Earth Planet. Sci. Lett.* **47**, 243–54.
Taylor, S. R. 1967. The origin and growth of continents. *Tectonophysics* **4**, 1–7.
Taylor, S. R. 1979. Chemical composition and evolution of the continental crust: the rare earth element evidence. In *The earth: its origin, structure and evolution*, M. W. McElhinny (ed.), 353–76. London: Academic Press.
Taylor, S. R. & S. McLennan 1985. *The continental crust: its composition and evolution*. Oxford: Blackwell.
Twist, D. 1985. Geochemical investigation of the Rooiberg silicic lavas in the Loskop Dam area, southeastern Bushveld. *Econ. Geol.* **80**, 1153–65.
Twist, D. & B. M. French 1983. Voluminous acid volcanism in the Bushveld Complex: a review of the Rooiberg Felsite. *Bull. Volcanol.* **46**, 225–42.
Twist, D. & R. E. Harmer 1987. Geochemistry of contrasting siliceous magmatic suites in the Bushveld Complex: genetic aspects and implications for tectonic discrimination diagrams. *J. Volcanol. Geotherm. Res.* **32**, 83–98.
Vermaak, C. F. 1977. The Merensky Reef – thoughts on its environment and genesis. *Econ. Geol.* **71**, 1270–98.

Von Gruenewaldt, G. 1973. The main and upper zones of the Bushveld Complex in the Roossenekal area, Eastern Transvaal. *Trans. Geol. Soc. S. Afr.* **76**, 207–27.

Von Gruenewaldt, G., M. R. Sharpe & C. J. Hatton 1985. The Bushveld Complex: introduction and review, *Econ. Geol.* **80**, 803–12.

Wager, L. R. & G. M. Brown 1968. *Layered igneous rocks.* Edinburgh: Oliver & Boyd.

Wanke, H., H. Baddenhausen, H. Palme & B. Spettel 1974. On the chemistry of the Allende inclusions and their origin as high temperature condensates. *Earth Planet. Sci. Lett.* **23**, 1–7.

8 Petrology and geochemistry of early Proterozoic high-Mg dykes from the Vestfold Hills, Antarctica

S. M. KUEHNER

Abstract

The oldest undeformed mafic dykes in the Vestfold Hills of the East Antarctic Shield are characterized by relatively high SiO_2 (51–57%), MgO (7–14%), Cr and Ni (up to 1600 ppm and 417 ppm, respectively) and low TiO_2 (0.49–0.70%) contents. The suite was emplaced at ~2400 Ma, directly following an amphibolite–granulite facies tectonothermal metamorphic event at ~2500 Ma. Chondrite-normalized plots of incompatible trace elements, major-element ratios and phase compositions divide the high-Mg suite into three distinct subgroups. Trace-element characteristics prohibit these subgroups from being related by crystal fractionation, and previous isotopic studies preclude crustal contamination as a significant source of chemical variations. Major- and trace-element evaluation also indicates that two of the chemically distinct subgroups were derived from primitive liquids extracted from separate 'chondritic' upper-mantle sources, leaving olivine + orthopyroxene residues. Some selective trace-element contamination has been superimposed upon the chondritic characteristics, possibly through a 'wallrock' reaction-type process with a plagioclase-bearing lherzolite. Comparison of an estimated parental liquid composition (Mg # = 0.75) with experimental melting studies indicates that magma extraction took place at pressures of ~10 kbar (35 km), consistent with the geochemical signature indicating partial melting within a plagioclase-bearing mantle.

8.1 Introduction

This paper discusses the petrology and geochemistry of the Vestfold Hills (Antarctica) high-Mg (HMG) tholeiite suite, one of five chemically distinct dyke suites emplaced in the Vestfold Hills (Fig. 8.1) during the early–middle Proterozoic. Unlike the dyke samples collected by Collerson & Sheraton (1986) as part of a regional geological–geochronological survey, the material discussed here forms part of a detailed study aimed specifically at the hundreds of mafic dykes outcropping in the Vestfold Hills. Fieldwork for this study focused on two aspects of the dyke occurrences. First, the near-continuous exposure allowed documentation of many dyke intersections. Thus, it was possible to detail the relative ages of individual dykes within a specific suite and so monitor the chemical evolution with time within the suite. Secondly, sampling emphasized the collection of 'quenched' margin material. As 'phenocrysts' in the chilled margins are typically < 1 mm long, the samples are considered to represent liquid compositions closely. Furthermore, recognition of liquidus phases and, therefore, possible fractionating assemblages is unequivocal in chilled-margin samples. This is not always the case in coarse-grained, slowly cooled rocks where differences in growth rates, crystal–liquid reactions or continual re-equilibration of the precipitating phases may obscure liquidus phase relations and compositions. In the HMG dykes, morphologically distinct chilled margins, as emphasized in weathering profiles, were generally not observed. Instead, the dykes vary continuously from very fine-grained, almost glassy, margins (with conchoidal fracture) to coarser-grained, often noritic, interiors.

Compositionally similar HMG dykes also outcrop in the Napier Complex of Enderby and Kemp Land, ~1000 km west of the Vestfold Hills (Fig. 8.1). Both the Vestfold Hills and the Napier Complex are high-grade granulite terrains with complex Archaean geological histories extending to at least ~3100 Ma (James & Black 1981, Collerson et al. 1983) and contain petrological evidence for prolonged (~2000 Ma) deep crustal residence times (Kuehner 1986, 1988). Emplacement of the HMG dykes in the Vestfold Hills (2424 ± 72 Ma; Collerson & Sheraton 1986) and the Napier Complex (2350 ± 48 Ma; Sheraton & Black 1981) immediately post-dates an upper amphibolite–granulite facies tectonothermal metamorphic event affecting both terrains at ~2400–2500 Ma (Black & James 1983, Collerson et al. 1983). The Napier Complex HMG dykes are undeformed, but show slight-to-extensive recrystallization and thus apparently post-date the emplacement of approximately coeval undeformed Fe-rich dykes which have granoblastic textures (Sheraton & Black 1981). The Vestfold Hills HMG dykes, in contrast, have retained their igneous textures and primary phenocryst compositions and were emplaced contemporaneously with an Fe-rich tholeiite suite (Kuehner 1986). Geochemical studies by Kuehner (1986) and Sheraton & Black (1981) have shown that there is no genetic relationship between the Fe-rich and

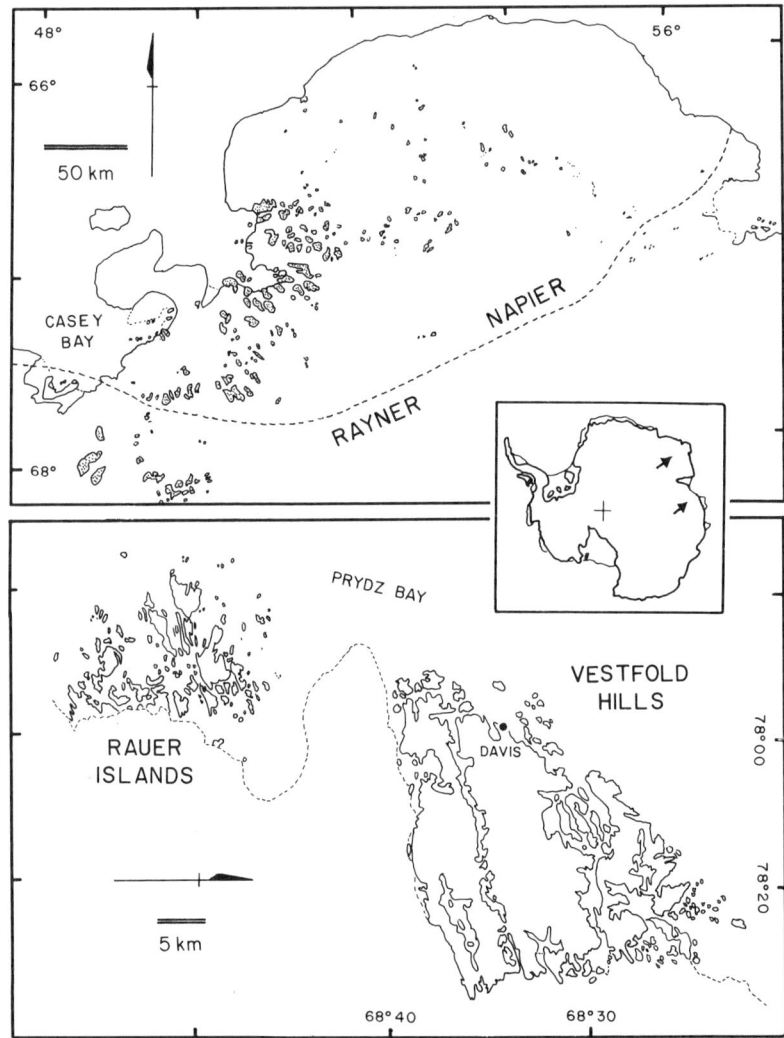

Figure 8.1 Location map showing the Archaean terrains of the Vestfold Hills and Napier Complex, and adjacent Proterozoic belts of the Rayner Complex and Rauer Islands. See James & Tingey (1983).

Mg-rich dyke associations of either terrain. Experimental reproduction of the HMG phenocryst assemblage by Kuehner (1986, 1988) has shown that these suites crystallized under a load pressure of 7–8 kbar.

8.2 Characterization of subgroups

Geochemical studies of 30 Vestfold Hills HMG dykes have delineated three distinct subgroups based on trace-element concentrations and major-element ratios. A convenient method for visualizing the chemical differences which define the groups is by comparing incompatible trace-element concentrations on normalized element variation diagrams (see Fig. 8.7). This plot subdivides the HMG suite into two distinct groups based on the presence or absence of a negative Sr anomaly (Sr depletion), a characteristic which distinguishes samples that cannot be related to one another through the fractionation of olivine, orthopyroxene or clinopyroxene. In general, these chemical differences are reflected petrographically, in that the 'normal Sr' subgroup is olivine + orthopyroxene phyric (SGP I), while the 'Sr-depleted' subgroup is dominated by olivine-free dykes. Exceptions to this generalization are the most LILE-enriched SGP I sample (see Fig. 8.7a), which is olivine-free, and the three Sr-depleted samples with the lowest LILE abundances, which are olivine-bearing.

A further subdivision of the Sr-depleted group can be made based on projections from the clinopyroxene apex of the basalt tetrahedron (Green 1970) onto the plane olivine–quartz–(jadite + Ca-tschermak) (see Fig. 8.9). Analyses of all olivine-bearing samples from each trace-element-defined subgroup plot along the Plag–Opx join, whereas the olivine-free samples are subdivided into two distinct groupings. Comparing these latter two populations with Figure 8.7b shows that the samples with the lowest normative olivine contents correspond to analyses with the most evolved LILE concentrations (note, for example, Rb_N). Characteristics of the subgroups are summarized below:

Subgroup SGP I :   normal Sr_N, olivine + orthopyroxene phyric (10 samples)
Subgroup SGP II : Sr_N-depleted, orthopyroxene phyric, Rb-poor (9 samples)
Subgroup SGP III: Sr_N-depleted, orthopyroxene phyric, Rb-rich (11 samples)

8.3 Age relationships of the HMG subgroups

Collerson & Sheraton (1986) found that Rb–Sr isotopic data from their 11 HMG samples did not produce an isochron within acceptable statistical limits (MSWD = 710) and four analyses were consequently excluded from the age determination. The resulting isochron (MSWD = 59) was defined by five samples with negative Sr anomalies (SGP II, III) and two SGP I samples, so that the emplacement age of 2424 ± 72 Ma does not coincide with a particular subgroup as defined here, but was obtained from samples belonging to each of the three subgroups.

Because dykes of the HMG suite outcrop infrequently, dyke intersections which define the relative emplacement age between members of the suite are uncommon. Thus, in order to generalize from the intersections that were observed it is necessary to assume that the dykes of each subgroup were intruded in separate, discrete events and did not overlap in time with the emplacement of dykes from other subgroups. Three dykes belonging to SGP I were found to be crosscut by SGP II or III dykes, and as such SGP I is inferred to be the oldest of the three subgroups. No intersections were found to indicate the relative ages of SGP II and SGP III. However, a compilation of estimated dyke widths shows that the average dyke width of each subgroup decreases in the order I, II, III (10.6 m, 6.4 m, 1.1 m respectively). As the oldest subgroup, based on dyke intersections, has the widest average dyke width and is also chemically the most primitive, the decrease in average dyke width may indicate that emplacement of the subgroups was influenced by a magma-induced stress field. It is envisaged that as the magmatic episode waned, dyke widths narrowed and thus the chemically evolved SGP III dykes are inferred to be the youngest subgroup.

8.4 Petrography

It has already been noted that the two trace-element-defined divisions of the HMG suite can in general be characterized on the basis of their phenocryst assemblage. In detail, the olivine-free samples (Sr-depleted) can be further subdivided texturally, a distinction also consistent with the Rb-rich, Rb-poor designations. The petrography and crystal chemistry of the three subgroups are discussed below.

Orthopyroxene phenocrysts in SGP I samples are stout, euhedral, colourless crystals rarely exceeding 1 mm in length and comprise less than 14% of the mode in the samples studied (Fig. 8.2a). The crystals display normal chemical zoning with a maximum of Mg# = 0.88 in cores and Mg# = 0.70 in rims in the more evolved dyke compositions. The composition of the most magnesian olivine phenocrysts is consistent with crystallization in equilibrium with orthopyroxene cores based on an Ol–Opx $K_D^{(Fe-Mg)}$ of 1.1 (Jaques & Green 1979). Texturally, however, the olivine grains (< 5% of the mode) have a highly irregular form and are typically rimmed by a fine-grained mosaic of groundmass pyroxene or are occasionally partially enclosed within a clot

Figure 8.2 Plain-light photomicrographs of Vestfold Hills HMG tholeiites. (a) Euhedral orthopyroxene phenocrysts with resorbed, pyroxene-rimmed olivine and lath-shaped Ca-clinopyroxene; SGP I sample 054A; length of frame 5.5 mm. (b) Laths of orthopyroxene phenocrysts and resorbed pyroxene-rimmed olivine in comb-textured groundmass; SGP II sample 214; length of frame 3.0 mm. (c) Quench-textured orthopyroxene phenocrysts with hollow cross sections and castellated terminations in a groundmass of swallowtailed, hollow pigeonite and calcic clinopyroxene; SGP III sample 027; length of frame 3.0 mm. (d) Flow-banded sample with quench orthopyroxene phenocrysts; SGP III sample 088; length of frame 2.5 mm.

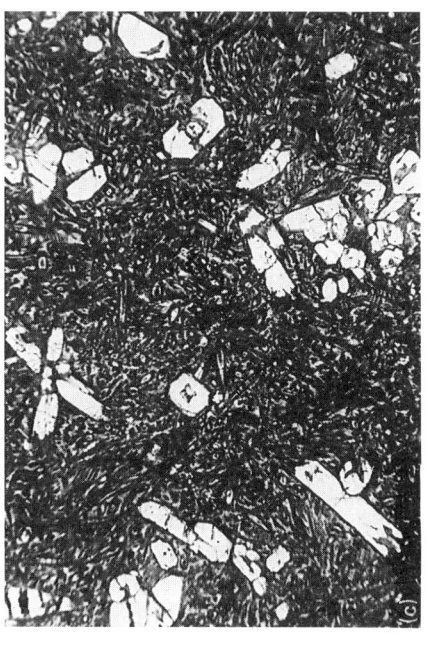
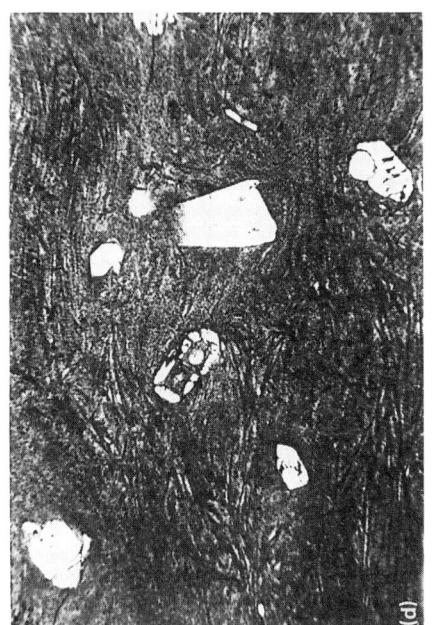
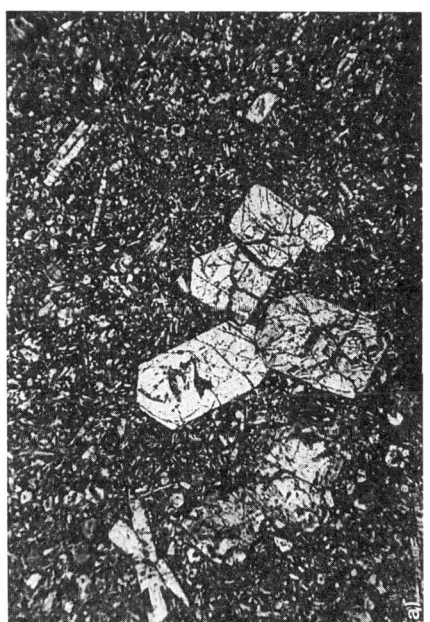

of phenocrystic orthopyroxene; these features indicate a subliquidus reaction relationship between olivine and coexisting liquid. The groundmass contains lath-shaped microphenocrysts of calcic clinopyroxene ($Wo_{38.0}En_{52.0}Fs_{10.0}$) and Mg-rich pigeonite enclosed within a much finer-grained groundmass of red-brown pyroxene and randomly orientated needles of plagioclase. Equilibrium crystallization of the pigeonite cores is indicated by near-constant CaO contents (4.5–5.0%) over the range of Mg# from 0.86 to 0.80, and contrasts with rim analyses where the CaO contents span the two-pyroxene immiscibility gap. All pyroxenes are Cr-rich; orthopyroxene and pigeonite contain about 0.5% Cr_2O_3, and calcic clinopyroxene about 1.0% Cr_2O_3. Only the most evolved SGP I sample lacks phenocrysts of olivine. In this rock, the phases formed earliest are orthopyroxene and pigeonite, which have ragged and resorbed margins and form the cores to many of the calcic clinopyroxene phenocrysts. Cr-rich spinel is found as rare, opaque inclusions only within the phenocryst phases of olivine-bearing samples. Cr# values (0.70–0.82) exceed those of spinel grains in primitive MORB (Dick & Bullen 1984) and overlap the compositional fields of Cr-spinels in komatiites, W Pacific boninites and those found in the Troodos Upper Pillow Lavas (Cameron *et al* 1979, Cameron & Nisbet 1982, Cameron 1985, Duncan & Green 1987). The SGP I spinels, however, are unusually Fe-rich (Mg# = 0.07–0.28) compared to spinel grains in komatiites, boninites and the Troodos Upper Pillow Lavas (Mg# = 0.30–0.80), but are similar in Cr# and Mg# to chromite grains in the Bushveld micropyroxenites (Sharpe & Hulbert 1985). The unusually Fe-rich character of spinel grains in the latter two plutonic suites is believed to be the result of the subsolidus Fe–Mg exchange between spinel and mafic silicates. Such a process is well documented (e.g. Ozawa 1983) and is supported by the occurrence, in both suites, of generally more Fe-rich spinel inclusions in olivine than in orthopyroxene (Kuehner 1986, Sharpe & Hulbert 1985).

The orthopyroxene phenocrysts of SGP II samples have a columnar habit with slightly tapered to rounded terminations and rarely exceed 0.5 mm in length (Fig. 8.2b). The most Mg-rich orthopyroxene compositions (up to Mg# = 0.877) are found in the three olivine-bearing samples and result in an Ol–Opx $K_D^{(Fe-Mg)} = 1.2$. Olivine comprises < 1% of the mode in each of these samples, has a highly irregular form and is enclosed by clots of orthopyroxene. Considering the small olivine population available for analysis, the marginally high K_D is considered evidence for co-precipitation of the olivine and orthopyroxene phenocrysts. In contrast to SGP I, plagioclase appears to have nucleated prior to pyroxene in the groundmass of SGP II dykes, and forms comb-textured arrays in moderately crystallized samples (Fig. 8.2b). Rare Cr-rich spinel inclusions are present in olivine crystals. The spinel grains are slightly less refractory (Cr# = 0.55–0.72) than SGP I spinel inclusions, and also have a slightly more Fe-rich range of compositions (Mg# = 0.4–0.22).

All subgroup III chilled-margin samples are characterized by quench-textured phenocrysts of orthopyroxene, about 2 mm in length. Longitudinal sections display irregularly shaped internal voids and many crystals have castellated terminations. Three samples have a very fine-grained, dark-brown groundmass with wisps of lighter material, defining what is interpreted as flowlines, about the phenocrysts (Fig. 8.2b). The groundmass in these three samples may have originally been a quenched glass. The groundmass in other SGP III samples has swallowtailed, hollow pigeonite and calcic clinopyroxene quench crystals, enveloped by fine, comb-textured plagioclase (Fig. 8.2c). Cumulate-textured norite samples collected from a ~50 m wide ring dyke outcropping in the Northern Peninsula region of the Vestfold Hills also have SGP III geochemical characteristics. The norites are dominated by 2–3 mm long orthopyroxene primocrysts (Mg# = 0.86–0.76), many of which have partially reacted with intercumulus liquid, resulting in broad rims of pigeonite (inverted). Dusty plagioclase, titanomagnetite (exsolved), biotite, K-feldspar + quartz granophyre and locally abundant Fe−Ni−S blebs compose the intercumulus material. Nodules of orthopyroxenite (5–10 cm in diameter) are locally abundant at the margin of the ring dyke. The single nodule examined is composed of multiple, optically continuous aggregates of stocky orthopyroxene crystals (up to Mg# = 0.912) with very minor intercumulus phlogopite, calcic clinopyroxene and K-feldspar + quartz granophyre.

Compositions of orthopyroxene phenocrysts from each subgroup, including the orthopyroxenite nodule, are compared in Figure 8.3. The phenocrysts of each subgroup have overlapping ranges in Fe/Mg, but are distinct in their Wo contents, which decrease in the order SGP I, II and III. As the HMG liquids are not saturated in both a Ca-rich and Ca-poor pyroxene, the Wo content of the orthopyroxene phenocrysts is not temperature-sensitive but reflects only the normative Di/(Di + Hy) of the liquid. The average value of this ratio also

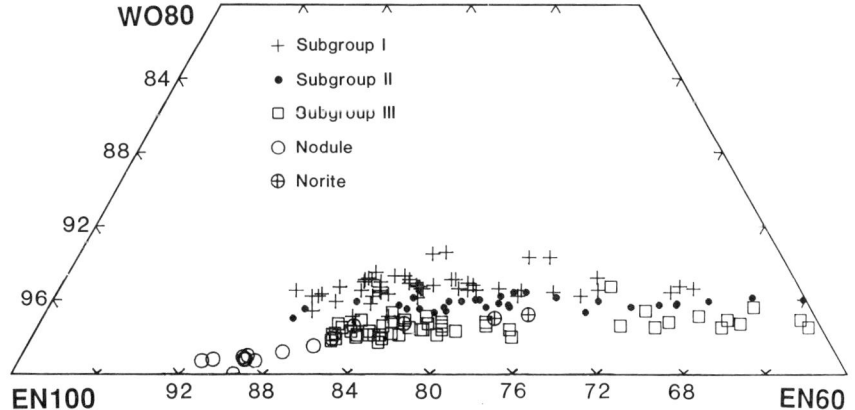

Figure 8.3 Comparison of orthopyroxene phenocryst compositions between subgroups, including analyses from the norite and pyroxenite nodule. Note that the diagram covers only the Mg-rich section of the pyroxene quadrilateral.

decreases from SGP I to III (0.37, 0.32, 0.25). These features clearly indicate that the three subgroups cannot be related to one another through fractionation of the observed phenocryst phases and infer their derivation from three distinct primary liquids. Orthopyroxenes in the nodule are Mg-rich and have low Wo contents (Fig. 8.3). With increasing Fe/Mg, their Wo contents increase, and merge with those of orthopyroxenes from the host norite and the SGP III trend, providing strong evidence for a cognate origin of the nodule.

8.5 Wholerock geochemistry

8.5.1 *Introduction*

Liquid compositions of the HMG suite, inferred from analyses of chilled margins (Table 8.1), are relatively siliceous (51–57% SiO_2) while having high MgO (7–14%), Cr and Ni contents (up to 1600 ppm and 417 ppm, respectively) and low TiO_2 (0.49–0.70%); the analysed samples range from 7% normative olivine to 10% normative quartz ($Fe^{2+} = Fe^T$). Similar chemical features characterize suites transitional between tholeiites and boninites, such as the micropyroxenites of the Bushveld Complex (Sharpe & Hulbert 1985) and the Troodos Upper Pillow Lavas (Duncan & Green 1987). Wholerock analyses from these suites and of Cape Vogel boninites are compared with the Vestfold Hills HMG dykes on a Jensen cation diagram (Fig. 8.4) and on Ni and Cr versus FeO^*/MgO plots (Fig. 8.5). Each of these suites displays minimal Fe enrichment during differentiation, contrasting sharply with the tholeiitic Fe-enrichment trend of the Skaergaard Complex and primitive MORB glasses (Fig. 8.4). This characteristic reflects the relatively minor role of olivine and dominant role of orthopyroxene ± clinopyroxene in controlling the compositional variation within the SiO_2-rich suites. The Vestfold Hills HMG dykes differ from the other three suites in having higher absolute concentrations of FeO and TiO_2, resulting in a higher Fe/Mg at comparable levels of the compatible elements Cr and Ni (Figs 8.5a & b).

It has been concluded from geochemical and experimental studies of Cape Vogel boninites and Troodos Upper Pillow Lavas that these suites are the result of low-pressure melting ($\leqslant 10$ kbar) of a refractory, harzburgitic source (Jenner 1982, Hickey & Frey 1982, Duncan & Green 1987). In light of the overall similarity in differentiation trends and major-element geochemistry between these two suites and the Vestfold Hills HMG dykes, especially with regard to their Mg-rich, yet relatively high-Si nature, it is possible that the Vestfold Hills suite may also have been derived by low-pressure melting of peridotite. Although none of the Vestfold Hills samples may have the composition of a primary liquid, an evaluation of the source characteristics of each subgroup can still be made as the dominant fractionating phase,

Table 8.1 Selected wholerock analyses[a] of high-Mg dykes from the Vestfold Hills, Antarctica.

Sample Subgroup	65054A I	65060 I	65069 I	65209 I	65034 II	65214 II	65265 II	65027 III	65072 III	65088 III
SiO_2	52.17	52.10	54.45	53.67	52.29	52.29	54.30	56.40	54.60	56.44
TiO_2	0.57	0.62	0.65	0.71	0.50	0.49	0.69	0.74	0.58	0.70
Al_2O_3	11.04	11.89	13.20	8.33	11.72	11.78	14.68	12.53	13.23	12.16
Fe_2O_3	11.54	12.00	10.62	12.19	10.99	11.03	10.87	11.56	10.88	11.52
MnO	0.18	0.18	0.16	0.18	0.18	0.18	0.17	0.16	0.17	0.16
MgO	13.00	11.22	7.52	13.45	12.96	13.29	6.84	8.89	9.07	8.70
CaO	9.24	9.93	9.99	9.23	9.21	9.14	9.73	7.79	8.55	7.74
Na_2O	1.86	1.78	2.22	1.65	1.27	0.96	1.74	1.52	1.40	1.21
K_2O	0.51	0.60	0.95	0.75	0.59	0.59	0.87	1.26	0.97	1.30
P_2O_5	0.08	0.10	0.10	0.09	0.05	0.06	0.10	0.10	0.09	0.11
LOI	−0.19	−0.60	−0.12	−0.26	0.0	−0.10	0.01	−0.21	−0.05	−0.22
H_2O^-	0.17	0.09	0.14	0.06	0.13	0.03	0.09	0.0	0.04	0.01
total	100.17	99.91	99.88	100.05	99.89	100.04	100.09	100.84	99.53	99.83
Zr	55	61	69	69	70	72	94	112	88	111
Sr	146	176	272	209	89	90	112	110	99	104
Nb	5	5	4	5	6	6	8	8	7	8
Y	15	15	13	13	16	16	20	22	18	21
Rb	20	23	39	31	18	18	42	65	51	67
Ba	196	254	311	174	208	222	249	288	219	277
V	219	221	229	218	206	206	237	217	205	218
Ni	333	250	106	377	417	409	130	205	186	201
Cr	1183	933	462	1600	1461	1397	143	681	679	715
La	8	9	12	9	17	15	16	22	15	19
Ce	13	16	20	17	25	27	32	39	31	37
Nd	7	9	13	10	10	12	16	19	13	17
Sc	36	33	32	31	37	34	51	29	34	35
100 Mg #	69.05	64.94	58.38	68.61	70.02	70.47	55.48	60.37	62.28	59.93

[a] All elements by XRF analysis at the University of Tasmania (Kuehner 1986).

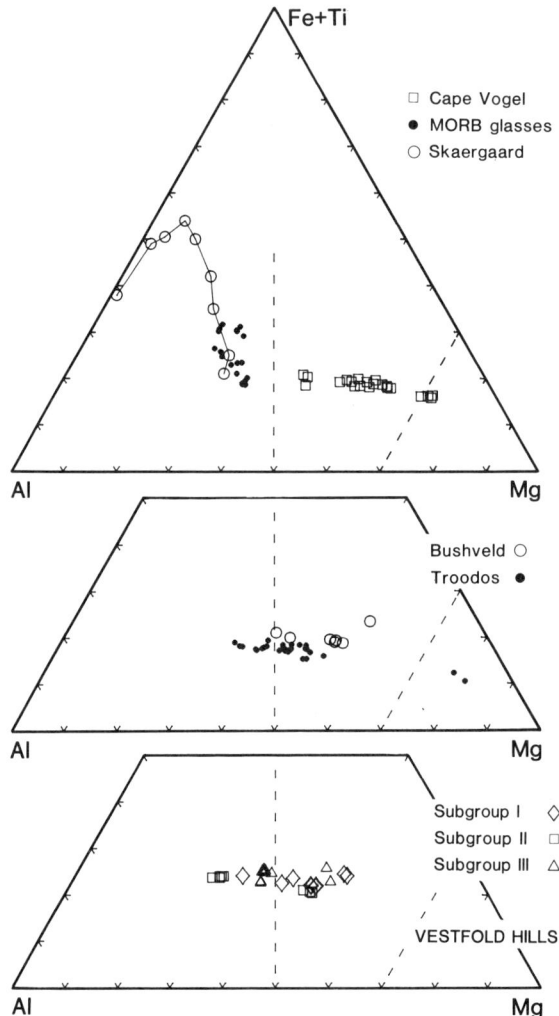

Figure 8.4 Comparison of the compositional trends of various Mg-rich tholeiite suites on a Jensen cation diagram (Jensen 1976). The boundaries from right to left are: ultramafic komatiite, basaltic komatiite and tholeiitic basalt (Jensen 1976). Data sources: this study, Jensen (1976), Sharpe & Hulbert (1985), Frey *et al.* (1980), Bryan & Moore (1977), Cameron (1985) and Jenner (1981).

orthopyroxene, does not significantly modify ratios of incompatible or moderately incompatible elements.

8.5.2 *Subgroup I*

Various elemental ratios of SGP I samples are shown normalized to chondritic ratios in Figure 8.6a. Ratios involving CaO, Al_2O_3 and TiO_2 are

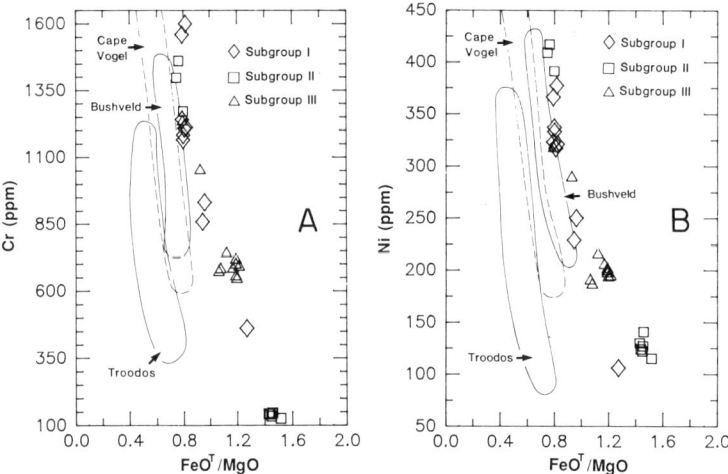

Figure 8.5 Plots of FeO*/MgO vs. (a) Cr and (b) Ni, illustrating the Fe-enriched nature of the Vestfold Hills HMG suite compared to the Bushveld, Cape Vogel and Troodos compositions. Data sources as in Figure 8.4.

essentially chondritic, indicating that Ca, Al and Ti behaved largely as incompatible elements during melting and subsequent fractionation. These ratios contrast with those of MORB, in which CaO/Al_2O_3 is generally chondritic (0.82), whereas CaO/TiO_2 displays a range of values all less than chondritic (<16; Sun & Nesbitt 1978) reflecting the retention of Ca–Al-bearing phase(s) in the residue (Cpx ± Gt ± Sp ± Pl) following MORB extraction. The Cape Vogel and Troodos suites, however, have Al_2O_3/TiO_2 and CaO/TiO_2 greater than chondritic, implying a source depleted in incompatible elements, including Ti, by extraction of basaltic liquids during a previous melting event (Nesbitt & Sun 1980). The chondritic nature of SGP I samples with respect to Ca–Al–Ti ratios indicates that the primary magma parental to this subgroup was derived from a peridotitic source which had not been modified by previous melting events. Furthermore, residual Ca–Al-rich phases are not indicated by the Ca–Al–Ti ratios, requiring that these phases were eliminated through fairly high degrees of partial melting. Chondritic Ti/Y values also indicate that garnet was not retained in the residue following extraction of SGP I parental liquids and imply that garnet was not involved in any geochemical enrichment process that may have modified the incompatible-element content of the source region.

Vanadium and Sc share similar geochemical characteristics but can be decoupled from one another by spinel (V-rich), garnet (Sc-rich) and, to a lesser extent, pyroxene fractionation, as Sc is a moderately compatible element in both clino- and orthopyroxene compared to V (Frey *et al.* 1978). Since garnet involvement is not indicated, the relatively broad range of Ti/Sc compared to Ti/V suggests that pyroxene fractionation has modified the Sc concentration

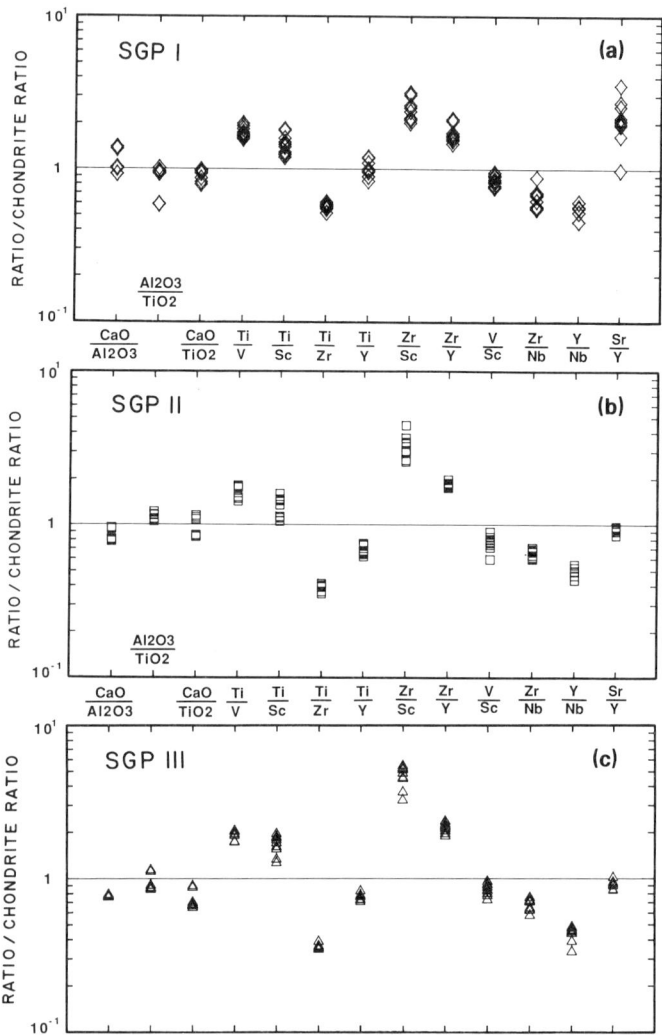

Figure 8.6 Various oxide and elemental ratios of (a) SGP I, (b) SGP II and (c) SGP III normalized with respect to chondritic ratios (Nesbitt & Sun 1980).

relative to Ti, consistent with the petrographic observations. Minor spinel fractionation is suggested by the greater-than-chondritic Ti/V, but the narrow range of values taken in conjunction with the restricted paragenesis of spinel implies that spinel fractionation took place only in the most primitive parental liquids.

Titanium, Zr and Nb behave incompatibly at all but very low degrees of partial melting. As Ti/Y is chondritic, the chemically similar Zr and Nb would also be expected to share this feature. However, Figure 8.7a illustrates that Zr

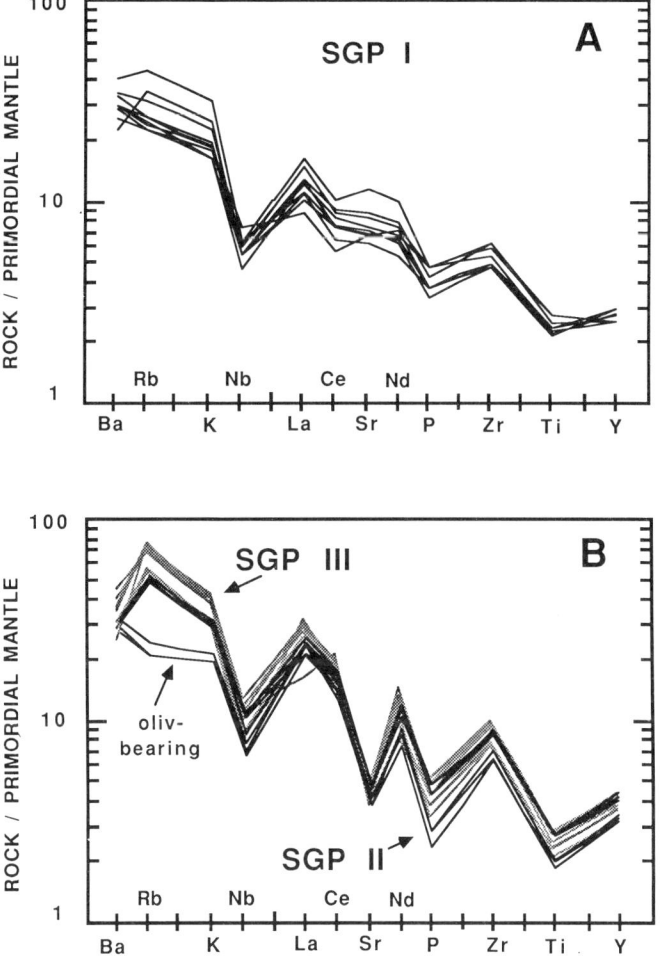

Figure 8.7 Normalized element variation plots of incompatible-element abundances from (a) olivine + orthopyroxene phyric samples and (b) olivine-free (except samples 229, 214, 034) orthopyroxene phyric dykes normalized to primordial mantle values (Wood *et al.* 1979).

and Nb have relative concentrations greater than that predicted from the enrichments of Ti and Y. The Ti/Zr for SGp I range from 67 to 56, considerably less than chondrite (110), and straddle the boundary suggested by Jenner (1981) to characterize boninites and high-Mg andesites generated at subduction zones (Ti/Zr < 60). The low Ti/Zr in Cape Vogel boninites is in part due to previous depletion in Ti with a subsequent enrichment in Zr, whereas in SGP I, the low Ti/Zr is attributed entirely to Zr enrichment upon an undepleted source. The less-than-chondritic Zr/Nb (Fig. 8.6a) further displays the decoupling of these geochemically similar elements and indicates relative enrichment in the order Nb > Zr > Ti. Even though Nb has been enriched relative to other HFS elements, it has not been enriched to the extent anticipated from elements of similar levels of incompatibility such as La or K (Fig. 8.7a).

The elements Rb, K, Ba, Sr, La and Ce are greatly enriched in subgroup I compared to primordial mantle (Fig. 8.7a) and would indicate low degrees of melting if considered independently of the geochemical features discussed above. Instead, the relative concentrations indicate that enrichment of these elements was superimposed onto a chondritic source prior to melting.

8.5.3 Subgroup II

Based on the younger relative age of SGP II compared to SGP I, and the lower CaO contents of the orthopyroxene phenocrysts in the former, the simplest model that might be considered for the origin of SGP II parental liquids is that they were derived by melting of the SGP I residue. This model would predict SGP II liquids to be severely depleted in incompatible elements and consequently to have high Al_2O_3/TiO_2 and CaO/TiO_2 with less-than-chondritic CaO/Al_2O_3. Figure 8.7b illustrates that SGP II, however, is highly enriched in incompatible trace elements. As the chondritic source of SGP I is inferred to have been enriched in incompatible elements prior to the melting event, it is possible that a second enrichment may have imposed a high incompatible-element character to a refractory SGP I residue (except Sr). However, the most primitive olivine-bearing samples of SGP II have only a marginally higher Al_2O_3/TiO_2 (23–24) compared to chondrite (20), which differs greatly from Troodos and Cape Vogel second-stage melts (~ 60 and ~ 33, respectively; Nesbitt & Sun 1980). Furthermore, the CaO/Al_2O_3 of the olivine-bearing SGP II samples is chondritic, which makes derivation from a clinopyroxene- (CaO-) poor source unlikely. These characteristics not only indicate separate sources for SGP I and II, but as for SGP I, also indicate an extensive degree of melting and the absence of Ca–Al-rich phases in the SGP II residue.

A comparison of the olivine-bearing samples of SGP II (034, 207, 214) with SGP I sample 054A reveals that, whereas these samples have similar MgO contents (12.6–13.3%), the SGP II samples are enriched in the compatible elements Cr and Ni (Figs 8.5a & b). Considering the undepleted nature of the SGP II source as indicated by the CaO/Al_2O_3 and Sr/Y, the primary liquid to SGP II probably resulted from a higher degree of melting than did the SGP I magma.

8.5.4 Subgroup III

The most mafic SGP III sample has a relatively low Mg# (0.622) and low Cr and Ni contents (670 ppm, 190 ppm) compared to the Mg# values of 0.69–0.70 for SGP I and II. If the primary melt to SGP III was derived from a chondritic source as inferred for SGP I and II, then the primitive characteristics of the source should be retained in the $CaO-Al_2O_3-TiO_2$ ratios of SGP III samples as long as orthopyroxene ($\pm$ olivine) was the only fractionating phase. Selected chondrite-normalized ratios are shown in Figure 8.6c. CaO/Al_2O_3 values for all samples are essentially identical and are consistent with fractionation of only orthopyroxene $\pm$ olivine from a primitive liquid. However, the absolute CaO/Al_2O_3 (~ 0.62) is significantly below chondrite (0.82; Sun & Nesbitt 1978) and suggests either that the SGP III source was depleted in CaO relative to the source regions of SGP I and II, or the fractionation of a Ca-rich, Al-poor phase modified the primary liquid composition. The latter possibility is unsupported by petrographic observations.

Although SGP III has lower CaO/Al_2O_3 values than SGP II and could thus represent the product of continued melting of the SGP II source, both suites have equivalent Sr/Y and SGP III is also enriched in Zr > Nb > Ti. These features are inconsistent with derivation of SGP III liquids from the SGP II residue unless the residue was enriched with incompatible elements (except Sr) prior to generation of the SGP III parental liquid. SGP III is also depleted in CaO (relative to Al_2O_3), as well as Ti and Sr (relative to Y), compared with SGP I ratios, but SGP III is enriched in Zr > Nb. These features are also inconsistent with the origin of SGP III parental liquids by melting the refractory SGP I residue without the prior introduction of incompatible elements.

As in SGP I, the normalized trace-element patterns of SGP II and III samples have negative Nb anomalies (Fig. 8.7), implying that the fluid or melt which introduced the incompatible elements to the SGP II and III source(s) was buffered by a phase which could decouple Nb from other incompatible elements (Saunders *et al.* 1980). A feature that distinguishes the normalized element variation diagrams of SGP II and III from those of SGP I is the existence of a negative Sr anomaly in the former. The Sr/Y is chondritic in both SGP II and III, whereas geochemically similar Nd and Ce are strongly enriched. By contrast, in modern island-arc petrogenesis a negative Nb anomaly is also characteristic but Nd, Ce and Sr are all enriched (Perfit *et al.* 1980).

If the Nb depletion is due to the buffering effect of a residual phase in the source region of the metasomatic fluid or melt, then the absence of Sr enrichment may also indicate that Sr was withheld from entering the enriching component by a residual phase. Of the major rock-forming minerals only plagioclase is able to decouple Sr from other LILE such as Ba, Rb and K. If plagioclase has played a role in modifying the source region(s) of SGP II and III, then melt extraction at pressures within the plagioclase peridotite stability field is required (< 12 kbar; Green & Hibberson 1970).

8.6 Rare-earth-element geochemistry

Rare-earth-element concentrations in a representative sample selected from each subgroup were determined following the ion-exchange X-ray fluorescence (XRF) procedure of Robinson *et al.* (1986). The three samples have generally parallel patterns (Fig. 8.8) characterized by flat HREE concentrations and strong LREE enrichment. The SGP I sample has a slight positive Eu anomaly, whereas SGP II and III samples have slight negative Eu anomalies as well as a higher Ce/Sm.

Based on the less-than-chondritic CaO/Al_2O_3 of SGP III samples, the origin of this subgroup by melting the residue of SGP I or II was considered possible if the clinopyroxene-poor source was enriched with incompatible elements prior to melting. However, the nearly identical REE patterns of 088 (SGP III) and 214 (SGP II) suggest that they were derived from a common parental liquid and differ in levels of REE enrichment due simply to the more fractionated composition of 088. To test this supposition, increments of equilibrium orthopyroxene were added to 088 until the resulting composition was equivalent in Mg# to 214 (0.705). Using partition coefficient data of Frey *et al.* (1978) and assuming Rayleigh fractionation, the Yb concentration in the hypothetical parental liquid (Mg# = 0.705) to SGP III sample 088 was then determined. The resulting Yb abundance (1.82 ppm) and, by inference from the parallel patterns, the other REE are indistinguishable from the concentrations in SGP II sample 214 (1.79 ± 0.07 ppm Yb). The calculation indicates

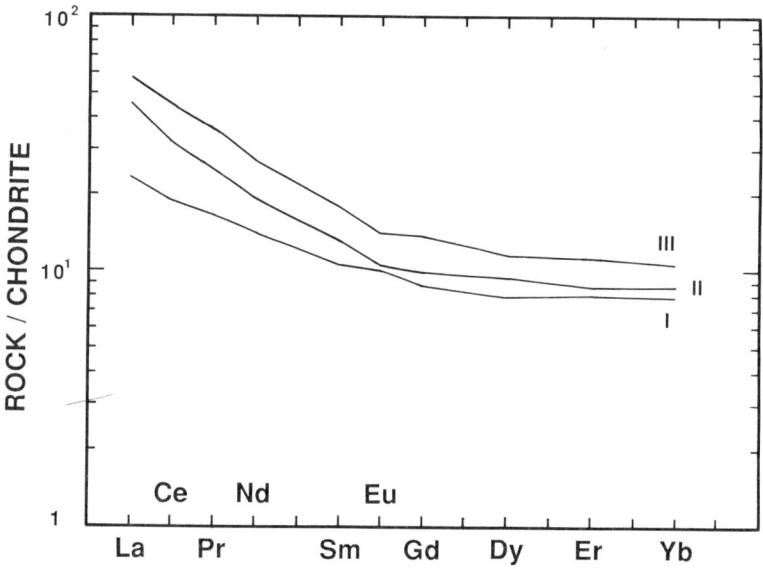

Figure 8.8 Chondrite normalized REE patterns (normalizing values of Masuda *et al.* (1973)) of samples representing the three HMG subgroups: SGP I, sample 054A; SGP II, 214; SGP III, 088.

that the source of SGP III was not a refractory residue, but had HREE (and Sr, Y) concentrations similar to the inferred primitive SGP II source.

8.7 Crustal contamination

The role of crustal contamination in modifying the compositions of the HMG suite is an important aspect to consider in view of the LILE-enriched nature of the suite, which up to now has been assumed to be a mantle characteristic. Sheraton & Black (1981) and Collerson & Sheraton (1986) have addressed this problem and considered the low and indistinguishable initial $^{87}Sr/^{86}Sr$ of the HMG suites from both the Vestfold Hills (0.7019 ± 0.0011) and the 1000 km distant Napier Complex (0.7020 ± 0.008) as evidence that extensive incorporation of old crust was not a major process in modifying the compositions of the HMG dykes. Furthermore, nearly identical trace-element ratios and enrichments are observed in the Napier Complex HMG dykes (see Sheraton & Black 1981) as in SGP II and III. These minor-element characteristics are unlikely to be the result of contamination by a young, homogeneous crust over such a large area and are better ascribed to a feature of the mantle source. These observations are considered good evidence for the absence of crustal assimilation as an important process in the evolution of the HMG suites in the Vestfold Hills and Napier Complex.

8.8 Estimation of the depth of magma genesis

The olivine-bearing samples 054A (SGP I) and 214 (SGP II) have Mg#, Ni and Cr contents consistent with their being unmodified melts from a mantle peridotite (Table 8.1). However, these liquid compositions are saturated in olivine and orthopyroxene at 7–8 kbar, the pressure at which the dykes crystallized (Kuehner 1986, 1988), which indicates that these liquid compositions could not be in equilibrium with mantle peridotite at higher pressure, thus requiring these samples to be differentiates of more mafic compositions. The only evidence presently recognized for more primitive liquids in this suite is the orthopyroxene (Mg# = 0.912) in the pyroxenite nodule, which requires the existence of a liquid with Mg# ~ 0.75 (K_D = 0.30; Jaques & Green 1979). The most primitive SGP III sample (the suite to which the pyroxenite nodule is cognate) is considerably differentiated compared to a primitive liquid composition. Thus in order to reconstruct a possible primary liquid composition sample 054A (Mg# = 0.71, Fe^{3+}/Fe^{2+} = 0.10) was used as a starting point because it is saturated in the phases thought to be residual following partial melting and has only experienced minor fractionation from the estimated Mg# ~ 0.75 parental liquid composition. Because the proportions of olivine + orthopyroxene theoretically fractionated to derive 054A from a primary liquid are not known, end-member compositions were estimated by adding increments of equilibrium olivine or orthopyroxene (assuming 2% Al_2O_3, 2.5% CaO) to 054A. The resulting compositions are listed in Table 8.2 and shown projected onto the plane (Jd + CaTs)–Ol–Qtz of the basalt tetrahedron (Fig. 8.9).

The pressure at which the calculated liquid compositions could be derived from a primitive mantle peridotite leaving a harzburgitic residue can be estimated by comparing these compositions with the experimental studies of Falloon & Green (1987, 1988), which are reversals of the anhydrous peridotitic melting studies of Jaques & Green (1980). The range of calculated compositions at Mg# = 0.75 plot between the 10 and 15 kbar olivine–orthopyroxene cotectics (Fig. 8.9). Two 10 kbar liquid compositions, saturated in olivine + orthopyroxene (1400°C and 1380°C; Falloon & Green 1988) are compared in Table 8.2 with the estimated parental liquid compositions of the HMG suite. Very good agreement exists between these compositions in all elements except TiO_2, CaO and FeO*. The much higher TiO_2 and marginally enriched CaO in the experimental glass compositions are due entirely to the very fertile Hawaiian pyrolite used in the experimental studies compared to the chondritic source demonstrated for the HMG suite. The higher FeO* content in the calculated primitive liquid is believed to indicate that the source for the HMG parental magma had a greater absolute concentration of FeO* than Hawaiian pyrolite at Mg# = 0.887. Similar inferences concerning the absolute Fe

Table 8.2 Comparison of calculated 'end-member' primitive liquid compositions with experimental 10 kbar glass analyses[a].

	054A + 8.7% Ol	054A + 11.9% Opx	10 kbar	
			1400°C Pt caps.[b]	1380°C Fe caps.[b]
SiO_2	51.67	53.04	52.29	51.44
TiO_2	0.53	0.51	2.11	2.12
Al_2O_3	10.21	10.16	10.22	10.53
Fe_2O_3	1.16	1.12	0.96	1.06
FeO	9.43	9.03	7.78	8.58
MnO	0.17	0.16	–	–
MgO	16.09	15.20	15.91	14.16
CaO	8.54	8.58	9.00	9.13
Na_2O	1.72	1.67	1.43	1.75
K_2O	0.47	0.46	0.41	0.37
P_2O_5	0.07	0.07	–	–
100Mg #[c]	75.3	75.0	78.5	74.6

[a]Glass analyses from Falloon & Green (1988).
[b]Fe and Pt caps. refer to iron and platinum capsules used in the experiments.
[c]100Mg # = $100Mg/(Mg + Fe^{2+})$; $Fe^{3+}/Fe^{2+} = 0.1$.

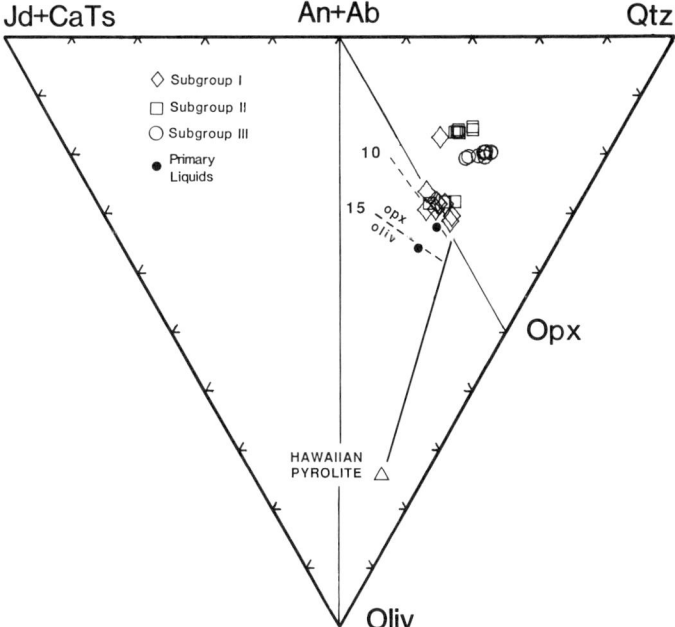

Figure 8.9 Compositions of the HMG dykes ($Fe_2O_3/FeO = 0.10$) projected from Di onto the plane (Jd + CaTs)–Ol–Qtz following Green (1970). Broken lines are Falloon & Green's (1988) 10 and 15 kbar Ol–Opx cotectics. Full circles represent the hypothetical Mg# = 0.75 primary liquid compositions to the HMG suite, calculated by adding olivine or orthopyroxene to sample 054A.

content of the Archaean mantle have been made by Glikson (1983) and Arndt (1977).

It is concluded from Figure 8.9 and the very close correspondence between the analyses presented in Table 8.2 that the parental liquid(s) to the HMG suite was(were) derived by anhydrous melting of a primitive and comparatively Fe-rich peridotite at ~10 kbar, leaving an olivine + orthopyroxene residue. This result is consistent with derivation from a plagioclase peridotite source (stable at $P < 12$ kbar; Green & Hibberson 1970), as suggested by the lack of Sr enrichment in SGP II and III patterns on normalized element variation diagrams (Fig. 8.7), and also with the absence of residual Ca–Al-rich phases in the source following melt extraction, as indicated by chondritic CaO–Al_2O_3–TiO_2 ratios in SGP I and II samples. Assuming a source with $2.2 \times$ chondrite HREE abundances (Ringwood 1979), 31% partial melting is required to produce the theoretical primary liquid.

8.9 Summary

The distinctive CaO contents in the orthopyroxenes of each subgroup indicate that the different 'activities' of the diopside molecule were intrinsic to the parental liquids of each subgroup. This is exemplified by the cognate pyroxenite nodule collected from a SGP III dyke, which requires the low normative Di/(Di + Hy) of this subgroup to be a feature of a parent liquid with Mg# ~ 0.75. Considering only these phenocryst characteristics, derivation of the primary liquids to each subgroup by removing batches of magma from a single source undergoing increasing degrees of partial melting of a clinopyroxene-bearing source could account for the variation in phenocryst compositions, as well as the inferred age relationships between the subgroups. However, successive extraction of liquids from a single source is inconsistent with many other geochemical features of the subgroups. The chondritic CaO/Al_2O_3 of SGP I and the olivine-bearing samples of SGP II rule out their derivation from a source retaining significant quantities of clinopyroxene, if it is assumed that the source initially had chondritic CaO/Al_2O_3. In addition, the absence of calcic clinopyroxene as a liquidus phase in high-Mg basalt compositions at any pressure (Kuehner 1986, Duncan & Green 1987, Jenner 1982) precludes models relying on high-pressure clinopyroxene fractionation to reduce the CaO content of a SGP I partial melt in order to derive the intermediate-CaO SGP II or low-CaO SGP III compositions.

It is inferred from the $CaO-Al_2O_3-TiO_2$ and trace-element abundances and ratios, and from REE patterns, that at least two separate source regions, each with primitive chondritic chemical characteristics modified by an incompatible-element enrichment event, were required to derive SGP I and II compositions. The more evolved SGP III samples have parallel, but enriched, trace-element patterns on normalized element variation diagrams compared with SGP II, and in the case of their REE profiles, SGP III can be simply modelled as differentiates of SGP II compositions. However, such a relationship cannot be accommodated by major-element (CaO/Al_2O_3) relations, which would require fractionation of a CaO-rich phase to derive SGP III from SGP II, and is also precluded by the differences in orthopyroxene phenocryst compositions.

The consistency of incompatible trace-element ratios between the olivine-bearing and olivine-free SGP II samples indicates that they were derived from a common parental liquid, which is in agreement with the CaO contents of the orthopyroxene phenocrysts in these samples. However, CaO/Al_2O_3 and Ba/Rb in olivine-free samples are significantly lower than in the olivine-bearing compositions, which cannot be explained by the fractionation of any likely phenocryst phase (olivine, orthopyroxene, pigeonite).

It is clear that simple models involving batch partial melting followed by closed-system crystal fractionation cannot adequately explain all the chemical features of the HMG suites, and more complex models are required. It is possible that some of the incompatible-element complexity was imparted to batches of liquid through a 'zone refining' or 'wallrock reaction' process (Harris 1974, Green & Ringwood 1967). Although different in mechanism, both models predict that mafic liquids will become enriched to a maximum of

$1/D$ in elements that are incompatible in the wallrock. These will typically be LILE (including LREE), but decoupling of these elements can be envisaged if the wallrock peridotite contains a small percentage of micas, amphiboles or other accessory phases. Such a process, if occurring in plagioclase peridotite, may explain the Sr depletion in SGP II and III.

This study has also shown that boninite-like liquids (high SiO_2, high MgO, low TiO_2) can be generated in a continental environment provided the crust is adequately thin, and in this case, also requires a sufficiently high heat flow for extensive ($\sim 30\%$) anhydrous partial melting, leaving a harzburgitic residue. The existence of these liquid compositions in the Vestfold Hills and the Napier Complex implies that the East Antarctic Shield was $\leqslant 35$ km thick at the end of the Archaean. Considering that the HMG dykes record emplacement pressures of 7–8 kbar (25–28 km; Kuehner 1986, 1988) and the present crust is ~ 30 km thick (Kadmina et al. 1983), a minimum of 20–23 km of crust had to be accreted to the base of the East Antarctic Shield since emplacement of the HMG dykes at ~ 2400 Ma.

Acknowledgements

This work forms part of my Ph.D. research, which was supported financially by a University of Tasmania Postgraduate Award. I would like to thank Professor D. H. Green for his guidance and discussions throughout this study, and the Australian Antarctic Division for logistical support in sample collection during the 1982–83 austral summer. Drs A. Crawford and J. Sheraton are thanked for their critical reviews of the manuscript.

References

Arndt, N. T. 1977. Ultrabasic magmas and high degree melting of the mantle. *Contrib. Mineral. Petrol.* **64**, 205–21.

Black, L. P. & P. R. James 1983. Geological history of the Archean Napier Complex of Enderby Land. In *Antarctic Earth science*, R. L. Oliver, P. R. James & J. B. Jago (eds), 11–15. Australian Academy of Science.

Bryan, W. B. & J. G. Moore 1977. Compositional variations of young basalts in the Mid-Atlantic Ridge rift valley near lat. 36°49′N *Geol. Soc. Am. Bull.* **88**, 556–70.

Cameron, W. E. 1985. Petrology and origin of primitive lavas from the Troodos ophiolite, Cyprus. *Contrib. Mineral. Petrol.* **89**, 239–55.

Cameron, W. E. & E. G. Nisbet 1982. Phanerozoic analogues of komatiitic basalts. In *Komatiites*, N. T. Arndt & E. G. Nisbet (eds), 29–50. London: George Allen & Unwin.

Cameron, W. E., E. G. Nisbet & V. J. Dietrich 1979. Boninites, komatiites and ophiolitic basalts. *Nature* **280**, 550–3.

Collerson, K. D. & J. W. Sheraton 1986. Age and geochemical characteristics of a mafic dyke swarm in the Archean Vestfold Block, Antarctica: inferences about Proterozoic dyke emplacement in Gondwanaland. *J. Petrol.* **27**, 853–86.

Collerson, K. D., E. Reid, D. Millar & M. T. McCulloch 1983. Lithological and Sr−Nd isotopic relationships in the Vestfold Block: implications for Archean and Proterozoic crustal evolution in the East Antarctic. In *Antarctic Earth science*, R. L. Oliver, P. R. James & J. B. Jago (eds), 77−84. Australian Academy of Science.

Dick, H. J. B. & T. Bullen 1984. Chromian spinel as a petrogenetic indicator in abyssal and Alpine-type peridotites and spatially associated lavas. *Contrib. Mineral. Petrol.* **86**, 54−76.

Duncan, R. A. & D. H. Green 1987. The genesis of refractory melts in the formation of oceanic crust. *Contrib. Mineral. Petrol.* **96**, 326−42.

Falloon, T. J. & D. H. Green 1987. Anhydrous partial melting of MORB pyrolite and other peridotite compositions at 10 kbar: implications for the origin of primitive MORB glasses. *Mineral. Petrol.* **37**, 181−219.

Falloon, T. J. & D. H. Green 1988. Anhydrous partial melting of pyrolite and peridotite from 8−35 kbar and petrogenesis of MORB. *J. Petrol.* (in press).

Frey, F. A., J. S. Dickey, G. Thompson, W. B. Bryan & H. L. Davies 1980. Evidence for heterogeneous primary MORB and mantle sources, NW Indian Ocean. *Contrib. Mineral. Petrol.* **74**, 387−402.

Frey, F. A., D. H. Green & S. D. Roy 1978. Integrated models of basalt petrogenesis: a study of quartz tholeiites to olivine melilites from southeastern Australia utilizing geochemical and experimental petrological data. *J. Petrol.* **19**, 463−513.

Glikson, A. Y. 1983. Geochemistry of Archaean tholeiitic basalt and high-Mg to peridotite komatiite suites, with petrogenetic implications. *Geol. Soc. India Mem.* **4**, 183−219.

Green, D. H. 1970. The origin of basaltic and nephelinitic magmas. *Trans. Leics. Lit. Phil. Soc.* **64**, 28−54.

Green, D. H. & W. Hibberson 1970. Instability of plagioclase in peridotite at high pressure. *Lithos* **3**, 209−21.

Green, D. H. & A. E. Ringwood 1967. The genesis of basaltic magmas. *Contrib. Mineral. Petrol.* **15**, 103−90.

Harris, P. G. 1974. Origin of alkaline magmas as a result of anatexis. In *The alkaline rocks*, H. Sorenson (ed.), 427−34. New York: Wiley.

Hickey, R. L. & F. A. Frey 1982. Geochemical characteristics of boninite series volcanics: implications for their source. *Geochim. Cosmochim. Acta* **46**, 2099−115.

James, P. R. & L. P. Black 1981. A review of the structural evolution and geochronology of the Archean Napier Complex of Enderby Land, Australian Antarctic Territory. Spec. Publ. Geol. Soc. Aust. no. 7, 71−83.

James, P. R. & R. J. Tingey 1983. The Precambrian geological evolution of the East Antarctic metamorphic shield − a review. In *Antarctic Earth science*, R. C. Oliver, P. R. James & J. B. Jago (eds), 5−10. Australian Academy of Science.

Jaques, A. L. & D. H. Green 1979. Determination of liquid compositions in high pressure melting of peridotite. *Am. Mineral.* **64**, 1312−21.

Jaques, A. L. & D. H. Green 1980. Anhydrous melting of peridotite at 0−15 kb pressure and the genesis of tholeiitic basalts. *Contrib. Mineral. Petrol.* **73**, 287−310.

Jenner, G. A. 1981. Geochemistry of high-Mg andesites from Cape Vogel, PNG. *Chem. Geol.* **33**, 307−32.

Jenner, G. A. 1982. Petrogenesis of high-Mg andesites. Ph.D.Thesis, University of Tasmania.

Jensen, L. S., 1976. *A new cation plot for classifying subalkalic volcanic rocks.* Ontario Div. Mines Misc. Pap. no. 66.

Kadmina, I. N., R. G. Kurinin, V. N. Masolov & G. E. Grikurov 1983. Antarctic crustal structure from geophysical evidence: a review. In *Antarctic Earth science*, R. L. Oliver, P. R. James & J. B. Jago (eds), 498−502. Australian Academy of Science.

Kuehner, S. M. 1986. Mafic dykes of the East Antarctic Shield, experimental, geochemical and petrological studies focusing on the Proterozoic evolution of the crust and mantle. Ph.D. Thesis, University of Tasmania.

Kuehner, S. M. 1988. Uplift history of the East Antarctic Shield; constraints imposed by high pressure experimental studies of Proterozoic mafic dykes. *Proc. Fifth Antarctic Earth Science Conf.* (in press).

Masuda, A., N. Nakamura & T. Tanaka 1973. Fine structures of mutually normalized rare-earth patterns of chondrites. *Geochim. Cosmochim. Acta* **37**, 239–48.

Nesbitt, R. W. & S.-S. Sun 1980. Geochemical features of some Archean and post-Archean high-magnesian–low-alkali liquids. *Phil. Trans. R. Soc. Lond. A* **297**, 365–81.

Ozawa, K. 1983. Evaluation of olivine–spinel geothermometry as an indication of thermal history for peridotites. *Contrib. Mineral. Petrol.* **82**, 52–65.

Perfit, M. R., D. A. Gust, A. E. Bence, R. J. Arculus & S. R. Taylor 1980. Chemical characteristics of island-arc basalts: implications for mantle sources. *Chem. Geol.* **30**, 227–56.

Ringwood, A. E. 1979. *Origin of the Earth and Moon*. New York: Springer.

Robinson, P., N. C. Higgins & G. A. Jenner 1986. Determination of rare-earth elements, yttrium and scandium in rocks by an ion exchange x-ray fluorescence technique. *Chem. Geol.* **55**, 121–137.

Saunders, A. D., J. Tarney & S. D. Weaver 1980. Transverse geochemical variations across the Antarctic Peninsula: implications for the genesis of calc-alkaline magmas. *Earth Planet Sci. Lett.* **46**, 344–60.

Sharpe, M. R. & L. J. Hulbert 1985. Ultramafic sills beneath the Eastern Bushveld Complex: mobilized suspensions of early lower zone cumulates in a parental magma with boninitic affinities. *Econ. Geol.* **80**, 849–71.

Sheraton, J. W. & L. P. Black 1981. Geochemistry and geochronology of Proterozoic tholeiite dykes of East Antarctica: evidence for mantle metasomatism. *Contrib. Mineral Petrol.* **78**, 305–17.

Sun, S.-S. & R. W. Nesbitt 1978. Geochemical regularities and genetic significance of ophiolitic basalts. *Geology* **6**, 689–93.

Wood, D. A., J.-L. Joron, M. Treuil, M. J. Norry & J. Tarney 1979. Elemental and Sr isotope variations in basic lavas from Iceland and the surrounding ocean floor: the nature of mantle source inhomogeneities. *Contrib. Mineral. Petrol.* **70**, 319–39.

9 Geological setting, petrology and chemistry of Cambrian boninite and low-Ti tholeiite lavas in western Tasmania

A. V. BROWN & G. A. JENNER

Abstract

Boninitic and spatially related low-Ti tholeiitic lavas form part of the Eocambrian–Cambrian volcano-sedimentary successions of W Tasmania. The boninitic lavas have many similarities with Tertiary boninite lavas which originate in supra-subduction zone environments. These similarities include their high SiO_2, MgO, Cr and Ni contents, Cr/Ni, Cr-rich spinels, presence of abundant orthopyroxene and clinoenstatite phenocrysts, concave to LREE-enriched REE patterns, and low HFSE abundances.

Two distinct geochemical groups of boninitic lavas from W Tasmania are recognized. The first group is very similar to Tertiary boninites, whereas the second group differs from Tertiary boninites in their HFSE ratios and have chemical similarities to the associated low-Ti tholeiitic lavas. The latter have low HFSE abundances, strong LREE depletion, variable ε_{Nd} (-2.2 to $+3.7$) compositions and an unusual ε_{Nd} versus Sm/Nd relationship.

The closest comparison to the Tasmanian boninite–low-Ti tholeiite sequences are the chemically similar Tertiary boninitic and low-Ti tholeiitic lavas which occur in the Yap–Mariana–Bonin forearc region and the Upper Pillow Lavas of the Troodos Ophiolite Complex. Geological, petrological and geochemical evidence obtained from the boninitic and associated volcanic rocks in W Tasmania indicate that this section of the extensive late Proterozoic–early Palaeozoic Lachlan Foldbelt in SE Australia was formed by the tectonic juxtaposition in the early Late Cambrian of two volcano-sedimentary successions, one (containing the boninites) formed in a supra-subduction zone setting and the other in an intra-continental rift environment. The most likely scenario for this juxtaposition involves an arc–continent collision.

9.1 Introduction

The Dundas Trough in W Tasmania (Fig. 9.1) is an elongate zone containing late Proterozoic to early Palaeozoic volcano-sedimentary and ultramafic–mafic rock successions. To the west, it is bounded by Precambrian successions of the Rocky Cape Region, which are multiply deformed but of low metamorphic grade. To the east, it is bounded by calc-alkaline lavas and

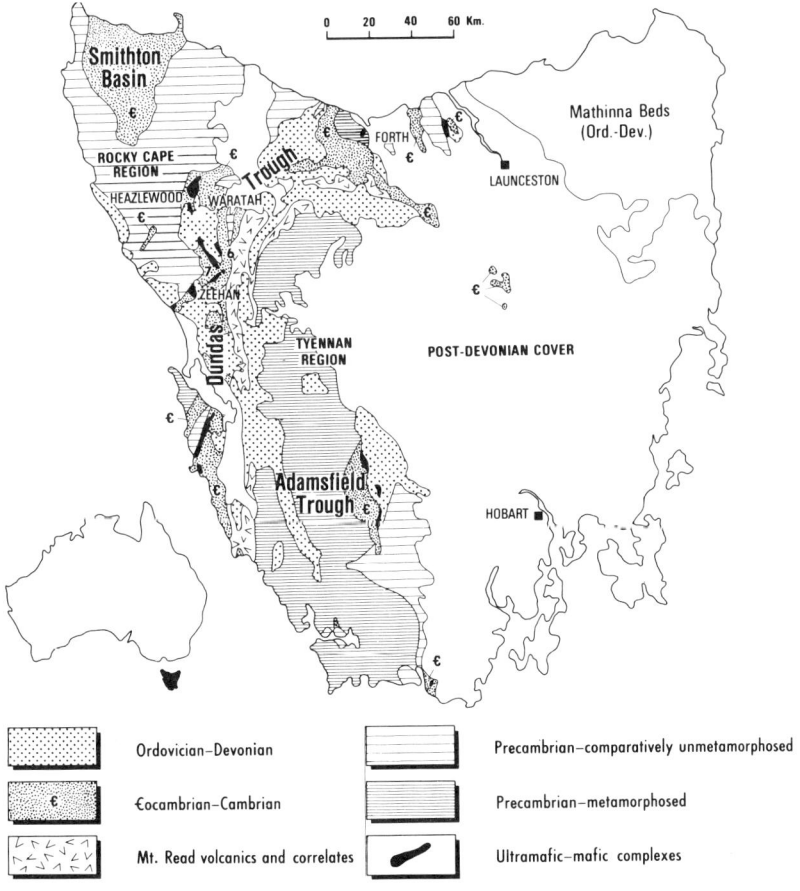

Figure 9.1 Schematic geological map of Tasmania. After Williams (1976).

associated sedimentary rocks of the Mt Read volcanic belt, which range in age up into the Late Cambrian. Within the Dundas Trough, a number of mafic and felsic volcanic sequences have been recognized as Eocambrian to Cambrian in age (Varne & Brown 1978, Brown *et al.* 1980, Brown 1986b, Varne & Foden 1987, Crawford & Berry 1989). Jenner & Brown (1989) have subdivided the mafic volcanic rocks on the basis of their geochemical characteristics and field relationships into four groups, rather than the two or three groups of previous authors.

The oldest of these sequences is composed of sub-alkaline to transitional basalts with a Within-Plate Basalt (WPB) geochemical signature. Lavas making up this sequence occur along the western side of the Dundas Trough and in the Smithton Basin (Fig. 9.1) and are part of the Crimson Creek Formation. The second group also consists of sub-alkaline basalts, but these lavas have Ocean-Floor Basalt (OFB) geochemical affinities; these occur in the Heazlewood–Waratah area and further south along the eastern side of the Dundas Trough to the east of the Huskisson Syncline and Meredith Granite. The third and fourth of the mafic volcanic sequences intrude and overlie the sub-alkaline basalt and the associated sedimentary rocks in the Heazlewood–Waratah area. Lavas belonging to the third group also occur in the Stonehenge area west of Zeehan (Fig. 9.1). The third sequence is a succession of boninitic rocks containing abundant Cr-spinel and pseudomorphs of low-Ca pyroxene phenocrysts. The fourth sequence consists of aphyric, low-Ti basalt–andesite (LOTI) lavas with tholeiitic characteristics and extreme LREE depletion. Associated with each of the latter three volcanic sequences is a succession of layered, cumulate ultramafic rocks (Brown 1986a,b).

The boninitic rocks are of particular petrological and tectonic interest. Based on limited major- and trace-element and petrographic data, it has been suggested that these lavas resemble present-day boninitic lavas from the Bonin Islands and Cape Vogel, generally assumed to have formed in a suprasubduction zone environment (Rubenach 1973, Varne & Brown 1978, Brown *et al.* 1980, Varne & Foden 1987). Berry & Crawford (1988) and Crawford & Berry (1989) have suggested recently that all the Cambrian (?) mafic and ultramafic complexes in W Tasmania represent fragments of a single, large allochthonous sheet emplaced during a Middle Cambrian arc–continent collision. These authors consider that the low-Ti basalt–boninite association of these mafic–ultramafic complexes formed in the forearc region of an oceanic arc which was overthrust onto thinned continental crust at the leading edge of a passive margin.

Despite the similarity of the Tasmanian boninitic rocks to Eocene–Recent boninite lavas, some tectonic models for W Tasmania suggest that the boninitic rocks were generated by second-stage melting in a failed continental rift (Brown *et al.* 1980, Brown & Waldron 1982), whereas in other models (Crook 1980, Corbett & Lees 1987) these rocks are overlooked.

This chapter presents new data on Tasmanian boninitic and LOTI lavas and a synthesis of geological field data pertinent to their occurrence. It aims to facilitate comparison of the Tasmanian boninitic rocks with potential recent analogues and with those of Cambrian age in Victoria (Crawford *et al.* 1984, Nelson *et al.* 1984, Crawford & Cameron 1985, Crawford & Keays 1987). The results place constraints on the tectonic setting of W Tasmania during the formation of the boninitic and LOTI lavas.

9.2 Regional geological setting

Detailed descriptions of the regional geology of W Tasmania can be found in Williams (1978), Corbett (1981) and Collins & Williams (1986). Specific details on the late Precambrian–Cambrian successions within the Dundas Trough appear in Brown (1986b) and Jenner & Brown (1989). The following summary is based on these references.

The Precambrian rocks of W Tasmania have been separated into two divisions. The first consists of multiply deformed, mainly upper greenschist facies–metamorphosed siltstone and orthoquartzite, which form the central Tyennan Region of W Tasmania, as well as a small outlier on the north coast near Forth (Fig. 9.1). The second division also consists of multiply deformed sedimentary rock sequences (siltstone and sandstone) but is mainly of lower to middle greenschist facies and locally contains mildly alkalic volcanic units. The rocks of the second division form the Rocky Cape Region and outliers to the south (Fig. 9.1). Part, or all, of the deformation affecting the rocks of these two divisions occurred during the Penguin Orogeny, dated at ~725 Ma. The earliest deposition in the Dundas Trough after the Penguin Orogeny was the Success Creek Group. This group consists of ~1000 m of shallow-water, fluviatile sedimentary rocks, which unconformably overlie the Precambrian basement along the western side of the Trough. The lower part of the group is dominated by siliceous sandstone and the upper part consists of siliceous siltstone, mudstone, dolomite and stromatolite clast-bearing, oolitic chert breccia (Brown 1986b).

The Success Creek Group is conformably followed by the Crimson Creek Formation, a sequence of turbiditic, volcaniclastic lithic wacke and siltstone, interbedded with alkaline to sub-alkaline basalt, mudstone and minor carbonate units. Chemically, the basalt suite has Within-Plate characteristics with LREE-enrichment REE patterns. The basalt suite also shows a progressive enrichment in Ti, Zr and Y up through the volcanic pile. A similar lithostratigraphy is also found in the Smithton Basin on the NW coast (Fig. 9.1).

In the Heazlewood–Waratah area, sub-alkaline basalts and associated sedimentary rocks occur, both with oceanic affinities (Jenner & Brown 1989). Intruding these rocks are successions of boninitic and LOTI lavas. Fault-bounded areas of three different successions of ultramafic rocks (each of which is considered to be associated with one of the three mafic lava sequences found in the Heazlewood–Waratah area) are also found within the Dundas Trough (Brown 1986a,b). One succession is a layered dunite, orthopyroxene-bearing dunite and minor harzburgite sequence, which contains a tectonic fabric formed by plastic deformation parallel to the cumulate layering. This succession is considered to be a magma chamber cumulate from the liquid which gave rise to the boninitic lavas and consists of Fo_{93-94} olivine, enstatite (Mg # = 0.93–0.94) and Cr-spinel with Cr # = 0.87–0.93. No clinopyroxene or feldspar has been recorded from these rocks. Electron probe microanalyses of enstatite from this succession indicate CaO, Cr_2O_3 and Al_2O_3 contents of less than 0.5%, suggesting that the original pyroxene formed as protoenstatite and did not quench-invert, but retained the high-temperature orthorhombic symmetry.

To the east of the Dundas Trough, a belt of acid to intermediate sub-aerial volcanics, the Mount Read volcanics, occurs (Fig. 9.1) (Corbett 1981, Corbett & Lees 1987). Biostratigraphic evidence indicates that the sedimentary rock sequences along the western part of this belt interdigitate with the lower part of the middle Middle Cambrian Dundas Group in the Dundas–Ring River area. The Dundas Group (Elliston 1954, Banks 1956) is now known to consist of two distinct parts: a stratigraphically lower part, associated with low-Ti basaltic lavas, which ranges in age from middle Middle Cambrian to early Late Cambrian, and a stratigraphically upper part which is of middle to late Late Cambrian age (Brown 1986b). The lower part contains mass flow and turbiditic conglomerate sequences separated by laminated mudstone–siltstone horizons and consists dominantly of chert, mudstone and carbonate with minor quartzite boulders, LOTI volcanic and ultramafic detritus (Rubenach 1974, Brown 1986b). The upper part of the group consists of conglomerate members interbedded with siltstone–sandstone members which grade upwards into siliceous fluviatile sequences containing upper Late Cambrian faunas.

From Late Cambrian to Early Ordovician, the southern part of the Dundas Trough and most of the Mt Read volcanics were covered by a terrestrial to shallow marine siliceous conglomerate and sandstone sequence, the Owen Conglomerate. Continuing deposition during the remainder of the Ordovician, Silurian and Early Devonian produced another 5000 m of shallow-water limestone, siliceous sandstone and siltstone–mudstone sequences, which constitute the Gordon Limestone and Eldon Group.

During late Early to early Middle Devonian two main phases of deformation extensively deformed the rock successions of W Tasmania, producing open fold structures with steep axial surfaces and associated cleavages, fault emplacement of blocks of Rocky Cape Region basement through overlying rock successions, and folding and re-emplacement of ultramafic and associated mafic rocks to produce their present spatial juxtaposition. Late in this orogeny, granitic batholiths intruded both the basement and overlying rock successions (Collins & Williams 1986).

9.3 Field relationships of the boninitic and low-Ti (tholeiitic) lavas

Boninitic lavas have been found in two main areas in W Tasmania: the Heazlewood–Magnet and the Stonehenge–Zeehan areas (Fig. 9.1). Geological maps showing sample localities are included as Figures 9.2 and 9.3. In the Heazlewood–Magnet area (Fig. 9.2), boninitic rocks occur (a) as dykes intruding the older oceanic sub-alkaline volcano-sedimentary successions, (b) as fault-bounded blocks along the margin of these successions and ultramafic rocks, and (c) within NE–SW orientated linear zones, apparently overlying the oceanic sub-alkaline volcano-sedimentary successions. The LOTI lavas intrude both the lower oceanic basalt and the boninitic lavas and, in places, overlie the boninitic lavas (Brown 1986b).

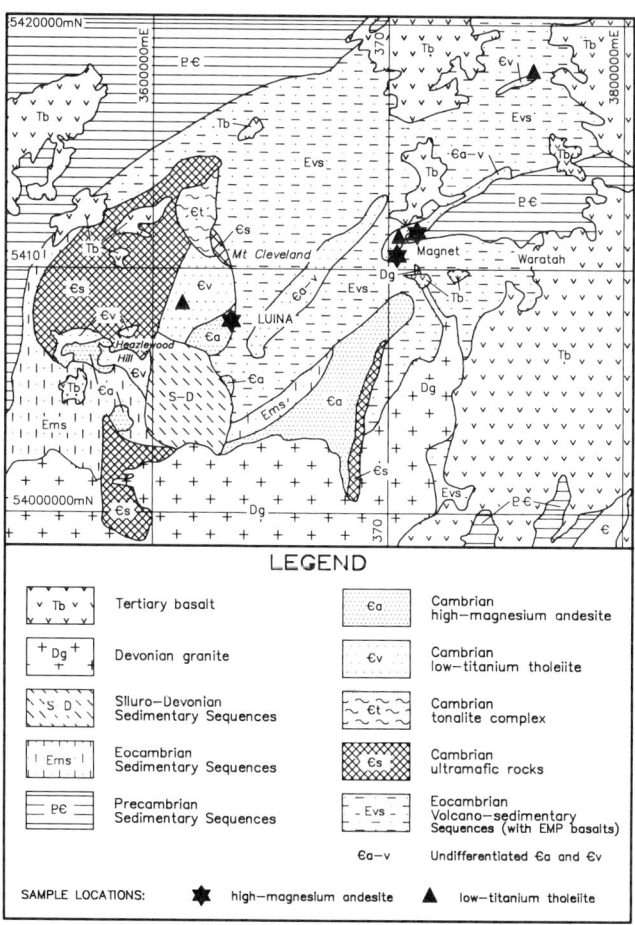

Figure 9.2 Schematic geological map of the Heazlewood–Waratah area. After Brown (1986b).

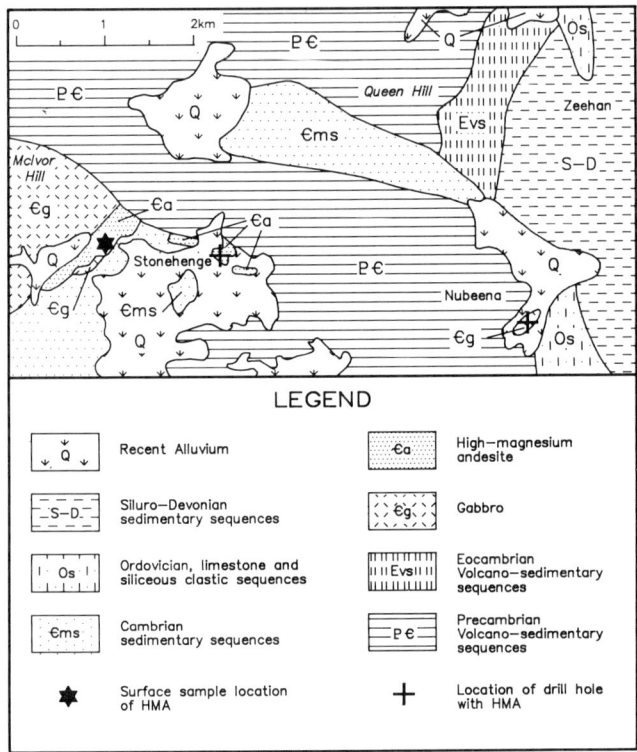

Figure 9.3 Schematic geological map of the Stonehenge–Zeehan area. After Blissett & Gulline (1962).

In the Stonehenge–Zeehan area (Fig. 9.3), boninitic lavas occur (a) as a fault-bounded block between gabbro (genetically related to the LOTI lavas) and sedimentary rocks correlated with the Middle Cambrian Razorback Conglomerate member of the lower Dundas Group, (b) as isolated outcrops in the Stonehenge area occurring in depressions within the Precambrian basement of the area and intruded by gabbro associated with the LOTI volcanics, and (c) tectonically intermixed with deformed black-grey mudstone–siltstone units in the Nubeena area. Samples from the latter two areas were obtained from diamond drill core.

In addition to the fault-bounded block of lavas in the Heazlewood–Waratah area, LOTI lavas have also been found in the Black Hill–Ring River area north of Dundas (Brown 1986b). In this area they are either overlain by, or interdigitate with, a basal conglomerate member of the Dundas Group (Red Lead Conglomerate) which also contains detritus of the lavas. To the west of Black Hill, LOTI lavas are fault juxtaposed with a plagioclase-bearing, layered ultramafic/gabbro succession which forms the Serpentine Hill Ultramafic Complex (Rubenach 1974, Brown *et al.* 1980, Brown 1986a,b).

9.4 Petrography

9.4.1 Boninitic rocks

HEAZLEWOOD–MAGNET AREA

Samples from this area come from highly weathered, vesicular pillow lava, breccia and crystal-mush flows. Pyroxene pseudomorphs up to 5 mm long and spinel grains (0.3 to 1.0 mm in diameter) are clearly visible in hand specimen. The original phenocryst mineralogy, excluding spinel, has been replaced by chlorite and amphibole, but original igneous textures are commonly preserved. Pseudomorphed pyroxene phenocrysts occur as single crystals, cruciform twins, or in glomeroporphyritic patches. The larger phenocrysts are generally euhedral and do not show remnant twinning. Smaller (1.0–1.5 mm) phenocrysts are subhedral, exhibit multiple twinning, and are interpreted to have been clinoenstatite (Cen). Originally, groundmass varied between vesicular and glassy, and felty texture of pyroxene microlites (0.07–0.15 mm long) with interstitial glass. Glass has been replaced by chlorite, and pyroxene by Cr-actinolite. In some amygdaloidal and breccia flows the groundmass has been replaced by calcite and quartz.

STONEHENGE–ZEEHAN AREA

In the coarser-grained parts of sheet flows, amphibole pseudomorphs after orthopyroxene and clinoenstatite (Figs 9.4a & b) occur. The pyroxene phenocrysts are 4–6 mm long, some enclosing Cr-spinel grains (up to 0.5 mm diameter). In all samples, primary minerals, with the exception of Cr-spinel, have been altered to actinolite–tremolite and chlorite.

Samples from a diamond drill hole near Stonehenge consist of chlorite and quartz with minor calcite after vesicular, fine- to medium-grained, and porphyritic lavas. The majority of these flows exhibit remnant flow banding with a quenched groundmass. Quench-textured lavas consist of 0.2 to 0.45 mm long acicular actinolite (after pyroxene) microlites, and interstitial glass. Phenocrysts are both euhedral and subhedral and up to 1.65 mm long, some enclosing Cr-spinel grains (0.02–0.05 mm diameter).

Most samples, including all those analysed, contain chlorite pseudomorphs with remnant multiple twinning, after clinoenstatite. These phenocrysts occur as cruciform twins or occasionally as rosettes. Most single-crystal, low-Ca pyroxene phenocrysts had quench overgrowths, which are now actinolite, but were probably originally strongly zoned Ca-rich pyroxenes.

The sample from the Nubeena drillhole is a medium-grained, porphyritic, flow-banded lava with stubby chlorite-pseudomorphed low-Ca pyroxene phenocrysts (0.35–0.55 mm) in a quenched groundmass consisting originally of spear-shaped pyroxene grains (up to 1.0 mm long), Cr-spinel and glass. The groundmass consists of a fine-grained quartz–chlorite intergrowth. Twinned pseudomorphs after Cen were not seen.

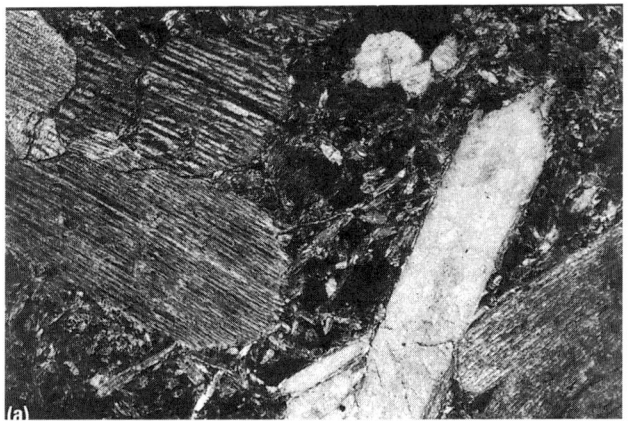

Figure 9.4 Amphibole pseudomorphs after clinoenstatite (twinned) and orthopyroxene (untwinned). Polarized light, Specimen 850031. Field of view 2.5 × 3.5 mm.

9.4.2 Low-Ti basalt (LOTI)

Petrographically, all samples of LOTI basalt are very similar, irrespective of whether they are from the Cleveland (Creenaune 1980), Waratah, or Black Hill area (Brown 1986b). Texturally, the LOTI lavas are aphyric, but in the thicker flows a considerable range in grain size occurs. These flows vary from pillowed flows to more massive flows with variolitic or amygdaloidal tops, to fine- to medium-grained ophitic- and subophitic-textured basalt in interior portions of flows. Rare coarse-grained (3–5 mm) granophyric areas occur in some thick flows in the Cleveland area (Creenaune 1980); these rapidly grade into a chilled basal zone.

Quenched and fine-grained samples of LOTI basalt contain variolitic sheaves of actinolite, chlorite and albitized plagioclase in a formerly glassy groundmass now replaced dominantly by quartz, albite, calcite and chlorite. Fine- to medium-grained samples contains fibrous actinolite and epidote alteration of subophitic intergrowths of pyroxene (0.5 mm) and plagioclase (0.2 mm) grains, secondary quartz, altered skeletal opaque grains and secondary pyrite. In some samples, the original mineralogy is totally replaced by an intergrowth of quartz and amphibole but the subophitic texture is still preserved. The coarser-grained ophitic-textured lavas are composed of actinolite, chlorite, albite, skeletal opaques and chlorite.

9.5 Mineral chemistry

Cr-spinel grains are the only relict primary phase in the samples studied. Microprobe analyses of the Cr-spinels were done at the University of Tasmania with a JEOL JX50A microprobe using an energy dispersive (EDAX) analytical system. Data reduction techniques and operating conditions used in this system are described in Griffin (1979). The composition of representative Cr-spinel grains from the Tasmanian boninitic rocks are presented in Table 9.1. They are compositionally very similar to chromite grains from Palaeozoic and Tertiary boninite lavas, as is shown in Figure 9.5a. Compositions of Cr-spinel grains from cumulate ultramafic rocks associated with the Tasmanian boninitic lavas are shown in Figure 9.5b. The Cr-spinel grains from the Cleveland–Magnet boninitic lavas are more Mg- and Cr-rich than those from the Stonehenge area. Chromite grains from the boninitic lavas within contact aureoles of Devonian granitoids always have Mg# < 0.40, presumably because of subsolidus re-equilibration with serpentine–amphibole–magnetite assemblages. The large compositional overlap between Cr-spinel grains from the boninitic lavas and those from associated layered dunite–harzburgite complexes in W Tasmania provides strong evidence for a genetic relationship between these suites (Brown 1986a).

9.6 Geochemistry

9.6.1 Methods

Major and trace elements were determined at the Analytical Laboratory of the Department of Mines, Tasmania. Major elements were determined by XRF using the heavy-absorber technique of Norrish & Hutton (1969). Zr, Y, Sc, V, Ni and Cr were determined by XRF on pressed powder pellets. Replicated analyses of samples and a comparison of samples analysed at both the University of Tasmania and the Department of Mines Laboratory verify that the analytical accuracy and precision of these analyses are similar to that

Table 9.1 Compositions of selected Cr-spinels from Tasmanian Cambrian boninite samples.

Sample no. Location	1 850023 Clev	2 850025 Mag	3 850026 Mag	4 850027 Mag	5 850028 SH	6 850029 SH	7 850030 SH	8 850055 SHDH1	9 850056 SHDH1	10 850249 SHDH2
TiO_2	–	–	–	0.27	0.22	–	0.29	–	–	–
Al_2O_3	6.49	3.82	4.74	4.48	6.97	7.14	7.11	5.02	7.39	4.16
Cr_2O_3	61.76	67.48	61.50	63.56	59.36	60.97	58.57	65.04	61.19	63.92
FeO	19.69	10.97	22.26	19.23	18.89	18.06	19.38	17.69	17.84	20.33
MnO	0.33	0.32	0.46	0.36	0.50	–	0.45	–	–	–
MgO	11.44	17.15	10.81	11.79	13.88	13.20	13.75	11.85	12.86	11.05
total	99.71	99.74	99.77	99.69	99.82	99.37	99.55	99.60	99.28	99.46
Ti	–	–	–	0.0543	0.0431	–	0.0569	–	–	–
Al	2.0342	1.1640	1.5014	1.4133	2.1385	2.2104	2.1870	1.5819	2.2934	1.3239
Cr	12.9807	13.7884	13.0632	13.4456	12.2130	12.6573	12.0810	13.7433	12.7339	13.6409
Fe^{3+}	0.9851	1.0476	1.4353	1.0868	1.6053	1.1323	1.6750	0.6748	0.9727	1.0352
Fe^{2+}	3.3929	1.3237	3.5667	3.2167	2.5062	2.8341	2.5539	3.2796	2.9549	3.5545
Mn	0.0743	0.0701	0.1047	0.0816	0.1102	–	0.0995	–	–	–
Mg	4.5328	6.6062	4.3286	4.7017	5.3836	5.1659	5.3467	4.7204	5.0451	4.4455
100Mg #	57.2	83.3	54.8	59.4	68.2	64.6	67.7	59.0	63.1	55.6
100Cr #	86.5	92.2	89.7	90.5	85.1	85.1	84.7	89.7	84.7	91.2

Total iron reported as FeO. Structural formulae calculated with oxygen = 32 assuming the general formulae $R^{2+}R^{3+}_2O_4$ and apportioning total iron as Fe^{2+} and Fe^{3+} such that $2(Mg + Mn + Ca + Fe^{2+}) = (Cr + Al + Ti + Fe^{3+})$.
Locations: Clev = Cleveland area; Mag = Magnet area; SH = Stonehenge area; SHDH1 = drillhole no. 1 near Stonehenge; SHDH2 = drillhole no. 2 near Stonehenge (see Figs 9.1 and 9.2).
Sample numbers are Department of Mines registered numbers.

quoted for international standards in Jenner (1981). Nine samples were analysed for REE, using two techniques: analyses where Tb and Ho are reported were analysed by radiochemical neutron activation (RNAA); those with Er but not Tb were analysed by an XRF ion-exchange technique. RNAA analyses were done at the University of Melbourne by Helen Waldron, using

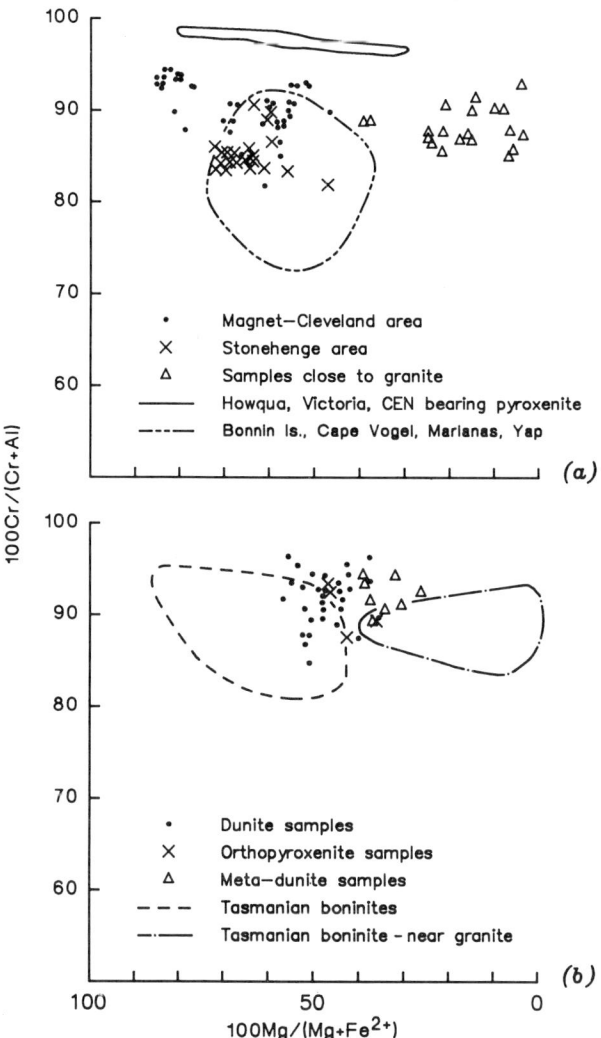

Figure 9.5 Variation in composition of Cr-spinel from Tasmanian boninitic and related ultramafic complexes and Tertiary boninite spinel: (a) Tasmanian boninite spinels; (b) spinels in Tasmanian ultramafics. Comparison fields for: Tertiary boninite samples from the Bonin–Mariana–Cape Vogel occurrences (Jenner 1982); Yap–Mariana Group A basaltic volcanics (Crawford et al. 1986); Howqua (Crawford 1980).

Table 9.2 Wholerock compositions of Tasmanian Cambrian boninites.

	Group I							Vic	Group II								Vic
	1	2	3	4	5	6	7	8	9	10	11	12	13	14	15	16	17
Sample no.	850023	850026	850027	850025	850028	850249	–	HGB-A	850238	850242	850241	850240	850029	850030	850031	–	–
Mineralogy	Cep	Cep	Cep	Cep	Cep	Pxp	Av.	Av.	Pxp	Pxp	Pxp	Pxp	Cep	Cep	Cep	Av.	How
					SH	SHD	(6)	(5)	SHD	SHD	SHD	SHD	SH	SH	SH	(7)	Av. (2)
Location	Clev	Mag	Mag	Mag													
SiO_2	56.31	55.91	61.29	61.03	54.73	60.56	58.31	56.95	57.62	61.30	60.46	61.67	53.59	55.00	54.47	57.73	56.10
TiO_2	0.08	0.03	0.09	0.10	0.04	0.14	0.08	0.30	0.20	0.19	0.22	0.22	0.13	0.11	0.10	0.17	0.08
Al_2O_3	9.24	7.13	12.03	11.77	4.88	12.29	9.56	9.34	8.89	8.66	9.21	9.04	5.77	5.18	5.05	7.40	5.05
FeO*	9.85	9.74	9.39	9.73	10.17	9.53	9.74	10.12	9.26	8.84	9.09	9.19	10.22	10.18	10.28	9.58	11.44
MnO	0.18	0.22	0.29	0.27	0.35	0.11	0.24	0.16	0.20	0.16	0.19	0.16	0.29	0.28	0.29	0.22	0.18
MgO	18.81	24.11	11.66	12.23	18.89	17.07	17.13	11.45	15.45	15.36	16.13	15.43	19.77	18.57	19.25	17.14	16.75
CaO	5.06	2.76	4.07	3.89	10.37	0.18	4.39	7.06	7.51	4.28	3.89	3.67	9.61	10.00	10.04	7.00	8.54
Na_2O	0.40	0.03	0.03	0.06	0.22	0.03	0.13	3.69	0.78	0.77	0.74	0.54	0.31	0.27	0.20	0.52	1.65
K_2O	0.70	0.02	1.10	0.88	0.20	0.03	0.38	0.35	0.04	0.43	0.04	0.03	0.29	0.39	0.28	0.21	0.17
P_2O_5	0.01	0.02	0.03	0.04	0.12	0.04	0.04	0.07	0.05	0.01	0.04	0.04	0.02	0.02	0.03	0.03	0.16
H_2O	7.8	8.1	6.3	6.4	–	8.7	7.05	–	6.1	5.4	6.3	6.3	4.3	4.0	4.2	5.23	–
CO_2	0.5	0.9	6.2	6.0	–	0.1	2.30	–	3.1	1.5	0.8	1.3	0.1	0.1	0.1	1.00	–
LOI	–	–	–	–	5.1	–	–	3.02	–	–	–	–	–	–	–	–	–
100Mg #	77	82	69	69	77	76	75	67	75	76	76	75	78	76	77	76	72
Cr	2045	2723	1009	930	3104	996	1801	1304	1167	1176	1197	1092	2719	3052	2832	1891	1580
Ni	756	667	190	184	610	229	451	238	256	208	239	229	596	610	598	391	406
Sc	29	26	29	29	22	22	26	31	34	34	35	34	27	25	24	31	27
V	107	105	150	126	105	104	116	204	289	176	261	273	120	111	110	191	129
Zr	7	17	23	18	5	35	18	28	10	13	10	10	5	3	5	8	29
Ti	462	200	554	617	257	862	492	1799	1199	1153	1305	1309	752	632	629	997	493
Y	8	4	4	6	5	8	7	9	7	7	7	7	5	3	4	6	14

FeO* = total Fe as FeO.
Analyses are recalculated volatile-free ($H_2O + CO_2$ or loss on ignition) to 100%. Cep = clinoenstatite pseudomorphs present; Pxp = pyroxene pseudomorphs.
Locations: Clev = Cleveland area; Mag = Magnet area; SH = Stonehenge area; SHD = drillholes at or near Stonehenge (see Figs 9.1 & 9.2).
Sample numbers are Department of Mines, Tasmania, registered numbers; HGB-A = Heathcote Greenstone Belt (Crawford & Cameron 1985); How = Howqua (Nelson et al. 1984); Vic = Victoria.

the technique described in Bishop & Hughes (1984). XRF analyses were done at the Memorial University of Newfoundland using the techniques described by Robinson *et al.* (1986).

Sm and Nd concentrations and isotopic compositions given in Table 9.4 were determined by G.A.J. at the Max-Planck-Institut für Chemie, Mainz, using slightly modified versions of the techniques described in White & Patchett (1984).

9.6.2 Wholerock geochemistry and alteration

Major- and trace-element analyses of the boninitic rocks are reported in Table 9.2, and of LOTI in Table 9.3. Aspects of the major- and trace-element chemistry are illustrated in Figures 9.6 to 9.9, and a comparison with other low-Ti volcanics is given in Table 9.4. REE data are given in Table 9.5 and illustrated in Figure 9.10. Note that in all figures in which abundances are plotted, data have been corrected for volatile loss.

The boninitic and LOTI volcanics have been subjected to regional greenschist facies metamorphism. Severe weathering has affected volcanic rocks, to the extent that it is sometimes not possible to find samples suitable for analysis. Post-eruptive metamorphic and weathering processes have altered the chemistry of these volcanics. The major elements, particularly Si, Na, Ca and K, may have undergone significant modification (Pearce 1975) and abundances of LILE such as Rb, Ba, Sr and Cs probably have also been significantly affected. The HFSE (Ti, P, Zr, Nb, Y, Hf, Ta) and REE are relatively immobile during low-grade alteration (Pearce 1975, Hellman *et al.* 1979, Saunders *et al.* 1980). The transition elements (Cr, Ni, Sc, V) are also relatively immobile during alteration (Pearce 1975, Shervais 1982).

The effects of alteration on the composition of the boninitic and LOTI lavas are illustrated in Figure 9.6, in which K, a potentially mobile element, does not behave in a coherent manner relative to immobile Ti. Based on this, and comparisons with other studies of altered high-Mg volcanics (e.g. Arndt & Jenner 1986), we have chosen not to report data for the low-field-strength elements. The immobile behaviour of Ti, V and Zr is illustrated in Figures 9.7 and 9.8, in which the behaviour of these elements is consistent and/or predictable when compared with indicators of the degree of fractionation (e.g. FeO^*/MgO).

Another indication of the degree of alteration the boninitic lavas have suffered is given by the lack of coherence between CaO content, CaO/Al_2O_3 and the occurrence of Cen pseudomorphs. In Tertiary Cen-bearing boninite, clinoenstatite only occurs in those boninitic rocks with $CaO/Al_2O_3 \leqslant 0.7$ (Jenner 1982). An increase in this ratio (and in the normative $Di/(Di + Hy)$ of the magma) in Tertiary Cen-bearing boninite results in cessation of protoenstatite crystallization (see also Komatsu 1980, Umino & Kushiro 1989). In

Table 9.3 Wholerock compositions of Tasmanian Cambrian low-Ti tholeiites (LOTI).

	Cleveland area									
Sample no.	1 60896	2 60898	3 60894	4 60897	5 60901	6 60892	7 60919	8 60900	9 60903	Av. (9)
SiO_2	50.36	52.52	48.71	52.79	55.06	53.98	50.31	52.64	49.82	51.80
TiO_2	0.21	0.20	0.22	0.23	0.25	0.31	0.25	0.25	0.13	0.23
Al_2O_3	13.62	14.62	15.80	14.18	15.22	13.83	14.95	14.95	14.69	14.65
FeO^*	10.42	8.91	10.14	9.30	11.28	10.64	9.68	10.08	8.66	9.90
MnO	0.09	0.17	0.19	0.18	0.19	0.23	0.18	0.18	0.22	0.18
MgO	11.96	9.18	11.48	9.24	7.92	8.15	10.65	9.07	10.18	9.76
CaO	13.09	12.23	10.68	12.92	8.74	9.15	11.90	9.90	13.74	11.37
Na_2O	0.01	1.97	2.56	0.99	0.66	3.52	1.46	2.28	1.73	1.69
K_2O	0.07	0.13	0.17	0.11	0.66	0.14	0.57	0.62	0.82	0.37
P_2O_5	0.15	0.05	0.04	0.05	0.02	0.04	0.04	0.03	0.01	0.05
H_2O	–	–	–	–	–	–	–	–	–	–
CO_2	–	–	–	–	–	–	–	–	–	–
LOI	1.4	1.5	2.3	1.0	2.0	2.8	1.7	1.6	2.0	1.81
100Mg #	67	65	67	64	56	58	66	62	68	64
Cr	–	–	–	–	–	–	–	–	–	–
Ni	–	123	130	106	91	68	175	84	107	111
Sc	–	–	–	–	–	–	–	–	–	–
V	400	219	285	233	275	264	217	267	107	437
Zr	11	10	13	11	12	11	15	11	7	11
Ti	1230	1212	1339	1394	1504	1876	1522	1526	794	1377
Y	16	12	11	9	13	18	14	16	7	13

Columns 1–9 = Cleveland area: 10–14 = Black Hill area; 15 = Magnet area; 16–18 = Waratah area (see Figs 9.1 and 9.2).
FeO^* = total Fe as FeO.
Analyses are recalculated volatile-free ($H_2O + CO_2$ or loss on ignition) to 100%.
Sample numbers 60... are University of Tasmania registered numbers, analyses from Creenaune (1980); sample numbers 85... are Department of Mines, Tasmania, registered numbers.

contrast, the Tasmanian boninitic rocks (and other Victorian and Newfoundland boninites; Crawford 1980, Jenner unpubl. data) with Cen pseudomorphs often have $CaO/Al_2O_3 > 1-2$ (see Table 9.2, columns 5, 13–15; Table 9.4, column 4), indicating substantial mobility and addition of CaO (see also Nelson et al. 1984).

9.6.3 *Major-element and trace-element characteristics*

The boninitic rocks have high SiO_2, Cr (> 900 ppm) and Ni (> 180 ppm) contents, high Mg # (> 0.69) and low HFSE abundances. Two groups of boninitic rocks are distinguished on the basis of their HFSE ratios (see Table 9.4; Figs 9.8 & 9.9). The first group (HMA-I) is characterized by low Ti/Zr (av. 28) and Ti/Y (av. 74), and high Zr/Y (av. 2.7). The REE patterns of

Table 9.3 (*continued*)

Sample no.	Black Hill Area						Waratah area				
	10 850032	11 850033	12 850034	13 850035	14 850036	Av. (5)	15 850052	16 850037	17 850038	18 850039	Av. (4)
SiO_2	59.33	57.79	62.84	51.71	55.72	57.48	61.45	54.98	52.68	52.63	55.44
TiO_2	0.29	0.63	0.46	0.51	0.29	0.44	0.23	0.57	0.41	0.26	0.37
Al_2O_3	13.40	19.26	14.15	15.65	15.36	15.56	15.67	13.84	15.04	14.00	14.64
FeO^*	12.47	9.63	9.09	13.88	12.09	11.83	7.13	12.93	11.16	10.95	10.54
MnO	0.14	0.10	0.15	0.21	0.21	0.16	0.15	0.20	0.19	0.19	0.18
MgO	4.57	4.32	5.04	7.07	5.43	5.29	6.85	5.22	8.03	7.01	6.78
CaO	7.90	3.65	4.83	8.23	8.89	6.70	3.70	6.62	9.20	10.61	7.53
Na_2O	1.66	1.55	3.18	2.32	1.81	2.10	4.72	5.40	3.00	3.95	4.27
K_2O	0.20	2.99	0.21	0.36	0.15	0.78	0.02	0.18	0.22	0.35	0.19
P_2O_5	0.04	0.06	0.05	0.05	0.05	0.05	0.07	0.07	0.06	0.05	0.06
H_2O	2.8	5.0	4.4	5.1	3.1	4.08	–	–	–	–	–
CO_2	0.1	3.6	3.7	7.9	0.2	3.10	–	–	–	–	–
LOI	–	–	–	–	–	–	6.3	2.7	4.8	3.4	4.3
100Mg#	40	44	50	48	44	45	63	42	56	53	54
Cr	25	205	168	177	57	126	189	85	276	84	158
Ni	60	90	89	94	57	78	108	66	111	75	90
Sc	49	63	57	60	70	60	21	56	60	34	43
V	270	543	307	325	406	370	124	386	308	434	313
Zr	8	18	15	17	8	13	19	23	18	9	17
Ti	1745	3783	2761	3059	1748	2274	1359	3440	2486	1548	2208
Y	8	12	10	16	9	11	8	19	15	8	12

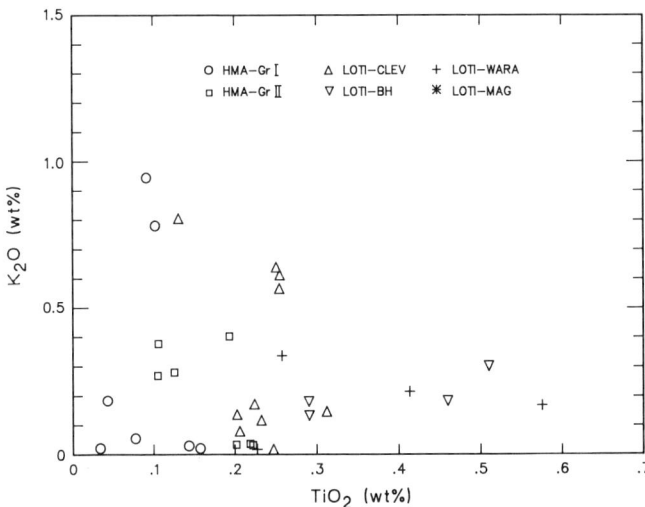

Figure 9.6 Potassium versus titanium variation in Tasmanian boninitic and LOTI lavas.

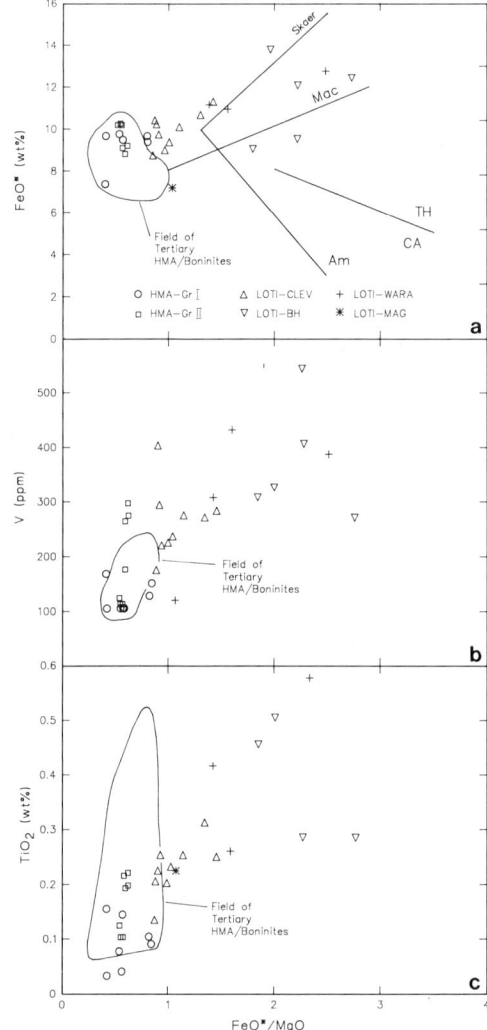

Figure 9.7 (a) FeO* versus FeO*/MgO in Tasmanian boninitic and LOTI samples. After Miyashiro (1974). Data sources for the field of Tertiary boninite are given in the text. Reference differentiation trends for tholeiitic rocks given by lines labelled Skaer (Skaergaard) and Mac (Macauley Island). Calc-alkaline differentiation trend defined by line Am (Amagi Island). The full line TH/CA divides the diagram into fields for tholeiitic (TH) and calc-alkalic (CA). FeO* = total iron as FeO. (b) V versus FeO*/MgO in Tasmanian boninitic and LOTI samples. The increase in V with FeO*/MgO indicates a tholeiitic affinity for the volcanics (Miyashiro & Shido 1975). Ti/V in these volcanics (not shown) are less than 10, and plot in the arc field of Shervais (1982). (c) TiO$_2$ versus FeO*/MgO in Tasmanian boninitic and LOTI samples. Note the extremely low concentrations of TiO$_2$ in the boninitic samples, which are amongst the lowest reported for any volcanic. The enrichment in TiO$_2$ with increasing value of the differentiation index in the LOTI suggests a tholeiitic affinity for these volcanics.

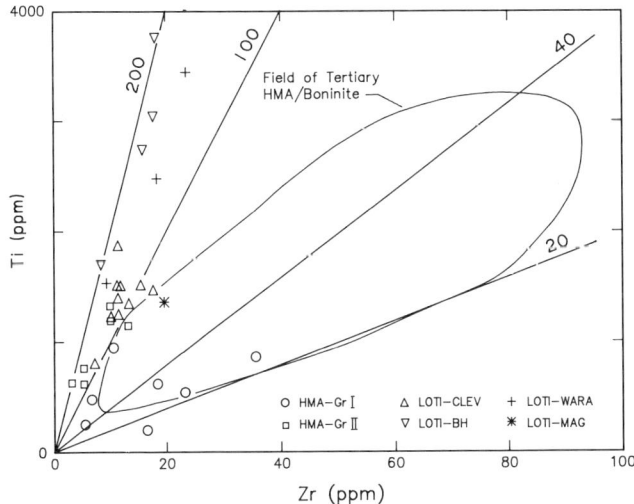

Figure 9.8 Ti versus Zr for Tasmanian boninitic and LOTI samples. The full lines numbered 20, 40, 100 and 200 are reference lines for constant Ti/Zr. Note the low Ti/Zr values in HMA-I, similarity in HMA-II and LOTI, coherent behaviour of Ti and Zr (especially within geochemical groups) and the low concentrations of Zr (< 40 ppm) in all of these volcanics. Compared to the reference fields (for oceanic and arc volcanics) of Pearce & Cann (1973), usually shown on this type of diagram, the boninitic and LOTI samples plot outside the fields at lower concentrations.

HMA-I samples (Fig. 9.10a) vary from concave to moderately LREE-enriched. Note the extremely low abundances of the HREE (<3 × primitive mantle abundances) and the depletion of Ti relative to the REE. The second group (HMA-II) have similar Mg#, SiO_2 and Al_2O_3 contents to those of HMA-I; however, they are higher in Ti, Sc, V and have lower Zr abundances. These differences are clearly shown by immobile-element ratios (Table 9.4), with the HMA-II lavas being characterized by high Ti/Zr (av. 125), Ti/Y (av. 175) and low Zr/Y (1.4) and Al_2O_3/TiO_2 (44). Unfortunately, there are no REE patterns available for HMA II samples.

The LOTI volcanics have been grouped in Table 9.3 according to their geographic areas – Cleveland (Clev), Black Hill (BH) and Waratah (Wara). The LOTI at Cleveland and Waratah are very closely associated with boninitic lavas, while those at Black Hill are not. The LOTI-Clev range in Mg# from 0.56 to 0.68, have 65–130 ppm Ni and SiO_2 contents of 49–55%. They are the most primitive LOTI group (Fig. 9.7), consisting of basalt and basaltic andesite. The enrichment in FeO^*, TiO_2 and V with FeO^*/MgO (Figs 9.8–9.10) indicates a tholeiitic affinity for these rocks. The average Ti/Zr in LOTI-Clev is about 125 (Fig. 9.8, Table 9.4). REE and HFSE patterns for LOTI-Clev are shown in Figure 9.10c. They are characterized by low abundances (<5 × primitive mantle), relatively coherent HFSE–REE behaviour, and strong LREE depletion. Two samples (60903 and 60900) show

Table 9.4 Average compositions and selected ratios of Tasmanian Cambrian boninites (HMA) and low-Ti tholeiites (LOTI), Victorian Cambrian boninites and W Pacific Tertiary boninites.

	1 Tas HMA-I	2 Tas HMA-II	3 Vic HGB-A	4 Vic How	5 Tert Bon Type C	6 Tert Bon Type E	7 Tas LOTI Clev	8 Tas LOTI BH	9 Tas LOTI Mag	10 Tas LOTI War
SiO_2	58.31	57.73	56.95	56.10	57.67	59.50	51.80	57.48	61.45	53.43
TiO_2	0.08	0.17	0.30	0.08	0.15	0.29	0.23	0.44	0.23	0.41
Al_2O_3	9.56	7.40	9.43	5.05	11.04	9.78	14.65	15.56	15.67	14.29
FeO^*	9.74	9.58	10.12	11.44	8.60	8.19	9.90	11.83	7.13	11.68
MnO	0.24	0.22	0.16	0.18	0.17	0.19	0.18	0.16	0.15	0.19
MgO	17.13	17.14	11.45	16.75	12.91	15.36	9.76	5.29	6.85	6.75
CaO	4.39	7.00	7.06	8.54	7.26	4.43	11.37	6.70	3.70	8.81
Na_2O	0.13	0.52	3.96	1.65	1.50	1.75	1.69	2.10	4.72	4.12
K_2O	0.38	0.21	0.35	0.17	0.63	0.47	0.37	0.78	0.02	0.25
P_2O_5	0.04	0.03	0.07	0.06	0.03	0.05	0.05	0.05	0.07	0.06
Cr	1890	1891	1304	1580	942	1822	–	126	189	148
Ni	451	391	238	406	214	444	111	78	108	84
Sc	26	31	31	27	28	32	–	60	21	50
V	116	191	204	129	138	157	437	370	124	376
Zr	18	8	28	29(15)	21	65	11	13	19	17
Y	7	6	9	14(7)	4	8	13	11	8	14
100Mg#	75	76	67	72	73	77	64	45	63	51

	1	2	3	4	5	6	7	8	9	10
CaO/Al$_2$O$_3$	0.46	0.95	0.75	1.69	0.62	0.46	0.78	0.43	0.24	0.62
CaO/TiO$_2$	55	41	24	107	38	15	49	15	16	21
Al$_2$O$_3$/TiO$_2$	120	44	31	63	61	34	64	35	68	35
Cr/Ni	3.6	4.8	5.5	3.9	4.4	4.1	–	1.6	1.8	1.8
V/Sc	4.5	6.3	6.6	4.8	5.0	4.9	–	6.2	5.9	7.5
Ti/V	4.2	5.2	8.8	3.8	6.5	11	3.2	6.1	11	6.6
Ti/Sc	19	33	58	18	32	54	–	38	65	50
Ti/Zr	28	125	64	17	44	27	125	175	72	147
Zr/Y	2.7	1.4	3.1	2.1	5.3	8.2	0.85	1.9	2.4	1.2
Ti/Y	74	175	200	35	225	214	106	207	170	178
La/Sm	3–7	–	3	2–4	2.6	3.7	0.4–1.0	0.4–1.5	–	–
La/Yb	1.7–7	–	3	0.8–8	2.1	5.7	0.1–0.3	0.1–0.5	–	–
Zr/Sm	21	–	22	24–100	60	51	28	21	–	–

References: column 1, this study (Table 9.1, columns 1–6); 2, this study (Table 9.1, columns 7–13); 3, Crawford & Cameron (1985); 4, Nelson *et al.* (1984); 5 and 6, Jenner (1981, 1982), Hickey & Frey (1982), Cameron *et al.* (1983) and Umino (1986); 7–10, this study. See text for discussion of bracketed Zr and Y contents in column 4.

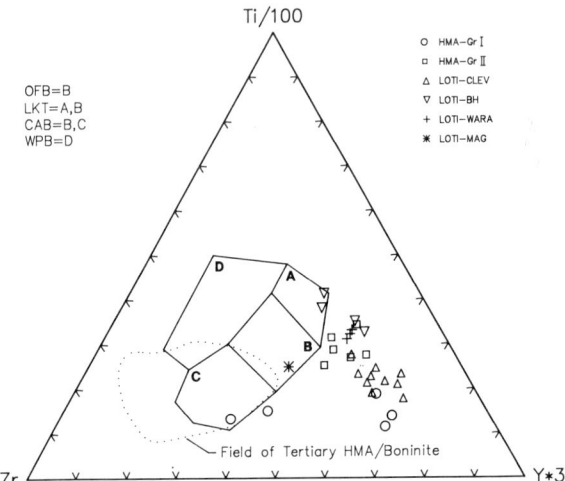

Figure 9.9 Ti–Zr–Y tectonic discrimination diagram. After Pearce & Cann (1973). The dotted curve shows field for Tertiary boninites (data sources given in the text). Note that the Tasmanian boninitic and LOTI samples generally plot outside of the reference fields, and are displaced towards the Y corner of the diagram. Two HMA-I samples plot close to or within the field defined by Tertiary boninite, emphasizing the similarity in HFSE between the two groups. However, the HFSE ratios in HMA-II differ notably from those of the Tertiary boninite and are more similar to that observed in the LOTI or very depleted supra-subduction zone volcanics.

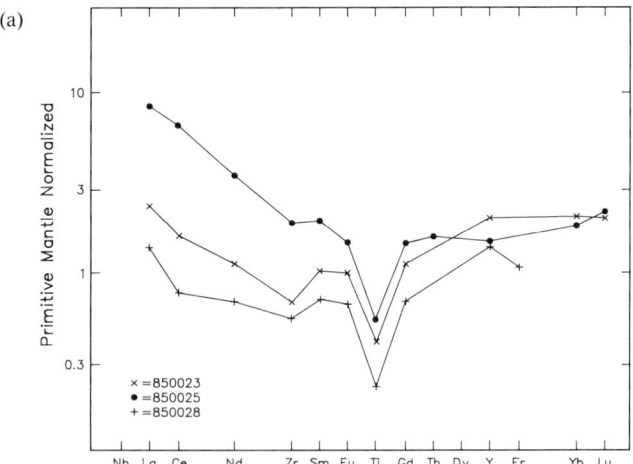

Figure 9.10 Primitive mantle normalized HFSE and REE abundances in (a) boninitic and (c) LOTI volcanics. In (b) the reference field for Bonin Island boninite is based on data from Cameron *et al*. (1983), Jenner (1982) and Hickey & Frey (1982). The Nepoui data are from Cameron *et al*. (1983). Two of the Tasmanian boninitic rocks (850023 and 850028) have the concave pattern typical of Type C boninite, while the other (850025) has the LREE enrichment typical of Type E boninite (Jenner 1981). Note, in particular, the lack of Zr enrichment relative to Sm and Nd in the Tasmanian boninitic samples, which is a characteristic geochemical signature found in Tertiary boninite. Normalizing values used are given in Jenner *et al*. (1987) and Jenner & Brown (1989).

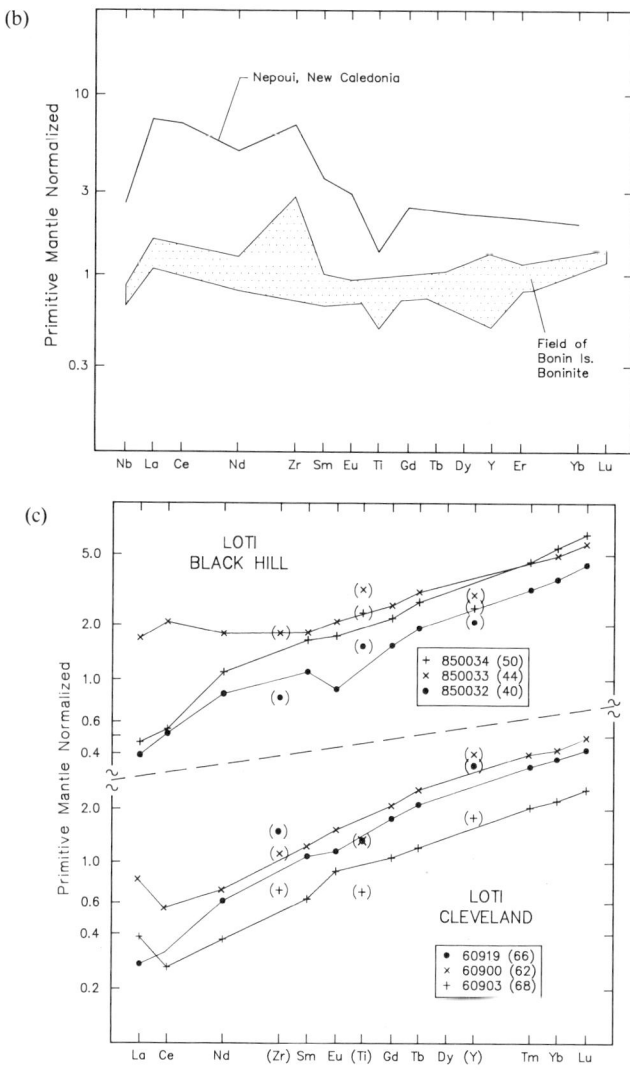

an abrupt enrichment in La relative to Ce. This may be a result of alteration; however, La enrichments relative to Ce have been noted in fresh suprasubduction zone volcanics (Rautenschlein *et al.* 1985, White & Patchett 1984).

The LOTI-Wara are generally more fractionated (Fig. 9.7) than LOTI-Clev, but have variably higher HFSE abundances (Figs 9.7 and 9.8). In particular, the average TiO_2 content in the LOTI-Wara is twice as high as that in the LOTI-Clev, whereas Zr is only slightly higher and Y is essentially the same. These differences in relative HFSE abundances are illustrated in Fig. 9.9 and by the HFSE ratios in Table 9.4. In general, the LOTI-Wara also appear to be

Table 9.5 Rare-earth-element concentrations (ppm) in selected Tasmanian Cambrian boninite (HMA) and low-Ti tholeiite (LOTI) samples.

	1	2	3	4	5	6	7	8	9
Sample no.	850023	850025	850028	+60919	+60900	+60913	850032	850033	850034
Rock type	HMA	HMA	HMA	LOTI	LOTI	LOTI	LOTI	LOTI	LOTI
Location	Clev	Mag	SH	Clev	Clev	Clev	BH	BH	BH
La	1.52	5.34	0.86	0.18	0.51	0.24	0.25	1.07	0.29
Ce	2.57	10.71	1.21	0.50	0.89	0.41	0.84	3.34	0.88
Nd	1.34	4.14	0.83	0.74	0.85	0.45	1.03	2.16	1.34
Sm	0.41	0.77	0.28	0.44	0.50	0.25	0.44	0.73	0.68
Eu	0.15	0.22	0.10	0.18	0.24	0.14	0.14	0.33	0.27
Gd	0.59	0.77	0.37	0.96	1.12	0.57	0.83	1.41	1.17
Tb	0.14	0.15	0.14	0.21	0.25	0.12	0.19	0.31	0.27
Ho	0.25	0.18	–	0.44	0.48	0.26	0.36	0.56	0.58
Er	–	–	0.46	–	–	–	–	–	–
Tm	0.13	0.10	–	0.23	0.27	0.14	0.22	0.23	0.31
Yb	0.89	0.79	–	1.65	1.88	1.00	1.61	2.24	2.37
Lu	0.13	0.14	–	0.28	0.23	0.17	0.29	0.38	0.43

Rock types: HMA = boninite; LOTI = low-titanium basalt.
Locations: Clev = Cleveland; Mag = Magnet; SH = Stonehenge; BH = Black Hill.
Sample 3 analysed by XRF–ion exchange, Memorial University of Newfoundland. All other samples analysed by RNAA at the University of Melbourne by H. M. Waldron.

of tholeiitic affinity (Fig. 9.7). LOTI-Mag is an extensively altered sample, characterized by high SiO_2, Na_2O and LOI, and low FeO^*/MgO (Table 9.3, column 15).

The LOTI-BH are the most fractionated group of LOTI (Fig. 9.7) and show marked variations in SiO_2, TiO_2 and V within a small range of FeO^*/MgO ratios (Fig. 9.7, Table 9.3). REE and HFSE patterns are illustrated in Fig. 9.9c. While there is an overall similarity in the patterns (i.e. overall LREE depletion and similar HREE pattern slopes), there is a marked variation in the LREE abundances, particularly for sample 850033. Ti/Zr remain relatively constant (Fig. 9.8); however, there is some variation in ratios involving Y (see Fig. 9.9).

9.6.4 Nd isotopic compositions

Concentrations of Sm and Nd, and Nd isotopic compositions determined on two boninitic and two LOTI samples are reported in Table 9.6. The REE concentrations in these samples are very low, which results in relatively large but acceptable errors on their isotopic compositions. Good to excellent agreement exists between the isotope dilution measurements and the radiochemical neutron activation determinations (Tables 9.5 & 9.6). The age uncertainty on these samples is about ± 50 Ma; however, this introduces

Table 9.6 Sm–Nd isotopic data for Tasmanian Cambrian boninite (HMA) and low-Ti tholeiite (LOTI) samples.

	1	2	3	4
Sample no.	850023	850028	+60919	850032
Rock type	HMA	HMA	LOTI	LOTI
Location	Clev	SH	Clev	BH
Sm (ppm)	0.361	0.271	0.439	0.468
Nd (ppm)	1.275	0.991	0.747	1.002
$^{117}Sm/^{144}Nd$	0.1713	0.1654	0.3555	0.2825
$(^{143}Nd/^{144}Nd)_{(M)}$	0.512423	0.512375	0.513089	0.513137
	33	20	52	33
$(^{143}Nd/^{144}Nd)_{(I)}$	0.51182	0.51180	0.51184	0.51215
$\varepsilon_{Nd(T)}$	−2.6	−3.0	−2.2	+3.7

$^{143}Nd/^{144}Nd$ normalized to $^{146}Nd/^{144}Nd = 0.7219$.
Errors represent ± 2 standard errors of the mean.
ε_{Nd} calculated for $T = 535$ Ma using present-day $(^{147}Sm/^{144}Nd)_{CHUR} = 0.1967$ and $(^{143}Nd/^{144}Nd)_{CHUR} = 0.51264$.
$\lambda(Sm) = 6.54 \times 10^{-12}$.
M = measured ratio, I = calculated initial ratio.

errors in ε_{Nd} which are either insignificant or generally irresolvable from the analytical errors (see also Nelson *et al.* 1984).

The two boninitic samples are from HMA-I (Table 9.2), from the Cleveland and Stonehenge areas. Although from widely separated localities, the initial ε_{Nd} compositions of these samples are similar at −2.8. The two LOTI samples from the Cleveland and Black Hill areas, however, show strikingly different initial ε_{Nd} compositions, LOTI-Clev = −2.2 and LOTI-BH = +3.7. It is perhaps significant that the LOTI-Clev sample is spatially associated with boninitic lavas having a similar initial ε_{Nd} composition (although very different major- and trace-element characteristics), whereas the LOTI-BH not associated with boninitic lavas have a different Nd isotopic composition.

9.7 Discussion

9.7.1 Major- and trace-element chemistry

Reference fields for Tertiary boninite are included in Figures 9.7–9.10, and average boninite compositions are given in Table 9.4 (columns 5 and 6). The boninite localities used in this compilation are: Cape Vogel (Jenner 1981, 1982, Cameron *et al.* 1983); New Caledonia (Sameshima *et al.* 1983, Cameron *et al.* 1983); Bonin Islands (Shiraki *et al.* 1980, Komatsu 1980, Cameron *et al.* 1983, Jenner 1982, Hickey & Frey 1982, Umino 1986); Mariana Trench (Dietrich *et al.* 1978, Sun & Nesbitt 1978, Hickey & Frey 1982, Jenner 1982). Type C boninites are those with concave REE patterns, while Type E are characterized by LREE-enriched patterns (see Jenner 1981). Type E occurs at Cape Vogel and in New Caledonia; Type C occurs at the Bonin, Marianas and Cape Vogel localities.

The Tasmanian boninitic rocks are similar to the Tertiary boninite in having high SiO_2, MgO, Cr, Ni, Cr/Ni > 3, and low HFSE and HREE abundances. The CaO/Al_2O_3, in spite of its susceptibility to alteration processes, is very similar in the HMA-I and Tertiary samples. The similarity between HMA-I and Tertiary boninite extends to the low Ti/Zr, often considered a distinctive feature of these rocks. However, the low Ti/Zr in the Tasmanian boninitic rocks is due more to a depletion in Ti, rather than an enrichment in Zr (see Fig. 9.10a), clearly shown by the much lower Zr/Sm in the Tasmania boninitic rocks (21) compared to Tertiary boninites (51, 60). The Tasmanian HMA-I samples, while showing a trend towards Tertiary boninite, plot very much towards the Ti/Y join in the Ti–Zr–Y plot (Fig. 9.9). Based on REE patterns, both Type C and Type E boninites are present in Tasmania (see Fig. 9.10a), similar to the Cape Vogel occurrence. Tasmanian HMA-I show an overall similarity with the Howqua Cen-bearing volcanics described by Crawford (1980) (see Table 9.4, column 4), and this extends to the REE patterns given for the Howqua boninite in Nelson *et al.* (1984). The bracketed concentrations of Zr and Y shown in Table 9.4 (column 4) are those which would be required for the Y (primitive mantle normalized) abundance to be consistent (non-anomalous) with respect to the HREE abundances. If the concentrations reported for the Howqua samples are somewhat high, then the comparison between the Victorian and Tasmanian HMA-I would be quite strong.

The Tasmania HMA-II differ greatly from Tertiary boninites (and HMA-I) with respect to their HFSE ratios (see Table 9.4, and Figs 9.8 & 9.9). In particular, the Tasmanian HMA-II have a high Ti/Zr (125), much more like that of modern MORB or the LOTI volcanics. Although there are some similarities between the HMA-II and boninites in the Victorian Cambrian Heathcote Greenstone Belt (see Table 9.4, column 3), there is a substantial difference in Ti/Y. We suspect that the HMA-II may have a LREE-depleted or concave REE pattern (more like those of the LOTI), in contrast to the LREE-enriched and HREE-depleted Heathcote boninites (see Nelson *et al.* 1984).

The tholeiitic affinity, LREE-depleted patterns and low HFSE abundances in the LOTI suggest that these volcanics are akin to island-arc tholeiite (IAT) series volcanics (Jakes & Gill 1970, Gill 1981). Most IAT are, however, not as depleted in LREE and HFSE as those reported here. Note that, in Fig. 9.9, the field for IAT is A and B, and that most of the Tasmanian LOTI plot outside these fields, trending from field A towards the Y apex. A survey of the literature revealed only a very limited subset of island-arc or supra-subduction zone volcanics which closely resemble the LOTI. In particular, strongly

LREE-depleted, low-Ti basalts from the Mariana forearc (1438C of Crawford *et al.* 1986), the Troodos Complex (103A of Kay & Senechal 1976; 315 of Rautenschlein *et al.* 1985) and the Lau Basin (95-1 from Gill 1976). Some samples from the Palau–Kyushu and West Mariana Ridge, reported in Bloomer (1983), plot in similar areas on the Ti–Zr–Y diagram.

9.7.2 Nd isotope geochemistry

Variation in the Nd isotopic composition of the Tasmanian boninitic and LOTI lavas is illustrated in Fig. 9.11; also shown are Nd isotopic compositions for the Victorian Howqua boninites, some Ordovician boninites and LOTI from Newfoundland, and Cambrian crust and mantle sources. As noted in Table 9.6, ε_{Nd} values for Tasmanian boninitic rocks are calculated for an age of 535 Ma, and the same age has been used in calculating the Cambrian mantle and crustal sources. Since different ages have been used in calculating the ε_{Nd} values for the Howqua and Newfoundland samples, these ages are shown on Fig. 9.11.

Interpretation of the major, trace and isotopic geochemistry of Tertiary boninite has emphasized that the mantle source areas for these volcanics has a

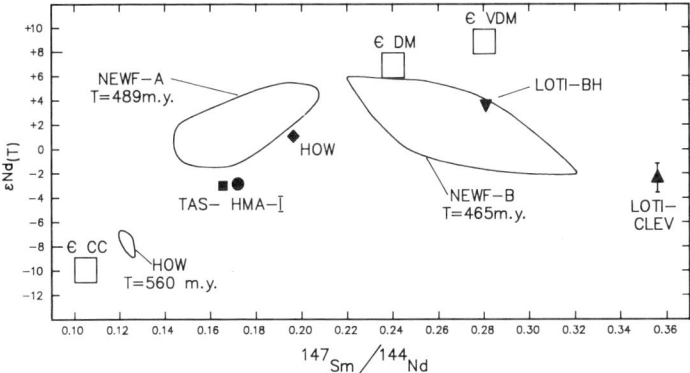

Figure 9.11 $\varepsilon_{Nd(T)}$ versus $^{147}Sm/^{144}Nd$ for selected Cambrian–Ordovician high-Mg volcanics and related rocks. Tasmanian boninitic and LOTI samples: full square, 850023 (HMA-I, Clev); full circle, 850028 (HMA-I, SH); full triangle, +60919 (LOTI-Clev) (error bar shown because of the large analytical uncertainty in the Nd isotopic composition); full inverted triangle, 850032 (LOTI-BH), data in Tables 9.2 and 9.5. Fields for potential mantle source areas and contaminants are given as CC, DM and VDM. CC is based on average upper continental crust (Taylor & McLennan 1985) and a present-day ε_{Nd} of -16 for this source material. CC represents the value both for contamination during eruption and for average Cambrian subducted sedimentary rocks. DM = depleted mantle and VDM = very depleted mantle Newf-A field is based on unpublished data for Ordovician volcanics from the Pacquet Harbour Group and Betts Cove Ophiolite, both of which contain boninite with probable clinoenstatite pseudomorphs. Newf-B is based on very depleted arc-related volcanics from the Ordovician Wild Bight Group of Central Newfoundland (Swinden 1987). See text for further discussion. How data are the Victorian Howqua data from Nelson *et al.* (1984), where the full curve encloses their samples 26324 and 26322 (their Table 3) and the full diamond is their sample 26313.

complex history of depletion (caused by previous partial melting episodes), followed by enrichment in LREE and LFSE (Sun & Nesbitt 1978, Jenner 1981, Hickey & Frey 1982, Cameron *et al.* 1983). The ultimate source for the enriched component in Tertiary boninite is still not completely understood. The two most likely sources are either mantle sources such as those responsible for the enrichment observed in oceanic islands, or subducted sedimentary material derived from continental crust. In either case, the enriched component is characterized by a low ε_{Nd}.

In Fig. 9.11, the depleted Cambrian mantle source is represented by the fields for DM (depleted mantle) and VDM (very depleted mantle). Based on comparisons with modern volcanics, the DM source would be similar to that for MORB, while VDM source mantle could be found in oceanic spreading-centre or supra-subduction zone environments (Duncan & Green 1987). The Cambrian continental crust/sedimentary source shown (Fig. 9.11) was calculated using average upper continental crust (CC) REE patterns and Nd isotopic compositions (Taylor & McLennan 1985). Note that an enriched Cambrian mantle source would probably not be as extreme in either its degree of LREE enrichment or low ε_{Nd}, and could lie on a line between DM/VDM and CC (Fig. 9.11).

The isotopic composition of the Tasmanian boninitic rocks lies between those of samples from Howqua and the DM/VDM and CC sources (Fig. 9.11). This suggests that, like Howqua, the enriched source for the Tasmanian boninitic rocks was characterized by an extremely low ε_{Nd}. The ε_{Nd} of about -2.8 for the Tasmanian boninitic rocks is lower than that observed in modern boninites and Ordovician boninite in Newfoundland. However, the mixing line defined by the Victorian Howqua and Tasmanian boninitic rocks on the ε_{Nd} versus $^{147}Sm/^{144}Nd$ plot is similar to that observed for Bonin Islands Tertiary boninites (Nelson *et al.* 1984, Hickey & Frey 1982).

Whereas the isotopic compositions of the Tasmanian boninitic rocks are somewhat more extreme than those observed in modern analogues, they are nonetheless easily explained by mixing between time-integrated LREE-enriched and LREE-depleted sources. In either modern or ancient volcanics, it is usual or 'normal' to find Nd isotopic and Sm/Nd relationships lying along positive trends between ε_{Nd} and Sm/Nd (i.e. from lower left to upper right in Fig. 9.11). However, the relationship between Nd isotopic compositions and Sm/Nd shown by the LOTI is relatively uncommon and requires a more complex explanation. The trend shown by the LOTI volcanics requires a source with low or negative ε_{Nd}, i.e. the Nd isotopic signature expected to be associated with a time-integrated LREE-enriched source, and extreme LREE depletion. One explanation for the origin of these contradictory characteristics is as follows: a depleted mantle source is mixed with, or metasomatized by, material from an enriched mantle or crustal source; at some later time, a partial melting episode leaves a more refractory, depleted residue, whose isotopic signature is still that of an enriched source; later partial melting of this source gives rise to LOTI-like volcanics. Although this scenario could work for LOTI-BH, it is intuitively difficult to envisage (or model) how the extreme LREE-depletion seen in LOTI-Clev could form in a source which had been LREE-enriched at some stage in its history.

The field for Newfoundland Ordovician volcanics (Newf-B) with similar, but less extreme, ε_{Nd} and Sm/Nd relationships has been included in Fig. 9.11 for two reasons; first, to show that whatever the process responsible for the negative ε_{Nd} versus Sm/Nd trend, it does occur in other areas; secondly, in the Newfoundland case, there is good geochemical and field evidence to suggest that this process occurs during backarc-basin formation or intra-arc rifting (Swinden 1987). This comparison, in conjunction with the geochemical comparisons discussed above, suggests that the Tasmanian LOTI volcanics are formed in a supra-subduction zone environment. The tectonic implications of this are more fully explored in Jenner & Brown (1989), and here we simply note that this is consistent with the 'allochthon' tectonic model for the Cambrian of W Tasmania proposed by Berry & Crawford (1988).

9.7.3 *Relationship between Tasmanian boninitic and LOTI lavas*

Field relationships discussed earlier in this paper and in Brown (1986b) have established that the LOTI are both temporally and spatially closely associated with the boninitic rocks, particularly in the Heazlewood–Waratah area (Fig. 9.3). This close association is also shown by the geochemical and isotopic data. In particular, the HMA-II and LOTI volcanics share similar HFSE characteristics. However, the high SiO_2 and MgO contents of the HMA-II and the presence of pseudomorphs after Cen and low-Ca pyroxene contrasts with the aphyric nature and lower SiO_2 contents of the LOTI. One possible explanation of these features is that the boninitic and LOTI lavas represent hydrous (almost water-saturated) and anhydrous (or markedly water-undersaturated) melts, respectively, of similar refractory mantle sources. This suggestion is based on comparisons between the petrogenesis of Tertiary boninite (Jenner 1982, Umino & Kushiro 1989) and second-stage melts (Duncan & Green 1980, 1987). The degree of similarity in the mantle sources for the boninitic and LOTI lavas requires further study; however, the range in Nd isotopic compositions between the LOTI and boninitic rocks does overlap, and the origin of the Sm/Nd isotopic characteristics in the LOTI suggests some link to the boninitic rocks via a complex, partial melting/source recycling mechanism.

9.8 Conclusions

Cambrian boninitic rocks in W Tasmania have many similarities with Tertiary boninites, which originate in supra-subduction zone environments. Differences, nonetheless, exist between the Tertiary and Tasmanian boninites. Two distinct geochemical groups of boninitic rocks are recognized in the Tasmanian. HMA-I are characterized by low Ti/Zr and in this respect are similar to Tertiary boninite. However, in contrast to the Tertiary boninite, the Tasmanaian HMA-I do not show the characteristic enrichment in Zr relative to the REE. HMA-II differ significantly from Tertiary boninite in their HFSE ratios, and show more similarity to the Tasmanian LOTI volcanics. The Tasmanian LOTI volcanics are characterized by an overall tholeiitic affinity, strong LREE depletion, low HFSE abundances, variable Nd isotopic compositions and a peculiar Nd isotopic versus Sm/Nd relationship. A link with some of the Tasmanian boninitic rocks is suggested by their close temporal, spatial and certain trace-element and isotopic similarities. Close geochemical analogues to the Tasmanian LOTI volcanics can be found; however, they are rare and occur in a variety of modern supra-subduction zone settings (e.g. Mariana forearc, north Tongan forearc, Lau Basin) or in ophiolites believed to have formed in such settings (e.g. Troodos Complex). For both the Mariana forearc and Troodos examples, this comparison is strengthened by the close association in both areas of these volcanics with low-Ca pyroxene-dominated ultramafic–mafic complexes. The geochemical and petrological characteristics of the Tasmanian boninitic and LOTI lavas (and their associated ultramafic–mafic complexes) are consistent with their formation in a supra-subduction zone environment. In fact, we know of no other tectonic environment of formation which is characterized by documented occurrences of volcanics similar in composition to the Tasmanian boninitic and LOTI lavas.

Acknowledgements

This research was supported by a NSERC operating grant to G.A.J. and by the Tasmanian Department of Mines. We are grateful to Dr A Hofmann and I. Raczek of the Max-Planck-Institut (Mainz) and H. Waldron (Melbourne University) for their invaluable analytical assistance. At various stages during this study, the authors have benefited from discussions with D. H. Green, A. J. Crawford, R. Varne, R. Berry and J. Foden. We are particularly grateful to Professor Green, who offered moral and financial support to G.A.J. A.V.B. publishes with the permission of the Tasmanian Director of Mines.

References

Arndt, N. T. & G. A. Jenner 1986. Crustally contaminated komatiites and basalts from Kambalda, Western Australia. *Chem. Geol.* **56**, 229–55.

Banks, M. R. 1956. The Middle and Upper Cambrian series (Dundas Group and its correlates) in Tasmania. In *El Sistema Cambrico* 2, J. Rodgers (ed.), 165–212. 20th Int. Geol. Congr., Mexico.

Berry, R. F. & A. J. Crawford 1988. The tectonic significance of Cambrian allochthonous mafic–ultramafic complexes in Tasmania. *Austr. J. Earth Sci.* **35**, 72–90.

Bishop, J. G. & T. C. Hughes 1984. Radiochemical neutron activation analysis determination of rare earth elements in geological materials with ppb sensitivity. *J. Radioanal. Chem.* **84**, 213–21.

Blissett, A. H. & A. B. Gulline 1962. *Geological atlas 1 mile series*, Zone 7, Sheet No. 50, *Zeehan*. Dept Mines, Tasmania.

Bloomer, S. H. 1983. Distribution and origin of igneous rocks from the landward slopes of the Mariana Trench: implications for its structure and evolution. *J. Geophys. Res.* **88**, 7411–28.

Brown, A. V. 1986a. Geology of the Dundas–Mt. Youngbuck Area, Western Tasmania. Ph.D. Thesis, University of Tasmania.

Brown, A. V. 1986b. *Geology of the Dundas–Mt. Lindsay–Mt. Youngbuck region*. Dept of Mines, Tasmania, Geol. Surv. Bull. no. 62.

Brown, A. V. & H. Waldron 1982. *Preliminary review of the Eocambrian–Cambrian basaltic association and tectonic setting within western and north-western Tasmania*. Dept of Mines, Tasmania, Rep. no. 1982/10.

Brown, A. V., M. J. Rubenach & R. Varne 1980. Geological environment, petrology and tectonic significance of the Tasmanian Cambrian ophiolitic and ultramafic complexes. In *Proc. Int. Ophiolite Symp.*, A. Panayiotou (ed.), 649–59. Cyprus.

Cameron, W. E., M. T. McCulloch & D. A. Walker 1983. Boninite petrogenesis: chemical and Nd–Sr isotopic constraints. *Earth Planet. Sci. Lett.* **65**, 75–89.

Collins, P. L. F. & E. Williams 1986. Metallogeny and tectonic development of the Tasman fold belt system in Tasmania. *Ore Geol. Rev.* **1**, 153–201.

Corbett, K. D. 1981. Stratigraphy and mineralization in the Mt Read volcanics, western Tasmania. *Econ. Geol.* **76**, 209–30.

Corbett, K. & T. C. Lees 1987. Stratigraphic and structural relationships and evidence for Cambrian deformation at the western margin of the Mt Read Volcanics, Tasmania. *Austr. J. Earth Sci.* **34**, 45–68.

Crawford, A. J. 1980. A clinoenstatite-bearing cumulate olivine pyroxenite from Howqua, Victoria. *Contrib. Mineral. Petrol.* **75**, 353–67.

Crawford, A. J. & R. F. Berry 1989. Tectonic implications of Late Precambrian–Early Palaeozoic igneous rock associations in Tasmania. In preparation.

Crawford, A. J. & W. E. Cameron 1985. Petrology and geochemistry of Cambrian boninites and low-Ti andesites from Heathcote, Victoria. *Contrib. Mineral. Petrol.* **91**, 93–104.

Crawford, A. J. & R. R. Keays 1987. Petrogenesis of Victorian Cambrian tholeiites and implications for the origin of associated boninites. *J. Petrol.* **28**, 1075–109.

Crawford, A. J., L. Beccaluva, G. Serri & J. Dostal 1986. Petrology, geochemistry and tectonic implications of volcanics dredged from the intersection of the Yap and Mariana trenches. *Earth Planet. Sci. Lett.* **80**, 265–80.

Crawford, A. J., W. E. Cameron & R. R. Keays 1984. The association boninite low-Ti andesite–tholeiite in the Heathcote Greenstone Belt, Victoria; ensimatic setting for the early Lachlan Fold Belt. *Austr. J. Earth Sci.* **31**, 161–75.

Creenaune, P. 1980. The volcanics of the Heazlewood River complex. B.Sc. (Hons.) Thesis, University of Tasmania.

Crook, K. A. W. 1980. Fore-arc evolution in the Tasman Geosyncline: the origin of the southeast Australian continental crust. *J. Geol. Soc. Austr.* **27**, 215–32.

Dietrich, V., R. Emmerman, R. Oberhansli & H. Puchelt 1978. Geochemistry of basaltic and gabbroic rocks from the west Mariana Basin and Mariana Trench. *Earth Planet. Sci. Lett.* **39**, 127–44.

Duncan, R. A. & D. H. Green 1980. Role of multistage melting in the formation of oceanic crust. *Geology.* **8**, 22–6.

Duncan, R. A. & D. H. Green 1987. The genesis of refractory melts in the formation of oceanic crust. *Contrib. Mineral. Petrol.* **96**, 326–42.

Elliston, J. 1954. Geology of the Dundas district, Tasmania. *R. Soc. Tasmania, Papers Proc.* **88**, 161–83.

Gill, J. B. 1976. Composition and age of Lau Basin and ridge volcanic rocks: implications for evolution of an inter-arc basin and remanent arc. *Geol. Soc. Am. Bull.* **87**, 1384–95.

Gill, J. B. 1981. *Orogenic andesites and plate tectonics.* New York: Springer.

Griffin, B. J. 1979. *Energy dispersive analysis system calibration and operation with TAS-SUEDS, an advanced interactive data reduction package.* Univ. Tasmania, Dept Geol. Publ. no. 343.

Hellman, P. L., R. E. Smith & P. Henderson 1979. The mobility of the rare earth elements: evidence and implications from selected terrains affected by burial metamorphism. *Contrib. Mineral. Petrol.* **71**, 23–44.

Hickey, R. L. & F. A. Frey 1982. Geochemical characteristics of boninite series volcanics: implications for their source. *Geochim. Cosmochim. Acta.* **46**, 2099–115.

Jakes, P. & J. B. Gill 1970. Rare earth elements and the island arc tholeiitic series. *Earth Planet. Sci. Lett.* **9**, 17–28.

Jenner, G. A. 1981. Geochemistry of high-Mg andesites from Cape Vogel, Papua New Guinea. *Chem. Geol.* **33**, 307–32.

Jenner, G. A. 1982. Petrogenesis of high-Mg andesites. Ph.D. Thesis, University of Tasmania.

Jenner, G. A. & A. V Brown 1989. Geological setting, trace element and Sm/Nd isotopic composition of Late Precambrian–Cambrian volcanics in western Tasmania: implications for tectonic models. In preparation.

Jenner, G. A., P. A. Cawood, M. Rautenschlein & W. M. White 1987. Composition of back-arc basin volcanics, Valu Fa Ridge, Lau Basin: evidence for a slab-derived component in their mantle source. *J. Volcanol. Geotherm. Res.* **32**, 209–22.

Kay, R. W. & R. G. Senechal 1976. The rare earth geochemistry of the Troodos Ophiolite Complex. *J. Geophys. Res.* **81**, 964–70.

Komatsu, M. 1980. Clinoenstatite in volcanic rocks from the Bonin Islands. *Contrib. Mineral. Petrol.* **74**, 329–38.

Miyashiro, A. 1974. Volcanic rock series in island arcs and active continental margins. *Am. J. Sci.* **274**, 321–55.

Miyashiro, A. & F. Shido 1975. Tholeiitic and calc-alkalic series in relation to the behaviours of titanium, vanadium, chromium, and nickel. *Am. J. Sci.* **275**, 265–77.

Nelson, D. R., A. J. Crawford & M. T. McCulloch 1984. Nd–Sr isotopic and geochemical systematics in Cambrian boninites and tholeiites from Victoria, Australia. *Contrib. Mineral. Petrol.* **88**, 164–72.

Norrish, K. & J. T. Hutton 1969. An accurate x-ray fluorescence spectrographic method for the analysis of a wide range of geologic samples. *Geochim. Cosmochim Acta* **33**, 431–51.

Pearce, J. A. 1975. Basalt geochemistry used to investigate past tectonic environment on Cyprus. *Tectonophysics* **25**, 41–67.

Pearce, J. A. & J. R. Cann 1973. Tectonic setting of basic volcanic rocks determined using trace element analysis. *Earth Planet. Sci. Lett.* **19**, 290–300.

Rautenschlein, M., G. A. Jenner, J. Hertogen, A. W. Hofmann, R. Kerrich, H-U. Schmincke & W. M. White 1985. Isotopic and trace element composition of volcanic glasses from the Akaki Canyon, Cyprus: implications for the origin of the Troodos Ophiolite. *Earth Planet. Sci. Lett.* **75**, 369–83.

Robinson, P., N. C. Higgins & G. A. Jenner 1986. Determination of rare earth elements, yttrium and scandium in international standard rocks using an ion-exchange–x-ray fluoresence technique. *Chem. Geol.* **55**, 121–37.

Rubenach, M. 1973. The Tasmanian ultramafic–gabbro and ophiolite complexes. Ph.D. Thesis, University of Tasmania.

Rubenach, M. 1974. The origin and emplacement of the Serpentine Hill complex, western Tasmania. *J. Geol. Soc. Austr.* **21**, 91–106.

Sameshima, T., J-P. Paris, P. M. Black & R. F. Heming 1983. Clinoenstatite-bearing lava from Nepoui, New Caledonia. *Am. Mineral.* **68**, 1076–82.

Saunders, A. D., J. Tarney, N. G. Marsh & D. A. Wood 1980. Are ophiolites ocean crust or marginal basin crust: a geochemical approach. In *Proc. Int. Ophiolite Symp.*, A. Panayiotou (ed.), 193–204. Cyprus.

Shervais, J. W. 1982. Ti–V plots and the petrogenesis of modern and ophiolitic lavas. *Earth Planet. Sci. Lett.* **59**, 101–18.

Shiraki, K., N. Kuroda, N. Urano & S. Maruyama 1980. Clinoenstatite in boninites from the Bonin Islands, Japan. *Nature* **285**, 31–2.

Sun, S-S. & R. W. Nesbitt 1978. Geochemical regularities and genetic significance of ophiolitic basalts. *Geology* **6**, 689–93.

Swinden, S. 1987. Ordovician volcanism and mineralization in the Wild Bight Group, central Newfoundland: a geological, petrological, geochemical and isotopic study. Ph.D. Thesis, Memorial University of Newfoundland.

Taylor, S. R. & S. M. McLennan 1985. *The continental crust: its composition and evolution*. Oxford: Blackwell.

Umino, S. 1986. Magma mixing in boninite sequence of Chichi-jima, Bonin Islands. *J. Volcanol. Geotherm. Res.* **29**, 125–57.

Umino, S. & I. Kushiro 1989. Experimental studies on boninite petrogenesis (this volume).

Varne, R. & A. V. Brown 1978. The geology and petrology of the Adamsfield Ultramafic Complex, Tasmania. *Contrib. Mineral. Petrol.* **67**, 195–207.

Varne, R. & J. D. Foden 1987. Tectonic setting of Cambrian rifting, volcanism and ophiolite formation in western Tasmania. *Tectonophysics* **140**, 1–19.

White, W. M. & J. Patchett 1984. Hf–Nd–Sr and incompatible element abundances in island arcs: implications for magma origins and crust–mantle evolution. *Earth Planet. Sci. Lett.* **67**, 167–85.

Williams, E. 1976. *Structural map of pre-Carboniferous rocks of Tasmania*. Dept of Mines, Tasmania.

Williams, E. 1978. Tasman Fold Belt System in Tasmania. *Tectonophysics* **48**, 159–206.

10 Boninitic lavas in Appalachian ophiolites: a review

R. A. COISH

Abstract

Early Palaeozoic lavas in three volcanic terrains in the northern Appalachians are similar to modern boninites. Samples are described from the Betts Cove ophiolite and Pacquet Harbour Group, Newfoundland, and from the Thetford Mines ophiolite, Quebec, Canada. The lavas are mostly quench textured pillows that have very low contents of TiO_2, Zr, Y, REE, low Ti/Zr, high MgO, Ni, Cr, U-shaped REE patterns and initial $\varepsilon_{Nd} = +0.4$ to $+4.7$. All chemical features match modern boninites but are distinct from Archaean basaltic komatiites. The boninitic lavas are invariably associated with other lavas exhibiting chemical features of island-arc tholeiites and/or MORB. The occurrence of boninitic lavas in some Appalachian ophiolites suggests that these ophiolites probably formed in the vicinity of a subduction zone. Furthermore, the presence in some of the ophiolites of associated lavas that are like MORB and island-arc tholeiites implies that igneous processes involved in forming the ophiolites were complicated and included melting a variety of source rocks in a subduction-zone environment. Plate models for the development of the Appalachians can accommodate the formation of ophiolites above subduction zones. It is still not clear whether all ophiolites in the northern Appalachians were formed near subduction zones.

10.1 Introduction

Boninites are vitrophyric lavas, usually containing phenocrysts of low-Ca pyroxene and olivine, found in forearc regions of some W Pacific island arcs (Cameron et al. 1979, Meijer 1980, Crawford et al. 1981, Bloomer & Hawkins 1987). Their unusual chemical features include high MgO ($> 9\%$), high SiO_2 ($> 55\%$) and extremely low TiO_2 contents (Hickey & Frey 1982). Boninites in

W Pacific arcs are all Cenozoic in age. In recent years, workers have found that some lavas in older volcanic sequences, including ophiolites, have petrographic and chemical features similar to boninites (Coish *et al.* 1982, Crawford & Cameron 1985). The occurrence of boninites in these older ophiolites has important implications for the tectonic environment of formation of the ophiolites.

Boninitic lavas are found in association with several northern Appalachian ophiolitic terrains (Laurent 1980, Coish *et al.* 1982, Upadhyay 1982, Oshin & Crockett 1986). All of the ophiolites are of early Palaeozoic age (505 to 485 Ma old). This chapter focuses on occurrences of boninitic lavas in the Betts Cove ophiolite, Newfoundland, in the Pacquet Harbour Group near Betts Cove, and in the Thetford Mines ophiolite, Quebec (Fig. 10.1). The lavas are referred to as boninitic because they have many chemical characteristics similar to boninites from the type localities in the W Pacific but they are not *exactly* like boninites. It should also be noted that these early Palaeozoic boninitic lavas are invariably associated with 'island-arc' or 'mid-ocean-ridge' tholeiites. In this chapter, I review field, petrographic and chemical data including major and trace elements, and Nd isotopes on the boninitic lavas, compare these data to modern boninites and Archaean basaltic komatiites, and show the importance of the lavas in interpreting tectonic environments of formation of the Appalachian ophiolites.

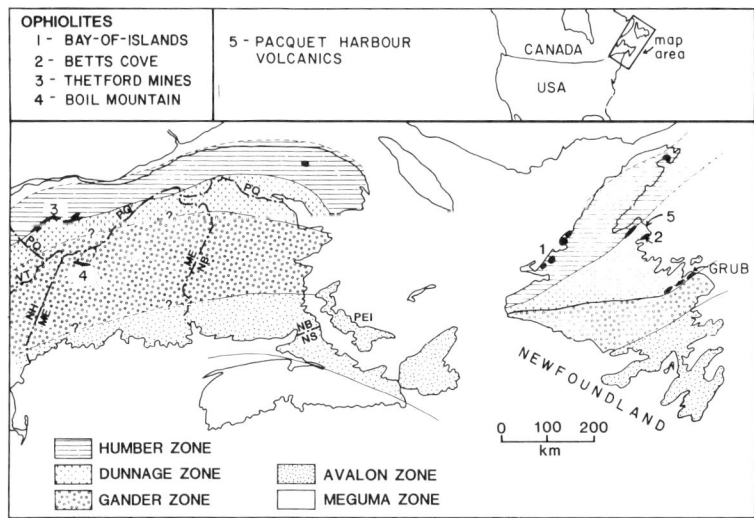

Figure 10.1 Tectono-stratigraphic map of the northern Appalachians, after Williams (1978). The locations of the principal ophiolites and the Pacquet Harbour volcanics are shown. GRUB refers to the Gander River Ultrabasic Belt (Blackwood 1982).

10.2 Regional setting

The northern Appalachians are divided into five major tectono-stratigraphic zones (Williams 1978) (Fig. 10.1). The Humber, Dunnage and Gander Zones represent, respectively: remnants of the ancient North American continental margin; remnants of the Iapetus (proto-Atlantic) ocean; and remnants of the eastern margin of Iapetus. The Avalon Zone represents an exotic terrain accreted to North America during final closing of Iapetus in the Siluro-Devonian. The best-known ophiolite localities occur in the western part of the Appalachians, in the Humber Zone and near the western edge of the Dunnage Zone (Fig. 10.1). An eastern line of ophiolitic remnants occurs in the eastern Dunnage Zone (Strong 1979, Coish & Rogers 1987), but these are not part of this study.

10.3 Local geological setting

All three localities for boninitic lavas in the N Appalachians are in or close to ophiolitic sequences. The Betts Cove ophiolite has been dated at 488 Ma (Dunning & Krogh 1985). It forms the base of the Snooks Arm Group (Snelgrove 1931, Upadhyay *et al.* 1971). Above the ophiolite proper is a sequence of siltstone, epiclastic sediments and basaltic lava flows, interpreted as a volcanic arc (Upadhyay *et al.* 1971, Dewey & Bird 1971) or an oceanic island/backarc basin sequence (Jenner & Fryer 1980). From bottom to top, the stratigraphic sequence of the ophiolite comprises layered peridotite, layered and massive gabbro, sheeted dyke complex and mafic pillow lavas (Fig. 10.2). A tectonite ultramafic unit is notably absent; the ultramafic cumulate sequence is in fault contact with country rock. Another notable feature of the stratigraphy is the thinness of the gabbro layer, and the relatively large amount of pyroxenite. The Betts Cove ophiolite was interpreted as a section of typical ocean crust by Upadhyay *et al.* (1971), Church & Stevens (1971) and Dewey & Bird (1971). Later detailed work on the geochemistry of the lava sequences has demonstrated unusual chemical features (low Ti, Zr, Y and Nb, and U-shaped REE patterns) not found in modern ocean-ridge basalts (Coish & Church 1979, Upadhyay 1978, 1982, Upadhyay & Neale 1979, Coish *et al.* 1982). This has led to the suggestion that the ophiolite formed in the vicinity of a subduction zone rather than at a mid-ocean ridge.

Coish *et al.* (1982) divided the mafic volcanic sequence at Betts Cove into three chemical units. These units were termed the *lower lavas* (< 0.30% TiO_2), *intermediate lavas* (0.30–0.75% TiO_2) and *upper lavas* (> 0.75% TiO_2). It is emphasized that all three units are mafic in composition. Thus, to avoid confusion with intermediate-SiO_2 lavas, the intermediate lavas are renamed *intermediate-Ti lavas*. These chemical units correspond to strati-

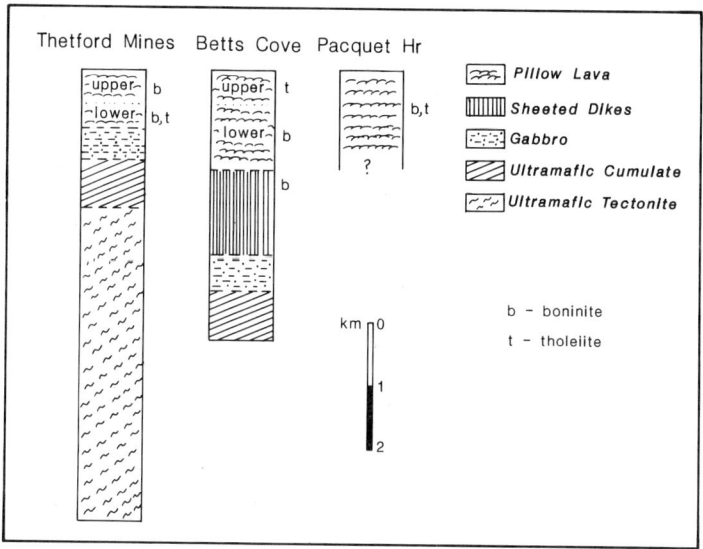

Figure 10.2 Stratigraphic columns for the Thetford Mines ophiolite (after Laurent 1975), the Betts Cove ophiolite (after Coish & Church 1979) and the Pacquet Harbour Group (after Hibbard 1983). Approximate stratigraphic locations of boninites (b) and tholeiites (t) are shown. Although not indicated in this diagram, within the lower unit at Thetford Mines, upper portions are Type II (boninitic) and lower portions are Type I (tholeiitic) (Oshin & Crockett 1986). In the Betts Cove ophiolite, intermediate-Ti lavas occur within the lower stratigraphic unit.

graphic units inasmuch as the lower lavas are below the upper lavas; however, the intermediate-Ti lavas are not clearly separable stratigraphically from the lower lavas. The lower- and intermediate-Ti lavas include samples from the sheeted dyke unit. Boninites are found in the lower lavas. However, the upper lavas have no boninitic affinities, but rather are tholeiites similar to present-day MORB.

The Pacquet Harbour Group occurs in the north-central part of the Baie Verte Peninsula (Fig. 10.1), and is defined as a moderately to steeply northerly dipping sequence of variably deformed and metamorphosed mafic volcanic and volcaniclastic rocks, and mafic dykes (Hibbard 1983). The rocks have been correlated with the upper parts of the Betts Cove ophiolite, which outcrops to the southeast. Thus, Hibbard (1983) interpreted the volcanic sequence of the Pacquet Harbour Group to be the top of an ophiolite sequence. Volcanic rocks in this group were analysed by Gale (1973) and the mafic portions were divided into theoleiitic and komatiitic basalts. Some samples thought to be basaltic komatiites by Gale (1973) are now interpreted to have boninitic affinities (Hibbard 1983). Tholeiitic rocks were also recognized by Hibbard (1983) but were not separated stratigraphically from the boninites. All boninitic analyses from the Pacquet Harbour Group presented in this chapter (from Hibbard 1983) are of rocks south of the area described by Gale (1973).

The Thetford Mines Complex is one of several mafic–ultramafic bodies that form a discontinuous ophiolite belt within the Quebec Appalachians (Laurent 1975, 1977). The Complex exposes a near-complete ophiolite sequence; a basal unit of tectonized harzburgite, layered ultramafic and gabbroic rocks, and upper layers of mafic volcanic rocks interbedded with fine-grained cherty, argillaceous sediments (Fig. 10.2). There is no well developed sheeted dyke unit, although a zone of massive gabbro and metadiabase intruded by diabase dykes occurs (Laurent 1975). The mafic volcanic unit was divided into a lower and upper unit, separated in some places by a thin cherty layer (Seguin & Laurent 1975). Oshin & Crockett (1986) subdivided the lower unit into Type I and Type II. Type I samples have tholeiitic affinities whereas Type II rocks have boninitic features, such as low Ti, P, Zr, Hf and total REE. Oshin & Crockett (1986) imply that Type II samples are stratigraphically above Type I samples. The upper volcanic unit has boninitic affinities and is similar in chemistry to Type II samples from the lower unit (Oshin & Crockett 1986). Thus, all boninitic samples from the Thetford Mines ophiolite are in the upper parts of the volcanic pile (Fig. 10.2).

In summary, boninitic samples considered here are those from (a) the lower mafic volcanics from the Betts Cove ophiolite, (b) the lower unit, Type II and the upper mafic unit from the Thetford Mines ophiolite, and (c) low-Ti mafic volcanics from the Pacquet Harbour Group (Fig. 10.2).

10.4 Petrography of boninitic lavas

In all three localities in the northern Appalachians, boninitic rocks occur both as pillow lavas and as dykes. All samples are metamorphosed to the greenschist facies; however, igneous textures and some igneous mineralogy are preserved. The dominant minerals are chlorite, albite, actinolite, clinopyroxene, epidote, calcite and quartz. Phenocryst phases include pseudomorphs after olivine and probably low-Ca pyroxene. In Betts Cove boninites, plagioclase and clinopyroxene are groundmass constituents. Also, Fe-Ti oxides are in low abundance and restricted to the groundmass. Chromite is a common phase in all of the boninitic samples. Quench textures showing elongate, skeletal crystals of clinopyroxene are common in pillow lavas at Betts Cove. These were interpreted as spinifex-textured basaltic komatiites by Upadhyay (1978). Varioles and spherulites are also present in many samples.

Can petrography be used to classify these early Palaeozoic rocks as boninites? Fresh boninites are glassy rocks with phenocrysts of predominantly low-Ca pyroxene and some olivine. Plagioclase is conspicuously absent. Magnesiochromite occurs as small grains included in olivine and low-Ca pyroxenes. The order of crystallization in modern boninites is magnesiochromite, olivine, low-Ca pyroxene, calcic pyroxene, + amphibole + plagioclase (Cameron et al. 1979). Olivine is always subordinate to low-Ca pyroxene, which can be clinoenstatite, orthopyroxene, or magnesian pigeonite (Crawford et al. 1984). Furthermore, most boninites have abundant brownish glass riddled with skeletal grains of pyroxene. The igneous mineralogy of the Appalachian boninitic rocks is difficult to know because of metamorphism and, thus, petrographic comparisons with modern boninites are tenuous. However, certain features deserve comment. The boninitic lavas at Betts Cove clearly exhibit phenocrystic and quench textures similar to boninites; however, similar textures are seen in many mid-ocean-ridge basalts and in Archaean basaltic komatiites. So, *texture is not a diagnostic* characteristic. Some features of the inferred igneous mineralogy of the Betts Cove samples are similar to modern boninites. First, plagioclase is not a phenocryst phase, implying that it crystallized late in the Betts Cove lavas. Secondly, the samples at Betts Cove exhibit large clots of chlorite/serpentine interpreted as pseudomorphs of olivine and low-Ca pyroxene (Coish & Church 1979). In fact, some pyroxene pseudomorphs have relict polysynthetic-type twinning, suggesting that it may have been clinoenstatite, a characteristic phenocryst in modern boninites. Appalachian boninitic samples do apparently contain a greater abundance of clinopyroxene than modern boninites, although the bulk of the clinopyroxene is in quench-textured groundmass. Supporting mineralogical evidence for the boninitic affinities of the Betts Cove lavas is found in the cumulate rocks associated with the lavas. The order of crystallization in these cumulates is chromite, olivine, orthopyroxene, clinopyroxene, plagioclase, similar to the order derived for modern boninites (Cameron et al. 1979, Howard & Stolper 1981). Thus, the petrography of the Appalachian boninitic rocks is consistent with modern boninites, but cannot be used to classify them as such. The chemistry of these samples is perhaps a better method of classification.

10.5 Mineral chemistry

Relict igneous phases in boninitic samples from the Appalachians include clinopyroxene and Cr-rich spinel. The most notable feature of the clinopyroxene composition is its low Ti content, a reflection of the low Ti content of the magma from which it crystallized (Coish & Church 1979). The Mg# ranges from 0.80 to 0.85, similar to clinopyroxene from modern boninites (Umino 1986). The composition of Cr-spinel shows high Cr# over a wide range of Mg# (Table 10.1, Fig. 10.3). The samples from Betts Cove plot in the range for modern boninites (Umino 1986, Bloomer & Hawkins 1987), and clearly have much higher Cr# than MORB. The high Cr content in the spinels is a reflection of the high Cr_2O_3 and low Al_2O_3 of the magma from which they crystallized (Crawford 1980). Low Al and high Cr contents are characteristic features of modern boninites.

Table 10.1 Representative analyses of chromite in boninitic rocks from the Betts Cove ophiolite.

Sample no.	7347a	7347b	7347c	7347d	7346a	73201	7346b
SiO_2	0.16	0.18	0.12	0.20	0.13	0.16	0.13
Al_2O_3	7.33	7.71	8.33	9.70	10.05	8.30	9.20
Cr_2O_3	61.87	57.53	60.59	56.18	55.88	56.85	55.90
TiO_2	0.00	0.04	0.04	0.01	0.51	0.02	0.43
Fe_2O_3[a]	4.67	5.07	4.71	4.77	5.43	5.24	6.37
FeO	11.39	18.60	11.31	15.10	14.94	17.63	15.54
MnO	1.20	0.98	1.19	1.36	0.79	0.91	1.34
MgO	14.12	9.25	14.27	11.57	12.30	9.81	11.79
total	100.74	99.36	100.56	98.89	100.03	98.92	100.70
cations per 32 oxygen							
Al	2.26	2.48	2.56	3.07	3.11	2.67	2.87
Cr	12.81	12.42	12.49	11.93	11.61	12.25	11.69
Ti	0.00	0.02	0.02	0.00	0.10	0.00	0.09
Fe^{3+}	0.92	1.04	0.92	0.96	1.07	1.08	1.27
Fe^{2+}	2.49	4.27	2.74	3.39	3.28	4.02	3.44
Mg	5.51	3.76	5.55	4.63	4.82	3.98	4.65

[a]Fe_2O_3 calculated assuming stoichiometry.

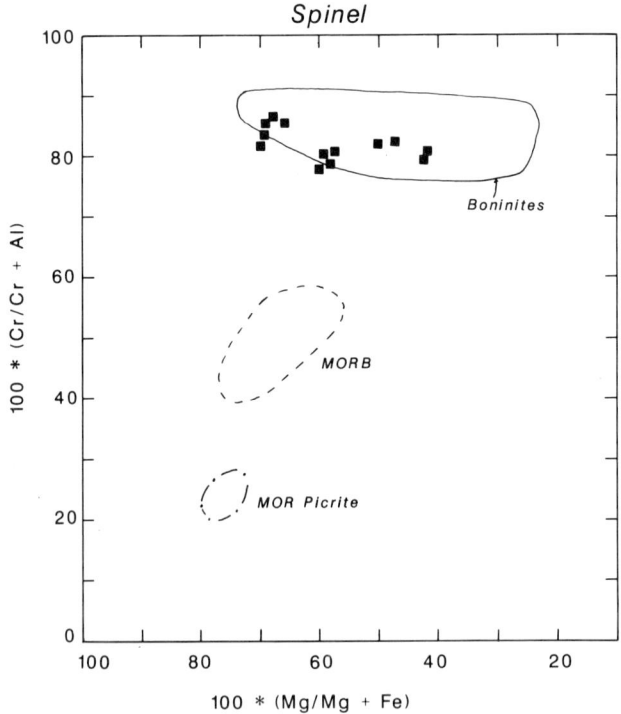

Figure 10.3 Cr# versus Mg# variations in Cr-spinel in boninitic samples from Betts Cove (■) compared to modern boninites from the Bonin Islands (Umino 1986) and MORB (Sigurdsson & Schilling 1976).

10.6 Wholerock geochemistry

10.6.1 Introduction

Analyses of boninitic rocks from Oshin & Crockett (1986), Hibbard (1983) and Coish et al. (1982) are used in the following discussions. Several questions are addressed. Are the Appalachian boninites compositionally similar to modern boninites? Can they be distinguished from basaltic komatiites? What is the petrogenesis of the boninitic samples? What are the tectonic implications of the occurrence of boninites in the Appalachians?

Table 10.2 Representative analyses of boninitic samples from the Betts Cove ophiolite[a].

Sample no.	73230	73256	7359	73149	7461a	7461b	73101a	73101b
SiO_2	53.54	50.87	55.79	53.29	46.66	50.99	42.92	60.39
TiO_2	0.27	0.14	0.16	0.11	0.19	0.17	0.14	0.12
Al_2O_3	15.44	12.64	12.85	11.38	12.66	11.29	11.04	11.68
Fe_2O_3	3.13	8.56	0.73	2.91	3.07	1.07	3.34	1.44
FeO	4.23	na[b]	6.62	4.96	5.77	6.60	8.06	4.16
MnO	0.13	na	0.14	0.17	0.16	0.14	0.20	0.15
MgO	6.53	13.40	9.31	11.82	14.36	12.89	15.57	7.60
CaO	9.01	6.11	5.27	8.74	13.17	11.86	11.67	6.71
Na_2O	4.14	1.63	3.16	2.52	0.17	0.72	0.52	5.03
K_2O	0.04	0.36	0.04	0.12	0.04	0.08	0.01	0.01
P_2O_5	0.03	0.01	na	0.02	0.01	0.01	0.03	na
LOI	2.23	6.26	5.37	3.06	3.97	3.27	6.09	2.19
total	98.72	99.98	99.44	99.10	100.23	99.09	99.59	99.48
Sc	36	41	41	46	34	34	na	na
Cr	107	670	274	778	918	787	730	756
Ni	61	238	95	155	387	335	231	268
Rb		4	5	4	5	2	1	na
Sr	52	71	102	196	160	237	74	60
Y	11	9	8	8	1	1	5	8
Zr	21	13	15	8	na	10	14	16
Ba	11	30	16	66	5	11	12	28
La	1.25	0.96	0.66	0.48	1.73	1.59	0.92	0.79
Ce	2.98	2.04	1.84	1.20	4.20	3.88	2.48	2.16
Nd	1.68	0.99	0.98	0.32	1.69	1.65	1.29	1.10
Sm	0.48	0.28	0.23	0.15	0.48	0.44	0.37	0.23
Eu	0.21	0.07	0.04	0.06	0.27	0.23	0.15	0.07
Tb	0.17	0.11	0.10	0.06	0.14	0.14	0.12	0.07
Yb	1.18	0.97	0.77	0.68	0.77	0.81	0.95	0.58
Lu	0.18	0.21	0.12	0.20	0.19	0.14	0.21	0.09

[a] Analyses from Coish et al. (1982).
[b] na = Analysis not available.

Table 10.3 Representative analyses of boninitic rocks from Thetford Mines[a].

(a) Lower unit, Type II.

Sample no.	II-A	II-B	II-C	II-D	II-E	II-F	II-G	II-H
SiO_2	52.49	54.93	48.12	52.17	56.18	49.15	53.67	53.22
TiO_2	0.22	0.31	0.23	0.30	0.31	0.18	0.24	0.21
Al_2O_3	14.59	16.03	15.04	15.16	15.10	9.88	14.09	11.91
FeO	11.28	7.33	9.77	9.95	7.68	4.82	8.80	7.86
MnO	0.25	0.11	0.25	0.16	0.11	0.10	0.17	0.19
MgO	9.97	7.80	7.92	9.04	8.57	6.11	8.85	10.46
CaO	2.40	4.75	5.91	4.21	4.35	14.89	5.33	7.57
Na_2O	4.61	4.49	4.30	3.36	3.82	2.12	5.01	3.79
K_2O	0.19	0.88	0.19	2.29	0.75	0.77	1.08	0.19
P_2O_5	0.01	0.02	0.01	0.01	0.02	0.01	0.01	0.01
LOI	3.99	3.35	8.26	3.36	3.10	11.97	2.76	4.58
total	100.00	100.00	100.00	100.01	99.99	100.00	100.01	99.99
Cr	253	213	335	243	339	335	119	897
Ni	63	84	58	63	102	80	55	109
Y	10	9	8	8	12	6	8	8
Zr	18	42	21	21	40	19	20	23
La	1.23	3.24	1.35	0.89	2.05	1.25	1.12	1.42
Ce	3.11	5.37	3.10	2.29	4.90	3.22	2.77	3.29
Nd	2.17	3.13	1.86	1.57	2.98	2.15	1.72	1.63
Sm	0.39	0.86	0.48	0.39	0.80	0.59	0.47	0.44
Eu	0.10	0.29	0.17	0.13	0.25	0.19	0.14	0.19
Tb	0.08	0.14	0.11	0.09	0.16	0.13	0.09	0.11
Yb	0.71	0.83	0.77	0.75	0.69	0.71	0.76	0.78
Lu	0.12	0.12	0.11	0.12	0.09	0.11	0.12	0.13

[a]Analyses from Oshin & Crockett (1986).
Trace elements in ppm.

10.6.2 Comparison with modern boninites and basaltic komatiites

Upadhyay (1982) suggested that both basaltic komatiites and boninites were found in the Betts Cove ophiolite. He used the high SiO_2 content of some samples as indicative of boninite, whereas other samples with high MgO but lower SiO_2 content he called komatiite or basaltic komatiite. However, SiO_2 is mobile during greenschist metamorphism, and is perhaps unreliable for use as a classification parameter. High-field-strength elements (HFSE) are less susceptible to alteration. Using variations in the HFSE, it is shown in the following sections that low-TiO_2 lavas, regardless of SiO_2 content, are similar to boninites but differ from basaltic komatiites.

The low-Ti samples from the N Appalachians are extremely depleted in P, Zr, Hf, Sm, Ti and Y compared to MORB and slightly depleted in those elements compared to basaltic komatiite from Munro Township, Ontario, Canada (Tables 10.2–10.4, Fig. 10.4). On the other hand, they compare

Table 10.3 *Continued*

(b) Upper lava unit.

Sample no.	A	B	C	D	E	F	G	H
SiO_2	52.79	51.13	52.92	49.12	51.77	55.21	48.28	52.18
TiO_2	0.24	0.24	0.21	0.20	0.25	0.18	0.22	0.23
Al_2O_3	12.98	13.36	11.81	10.93	10.77	11.05	10.48	12.16
FeO	8.50	9.47	8.80	9.49	9.08	6.75	10.29	9.08
MnO	0.21	0.18	0.17	0.18	0.17	0.15	0.20	0.21
MgO	10.64	13.07	12.75	12.56	13.68	10.05	15.74	12.65
CaO	5.14	4.44	5.80	9.82	5.81	5.78	8.38	6.43
Na_2O	3.66	4.00	3.05	1.98	2.49	3.65	1.26	3.56
K_2O	0.62	0.22	0.23	0.33	0.21	0.56	0.45	0.17
P_2O_5	0.01	0.01	0.01	0.01	0.01	0.01	0.01	0.01
LOI	5.22	3.87	4.25	5.38	5.76	5.61	4.68	3.33
total	100.01	99.99	100.00	100.00	100.00	99.00	99.99	100.01
Cr	2598	2132	2014	2853	3541	1786	2663	1949
Ni	283	274	271	322	317	250	357	294
Y	9	7	8	8	11	4	–	10
Zr	23	25	24	22	29	18	–	26
La	1.45	1.62	2.15	1.16	1.30	1.83	1.44	1.32
Ce	3.39	3.79	5.21	2.73	3.06	4.13	2.88	2.88
Sm	0.57	0.69	0.54	0.43	0.62	0.47	0.59	0.58
Eu	0.16	0.20	0.16	0.16	0.23	0.16	0.19	0.17
Tb	0.15	0.15	0.14	0.09	0.16	0.10	0.13	0.15
Yb	1.22	1.29	0.82	0.61	1.13	0.73	0.99	1.03
Lu	0.18	0.20	0.12	0.10	0.16	0.11	0.15	0.16

favourably to boninite from Bonin Islands. A characteristic feature of the REE pattern of many boninites is the U-shape shown by elements from Ce through Yb, a feature that will be discussed more thoroughly later. The elements Sr to Ba in Fig. 10.4 show erratic behaviour, indicating their mobility during metamorphism consistent with results from many studies in low-grade metamorphic terrains (e.g. Coish 1977). It is noteworthy that the Sc abundances in the Appalachian samples, modern boninites, Archaean basaltic komatiites and MORB are comparable. This may indicate that clinopyroxene and/or garnet are not important in the genesis of all these rocks, since Sc is readily incorporated into clinopyroxene and garnet.

The Appalachian samples are also compared to modern boninites and Archaean basaltic komatiites in Figures 10.5A and B. The samples show a general trend of decreasing Al_2O_3 with increasing MgO (Fig. 10.5A). This is similar to the trends for modern boninites and basaltic komatiites. This trend is presumably a reflection of the absence of plagioclase during early stages of crystal fractionation leading to an increase in Al_2O_3 in fractionated liquids. A clear difference between the Appalachian boninites and Archaean basaltic komatiites is the much lower TiO_2 content of the Appalachian samples at all

Table 10.4 Representative analyses of boninitic rocks from Pacquet Harbour Group[a].

Sample no.	1	5	10	13	14	15	16	17
SiO_2	52.60	51.40	49.90	53.80	49.70	51.70	52.00	51.30
TiO_2	0.17	0.07	0.12	0.17	0.18	0.10	0.09	0.07
Al_2O_3	11.95	9.40	10.55	9.90	10.00	8.80	9.25	7.90
Fe_2O_3	1.07	0.89	12.75	1.87	1.57	1.06	1.42	1.24
FeO	7.19	7.32	5.27	7.68	7.36	7.47	7.68	7.84
MnO	0.13	0.17	0.18	0.13	0.18	0.12	0.12	0.16
MgO	14.85	17.03	13.35	12.79	16.75	16.88	15.00	16.58
CaO	9.12	8.54	4.17	9.18	9.42	9.53	10.80	11.86
Na_2O	2.13	1.53	0.21	3.28	1.57	1.73	1.92	1.08
K_2O	0.14	0.06	0.05	0.09	0.06	0.08	0.18	0.33
P_2O_5	0.04	0.02	0.05	0.07	0.07	0.04	0.05	0.06
LOI	2.16	3.28	4.80	2.40	3.36	2.75	1.88	2.01
total	101.55	99.71	101.40	101.36	100.22	100.26	100.39	100.43
Cr	1048	1894	1158	1543	1816	1885	1571	1551
Ni	232	314	212	342	346	367	317	407
Rb	3	na[b]	na	1	na	1	3	6
Sr	5	68	10	69	91	57	103	62
Y	6	4	5	8	5	6	4	3
Zr	21	11	25	14	17	11	8	8
Ba	37	16	21	17	17	10	22	65

[a]Analyses are taken from Hibbard (1983).
[b]na = Analysis not available.

MgO levels (Fig. 10.5B). TiO_2 contents in the range of 0.15 to 0.35% are clearly in the field for modern boninites but outside the field for basaltic komatiite. Furthermore, the fact that the TiO_2 contents are lower for a given MgO content indicates that the low Ti content is not simply a reflection of the degree of fractionation in the magma.

A plot of Ti versus Zr also indicates that the Appalachian samples are more like modern boninites than Archaean basaltic komatiites (Fig. 10.6). Not only are the Appalachian samples lower in Ti than basaltic komatiites but they also have Ti/Zr from ~ 20 to 80, similar to the range for modern boninites (Hickey & Frey 1982), but unlike those in basaltic komatiites and MORB (100–120). In magnesian sub-alkaline lavas, the low Ti/Zr are unique to boninites and this is perhaps the strongest evidence that the Appalachian very low-Ti samples are indeed boninites.

The rare-earth-element patterns of the Appalachian samples have an unusual concave-upward or U shape (Figs 10.7A, B and C). All of the very low-TiO_2 samples from Betts Cove and Thetford Mines exhibit this pattern; no REE analyses were available for the Pacquet Harbour samples. U-shaped REE patterns are also found in modern boninites from the Bonin Islands, Mariana Trench and Cape Vogel, Papua New Guinea (Fig. 10.7D) and in Cambrian boninites from SE Australia (Crawford & Cameron 1985). The Appalachian

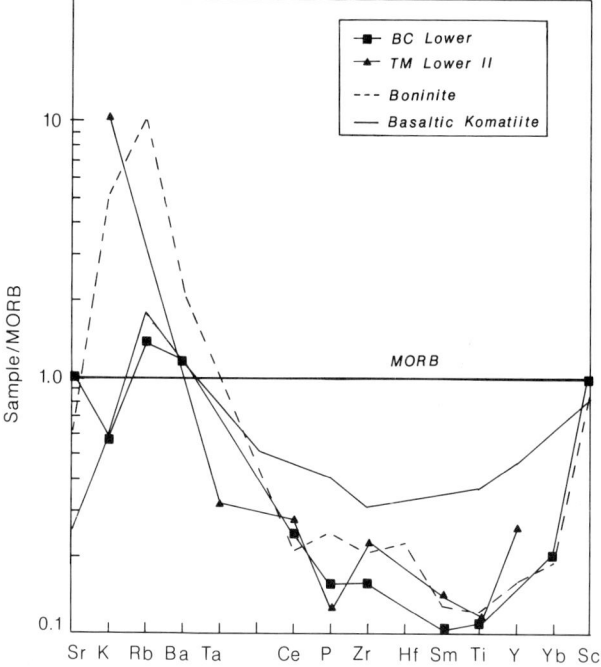

Figure 10.4 Element abundances normalized to average MORB (Pearce 1980) for typical Palaeozoic boninites from Betts Cove (BC Lower) and Thetford Mines (TM Lower II) ophiolites, modern boninite (sample 2983; Hickey & Frey 1982) and Archaean basaltic komatiite (Arndt & Nisbet 1982). Note the extremely low abundances of the HFSE in the Betts Cove and Thetford Mines samples and their favourable comparison with modern boninite.

samples are most like boninites from the Bonin Islands and the Mariana Trench in terms of their La/Yb; boninites from Cape Vogel and SE Australia have higher La/Yb while retaining the U-shaped pattern. Note that boninites from DSDP Site 458 do not exhibit U-shaped patterns; rather they are LREE-depleted (Hickey & Frey 1982). Nevertheless, boninites are the only *volcanic* rock type to show the unusual U-shaped REE patterns. In contrast to the Appalachian boninites, MORB and basaltic komatiites have REE patterns that are usually either LREE-depleted or flat (BVSP 1981, Arndt & Nisbet 1982).

Nd isotopes have been reported on 12 samples of volcanic rocks from the Betts Cove and Thetford Mines ophiolite (Table 10.5). For boninitic lavas, represented by the lower lavas at Betts Cove and all of the Thetford Mines samples, ϵ_{Nd} ranges from $+0.4$ to $+4.7$. This is within the range for boninites from the W Pacific (-0.3 to $+6.2$; Hickey & Frey 1982), and below values for calculated early Palaeozoic MORB (~ 6–10; Jacobsen & Wasserburg 1979). These Nd isotope values are also similar to values reported for some early Palaeozoic boninitic lavas from SE Australia (Nelson *et al.* 1984). ϵ_{Nd} values

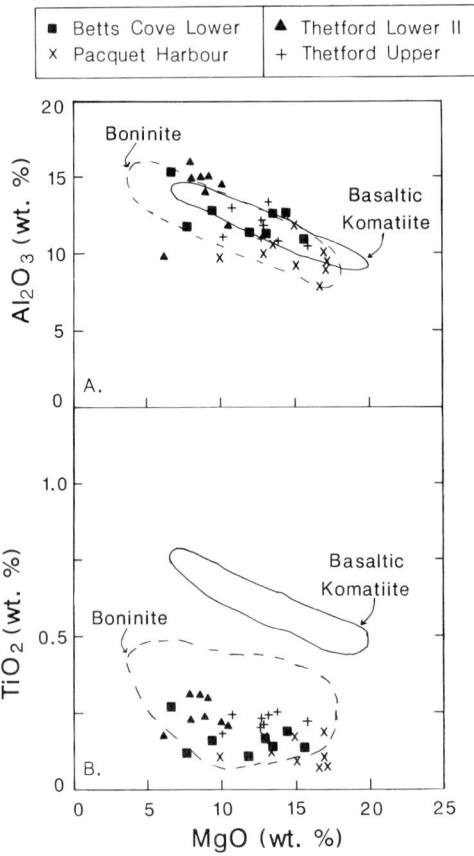

Figure 10.5 (A) Al_2O_3 versus MgO and (B) TiO_2 versus MgO variations for boninitic samples in the Appalachians. Note the similarity of the Appalachian samples with modern boninites (Hickey & Frey 1982) and their variance with Archaean basaltic komatiite in TiO_2 content.

for boninitic rocks at Betts Cove appear to grade into values for the intermediate-Ti lavas (+ 4.7 to + 6.9) but are distinctly lower than values for the upper lavas (+ 7.2 to + 7.6).

10.7 Volcanic rocks associated with the boninitic lavas

In all occurrences of low-Ti, boninitic lavas in the northern Appalachians, other volcanic rocks are found in close association. At Betts Cove, there are two chemical units above or interbedded with the boninites (Coish *et al.* 1982). The *intermediate-Ti* lavas are intermixed with the boninitic samples and have slightly higher TiO_2 (0.30 to 0.75%) with LREE-depleted REE patterns rather than U-shaped patterns (Coish *et al.* 1982). They also have slightly higher ε_{Nd}

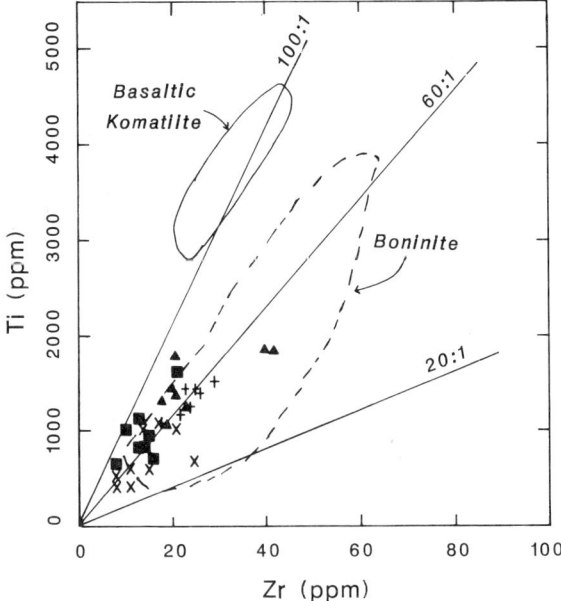

Figure 10.6 Variation of Ti with Zr. Appalachian samples have an average Ti/Zr around 60, similar to modern boninites. Data for boninites from Hickey & Frey (1982) and for basaltic komatiites from Arndt & Nisbet (1982). Symbols as in Figure 10.5.

values than the boninitic lavas, although the ε_{Nd} values do appear to grade from the lower (boninitic) lavas into the intermediate-Ti lavas (Table 10.5). The intermediate-Ti lavas are similar to island-arc tholeiites in their Cr–Y variations (Coish & Rogers 1987). A tentative conclusion is that the intermediate-Ti lavas are island-arc tholeiites, although their origin as boninites cannot be ruled out. Stratigraphically above both the lower and intermediate-Ti lavas are the upper lavas. Petrographically and chemically, the upper lavas are different from the lower units. They have a greater abundance of pyroxene and plagioclase phenocrysts and they have higher TiO_2, Zr, Y, REE and ε_{Nd}. Chemically, the upper lavas are like MORB or certain backarc-basin (BAB) basalts (Coish & Church 1979, Coish et al. 1982). Furthermore, a thick sequence of high-Ti lavas of the upper Snooks Arm Group overlies the upper lavas of the ophiolite (Jenner & Fryer 1980). These are enriched in LILE, and are similar chemically to BAB basalts from the Scotia Sea backarc basin (Jenner & Fryer 1980, Saunders & Tarney 1984). A similar stratigraphy of magma series with boninitic lavas overlain by BAB basalts is preserved in Cambrian greenstone belts in SE Australia (Crawford et al. 1984, Crawford and Cameron 1985, Crawford & Keays 1987).

In the Thetford Mines ophiolite, the boninitic rocks are represented by lower lavas Type II and the upper lavas (Oshin & Crockett 1986). In contrast to the stratigraphy at Betts Cove, higher-TiO_2 lavas (lower lavas Type I) in

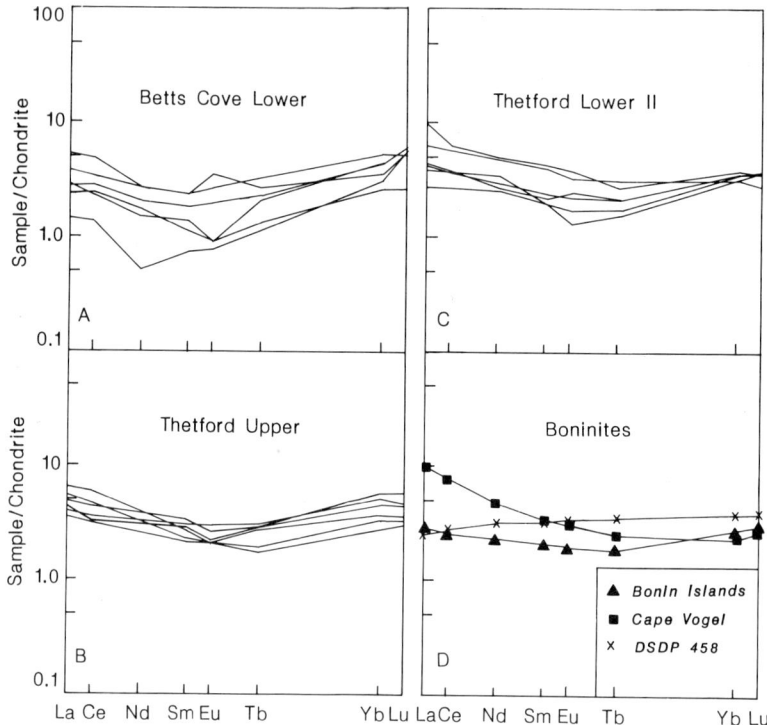

Figure 10.7 Rare-earth-element variations in (A–C) boninitic samples from Betts Cove and Thetford Mines ophiolites (Coish *et al.* 1982, Oshin & Crockett 1986) and (D) modern boninites (Hickey & Frey 1982).

the Thetford Mines ophiolite occur *below* the boninitic samples, similar to stratigraphy in the Troodos Ophiolite (Cameron 1985). The higher-TiO_2 samples are thought to be MORB-like lavas derived from an undepleted mantle source in a marginal or backarc basin (Oshin & Crockett 1986). However, there is some evidence that these high-Ti lavas are more like island-arc tholeiites (IAT). First, the TiO_2 content for a given FeO^*/MgO is low for MORB-like lavas and similar to IAT. Secondly, they have very low Ni contents. Thirdly, in a Cr–Y discriminant diagram, they clearly plot in the IAT field (Coish & Rogers 1987). Fourthly, the abundance of plagioclase phenocrysts is consistent with their derivation as IAT.

In the Pacquet Harbour Group, higher-TiO_2 basalts are also found, but their stratigraphic relationship to the boninitic samples is not clear from the report of Hibbard (1983). On the basis of four analyses presented by Hibbard (1983), the high-Ti rocks are similar to the Betts Cove upper lavas. Specifically, they fall along the same trend in a FeO^*/MgO versus TiO_2 diagram. They are unlike the high-Ti samples from Thetford Mines which have much lower TiO_2 contents at a given FeO^*/MgO. Thus, the higher-TiO_2 basalts in the Pacquet Harbour Group can be interpreted as MORB or BAB basalts.

Table 10.5 Nd isotopes in Betts Cove and Thetford Mines ophiolites.

	Sample	$^{143}Nd/^{144}Nd$	$^{147}Sm/^{144}Nd$	ε_{Nd}
Betts Cove[a]				
upper lavas	73-13b	0.513024 ± 18	0.2054	+7.2
	73-9	0.513014 ± 18	0.1948	+7.6
intermediate lavas	74-57	0.513140 ± 17	0.2475	+6.9
	75-4	0.512981 ± 18	0.2111	+6.0
	75-3	0.512843 ± 19	0.1887	+4.7
lower lavas	74-101a	0.512791 ± 16	0.1749	+4.5
	74-101b	0.512728 ± 19	0.1624	+4.1
	73-256	0.512782 ± 20	0.1816	+3.9
Thetford Mines[b]				
upper pillow	Qe-40	0.51197 ± 2	0.201	+2.2
lower pillow	Qe-13a	0.51181 ± 3	0.178	+0.4
lower pillow	Qe-18	0.51194 ± 2	0.178	+3.0
diabase	Qe-8b	0.51209 ± 2	0.205	+4.7

[a] Analyst for Betts Cove samples was R. Hickey. See Coish *et al.* (1982) for details of method. Note that the $^{143}Nd/^{144}Nd$ ratios are normalized to $^{146}Nd/^{144}Nd = 0.7219$ and $(^{143}Nd/^{144}Nd)_{BCR-1} = 0.51263$.

[b] Analyses from Shaw & Wasserburg (1984). Note that the $^{143}Nd/^{144}Nd$ ratios are normalized to $^{146}Nd/^{142}Nd = 0.636151$ and $(^{143}Nd/^{144}Nd)_{CHUR} = 0.511847$.

The occurrence of more 'normal' volcanic rocks closely associated with boninites is critical to any discussion of tectonic environment of formation. Boninites cannot be isolated and discussed without consideration of the significance of other volcanic rocks. In most examples in the Appalachian ophiolites, there is no reason to expect that the boninites were formed in a different tectonic environment from the other volcanics, and later juxtaposed. In some cases, the boninites are interbedded with the other volcanic rocks, whereas in other cases, boninites conformably overlie or underlie other volcanics. Thus, any tectonic environment envisioned must be able to produce boninitic lavas *and* higher-TiO_2 island-arc tholeiites and/or MORB-like lavas.

10.8 Petrogenesis of Appalachian boninites

Because of the demonstrated chemical similarity of boninitic samples in the Appalachian ophiolites with W Pacific boninites, ideas on the petrogenesis of modern boninites can be applied to those in the current study. The following discussion of the origin of modern boninites is based largely on Meijer (1980), Crawford *et al.* (1981), Hickey & Frey (1982), Coish *et al.* (1982), Cameron *et al.* (1983), Walker & Cameron (1983) and Hawkins *et al.* (1984).

The low Ti, Zr, Y and HREE concentrations in Appalachian and Cenozoic boninites can be explained by partial melting of a peridotite that is more depleted than the source for MORB (Sun & Nesbitt 1978, Coish & Church 1979, Jenner 1981, Coish *et al.* 1982, Hickey & Frey 1982, Cameron *et al.* 1983, Oshin & Crockett 1986). The source, then, was probably harzburgitic, although enough clinopyroxene must have been in the source to provide the necessary Al_2O_3 and CaO present in boninites. High MgO, Ni, Cr and low Al_2O_3 contents of the boninitic samples can be explained by melting this severely depleted peridotite. The U-shaped REE patterns, low Ti/Zr and low ε_{Nd} in the boninitic samples complicate this simple picture of progressive depletion of the mantle, and second- or third-stage melting of severely depleted mantle to form boninites. One would expect extreme depletion in the LREE relative the HREE and fairly constant Ti/Zr and ε_{Nd} if the only process involved in forming boninites were melting a severely depleted mantle. The LREE enrichment in boninites (Fig. 10.7d), resulting in the U-shaped REE pattern, has been explained by melting a source with a U-shaped REE pattern. The source is indeed a severely depleted peridotite, but one whose bulk chemistry has been altered by the introduction of an LREE-rich but Ti-poor fluid (Sun & Nesbitt 1978, Jenner 1981, Hickey & Frey 1982, Coish *et al.* 1982). Thus, the source for boninites has at least a two-stage evolutionary history; first, a partial melting stage to deplete the source in Ti, Zr, Y and REE (the melting stage may have involved two or three stages of small degrees of melting rather than a single large-degree melting event), and secondly, an enrichment stage to increase the abundance of LILE, including LREE. Possible support for this idea is the existence of harzburgite in several ophiolites and some spinel lherzolite inclusions in alkali basalt, with U-shaped REE patterns (Frey 1984). An alternative explanation is that, during partial melting, a minor residual phase in the source retains the MREE more strongly than either LREE or HREE, resulting in a U-shaped pattern in the liquid. However, no such phase has been clearly identified in harzburgites, or in peridotite melting studies.

The low ε_{Nd} values can also be explained by adding an enriched fluid component to the source rock. Hickey & Frey (1982) calculated that mixing about 0.5% enriched component with 99.5% depleted source could produce suitable Nd isotopic ratios in the source. The enriched component could either be a fluid or a melt phase. This enriched component was liberated from subducted oceanic crust, then migrated upwards and mixed with depleted mantle (see Figure 13 in Hickey & Frey 1982).

10.9 Tectonic models for boninite generation

Exactly when and where modern boninites form in volcanic arcs is unclear. They do appear to be restricted to intra-oceanic arcs. Meijer (1980) stated that boninites from the Mariana and Bonin forearc regions are always associated with but younger than island-arc tholeiites, and thus formed by melting the depleted mantle wedge above a subducting plate after the production of island-arc tholeiite. This scenario requires unusually high geothermal gradients beneath the arc and forearc.

Crawford et al. (1981) suggested that boninites form when arc volcanism ceases and backarc spreading is initiated. Boninites represent the early stages of backarc magmatism. They are generated by melting depleted subarc mantle, the heat for melting being supplied by rising diapirs of MORB-source mantle. The diapirs eventually initiate spreading in backarc basins and also melt to form backarc-basin basalts which cover the boninites formed earlier. As the backarc spreading centre migrates away from the trench, the chemistry of backarc-basin basalts is no longer influenced by the subducting slab and becomes more MORB-like. Crawford & Keays (1987) present a modified scheme to explain the occurrence of Cambrian boninites and tholeiites in Victoria, Australia. They suggest that a subducted spreading ridge which remains active after subduction could provide the necessary heat to melt overlying depleted mantle and produce boninites. In these Cambrian rocks, boninites overlie arc lavas, implying that arc volcanism *precedes* boninite eruption.

On the other hand, Hawkins et al. (1984) suggest that boninites are the first magmas produced in oceanic subduction zones. They speculate that the heat source may be 'mantle counterflow in the corner between the downgoing slab and overriding plate'. The mantle counterflow causes 'heating, up-arching, rupture and partial melting of the mantle wedge and crust' (Hawkins et al. 1984). Furthermore, their model indicates that boninites are produced at depths of 50 km or less and erupted in the forearc region. During this stage, the lithosphere is dilated and *extended*, due to trench roll-back, leading to the formation of boninitic ocean crust as older ocean crust is rifted. Hawkins et al. (1984) and Bloomer & Hawkins (1987) postulate that the bulk of island-arc volcanism *follows* this stage of boninite production, although they do suggest that some overlap between magma series can occur. Thus, boninite forms a substrate on which later arc volcanics are deposited. An important part of this model is that extension and backarc spreading occur before the development of the main arc. Alabaster et al. (1982) use a similar model to explain the occurrence of island-arc rocks in the Oman ophiolite.

10.10 Appalachian boninites and plate tectonics

Plate tectonic models of the Appalachian mountains have been based partly on the location of ophiolites and their interpretation as pieces of ocean crust formed at an oceanic spreading ridge. A generally accepted plate tectonic model for the Appalachians involves an ocean-episode in the late Precambrian or early Cambrian followed by ocean closing due to southeastward subduction during the mid-Ordovician. Southeastward subduction led to formation of an island arc; eventual collision of this arc with a western continental mass resulting in the Taconic Orogeny during the mid-Ordovician. A Devonian event perhaps related to continent–continent collision produced the Acadian Orogeny. In early plate models, ophiolitic remnants were assumed to have formed at a ridge as part of the ocean-opening phase, and then were subsequently caught up in a subduction or obduction zone (Church & Stevens 1971, Dewey & Bird 1971). Later, some ophiolites (e.g. Betts Cove) were interpreted to have formed in inter-arc or backarc basins after the onset or even cessation of island-arc volcanism (Nelson & Casey 1979, Coish *et al.* 1982) whereas others (e.g. Bay of Islands) were thought to have formed at an ocean ridge, presumably before initiation of island-arc volcanism.

Ideas on the generation of boninites as well as new age data for the Newfoundland ophiolites warrant further refinement of Appalachian plate tectonic models. Boninites are important because their unique chemical features and rare occurrence enable one to make strong inferences on the tectonic environment of formation. As modern boninites are found only in supra-subduction-zone settings, it is inferred that ancient boninites also formed above a subduction zone. The recognition of boninites in Appalachian ophiolites in turn suggests the ophiolites formed in a supra-subduction-zone environment rather than at an ocean ridge. Hence, the timing of some of plate processes has to be reassessed.

By analogy with modern boninites, Appalachian ophiolites containing boninites probably formed in the *early* stages of subduction, during an extensional event related to backarc spreading. In the interpretation of modern boninites, there is disagreement as to whether boninites preceded or followed island-arc volcanism (Crawford *et al.* 1981, Hawkins *et al.* 1984). The Appalachian examples indicate that both chronologies occur. In some Appalachian ophiolites (e.g. Thetford Mines), island-arc tholeiites (represented by Lower lavas I) occur below the boninites; in others (e.g. Betts Cove), boninitic samples are the lowest and, hence, first lavas and are overlain by and interbedded with possible island-arc tholeiites (intermediate-Ti lavas). In the case of Betts Cove, it also clear that significant extension took place during boninite formation because there is a well developed sheeted dyke complex. Furthermore, this extension took place before the onset of arc volcanism. It is obvious that this is not always the case, as shown, for example, by the Thetford Mines sequence. The order of eruption of boninite and island-arc tholeiite may be determined by the local geothermal gradient; a very high geothermal gradient would permit melting of severely depleted mantle to produce boninites whereas lower geothermal gradients could result in melting less depleted mantle to form island-arc tholeiites. Alternatively, boninitic and island-arc tholeiite lavas may be produced simultaneously in the mantle, and also erupted penecontemporaneously. Either lava sequence then could form the volcanic carapace on the intrusive sequence; at stratigraphically higher levels, these lava types would be interbedded. Whatever the correct scenario, it appears that extension leading to boninite generation occurred very early in the development of the subduction zone.

At Betts Cove, the upper lava sequence in the ophiolite and the upper Snooks Arm Group can be interpreted as backarc-basin basalts. Thus, boninitic and island-arc tholeiitic lavas form a substrate on which MORB-like marginal basin basalts are erupted. The later lavas are more like MORB, perhaps because of roll-back and migration of slab-influenced volcanism trenchwards, leaving only MORB-mantle sources available for melting in backarc regions. This does not mean that all ophiolites with boninitic lavas should be overlain by MORB-type lavas. Perhaps, what overlies boninitic lavas depends on the location of the ophiolite relative to the trench. Those farthest from the trench may be covered by MORB-type lavas; those in the forearc region may be covered by forearc sediment; those formed in the volcanic arc area may be covered by island-arc tholeiite and calc-alkaline lavas of the main arc sequence.

In Newfoundland, the ophiolitic sequences with boninitic lavas are located east of the well exposed and well known Bay of Islands massifs (Fig. 10.1). Lavas in the Bay of Islands sequences are all MORB-like, and hence the ophiolite has been cited as a good example of an ophiolite formed at an ocean ridge (Pearce et al. 1984, Casey et al. 1985). The age of the Bay of Islands ophiolite relative to the Betts Cove ophiolite may be critical to its interpretation as having formed at an ocean ridge. The Bay of Islands complex has been dated at ~505 Ma (U–Pb; Mattinson 1976; Nd–Sm: Jacobsen & Wasserburg 1979), and at 485 Ma (U–Pb: Dunning & Krogh 1985). The correct age of the Bay of Islands is currently being debated, and details of the arguments are beyond the scope of this chapter. The Betts Cove ophiolite has an accepted age of 488 Ma (Dunning & Krogh 1985). If the older age for the Bay of Islands ophiolite is correct, then it is quite reasonable to imagine that this ophiolite formed at a mid-ocean ridge at about 505 Ma; later, southeastward subduction within this ocean basin produced the Betts Cove ophiolite above a subduction zone at 488 Ma. The Bay of Islands ophiolite at this time may have been part of the upper ocean plate, and subsequent obduction of the upper ocean plate and island arc resulted in the Bay of Islands ophiolite emplaced in its present position west of the Betts Cove ophiolite. On the other hand, if the younger age for the Bay of Islands is accepted, then it is essentially the same age as the Betts Cove ophiolite, and it is less likely that the Bay of Islands ophiolite would have formed at an ocean ridge while the Betts Cove ophiolite formed above a subduction zone. While it is certainly possible for ocean-ridge processes to be happening at the same time as subduction processes, it is fortuitous to expect that the only obducted slice of ridge ophiolite would be exactly the same age as the obducted arc ophiolite. It seems more likely that both ophiolitic fragments would have formed in the same environment if they are the same age. Since the presence of boninites at Betts Cove indicates a subduction environment, then the Bay of Islands ophiolite would also have formed in this environment, probably at a marginal basin spreading centre in a zone not influenced by fluids from a subducting slab. The chemical data from the Bay of Islands ophiolite permit both interpretations, i.e. ocean ridge and marginal basin. Resolution of the problem of tectonic environment of formation of the Bay of Islands ophiolite awaits further work on age and perhaps on sedimentary rocks above the ophiolite.

10.11 Conclusions

The main conclusions of this study are as follows:

(a) Low-TiO_2 lavas from three ophiolitic terrains in the northern Appalachians have boninitic affinities. The low-TiO_2 lavas are closely associated with higher-TiO_2 lavas.
(b) The low Ti, Zr, Y, REE, ε_{Nd} and Ti/Zr, high MgO, Cr and Ni, and U-shaped REE patterns are chemical features that the Appalachian samples have in common with modern boninites from the W Pacific island arcs.
(c) The occurrence of the boninites in Appalachian ophiolites is strong evidence that the ophiolites formed in the vicinity of an early Palaeozoic island arc. The higher-TiO_2 lavas associated with the boninitic samples could have formed in a backarc basin (Betts Cover and Pacquet Harbour) or as island-arc tholeiites (Thetford Mines).
(d) Early Palaeozoic plate models of the Appalachians must take into account the formation of some ophiolites near subduction zones.

Acknowledgements

I thank Bill Church for much stimulating discussion on the origin of ophiolites. Reviews by A. J. Crawford and G. A. Jenner were very helpful. Part of this research was supported by NSF EPSCoR grant R11-8610679.

References

Alabaster, T., J. A. Pearce & J. G. Malpas 1982. The volcanic stratigraphy and petrogenesis of the Oman ophiolite complex. *Contrib. Mineral. Petrol.* **81**, 168–83.
Arndt, N. T. & E. G. Nisbet 1982. Geochemistry of Munro Township basalts. In *Komatiites*, N. T. Arndt & E. G. Nisbet (eds), London: George Allen & Unwin.
Blackwood, R. F. 1982. *Geology of the Gander Lake (2D/15) and Gander River (2E/2) area*. Newfoundland Dept Mines Energy, Min. Development Div., Rep. no. 82-56.

Bloomer, S. H. & J. W. Hawkins 1987. Petrology and geochemistry of boninite series, volcanic rocks from the Marianas trench. *Contrib. Mineral. Petrol.* **97**, 361–77.
BVSP (Basaltic Volcanism Study Project) 1981. *Basaltic volcanism on the terrestrial planets.* New York: Pergamon.
Cameron, W. E. 1985. Petrology and origin of primitive lava from the Troodos ophiolite, Cyprus. *Contrib. Mineral. Petrol.* **89**, 1–17.
Cameron, W. E., M. T. McCulloch & D. A. Walker 1983. Boninite petrogenesis: chemical and Nd–Sr isotopic constraints. *Earth Planet. Sci. Lett.* **65**, 75–89.
Cameron, W. E., E. G. Nisbet & V. J. Dietrich 1979. Boninites, komatiites and ophiolitic basalts. *Nature* **280**, 550–3.
Casey, J. F., D. L. Elthon, F. X. Siroky, J. A. Karson & J. Sullivan 1985. Geochemical and geological evidence bearing on the origin of the Bay of Islands and coastal complex ophiolites of western Newfoundland. *Tectonophysics* **116**, 1–40.
Church, W. R. & R. K. Stevens 1971. Early Paleozoic ophiolite complexes of Newfoundland Appalachians as mantle–ocean crust sequences. *J. Geophys. Res.* **76**, 1460–6.
Coish, R. A. 1977. Ocean floor metamorphism in the Betts Cove ophiolite, Newfoundland. *Contrib. Mineral. Petrol.* **60**, 255–70.
Coish, R. A. & W. R. Church 1979. Igneous geochemistry of mafic rocks in the Betts Cove ophiolite, Newfoundland: some unusual chemical characteristics. *Contrib. Mineral. Petrol.* **70**, 29–39.
Coish, R. A. & N. W. Rogers 1987. Geochemistry of the Boil Mountain ophiolitic complex, northwest Maine, and tectonic implications. *Contrib. Mineral. Petrol.* **97**, 51–65.
Coish, R. A., R. Hickey, & F. A. Frey 1982. Rare earth element geochemistry of the Betts Cove ophiolite, Newfoundland: complexities in ophiolite formation. *Geochim. Cosmochim. Acta* **46**, 2117–334.
Crawford, A. J. 1980. A clinoenstatite-bearing cumulate olivine pyroxenite from Howqua, Victoria. *Contrib. Mineral. Petrol.* **75**, 353–67.
Crawford, A. J. & W. E. Cameron, 1985. Petrology and geochemistry of Cambrian boninites and low-Ti andesites from Heathcote, Victoria. *Contrib. Mineral. Petrol.* **91**, 93–104.
Crawford, A. J. & R. R. Keays 1987. Petrogenesis of Victorian Cambrian tholeiites and implications for the origin of associated boninites. *J. Petrol.* **28**, 1075–109.
Crawford, A. J., L. Beccaluva & G. Serri 1981. Tectono-magmatic evolution of the west Philippine–Mariana region and the origin of boninites. *Earth Planet. Sci. Lett.* **54**, 346–56.
Crawford, A. J., W. E. Cameron & R. R. Keays 1984. The association boninite low-Ti andesite–tholeiite in the Heathcote greenstone belt, Victoria; ensimatic setting for early Lachlan Fold Belt. *Austr. J. Earth Sci.* **31**, 161–75.
Dewey, J. F. & J. Bird 1971. Origin and emplacement of the ophiolite suite: Appalachian ophiolites in Newfoundland. *J. Geophys. Res.* **76**, 3179–206.
Dunning, G. R. & T. E. Krogh 1985. Geochronology of ophiolites of the Newfoundland Appalachians. *Can. J. Earth Sci.* **22**, 1659–70.
Frey, F. A. 1984. Rare earth element abundances in upper mantle rocks. In *Rare earth element geochemistry*, P. Henderson (ed.), 153–203. Amsterdam: Elsevier.
Gale, G. H. 1973. Paleozoic basaltic komatiite and ocean floor type basalts from northeast Newfoundland. *Earth Planet. Sci. Lett.* **18**, 22–8.
Hawkins, J. W., S. H. Bloomer, C. A. Evans & J. T. Melchior 1984. Evolution of intra-oceanic arc–trench systems. *Tectonophysics* **102**, 175–205.
Hibbard, J. 1983. *Geology of the Baie Verte Peninsula, Newfoundland.* Newfoundland Dept Mines Energy, Min. Development Div., Mem. no. 2.
Hickey, R. L. & F. A. Frey 1982. Geochemical characteristics of boninite series volcanics: implications for their source. *Geochim. Cosmochim. Acta* **46**, 2099–115.
Howard, A. H. & E. Stolper 1981. Experimental crystallization of boninites from the Marianas Trench. *Trans. Am. Geophys. Union* **62**, 1091.

Jacobsen, S. B. & G. J. Wasserburg 1979, Nd and Sr isotopic study of the Bay of Islands ophiolite complex and the evolution of the source of midocean ridge basalts. *J. Geophys. Res.* **84**, 7429–45.

Jenner, G. 1981. Geochemistry of high-Mg andesites from Cape Vogel, Papua New Guinea. *Chem. Geol.* **33**, 307–22.

Jenner, G. & B. J. Fryer 1980. Geochemistry of the upper Snooks Arm Group Basalts, Burlington Peninsula, Newfoundland: evidence against formation in an island arc. *Can. J. Earth Sci.* **17**, 888–900.

Laurent, R. 1975. Occurrences and origin of the ophiolites of southern Quebec, northern Appalachians. *Can. J. Earth Sci.* **12**, 443–55.

Laurent, R. 1977. Ophiolites from the northern Appalachians of Quebec. In *North American ophiolites*, R. G. Coleman & W. P. Irwin (eds), 25–40. State of Oregon, Dept Geol. Min. Indust., Bull. no. 95.

Laurent, R. 1980. Environment of formation, evolution and emplacement of the Appalachian ophiolites from Quebec. In *Ophiolites: Proc. Int. Ophiolite Symp.*, A. Panayiotou (ed.), 628–636. Cyprus.

Mattinson, J. M. 1976. Ages of zircons from the Bay of Islands ophiolite complex, western Newfoundland. *Geology* **4**, 393–4.

Meijer, A. 1980. Primitive arc volcanism and a boninite series: examples from western Pacific island arcs. In *The tectonic and geologic evolution of Southeast Asian seas and islands*, D. E. Hayes (ed.), 269–82. Am. Geophys. Union Monogr. no. 23.

Nelson, D. R., A. J. Crawford & M. T. McCulloch 1984. Nd–Sr isotopic and geochemical systematics in Cambrian boninites and tholeiites from Victoria, Australia. *Contrib. Mineral. Petrol.* **88**, 164–72.

Nelson, K. D. & J. F. Casey 1979. Ophiolitic detritus in the upper Ordovician flysch of Notre Dame Bay and its bearing on the tectonic evolution of western Newfoundland. *Geology* **7**, 27–31.

Oshin, I. O. & J. H. Crockett 1986. The geochemistry and petrogenesis of ophiolitic volcanic rocks from Lac de l'Est, Thetford Mines Complex, Quebec, Canada. *Can. J. Earth Sci.* **23**, 202–13.

Pearce, J. A. 1980. Geochemical evidence for the genesis and eruptive setting of lavas from Tethyan ophiolites. In *Ophiolites: Proc. Int. Ophiolite Symp.*, A. Panayiotou (ed.), 261–72. Cyprus.

Pearce, J. A., S. J. Lippard & S. Roberts 1984. Characteristics and tectonic significance of supra-subduction zone ophiolites. In *Marginal basin geology*, B. P. Kokelaar & M. F. Howells (eds), 77–94. Geol. Soc. Lond. Spec. Publ. no. 16.

Saunders, A. D. & J. Tarney 1984. Geochemical characteristics of basaltic volcanism within back-arc basins. In *Marginal basin geology*, B. K. Kokelaar & M. F. Howells (eds), 59–76. Geol. Soc. Lond. Spec. Publ. no. 16.

Seguin, M. K. & R. Laurent 1975. Petrological features and magnetic properties of pillow lavas from the Thetford Mines Ophiolite (Quebec). *Can. J. Earth Sci.* **12**, 1406–20.

Shaw, H. F. & G. J. Wasserburg 1984. Isotopic constraints on the origin of Appalachian mafic complexes *Am. J. Sci.* **284**, 319–49.

Sigurdsson, H. & J.-G. Schilling 1976. Spinels in Mid-Atlantic Ridge basalts: chemistry and occurrence. *Earth Planet. Sci. Lett.* **29**, 7–20.

Snelgrove, A. K. 1931. Geology and ore deposits of the Betts Cove–Tilt Cove area, Notre Dame Bay, Newfoundland. *Can. Min. Met. Bull.* **24**, 477–519.

Strong, D. F. 1979. The enigmatic ultramafic and associated rocks of eastern Newfoundland. *Geol. Assoc. Can. Abstr. Progr.* **4**, 81.

Sun, S.-S. & R. W. Nesbitt 1978. Geochemical regularities and genetic significance of ophiolitic basalts. *Geology* **6**, 689–93.

Umino, S. 1986. Magma mixing in boninite sequence of Chichijima, Bonin Islands. *J. Volcanol. Geotherm. Res.* **29**, 125–57.

Upadhyay, H. 1978. Phanerozoic peridotitic and pyroxenitic komatiites from Newfoundland. *Science* **202**, 1192–5.

Upadhyay, H. 1982. Ordovician komatiites and associated boninite-type lavas from Betts Cove, Newfoundland. In *Komatiites*, N. T. Arndt & E. G. Nisbet (eds), 187–98. London: George Allen & Unwin.

Upadhyay, H. & E. R. W. Neale 1979. On the tectonic regimes of ophiolite genesis. *Earth Planet. Sci. Lett.* **43**, 93–102.

Upadhyay, H., J. F. Dewey & E. R. W. Neale 1971. The Betts Cove ophiolite complex, Newfoundland: Appalachian oceanic crust mantle. *Geol. Assoc. Can.* **24**, 27–34.

Walker, D. A. & W. E. Cameron 1983. Boninite primary magmas: evidence from the Cape Vogel Peninsula, PNG. *Contrib. Mineral. Petrol.* **83**, 15–158.

Williams, H. 1978 *Tectono-lithofacies map of the Appalachian orogen*. Memorial University of Newfoundland, Map no. 1.

11 Petrogenesis of boninitic lavas from the Limassol Forest Complex, Cyprus

N.W. ROGERS, C.J. MacLEOD & B.J. MURTON

Abstract

Major- and trace-element and Sr and Nd isotope analysis of six glassy lavas and two dykes from the Limassol Forest Complex (LFC), Troodos, Cyprus are presented. Major-element abundances bear many similarities with W Pacific boninites but lower SiO_2 (50–55%) and the predominance of olivine, as opposed to low-Ca pyroxene, control contrasts with boninites *sensu stricto*. Two magma groups are identified on the basis of REE and HFSE abundances, corresponding to Cameron's groups II and III. Group II lavas have REE and HFSE abundances 3–5 × chondritic while the group III lavas show 'U-shaped' normalized trace-element profiles. Age-corrected ε_{Nd} values range from +8.9 to +0.9 and $^{87}Sr/^{86}Sr$ from 0.70394 to 0.70752, but are not simply related to trace-element variations. A new model for the petrogenesis of these lavas is presented in which a depleted harzburgite mantle, residual after the extraction of MORB and/or Troodos axis sequence magmas, is first infiltrated by a fluid derived from the underlying subduction zone. This produces LILE enrichment and the development of the isotopic heterogeneity. Later migration of small-volume mafic silicate melts from the deeper asthenosphere, with $\varepsilon_{Nd} \sim +8$, reduces the isotopic heterogeneity to the values seen in the erupted lavas and produces the two styles of REE and HFSE enrichment. Melting of this complexly enriched but still harzburgitic mantle occurred during transtensional tectonism associated with the development of the LFC. Lavas were erupted rapidly, without retention in large magma chambers, thus recording the isotopic and trace-element heterogeneities in their source regions.

11.1 Introduction

Boninites and high-Mg andesites have recently attained a significance in the study of convergent plate margin magmatism that far outweighs their volumetric abundance. Their primary, unfractionated composition, an unusual feature in arc-related magmas, has encouraged inferences to be made concerning their source regions without the need to consider the effects of crustal contamination and similar processes that modify most magmas en route to the surface. In this way, boninites provide some of the most direct evidence for the nature of processes operating in the sub-arc mantle wedge. In particular, 'U-shaped' REE patterns and unusually low ε_{Nd} values have been interpreted as reflecting the addition of exotic components, possibly subducted sediment, to the magma source (Sun & Nesbitt 1978, Cameron et al. 1983). However, this model is perhaps difficult to reconcile with other geochemical and experimental evidence which is consistent with melting of depleted peridotite in the oceanic lithosphere at relatively low pressures (Duncan & Green 1980).

Magnesian lavas of boninitic affinity have long been recognized in the Troodos Ophiolite in Cyprus. Some of these are compositionally similar to the W Pacific boninites, being characterized by low TiO_2 abundances, 'U-shaped' REE profiles and variable ε_{Nd} values. However, they are closely associated with a second group of high-MgO, low-TiO_2 lavas which have slightly LREE-depleted profiles that cannot be modelled by the simple addition of trace-element-enriched subducted sediment to a depleted peridotite source. Yet this latter group still shows variation in ε_{Nd}.

The rocks discussed herein belong to these two magmatic groups and are from the Limassol Forest Complex (LFC), on the southern edge of the Troodos massif. Detailed remapping of this area, where primitive lavas predominate, has revealed that it is part of a fossil transtensional transform fault. This provides a well defined tectonic framework within which to constrain the petrogenesis of these lavas and assess the importance of local geological controls compared with broader-scale plate tectonic influences. Particular emphasis is placed on a single model to account for the trace-element characteristics of both magma groups, rather than appealing to models that require distinct components to explain these minor differences in two otherwise similar and closely associated magma groups. New trace-element and Nd and Sr isotope analyses are reported and used to evaluate the role of sediment subduction in the petrogenesis of the LFC lavas, and the implications for the origin of similar trace-element variations in the W Pacific boninites are briefly discussed.

11.2 Geology

The Troodos massif of Cyprus is arguably the most extensively studied ophiolite complex in the world. It preserves a complete and coherent Penrose stratigraphy (Anon. 1972), comprising well developed residual mantle, plutonic, sheeted dyke and volcanic sequences overlain by pelagic sediments. The lavas have traditionally been divided on volcanological, petrological and metamorphic grounds into two units (e.g. Wilson 1959, Bear 1960, Carr & Bear 1960, Gass 1960). The Lower Pillow Lavas, or 'axis sequence', are dominated by evolved andesites and dacites of island-arc tholeiite affinity (Robinson *et al.* 1983). They are, however, considered cogenetic with the underlying sheeted dyke complex (Bear 1960, Moores & Vine 1971, Gass & Smewing 1973) and to have formed by sea-floor spreading at a constructive plate margin (Gass *et al.* 1975). The Upper Pillow Lavas, in contrast, are of basalt–basaltic andesite composition and have been interpreted in terms of 'off-axis' volcanism (Gass & Smewing 1973).

The LFC lies south of the main part of the ophiolite and is separated from it by the Arakapas Fault Belt. This has been interpreted by Simonian & Gass (1978) as the northern wall of an E–W trending oceanic transform fault which, during the formation of the ophiolite, formed a major bathymetric depression progressively infilled by lava flows, volcaniclastic screes and turbidites. Recent work (Murton 1986b, Murton & Gass 1986, MacLeod 1988) has shown that much of the LFC also formed part of the transform and is also characterized by primitive mafic volcanism.

In the eastern part of the LFC the transform tectonized zone is 4–5 km wide, and is bounded to the south by a sliver of oceanic crust formed at a ridge on the opposite side of the transform. However, this relatively narrow, well defined zone becomes replaced towards the west by a much wider zone of dominantly transtensional strike–slip tectonism which is distributed across the entire width of the complex (> 10 km). In this area, the transcurrent slip along the transform became markedly transtensional, possibly in response to a change in instantaneous plate motions, giving rise to a localized pull-apart basin and rapid extensional disaggregation of the ridge axis crustal sequence. This was accompanied by uplift of the underlying lithospheric mantle to shallow structural levels and, at the same time, the intrusion of numerous ultramafic and mafic plutons and dykes into the mantle tectonites and disrupted axis sequence crust. These intrusive bodies are associated with the 'transform sequence' lavas (Murton 1986b) which were extruded into the transform trough, and also occur interbedded with lavas of the adjacent axis sequences. The transform sequence extrusives are the most depleted or 'boninitic' of all the lavas reported from Cyprus, and have been the subject of all previous work on boninites in this area (Simonian & Gass 1978, McCulloch & Cameron 1983, Cameron 1985, Murton 1986a,b, Duncan & Green 1987, Flower & Levine 1987). Significantly, the boninitic lavas are closely associated with the development of the extensional transform. Rapid uplift followed by limited melting of the underlying mantle was a direct consequence of localized extension across the transform zone and it is this local geological control that is considered to be most important in the petrogenesis of the boninitic lavas, rather than the effects of any underlying subduction zone.

11.3 Petrography

Primitive boninitic lavas from Cyprus are mostly olivine and olivine–orthopyroxene-phyric, although aphyric flows are not uncommon. In this they are distinct from W Pacific boninites, which have a low-Ca pyroxene-dominated mineralogy. The LFC lavas have a hyalopilitic texture in which groundmass clinopyroxene crystals (±olivine or orthopyroxene) with acicular quench morphologies are set in colourless fresh glass. Plagioclase is not present in the samples analysed but is present in the groundmass of some samples. Alteration in the samples selected is confined to partial replacement of olivine and interstitial glass by a yellow smectite.

Figures 11.1a and b show the tachylitic margin and centre respectively of a typical olivine-phyric pillow lava. Phenocrysts of olivine (Fo_{86-88}) are up to 2 mm long, displaying skeletal 'hopper' morphologies (Donaldson 1976) indicative of rapid cooling. They are present in the quenched pillow margin, but are in equilibrium with the whole rock (Roeder & Emslie 1970), suggesting that the analyses in Table 11.1 can be regarded as liquid compositions.

Olivine–orthopyroxene-phyric lavas contain olivine phenocrysts (Fo_{89}) of similar morphology to the above, with occasional large (up to 1 cm) euhedral xenocrysts ($Fo_{91.5}$) which may show kink banding. These are of similar composition to olivines from residual harzburgite tectonites from the LFC and suggest a near-primary nature for the lavas. The orthopyroxene ($En_{85}Wo_4Fs_{11}$) is most frequently observed forming glomerocrysts of 0.1–0.5 mm (rarely up to 2 mm) blocky, equant crystals associated with clinopyroxene of similar morphology. In some instances they form complex intergrowths (Figs 11.1c & d), with orthopyroxene cores surrounded by clinopyroxene rims. Identical jacketing of orthopyroxene has been observed in boninites from the Mariana forearc (Natland 1981), and of clinoenstatite from Bonin (Komatsu 1980) and Cape Vogel (Dallwitz et al. 1966). This appears to be characteristic of the boninite series, and is probably indicative of rapid disequilibrium crystallization. Also characteristic of the LFC lavas is the crystallization sequence Cr-spinel followed by olivine, orthopyroxene, clinopyroxene and finally plagioclase. This is in contrast to the olivine, plagioclase, clinopyroxene sequence typical of MORB, and may reflect the high SiO_2/Al_2O_3 and low CaO/Al_2O_3 of the LFC parent magmas (Natland 1981).

The transform sequence dykes are clinopyroxene–plagioclase-phyric dolerites with a hypidiomorphic granular groundmass. They have suffered lower greenschist facies metamorphism, with the replacement of groundmass clinopyroxene by epidote and actinolite. The euhedral pyroxene phenocrysts

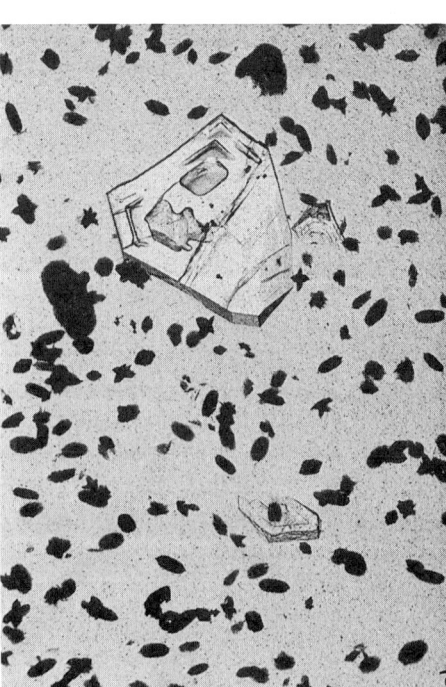

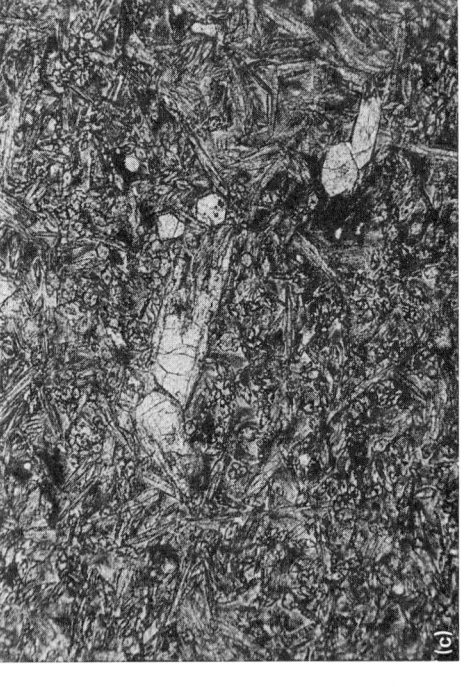

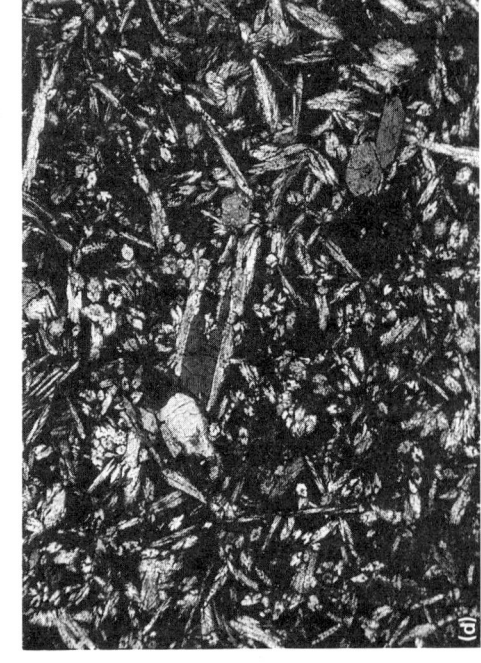

are only partially altered and the plagioclase is fresh. A fuller description of the boninitic dykes is given in Murton (1986b).

11.4 Results

11.4.1 Major and trace elements

Major and trace elements (V, Cr, Ni, Rb, Sr, Y and Zr) were determined by XRF at the Open University using techniques described by Potts *et al.* (1984). Sc, Th, Ta and Hf were analysed using INAA (Potts *et al.* 1985) and the REE by isotope dilution mass spectrometry. Representative analyses of six glassy lavas and two mafic dykes from the LFC are listed in Table 11.1. Other REE data plotted on some of the diagrams are by INAA and are available on request. The lavas are characterized by moderate to high MgO contents and SiO_2 generally $> 50\%$. Mg# values are high (0.65–0.75) as are Cr and Ni abundances (550–1140 and 150–430 ppm respectively), reflecting the relatively primitive, unfractionated nature of the lavas, an interpretation reinforced by the presence and composition of the olivine xenocrysts.

TiO_2 is low in all analyses, although it varies from 0.21 to 0.39%. In Figure 11.2, TiO_2 abundances in samples from identifiable stratigraphic units from the Kalavasos Mines in the eastern part of the LFC are plotted against MgO. Each of the units shows minimal variation in TiO_2 and defines short trends that are consistent with the fractionation of no more than 13% olivine, the dominant phenocryst phase. The greater variation in the picrites, which contain up to 48 modal% olivine, is more in accord with olivine accumulation. However, the trends for the five units are subparallel to one another rather than collinear, implying that they cannot be simply related by fractional crystallization. Rather, each unit represents a discrete magma batch and any link between them must be due to processes operating in their source region and not in a high-level magma chamber.

Al_2O_3 and CaO abundances are not markedly lower than those in most basalts and, while Al_2O_3 is comparable with W Pacific boninites, CaO is higher (9.9–11.7% compared with 4.9–10.8%). The combination of low TiO_2 with

---

Figure 11.1 (a) Tachylitic margin of olivine-phyric pillow: 'hopper' olivines are set in colourless isotropic glass, with subcrystalline spherulites of olivine and clinopyroxene. Sample 85/110; plane-polarized light (PPL); field of view (FOV) 2 mm. (b) Centre of olivine-phyric pillow: slower cooling rates give rise to a hyalopilitic texture in the groundmass, with acicular quench olivine and clinopyroxene microlites, the latter showing the incipient development of a curved branching 'cockscomb' form. Sample 85/110; PPL; FOV 3 mm. (c), (d) Centre of orthopyroxene-phyric pillow: composite phenocryst of orthopyroxene jacketed by clinopyroxene is set in a hyalopilitic matrix. Equant microphenocrysts and acicular microlites and olivine orthopyroxene are clinopyroxene, with minor chrome spinel. Sample 85/117; (c) PPL, (d) crossed polars; FOV 2.5 mm.

Table 11.1 Major- and trace-element concentrations in six lavas and two dykes from the LFC. All samples were crushed in agate to minimize the possibility of trace-element contamination.

Sample	85/29 LII	85/110 LII	84/199a LII	85/117 LIII	84/113 LIII	84/191 LIII	84/15b1 DIII	84/111 DIII
SiO_2	51.71	51.06	50.85	51.84	54.00	49.71	52.58	52.72
TiO_2	0.38	0.39	0.27	0.26	0.21	0.22	0.29	0.22
Al_2O_3	13.40	14.15	13.22	12.92	12.31	14.55	15.22	13.96
Fe_2O_3	8.31	8.21	8.86	8.46	7.90	7.66	9.58	9.32
MnO	0.16	0.15	0.15	0.15	0.22	0.17	0.16	0.16
MgO	9.62	8.51	10.27	9.91	10.70	12.11	6.51	10.41
CaO	10.54	10.55	11.69	11.47	9.94	10.06	8.19	10.75
Na_2O	1.50	1.65	0.62	0.46	0.68	0.41	2.81	0.94
K_2O	0.19	0.13	0.35	0.23	0.50	0.49	2.23	0.17
P_2O_5	0.06	0.08	0.06	0.08	<0.03	<0.03	0.03	<0.03
LOI	4.43	3.90	4.87	4.37	4.07	5.89	2.13	1.55
total	100.30	98.78	101.21	100.15	100.53	101.27	99.73	100.20
Sc	38.3	38.8	35.3	44.2	41.7	52.9	43.2	42.6
V	207	230	211	229	223	236	348	263
Cr	712	559	858	615	709	1137	–	610
Ni	262	157	412	156	254	434	106	199
Rb	6	5	5	7	14	11	12	1
Sr	214	88	88	101	196	240	91	28
Y	10	10	11	8	7	8	15	8
Zr	27	26	22	12	18	8	19	8
Nb	2.8	2.2	1.8	3.0	2.3	<1	5.1	3.5
Th	–	–	0.19	–	0.37	–	0.46	0.29
Ta	0.07	0.07	0.06	0.13	0.12	0.05	0.28	0.19
Hf	0.68	0.61	0.62	0.34	0.42	0.31	0.6	0.25
Ce	3.33	2.81	2.41	0.94	1.85	0.408	2.38	1.48
Nd	2.39	2.22	1.86	0.639	0.897	0.294	1.54	0.69
Sm	0.846	0.84	0.713	0.338	0.334	0.203	0.742	0.283
Eu	0.329	0.329	0.269	0.142	0.128	0.0914	0.252	0.109
Gd	1.29	1.31	–	0.819	0.656	–	1.45	–
Dy	1.52	1.78	1.55	–	1.07	–	2.06	1.36
Er	1.06	1.2	1.13	1.15	0.842	0.507	1.64	1.01
Yb	1.14	1.26	1.23	1.16	1.04	0.634	1.95	1.21

L = lava; D = dyke; II = Group II; III = Group III.

unexceptional CaO and Al_2O_3 leads to high and variable Al_2O_3/TiO_2 and CaO/TiO_2 which fall close to the trends defined by boninites but at slightly higher TiO_2 abundances (Fig. 11.3). Again CaO/TiO_2 are not as extreme as those from the Bonin Islands but they are closely comparable with those for boninites from the Mariana forearc. In addition, the LFC lavas divide naturally into two groups, those with low TiO_2 and high CaO/TiO_2 and a second group with $TiO_2 \sim 0.4\%$ and CaO/TiO_2 between 15 and 25; the latter ratio is similar to MORB and chondrites. This second group also shows some

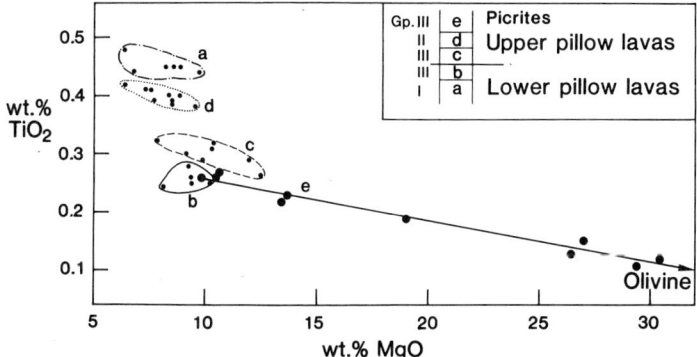

Figure 11.2 MgO–TiO₂ variation diagram for lava units from the Kalavasos Mines in the eastern Limassol Forest Complex. The letters refer to identifiable flow units, with their relative stratigraphic positions shown in the inset diagram. Variation within flow units can be related to fractionation of olivine. Between-unit variation is related to source processes. Units a, b and the picrites are Group III lavas, unit c is Group II and unit d is Group I (after Cameron 1985).

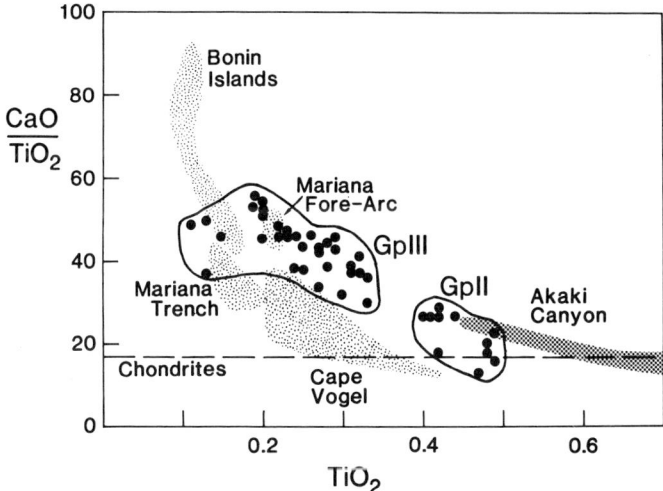

Figure 11.3 Variation of CaO/TiO₂ with TiO₂ for the Limassol Forest lavas in comparison with western Pacific boninites. Data from Hickey & Frey (1982), Umino (1986), Jenner (1981) and Cameron *et al.* (1983). The chondritic CaO/TiO₂ is shown for reference, as is the field for the volcanic glasses from the Akaki Canyon (Rautenschlein *et al.* 1985).

compositional overlap with the volcanic glasses from the Lower Pillow Lavas (LPL) in the Akaki Canyon in the main part of the Troodos Ophiolite, north of the Arakapas Fault Belt (Rautenschlein *et al.* 1985). However, despite this major-element similarity, incompatible-element variations suggest significant differences in their petrogeneses. The low- and high-TiO₂ groups in the LFC lavas correspond to Cameron's (1985) magma groups III and II respectively, and whereas stratigraphic variation of TiO₂ (Fig. 11.2) suggests that a more detailed classification might be more appropriate, it is convenient to conform to the original classification, particularly with reference to trace-element abundances, discussed below. However, it should be emphasized that, apart from TiO_2 the major-element abundances of the two groups are comparable and imply melting of similar source regions under similar $P-T$ conditions.

The above features are clearly comparable with those of the type boninites from the western Pacific arc localities. However, other compositional features, notably SiO_2 contents of $<55\%$ and the consequent dominance of olivine- as opposed to orthopyroxene-controlled fractionation, are distinct from the characteristics of type boninites.

As with all Troodos lavas, LILE are enriched compared with HFSE and REE, resulting in high Sr/Nd, which reflect the general supra-subduction zone environment of the Troodos Ophiolite (Pearce 1980, McCulloch & Cameron 1983). Abundances of the HFSE and REE are low and tend to be considerably lower than those in the LPL of the main part of the Troodos Ophiolite. This is illustrated in Figure 11.4 in which the REE are compared with the Akaki Canyon glasses (Rautenschlein *et al.* 1985), taken as representative of the Troodos LPL. The REE also reflect the division into two distinct geochemical

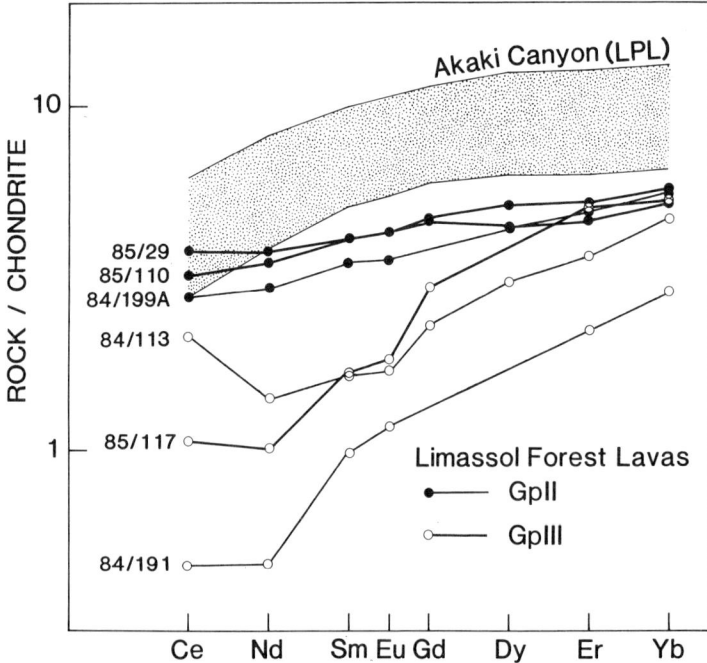

Figure 11.4 Chondrite-normalized abundances of the REE in six glassy LFC lavas. The Group III lavas tend to be characterized by slight up-turns in Ce_N abundances compared with Nd, reflecting the development of 'U-shaped' patterns. Note also that the Group II lavas are less LREE-depleted than the Akaki Canyon glasses despite the generally lower overall abundances.

groups. The Group II, higher-Ti samples are only mildly LREE-depleted, with chondrite-normalized abundances between 3 and 5. Significantly, these lavas are less LREE-depleted than the Akaki Canyon samples. In contrast the Group III lavas are characterized by more strongly LREE-depleted patterns but they also show a tendency towards the development of Ce enrichment over Nd. These variations are in accord with previous results for similar rocks from southern Troodos and trace-element abundances are comparable with those from W Pacific boninites (Hickey & Frey 1982, Cameron *et al.* 1983).

As with TiO_2, much of the variation in trace-element abundances cannot be accounted for by fractionation of phenocryst phases. For example, a plot of Zr versus Y (Fig. 11.5) shows that within the Group III, low-Ti lavas, the total range of Y concentrations (7–10 ppm) is small compared with that of Zr (8–18 ppm) resulting in a variation in Zr/Y from 1 to 2.6. Since both elements are of similar incompatibility in olivine, and olivine fractionation is limited to at most 13%, this enrichment in Zr must be attributed to source-related processes. In addition, although both Y and Zr are more abundant in the

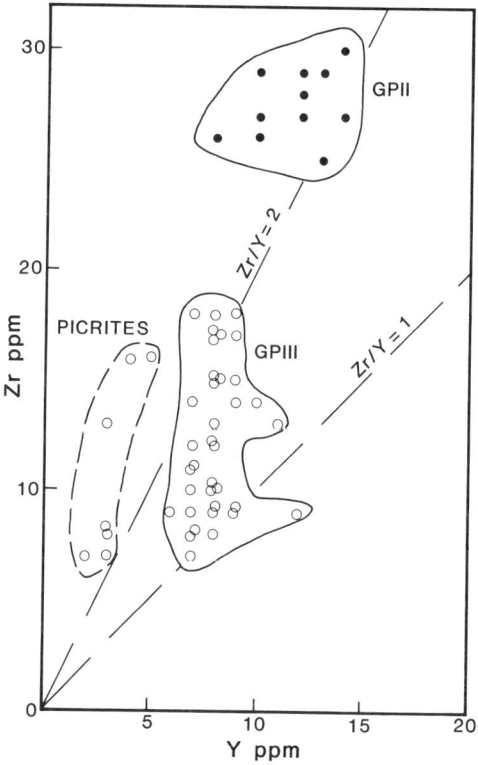

Figure 11.5 Zr–Y variation in the LFC lavas. Note the small variation in Y in the Group III lavas compared with Zr in the same samples.

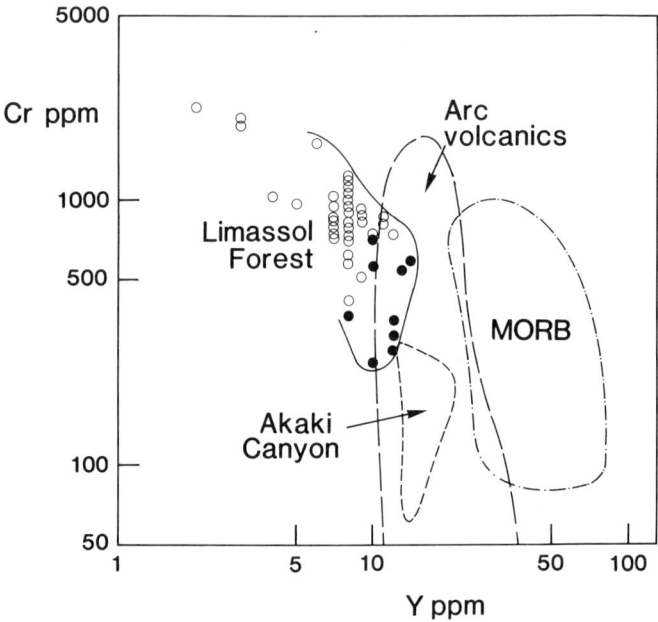

Figure 11.6 Variation of Cr with Y in the LFC lavas compared with the fields for MORB and arc-related volcanics generally. Y values for the Akaki Canyon glasses were estimated from Yb analyses assuming similar chondrite-normalized abundances for the two elements. The LFC lavas tend to plot at lower Y abundances than arc volcanics with only a few Group II samples lying within the lower limits of the arc field.

Group II lavas (Zr = 25–30 ppm, Y = 8–14 ppm), Zr/Y varies from 1.9 to 3.2, and this again must be attributed to source processes.

The low concentrations of Y are further evident in the Cr–Y diagram (Fig. 11.6) in which the Troodos lavas may be compared with MORB and arc-related volcanics. Pearce (1982) first used this diagram to discriminate between oceanic basalts from different tectonic environments noting that those from destructive plate margins plotted at lower Y values than MORB. This difference can be ascribed to a number of possible source-related features; for example, extent of source depletion, variable degrees of melting and the stability of Y-bearing phases during partial melting. Variation in Cr abundance is attributed to olivine and chromite or pyroxene fractionation.

The Limassol Forest lavas define a third field at even lower Y abundances than the arc volcanic field, with the Group III lavas in particular plotting at < 10 ppm compared with > 10 ppm in arc volcanics and > 20 ppm in MORB. Again this variation can be modelled in any of the three ways outlined above. However, if a MORB source is assumed, then the degree of melting required to reproduce the Group III Y abundances becomes unreasonably high ($\geqslant 30\%$). In addition, we agree with Hickey & Frey (1982) that low Ti/Sc and Ti/V argue against residual clinopyroxene, garnet or amphibole, the three

Table 11.2 Nd and Sr isotope analyses of six LFC lavas with corrected values assuming an age of 85 Ma.

Sample	^{143}Nd/^{144}Nd	^{147}Sm/^{144}Nd	(^{143}Nd/^{144}Nd)$_{85}$	$(\varepsilon_{Nd})_{85}$	^{87}Sr/^{86}Sr	^{87}Rb/^{86}Sr	(^{87}Sr/^{86}Sr)$_{85}$
Group II lavas							
84/199a	0.513115 ± 14	0.2313	0.512986	8.9	0.706009 ± 23	0.164	0.705811
85/29	0.512965 ± 17	0.2144	0.512846	6.2	0.706343 ± 16	0.081	0.706245
85/110	0.512995 ± 21	0.2288	0.512868	6.6	0.704134 ± 13	0.164	0.703936
Group III lavas							
84/113	0.512702 ± 10	0.2249	0.512577	0.9	0.707769 ± 19	0.207	0.707519
84/191	0.513028 ± 29	0.4166	0.512796	5.2	–	–	–
85/117	0.512957 ± 21	0.2544	0.512816	5.6	0.707397 ± 30	0.201	0.707155

most likely Y-bearing mantle phases, during magma genesis. Thus, supported by the high CaO/TiO_2 and Al_2O_3/TiO_2, we prefer to interpret Cr–Y variations in terms of a depleted source.

11.4.2 Radiogenic isotopes

The $^{143}Nd/^{144}Nd$ and $^{87}Sr/^{86}Sr$ of the analysed lavas are listed in Table 11.2 together with ε_{Nd} values and $^{87}Sr/^{86}Sr$ calculated at 85 Ma. These are plotted in Figure 11.7 together with data from McCulloch & Cameron (1983). The calculated ε_{Nd} values vary from $+0.9$ up to $+8.9$, slightly extending the range of values reported by McCulloch & Cameron (1983) for mafic Troodos lavas. All analyses lie on or to the right of the 'mantle array' and, whereas there is considerable overlap in both isotope ratios, there is a tendency for Group III lavas to plot at lower ε_{Nd} and higher $^{87}Sr/^{86}Sr$ values than those in Group II. The displacement to higher $^{87}Sr/^{86}Sr$ was interpreted by McCulloch & Cameron (1983) as reflecting the addition of a subduction-related component to the magma source. However, this isotopic displacement in Nd and Sr is not seen in Akaki Canyon glasses despite the strong evidence for the presence of such a component in the trace-element abundances and the Pb isotopes (Rautenschlein et al. 1985).

The broad range in ε_{Nd} in the LFC lavas corresponds to more than half of the total variation shown by present-day oceanic volcanics (Zindler & Hart 1986) and contrasts with the uniformity of the Akaki Canyon data. This

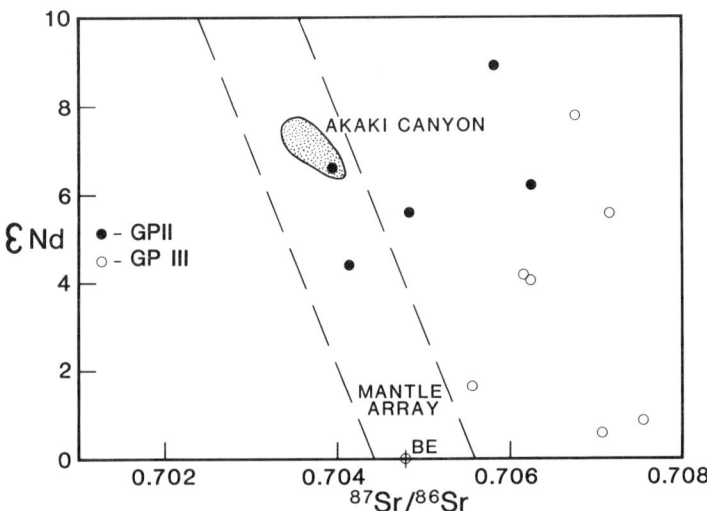

Figure 11.7 ε_{Nd} versus $^{87}Sr/^{86}Sr$ for the LFC lavas at 85 Ma. Included in this diagram are the data from McCulloch & Cameron (1983) and Rautenschlein et al. (1985) for the Akaki Canyon LPL glasses.

isotopic heterogeneity may be a reflection of the primitive, unfractionated nature of the LFC lavas, which appear little modified by magma chamber processes. The presence of olivine xenocrysts referred to above implies that residence times in magma chambers were very short. In contrast the evolved nature of the Akaki Canyon glasses requires a degree of fractional crystallization, and the spatial relation of the LPL to the sheeted dyke complex implies retention, fractionation and homogenization within relatively large, dynamic axial magma chambers. Since turbulent mixing in magma chambers is many orders of magnitude more effective at homogenizing isotopically distinct magmas than is diffusion in partially molten mantle (Kenyon & Turcotte 1987), the difference in degree of isotopic heterogeneity between the LFC lavas and the Akaki Canyon glasses might be related to residence times in magma chambers rather than to a real difference in the degree of isotopic heterogeneity of their respective sources.

11.5 Discussion

Many of the features described above, viz. high MgO, $SiO_2 > 50\%$ and low TiO_2, are consistent with hydrous ($<1\%H_2O$) partial melting of a major-element-depleted source (e.g. Duncan & Green 1987). The incompatible-element and isotopic data, however, clearly imply an enriched and isotopically heterogeneous source. These apparently contradictory features, which are also characteristic of W Pacific boninites, require the operation of contrasting source-related processes either prior to or associated with magma extraction; it is the nature of these processes that will now be discussed in relation to the evidence from trace elements.

In addition to the major-element characteristics, certain trace elements, notably Y and the HREE, also reflect the depleted nature of the sources of both Group II and Group III magmas and this can be more quantitatively investigated using the Cr–Y diagram (Fig. 11.8). On this diagram, melting produces subhorizontal variations as a result of the incompatibility of Y in garnet-free mantle assemblages. In contrast, fractional crystallization of olivine, chromite and clinopyroxene will lead to near-vertical tends due to the compatibility of Cr; hence the orientation of the MORB and arc volcanic fields. The variation in Cr abundance in the LFC lavas emphasizes the low degree of fractional crystallization that they have suffered.

Superimposed on this framework are the loci of a number of different melting models. S_1 represents a MORB source (after Pearce 1982) and L_1 is the locus of the liquids derived from it by 0–20% batch melting. S_2 is the residue after 10% batch melting of S_1 and L_2 the liquid locus for 0–10% melting of this depleted mantle. Similarly S_3 is the residue from 20% melting of S_1 and L_3 the locus of derived batch melts. Also shown on this diagram is the field for the

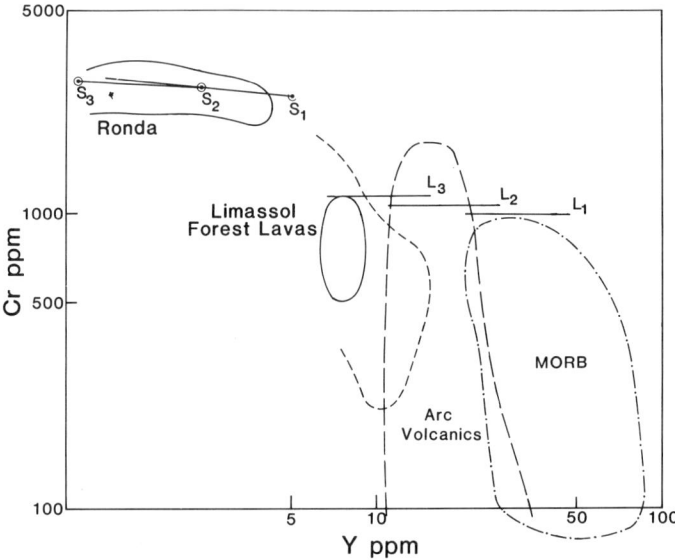

Figure 11.8 Partial melting models for Cr and Y. See text for description of S_1, S_2, S_3, L_1, L_2 and L_3. Delineated areas show the maximum extent of the LFC lava compositions and the area occupied by the majority of the analyses. Field for the Ronda peridotite taken from Frey *et al.* (1985).

Ronda peridotite, which has suffered depletion through basaltic melt extraction (Frey *et al.* 1985). Note the generally good agreement of this field with the predicted abundances of Cr and Y in our model residues.

From these calculations, it can be appreciated that MORB can be derived from its source by up to 20% batch melting followed by olivine and pyroxene fractionation. Arc lavas can also be derived from a MORB source but require a significantly greater degree of melting (>20%). Alternatively, melting the residue, S_2, can also produce an arc basalt in terms of Cr and Y abundances. The merits of these two alternatives have been discussed by Pearce (1982). Extracting the Limassol Forest magmas from a MORB source, however, requires yet higher degrees of melting (30–40%) and it is arguable whether such high melt volumes are feasible from both thermal considerations and in the light of recent models (e.g. McKenzie 1984, 1985) which favour the extraction of small-volume melts before large melt fractions are produced.

A more acceptable model involves remelting mantle residual after MORB and/or island-arc tholeiite extraction, as proposed by Duncan & Green (1987) from experimental evidence. The model illustrated here involves melting peridotite which is residual after 20% melt extraction (i.e. S_3) and clearly shows that Cr and Y abundances in the Group II and Group III lavas can both be successfully modelled by up to 10% melting of such a depleted source followed by a limited amount of olivine or pyroxene fractionation. It is important to emphasize here that the Cr–Y composition of this source is insensitive to the number of preceding batch melting depletion episodes. Thus the model is not inconsistent with the previous extraction of both MORB and the axis sequence basalts from the mantle prior to the final LFC magmatic episode.

The calculations can be extended to include Sc, V and TiO_2 (Fig. 11.9). Also included are the relative positions of L_1, L_2, L_3, S_1, S_2 and S_3 for reference, together with the data fields for the Limassol Forest lavas and the Akaki Canyon volcanic glasses as available. TiO_2 abundances are particularly well reproduced by the model although the high-Ti, Group II lavas can only be modelled by small degrees of melting. In contrast, V appears to be particularly insensitive to the degree of source depletion. Variations in Sc are qualitatively reproduced, the model predicting higher Sc in lavas derived from the more depleted source. This is in accordance with observation, Sc abundances in the

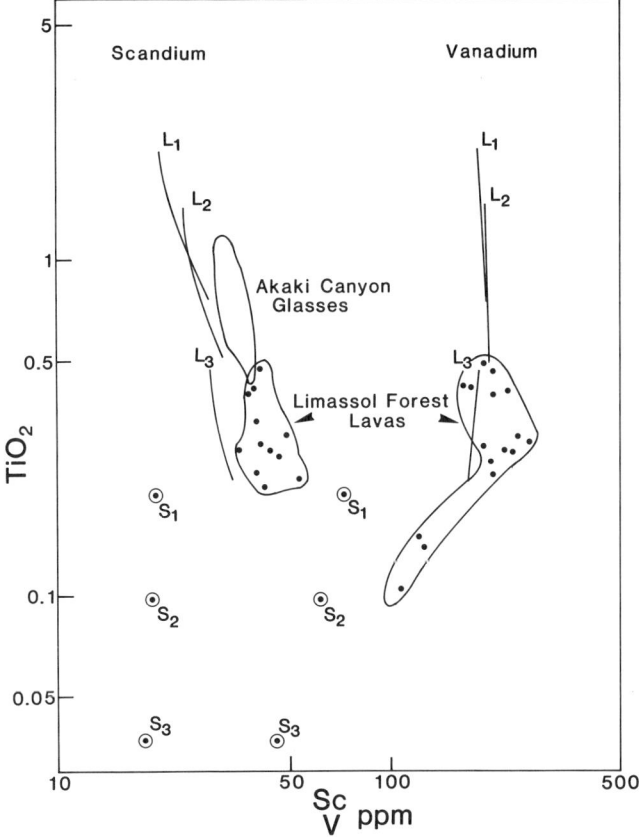

Figure 11.9 Partial melting models extended to TiO_2, Sc and V. Predicted Sc concentrations for the lavas are generally lower than the actual analyses. However, both models and analyses show the higher Sc abundances in the LFC lavas than in the Akaki Canyon glasses. Note the minimal variation in V predicted by the models compared with TiO_2.

Limassol Forest lavas being generally higher than those from the Akaki Canyon glasses. Quantitatively, model Sc concentrations are 10–30% lower than the actual abundances. This discrepancy may arise from poorly constrained orthopyroxene K_D values, the value used in this model (viz 1.1, taken from Frey *et al.* (1978)) being rather high when compared with the range of values given by Irving (1978). If it is reduced to 0.6, within the limits defined by Irving, then the results of the model agree very well with the analytical data. Despite this minor discrepancy, these calculations serve to illustrate that both Group II and Group III magmas were originally derived from sources that were similarly depleted in major elements, TiO_2, Sc, Y and the HREE, and imply that differences between the two groups must therefore be related to contrasting trace-element enrichment styles.

The depletion model described above predicts $(Ce/Yb)_N$ values of $\ll 0.1$ in any second-stage melt and, while observed values in the LFC lavas are low, most tend to be much greater than 0.1. This is particularly so in the high-TiO_2 Group II samples in which $(Ce/Yb)_N \sim 0.6$. Furthermore, even the low-TiO_2 Group III lavas show an enrichment of Ce over Nd and both elements are significantly more abundant in these samples than might be expected from a smooth extrapolation of their chondrite-normalized curves. Thus both magma groups show the effects of trace-element enrichment superimposed on a depleted peridotite mantle source, although the styles of enrichment are quite distinct. Nevertheless the close association of the two magma groups in the field and the similarity of the depleted component common to both imply a similarly close association of the two sources. At issue, therefore, is not only the nature of the enrichment process or processes but, in addition, the development of two contrasting but closely associated magma sources.

Previous studies of boninitic lavas, including those from Cyprus, have used correlations between trace elements and ε_{Nd} to characterize the source of the LREE enrichment. Preferred models involve deriving an added component from a low ε_{Nd} subducted sediment. For example, Hickey & Frey (1982) appealed to mantle metasomatism caused by a fluid derived from either subducted sediment or LREE-enriched mantle. Cameron (1985) and Cameron *et al.* (1983) have also suggested a subduction-related origin for the LREE enrichment, which was ascribed to the migration of a slab-derived hydrous fluid. However, these studies have maintained that LREE enrichment is correlated with an increase in Zr whereas Pearce (1983) suggests that HFSE such as Zr are not mobilized by slab-derived hydrous fluids. Moreover, in other volcanic provinces where sediments have been less ambiguously shown to contribute to the magma source (Rogers *et al.* 1985, 1987), an increase in sediment input does not lead to an increase in HFSE. Furthermore, present models of boninite petrogenesis are unsatisfactory for the LFC in that they only account for the LREE enrichment in the Group III lavas, ignoring their spatial and temporal association with Group II lavas whose trace-element characteristics also require a degree of source enrichment.

Chondrite-normalized abundances of the HFSE and REE are illustrated in Figure 11.10, and further emphasize the separation between the two groups defined on the basis of TiO_2. The high-TiO_2 group has smooth abundance patterns $3-5\times$ chondrite whereas the low-TiO_2 lavas have very low $(Sm/Yb)_N$ but high Ta/Hf and Ta/Sm and define roughly 'U-shaped' normalized curves with minima at Nd. Similar chondrite-normalized curves for the Akaki Canyon glasses are included in Figure 11.10. They define patterns which show a general depletion in the increasingly incompatible elements, although most are more abundant than in the LFC lavas. The exception is Ta, which has comparable abundances in the Group II lavas and is *more* abundant in the Group III samples.

The relatively high abundance of Ta compared with the REE and the association of high Ta concentrations with the 'U-shaped' REE patterns have implications for the origin of the REE enrichment. This can be most conveniently discussed on a diagram of Ce/Yb against Ta/Yb (Fig. 11.11) as originally used by Pearce (1982) to discriminate between MORB and within-plate basalts from those erupted above subduction zones. MORB and within-plate basalts define a positive trend (see Fig. 11.11) due to the concomitant increase of Ce with Ta relative to Yb, a feature most easily understood as resulting from migration of small-volume partial melts within the magma

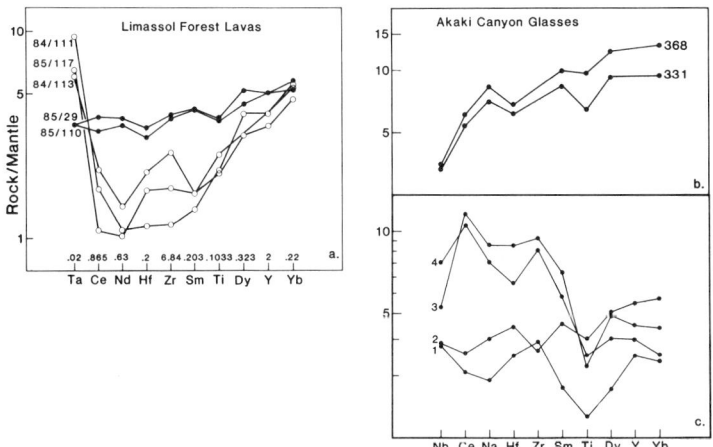

Figure 11.10 (a) Chondritic mantle-normalized trace-element abundance diagrams for the LFC lavas. The division between the Group II and III lavas, originally based on TiO_2 and Zr abundances, is also reflected in the abundances of the REE and Ta. Note the high Ta concentrations in the Group III lavas and the lack of significant negative Ta anomalies in the Group II samples. (b) Normalized trace-element diagram for two Akaki Canyon volcanic glasses (Rautenschlein *et al.* 1985). These have lower Ta/Ce than both LFC lava groups. (c) Normalized trace-element diagram for four boninites from the W Pacific: 1, Bonin Islands; 2, New Zealand; 3, New Caledonia; and 4, Papua New Guinea. All analyses taken from Cameron *et al.* (1983) and correspond to their sample nos. 1, 8, 5 and 3 respectively. Nb normalizing value 0.35 ppm.

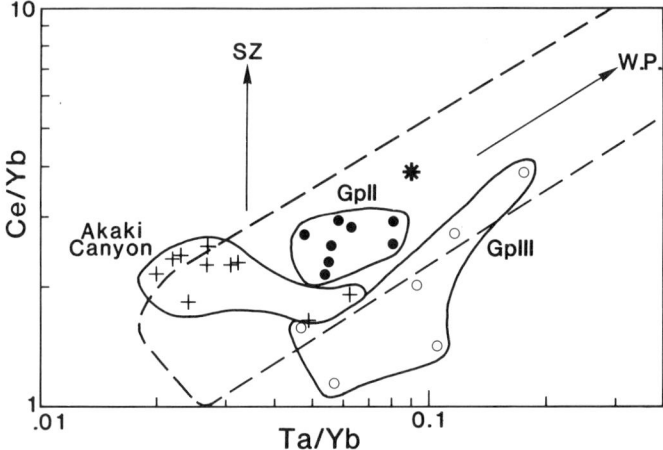

Figure 11.11 Plot of Ce/Yb against Ta/Yb. After Pearce (1982). Basalts from mid-ocean-ridge and within-plate environments define a diagonal trend indicating concomitant enrichment of both Ce and Ta over Yb (arrow marked WP). The lower bound of the MOR-WP field is outlined by the broken curve. In contrast, lavas from subduction-related settings tend to show an enrichment of Ce over both Yb and Ta, producing a vertical displacement, designated by the arrow marked SZ. The Group II and III lavas from the LFC define two separate fields within and below the main trend but significantly do not show the vertical displacement of typical subduction-related trace-element enrichment. The slight upward displacement of the Akaki Canyon LPL field above the main trend at lower Ta/Yb values is more consistent with a subduction-related enrichment style.

source. In contrast, subduction-zone basalts tend to be displaced above this trend to higher Ce/Yb values for a given value of Ta/Yb, resulting in high Ce/Ta, which are considered to reflect element mobilization by fluids or siliceous melts derived from the subducted lithospheric slab. The high-field-strength properties of both Ta and Yb render them immobile relative to the LILE, Th and, possibly, LREE in such an environment.

The most obvious feature of Figure 11.11 is the separation of the LFC and Akaki magma groups into three distinct fields. The majority of the Akaki Canyon glasses show a slight enrichment of Ce over Ta, plotting just above the main trend, and are interpreted as reflecting the typical subduction-zone signature of the axis sequence lavas (Rautenschlein *et al.* 1985). In contrast, both groups of LFC lavas plot at higher Ta/Yb values than the majority of the Akaki samples, approaching and occasionally in excess of the chondritic (bulk Earth) value. This implies that, in terms of these elements at least, the LFC magma source is more enriched than that of the Akaki Canyon glasses, a conclusion in marked contrast to inferences based on variations in CaO/TiO_2. However, neither group is displaced above the MORB − within-plate basalt trend, both groups having lower Ce/Ta than the Akaki Canyon glasses with Ce/Ta in Group II lavas being similar to MORB. More significantly, although

Group III lavas fall only just within or even below the lower bound of the main trend, they generally show a sympathetic increase of Ce/Yb with Ta/Yb. Neither group of LFC lavas shows the vertical displacement that might be expected if trace-element enrichment was the result of subduction-related processes. Rather, the approximately chondritic Ce/Ta values of the Group II lavas and the positive correlation and low Ce/Ta defined by the Group III lavas suggest a trace-element enrichment style more similar to that seen in MORB and within-plate basalts. Thus, in contrast to previous models, we conclude that LREE enrichment in the sources of both Group II and Group III magmas was controlled by the migration of small-volume mafic silicate melts and not by the addition of sediment or other components derived from a subduction zone.

However, the simple addition of a small volume melt to a depleted peridotite matrix is unable to explain adequately the differing incompatible-trace-element abundances in both magma groups, and a more sophisticated model is required if the two groups are to be related to the same process. One such model of direct relevance to this problem has recently been proposed by Navon & Stolper (1987), who describe the interaction between a mantle matrix and a migrating melt as being analogous to an ion-exchange process. As a result of continuous chemical exchange between the migrating melt and the matrix, the trace-element abundances of both can be modified quite markedly and in a way that is distinct from the effects of simple binary mixing. Moreover, this effect is most pronounced for the more incompatible elements and it is these that show the greatest variation in the LFC lavas. Navon and Stolper (1987) illustrate how 'U-shaped' REE profiles can be generated as transient features in depleted peridotite as an LREE-enriched melt passes through and interacts with the mantle matrix. The peridotite is originally characterized by an LREE-depleted pattern which gradually evolves through time to a transient 'U-shape' and finally to a flat profile when the matrix finally reaches equilibrium with the migrating melt. Significantly for the petrogenesis of the LFC lavas, both types of REE patterns can develop as a result of this one process of melt migration. It is suggested therefore that the 'U-shaped' profiles of the Group III lavas developed through limited interaction between migrating melt and depleted peridotite matrix while the more chondritic patterns of the Group II lavas are the result of complete equilibration with the migrating melt.

Such marked deviations from results of simple binary mixing effected by melt percolation have implications for the interpretation of combined isotope and trace-element variations. The model predicts that the source of the Group II lavas has more closely approached equilibrium with the migrating melt than that of the Group III lavas. Since there is a tendency for Group II to have higher ε_{Nd} than Group III (Fig. 11.7), this would imply that the migrating melt must also have been characterized by high ε_{Nd} relative to the depleted peridotite matrix. This is the opposite of inferences based on the simple mixing model (Cameron 1985) in which depleted mantle with $\varepsilon_{Nd} \sim +8$ to $+10$ is infiltrated by an exotic component with $\varepsilon_{Nd} < 0$; the feasibility of the new model should be examined in the light of evidence from elsewhere in the geological record.

LREE-enriched melts with high ε_{Nd} are not uncommon, most alkali basalt lavas from ocean islands being both enriched in LREE and possessing positive ε_{Nd}. Moreover, basaltic rocks from currently active oceanic fracture zones are also LREE-enriched, resembling E-type MORB in their trace-element and isotopic characteristics (Langmuir & Bender 1984). Thus, this aspect of the model is consistent with geological evidence and, more significantly, with the interpretation of the Limassol Forest Complex as part of a 'leaky' transform fault.

Depleted peridotite with $\varepsilon_{Nd} \sim 0$, however, seems intuitively less likely, yet such values have been recorded from both Alpine and ophiolitic peridotites (Richard & Allegre 1980, McCulloch & Cameron 1983, Menzies 1984, Reisberg & Zindler 1987). More specifically, in a detailed study of the Ronda peridotite, Reisberg & Zindler (1987) demonstrated a spread in ε_{Nd} greater than that recorded from all oceanic volcanics so far analysed, values ranging from $+13$ to -8. However, correlations of ε_{Nd} with indices of trace-element enrichment were not apparent. Furthermore, in an equally detailed study of trace-element variations in the same massif, Frey *et al.* (1985) found no evidence for incompatible element enrichment, all trace-element variations being adequately accounted for by depletion through melt extraction. Indeed, we have already demonstrated above how, in terms of Y and HREE, the Ronda peridotite is a good analogue for the depleted component of the LFC magma source. Thus in Ronda we have a well documented example of small- to medium-scale Nd isotopic heterogeneity in depleted mantle material without any clear evidence for REE enrichment, characteristics that closely match the requirements for the peridotite end-member of the proposed model for the LFC magma source.

Despite the apparent lack of REE enrichment, pyroxene separates from Ronda are characterized by high Sr/Nd, suggesting interaction with a hydrous fluid. Indeed, Reisberg & Zindler (1987) have proposed that chemical exchange between the peridotite and fluid could effect the observed isotopic change without net addition of Nd, providing that the fluid was characterized by low ε_{Nd} values. The ubiquitous LILE enrichment characteristic of all Troodos lavas, including those from the LFC, is also attributed to fluid interaction. In this case the fluid is related to subduction and carries the LILE from the subducting slab into the overlying mantle. It is therefore suggested, by analogy with Ronda, that isotopic heterogeneities in depleted oceanic lithosphere developed as a result of a pervasive and regional migration of subduction-related fluids which were characterized by low ε_{Nd} values, probably derived from subducted sediment. Thus although we would agree with previous workers that the low ε_{Nd} in the LFC lavas possibly reflect a component derived from subducted sediment, the latter was not responsible for the development of the 'U-shaped' REE patterns. These developed as a result of the migration of a mafic melt from deeper within the mantle, into the depleted but isotopically heterogeneous oceanic lithosphere (cf. Murton 1986a).

11.6 Possible re-evaluation of Western Pacific boninites?

The preceding discussion emphasizes the importance of melt-related processes in the development of the HFSE and REE abundance patterns of the Limassol Forest lavas. Since the REE abundances of the LFC lavas are comparable with those of W. Pacific boninites, this raises some doubts regarding earlier interpretations of the nature of both processes operating in, and components added to, their sources.

Chondrite-normalized trace-element patterns of four typical boninites from W Pacific localities (Cameron *et al.* 1983) are shown in Figure 11.10, with Nb substituted for geochemically similar Ta. As with the LFC lavas, normalized abundances of the more compatible elements (Ti–Yb) are $1-5\times$ chondrite, reflecting the depleted peridotite source of the boninites and confirming the notion that these lavas represent second-stage mantle melts. In contrast, abundances of the more incompatible elements show much greater variation, demonstrating the chemical diversity of this group of igneous rocks. The two most incompatible-element-enriched samples have Ce/Nb > chondritic values and probably reflect the influence of subduction-related processes in their petrogenesis. The two remaining samples, including one from the Bonin Islands, are less enriched in the incompatible elements but tend to have low Ce/Nb, similar to or slightly greater than chondritic. The New Zealand boninite in particular resembles the Group II lavas from the Limassol Forest. None of the examples, however, shows a marked depletion in Zr and Hf as might be expected for subduction-related magmas. Indeed previous studies have shown that the degree of LREE enrichment, expressed in terms of a percentage of a hypothetical added component, is most clearly correlated with Nb and Zr and shows no significant relation to variation in LILE (Cameron *et al.* 1983). Since enrichment in the latter is most unequivocally characteristic of fluid-related subduction processes, whereas Zr and Nb are generally thought to be more mobile in mafic silicate melts, this suggests that enrichment is melt-related and our conclusions regarding the LFC boninitic lavas may be of some relevance to the interpretation of other boninites.

11.7 Conclusions

High-Mg, high-Si lavas from the LFC, Cyprus, share some compositional characteristics with W Pacific boninites, notably low TiO_2, low incompatible-trace-element abundances, 'U-shaped' REE patterns, and relative enrichment of the LILE compared with both REE and HFSE. However, they are characterized by lower SiO₂ contents than true boninites (50–55%) and this is reflected in fractionation trends controlled by olivine, the dominant phenocryst phase, rather than low-Ca pyroxene. The LFC lavas are therefore more correctly termed boninitic, being too silica-poor to be termed boninites *sensu stricto*.

Major and compatible-trace-element abundances are consistent with derivation as a result of 5–10% melting of a depleted, probably harzburgitic, mantle source. Primitive lava compositions and the presence of forsteritic olivine xenocrysts also imply eruption directly from the mantle source region without retention and homogenization in longlived magma chambers.

Two magmatic groups are recognized on the basis of TiO_2, HFSE and REE abundances, corresponding to the compositional groups II and III of Cameron (1985). Both groups show preferential enrichment of LILE compared with HFSE, reflecting the general supra-subduction zone setting of the Troodos Ophiolite as a whole. Neither group, however, shows enrichment of REE compared with HFSE, Ce/Ta, for example, being comparable with MORB. This implies that REE abundances were not controlled by subduction zone-derived fluids, but by the migration of small-volume mafic melts prior to or during magma generation.

Both Group II and III magmas are isotopically heterogeneous with ε_{Nd} varying from +8.9 to +0.9 and age-corrected $^{87}Sr/^{86}Sr$ from 0.70394 to 0.70752, confirming results from previous investigations of similar lavas from Troodos. This heterogeneity in such primitive lavas is interpreted as reflecting the composition of the mantle source region. However, no simple correlations between isotope and trace-element variations can be recognized, conflicting with previous models in which 'U-shaped' REE patterns are related to the addition of sedimentary material with low ε_{Nd} to the magma source. A new model is presented in which a small-volume, LREE-enriched melt with high ε_{Nd} (~ +8 to +10) percolates through a depleted harzburgite matrix with variable ε_{Nd}. Gradual re-equilibration between matrix and introduced melts produces first 'U-shaped' and subsequently flat normalized trace-element patterns in the harzburgite. The boninitic lavas were produced by partial melting of this variably enriched harzburgite during the rapid uplift accompanying the strike–slip and transtensional tectonics associated with the development of the Limassol Forest Complex.

Acknowledgements

We would like to thank Ian Gass and Julian Pearce for their comments on an earlier version of this chapter and Shen-Su Sun and Tony Crawford for thoughtful reviews. Thanks are also due to John Taylor for the artwork and

Carol Whale for secretarial assistance. C. J. MacLeod and B. J. Murton acknowledge funding by NERC and the Royal Society, and isotope analyses at the Open University are funded by NERC.

References

Anon. 1972. Penrose field conference on ophiolites. *Geotimes* **17**, 24–5.
Bear, L. M. 1960. *The geology and mineral resources of the Akaki Lythrodondha area, Cyprus.* Geol. Surv. Dept Mem. No. 7.
Cameron, W. E., 1985. Petrology and origin of primitive lavas from the Troodos ophiolite, Cyprus. *Contrib. Mineral. Petrol.* **89**, 239–55.
Cameron, W. E., M. T. McCulloch & D. A. Walker 1983. Boninite petrogenesis: chemical and Nd–Sr isotopic constraints. *Earth Planet. Sci. Lett.* **65**, 75–89.
Carr, J. M. & L. M. Bear 1960. *The geology and mineral resources of the Peristerona–Lagoudhera area.* Cyprus Geol. Surv. Dept Mem. no. 2.
Dallwitz, W. B., D. H. Green & J. E. Thompson 1966. Clinoenstatite in a volcanic rock from the Cape Vogel area, Papua. *J. Petrol.* **7**, 375–403.
Donaldson, C. H. 1976. An experimental investigation of olivine morphology. *Contrib. Mineral. Petrol.* **57**, 183–213.
Duncan, R. A. & D. H. Green 1980. Role of multi-stage melting in the formation of oceanic crust. *Geology* **8**, 22–6.
Duncan, R. A. & D. H. Green 1987. The genesis of refractory melts in the formation of oceanic crust. *Contrib. Mineral. Petrol.* **96**, 326–42.
Flower, M. F. J. & H. M. Levine 1987. Petrogenesis of a tholeiite–boninite sequence from Ayios Mamas, Troodos ophiolite: evidence for splitting of a volcanic arc? *Contrib. Mineral. Petrol.* **97**, 509–24.
Frey, F. A., D. H. Green & S. D. Roy 1978. Integrated models of basalt petrogenesis: a study of quartz tholeiites to olivine melilitites from south eastern Australia utilising geochemical and experimental petrological data. *J. Petrol.* **19**, 463–513.
Frey, F. A., C. J. Suen & H. W. Stockman 1985. The Ronda high temperature peridotite: geochemistry and petrogenesis. *Geochim. Cosmochim. Acta* **49**, 2469–91.
Gass, I. G. 1960 *The geology and mineral resources of the Dali area.* Cyprus Geol. Surv. Dept Mem. no. 4.
Gass, I. G. & J. D. Smewing 1973. Intrusion, extrusion and metamorphism at constructive margins: evidence from the Troodos Massif, Cyprus. *Nature* **243**, 26–9.
Gass, I. G., C. R. Neary, J. Plant, A. H. F. Robertson, K. O. Simonian, J. D. Smewing, E. T. C. Spooner & R. A. M. Wilson 1975. Comments on 'The Troodos ophilitic complex was probably formed in an island arc' by A. Miyashiro and subsequent correspondence by A. Hynes and A. Miyashiro. *Earth Planet. Sci. Lett.* **25**, 236–8.
Hickey, R. L. & F. A. Frey 1982. Geochemical implications of boninite series volcanics: implications for their source. *Geochim. Cosmochim. Acta* **46**, 2099–115.
Irving, A. J. 1978. A review of experimental studies of crystal/liquid trace element partitioning. *Geochim. Cosmochim. Acta* **42**, 743–70.
Jenner, G. A. 1981. Geochemistry of high-Mg andesites from Cape Vogel, Papua New Guinea. *Chem. Geol.* **33**, 307–32.
Kenyon, P. M. & D. L. Turcotte 1987. Along-strike magma mixing beneath mid-ocean ridges: effects on isotope ratios. *Earth Planet. Sci. Lett.* **84**, 393–405.
Komatsu, M. 1980. Clinoenstatite in volcanic rocks from the Bonin Islands. *Contrib. Mineral. Petrol.* **74**, 329–38.

Langmuir, C. H. & J. F. Bender 1984. The geochemistry of oceanic basalts in the vicinity of transform faults: observations and implications. *Earth Planet. Sci. Lett.* **69**, 107–27.

McCulloch, M. T. and W. E. Cameron 1983. Nd–Sr isotopic study of primitive lavas from the Troodos ophiolite, Cyprus: evidence for a subduction-related setting. *Geology* **11**, 727–31.

McKenzie, D. P. 1984. The generation and compaction of partially molten rock. *J. Petrol.* **25**, 715–65.

McKenzie, D. P. 1985. The extraction of magma from the crust and mantle. *Earth Planet. Sci. Lett.* **74**, 81–91.

MacLeod, C. J. 1988. Upper Cretaceous evolution of the eastern Limassol Forest Complex: the role of the Southern Troodos Transform Fault in the rotation of the Cyprus microplate. In *Troodos '87; Ophiolites and Oceanic Lithosphere*. Geol. Survey Dept. Nicosia, Cyprus.

Menzies, M. A. 1984. Chemical and isotopic heterogeneities in orogenic and ophiolitic peridotites. In *Ophiolites and oceanic lithosphere*, I. G. Gass, S. J. Lippard & A. W. Shelton (eds), 231–40. Geol. Soc. Lond. Spec. Publ. no. 13.

Moores, E. M. & F. J. Vine 1971. The Troodos Massif, Cyprus and other ophiolites as oceanic crust: evaluation and implications. *Phil. Trans. R. Soc. Lond. A* **268**, 443–66.

Murton, B. J. 1986a. Anomalous oceanic lithosphere formed in a leaky transform fault: evidence from the Western Limassol Forest complex, Cyprus. *J. Geol. Soc. Lond.* **143**, 845–54.

Murton, B. J. 1986b. The tectonic evolution of the western Limassol Forest Complex, Cyprus. Ph.D. Thesis, Open University.

Murton, B. J. & I. G. Gass 1986. Western Limassol Forest Complex, Cyprus. Part of an Upper Cretaceous leaky transform fault. *Geology* **14**, 255–8.

Natland, J. H. 1981. Crystal morphologies and pyroxene compositions in boninites and tholeiitic basalts from Deep Sea Drilling Project holes 458 and 459B in the Mariana fore-arc region. Init. Rep. DSDP Leg 60, 681–707.

Navon, O. & E. Stolper 1987. Geochemical consequences of melt percolation: the upper mantle as a chromatographic column. *J. Geol.* **95**, 285–307.

Pearce, J. A. 1980. Geochemical evidence for the genesis and eruptive setting of lavas from Tethyan ophiolites. In *Proc. Int. Ophiolite Symp.*, A. Panayiotou (ed.), 261–72. Cyprus.

Pearce J. A. 1982. Trace element characteristics of lavas from destructive plate boundaries. In *Andesites*, R. S. Thorpe (ed.), 525–47. New York: Wiley.

Pearce, J. A. 1983. Role of the sub-continental lithosphere in magma genesis of active continental margins. In *Continental basalts and mantle xenoliths*, C. J. Hawkesworth & M. J. Norry (eds). Nantwich: Shiva.

Potts, P. J., P. C. Webb & J. S. Watson, 1984. Energy dispersive x-ray fluorescence analysis of silicate rocks for major and trace elements. *X-Ray Spectr.* **13**, 123–44.

Potts, P. J., O. Williams-Thorpe, M. C. Isaacs & D. W. Wright 1985. High-precision instrumental neutron activation analysis of geological samples employing simultaneous counting with both planar and coaxial detectors. *Chem. Geol.* **48**, 145–55.

Rautenschlein, M., G. A. Jenner, J. Hertogen, A. W. Hofmann, R. Kerrich, H. U. Schminke & W. M. White 1985. Isotopic and trace element composition of volcanic glasses from the Akaki Canyon, Cyprus: implications for the origin of the Troodos ophiolite. *Earth Planet. Sci. Lett.* **75**, 369–83.

Reisberg, L. & A. Zindler 1987. Extreme isotopic variations in the upper mantle: evidence from Ronda. *Earth Planet. Sci. Lett.* **81**, 29–45.

Richard, P. & C. J. Allegre 1980. Nd and Sr isotope study of ophiolite and orogenic lherzolite petrogenesis. *Earth Planet. Sci. Lett.* **47**, 65–74.

Robinson, P. T., W. G. Melson, T. O'Hearn & H.-U. Schminke 1983. Volcanic glass compositions of the Troodos ophiolite, Cyprus. *Geology* **11**, 400–4.

Roeder, P. L. & R. F. Emslie 1970. Olivine–liquid equilibrium. *Contrib. Mineral. Petrol.* **29**, 275–89.

Rogers, N. W., C. J. Hawkesworth, D. P. Mattey & R. S. Harmon 1987. Sediment subduction and the source of orogenic leucitites. *Geology* **15**, 451–3.

Rogers, N. W., C. J. Hawkesworth, R. J. Parker & J. S. Marsh 1985. The geochemistry of potassic lavas from Vulsini, central Italy and implications for mantle enrichment processes beneath the Roman region. *Contrib. Mineral. Petrol.* **90**, 224–57.

Simonian K. O. & I. G. Gass 1978. Arakapas fault belt, Cyprus: a fossil transform fault. *Geol. Soc. Am. Bull.* **89**, 1220–30.

Sun, S.-S. & R. W. Nesbitt 1978. Geochemical regularities and genetic significance of ophiolitic basalts. *Geology* **6**, 689–93.

Umino, S. 1986. Magma mixing in boninite sequence of Chichijima, Bonin Islands. *J. Volcanol. Geotherm. Res.* **29**, 125–57.

Wilson, R. A. M. 1959. *The geology of the Xeros–Troodos area, Cyprus.* Cyprus Geol. Surv. Dept Mem. no. 1.

Zindler, A. & S. R. Hart 1986. Chemical geodynamics. *Annu. Rev. Earth Planet. Sci.* **14**, 493–571.

12 Contrasting boninite–tholeiite associations from New Caledonia

W.E. CAMERON

Abstract

In New Caledonia there are two boninite–tholeiite sequences, one of Permo-Triassic age which outcrops near Koh in the centre of the island, and a second outcropping northwest of Nepoui on the west coast which is of early to mid-Tertiary age. Both are probably parts of dismembered ophiolites. The Koh suite has been metamorphosed in the greenschist facies and consists of evolved tholeiitic basalts and andesites ($\varepsilon_{Nd} \sim +8$) interbedded with and intruding LREE-depleted boninites with ε_{Nd} values between $+2.7$ and $+5.5$. One tholeiitic andesite shows extreme LREE and Zr depletion, has a ε_{Nd} of $+6.7$ and provides a possible link with boninites. The Nepoui boninites are extremely fresh and overlie tholeiitic basalts with prehnite–pumpellyite to lower greenschist facies mineralogy. The boninites are olivine-phyric and enriched in LREE, SiO_2, Na_2O, K_2O and Zr but severely depleted in CaO, Ti, V, Sc, Nb and P. They have low initial $^{87}Sr/^{86}Sr$ (0.7034–0.7035) and unusually high ε_{Nd} values ($+8$ to $+10$). This implies that the Nepoui boninites bear no petrogenetic relationship to the tholeiitic basalts ($\varepsilon_{Nd} = +3$ to $+4.6$). They are more enriched in LILE at a given Mg# than Cape Vogel boninites.

Both sets of tholeiites probably formed in backarc basins at an early stage of their development. The only geochemical characteristics they share with arc tholeiites found in association with boninites in the Mariana forearc and at Cape Vogel, PNG, are very low Nb concentrations and, in the case of the west coast tholeiites, low initial $^{143}Nd/^{144}Nd$.

12.1 Introduction

Phanerozoic boninites are often associated with tholeiitic basalts or tholeiitic andesites. Major clues to the tectonic settings in which boninites may be generated are given by the geochemistry of the tholeiitic rocks, which vary from island-arc type (e.g. Mariana forearc; Wood *et al.* 1981) to backarc-basin type (e.g. Cambrian of Victoria; Crawford & Keays 1987). This 'supra-subduction'-zone setting is well accepted (Sun & Nesbitt 1978) and allows the generation of boninite at any stage in the early development of backarc basins (e.g. Crawford *et al.* 1981).

The island of New Caledonia in the SW Pacific has two boninite suites. The younger boninites are remarkably fresh and overlie the altered 'Formation des Basaltes' which extends along the west coast of the island. Boninite outcrops are very limited, however, and occur along the western margin of the peridotite Kopeto Massif near Route 1 between Nepoui and Pouembout, at about 21°15'S, 164°55'E. Avias (1977) gave chemical analyses of these rocks and they were later recognized as boninites by Sameshima *et al.* (1983) and Cameron *et al.* (1983). As yet, the boninites have not been dated, but they are clearly younger than the bulk of the Formation des Basaltes (Palaeocene or possibly Cretaceous; Paris 1981) and thought to be older than the early Eocene emplacement age of the peridotite sheet which once covered most of the island.

The older boninite–tholeiite sequence occurs near the centre of New Caledonia, south of the village of Koh ('Massif de Koh', 21°33'S, 165°50'E), and was recognized by the present author from thin sections in the Geology Department, University of Auckland. Analyses 4 and 8 in Table 5 of Paris (1981) provided confirmation. This chapter presents chemical and isotopic data on both suites and compares them with other relevant occurrences, particularly that at nearby Cape Vogel, Papua New Guinea (Dallwitz *et al.* 1966, Jenner 1981).

12.2 Massif de Koh

12.2.1 Geology

Metaboninites interbedded with tholeiitic metabasalts and andesites occur on the NE margin of the southernmost outcrop of the Palaeozoic greenschist 'core' of the island. The contact is one of a series of NW-trending major faults which contain discontinuous serpentinite bodies along their length. The other two massifs, Tarouimba–Sphinx and Cantaloupaï (Fig. 12.1), are structurally similar and all three consist of gabbros, dolerites, basalts and minor acid rocks, some of which host small Cu deposits (Paris, 1981).

An isolated fourth outcrop near Pacquereux consists mainly of extrusives. It is possible that they form part of a Permian or Triassic ophiolite which is covered on the northern side by Triassic–Jurassic sediments.

This field study was limited to the excellent exposures along the portion of the new road from Sarraméa to Canala on the Canala–La Foa 1:50 000 sheet

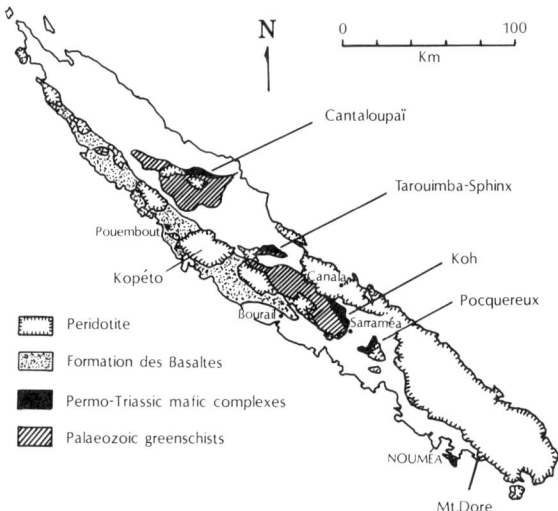

Figure 12.1 Geological setting of the units relevant to this chapter, New Caledonia. Most of the remainder of the island consists of Permian and Mesozoic sediments and metasediments.

(Fig. 12.2). The terrain is rugged and heavily timbered, with outcrop limited in the forests to huge float blocks, up to 2 m across, choking the creeks. Boninites can be identified easily by their pale-green colour on fractured surfaces and yellow weathering rinds, whereas the tholeiites are grey-to-black and weather various shades of brown. Both occur mainly as pillows.

Tholeiitic dolerite dykes are fairly common, particularly at the southern end of the section, and in several places cut boninites. The pillows are the right way up and the sequence appears to young to the NE. Near the top are acid tuffs, cherts, andesite and dacite. From the road section, it could be concluded that boninites and tholeiitic basalts–dolerites are in roughly equal proportions, with about 5% of andesite and dacite. In many places faults separate rock types but sufficient exposures provide evidence that they interfinger.

12.2.2 Petrography

The volcanic rocks are metamorphosed to lower greenschist facies. Igneous textures are well preserved but only spinel and clinopyroxene remain unaltered. Dolerites are generally of prehnite–pumpellyite grade and tend to have some unalbitized plagioclase. Boninites were originally glassy with a quench-textured groundmass and phenocrysts (typically 0.5 mm across) of olivine and/or orthopyroxene, now pseudomorphed by serpentine or quartz and chlorite respectively. There is no compelling evidence for the former presence of clinoenstatite. Mineral compositions were determined on the CAMECA electron microprobe housed in the Research School of Earth Sciences, ANU.

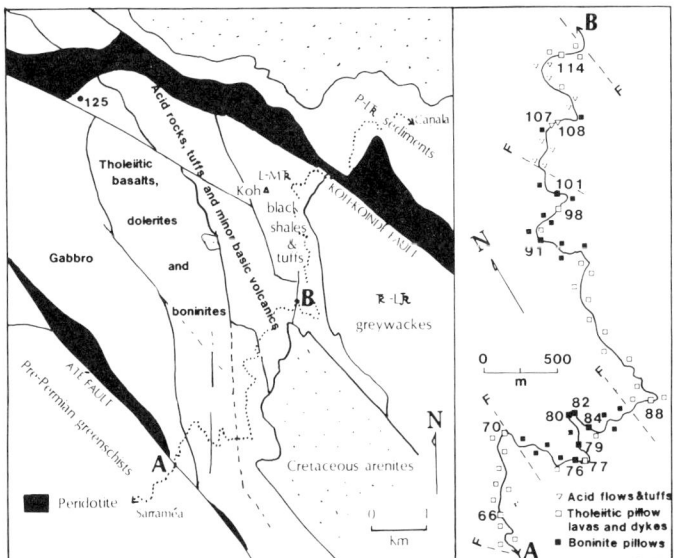

Figure 12.2 Locality maps for tholeiites and boninites from the Koh Massif, central New Caledonia. Base geology from Paris (1981).

Augite is typically acicular and ranges in composition from $En_{51}Wo_{32}Fs_{17}$ to $En_{46}Wo_{38}Fs_{16}$ with Al_2O_3 contents between 1.5 and 4%. In about half the samples, augite is pseudomorphed by actinolite. Magnesiochromite occurs as inclusions in former olivine or more commonly, as discrete grains in the groundmass. Spinel Cr# span the narrow range 0.90–0.84, Mg# = 0.63–0.45. Original glass has been replaced by quartz + albite and some or all of the secondary minerals pumpellyite, quartz and smectite.

Tholeiites also show perfect textural preservation. Most are aphyric, non-vesicular, augite–plagioclase rocks, but some samples also contain pseudomorphs of chlorite after groundmass orthopyroxene. The least-altered dolerites have fresh plagioclase feldspar and brown hornblende rims to augite (Table 12.1, sample 98). Clinopyroxene compositions are less magnesian than those in boninites, ranging from $En_{47}Wo_{38}Fs_{15}$ to $En_{38}Wo_{35}Fs_{27}$. Some pillow lavas have relict augite but plagioclase is altered to albite or hydrogrossular. Olivine pseudomorphs are very rare; in fact, many dolerites and thick flows have interstitial quartz which, from its texture, is of primary origin. The freshest gabbro sample, from a tributary of the Ouen Ouaï River, 4 km WNW of Koh, has olivine (Fo_{63}) and two pyroxenes and, because of its evolved nature, has affinity with the tholeiites.

Dacites and high-SiO_2 andesites are porphyritic rocks which weather a reddish or pale-green colour. Andesites have phenocrysts of plagioclase feldspar, augite and altered orthopyroxene, whereas dacites contain plagio-

Table 12.1 Microprobe analyses of selected minerals from New Caledonian boninites.

	Spinel	Olivine		Orthopyroxene		Hornblende	
Sample	Groundmass 173	Phenocryst 173	Skeletal phenocryst 184	Phenocryst 173	Resorbed phenocryst 184[a]	Rims augite 98	Quench groundmass 155
SiO_2	–	40.92	40.29	56.62	55.67	44.14	40.46
TiO_2	0.19	–	–	0.04	–	2.02	0.64
Al_2O_3	7.06	–	0.04	0.86	0.38	8.79	15.19
Cr_2O_3	61.78	0.14	0.12	0.65	0.14	–	–
FeO	15.59	6.27	9.77	7.19	13.20	18.14	23.31
MnO	0.14	0.10	0.15	0.18	0.23	0.37	0.20
NiO	0.09	0.52	0.31	0.04	–	–	–
MgO	13.61	51.84	49.15	33.66	28.94	11.87	6.89
CaO	–	0.07	0.07	0.92	0.89	10.17	8.78
Na_2O	–	–	–	–	–	2.23	1.33
K_2O	–	–	–	–	–	0.22	0.09
total	98.46	99.86	99.90	100.15	99.45	97.95	96.89
Mg #	0.671	0.936	0.900	0.892	0.796	0.538	0.345

[a]This analysis by TPD energy-dispersive microprobe operated at 15 kV, 3 nA. Other analyses by CAMECA microprobe operated at 25 kV, 50–60 nA and long counting times.

clase, quartz and hornblende phenocrysts in order of abundance. The fine-grained groundmass in acid rocks consists of albitized, acicular plagioclase and quartz.

12.2.3 Chemistry

Sixteen representative and least-altered samples were chosen for chemical analysis by X-ray fluorescence (XRF) and instrumental neutron activation (INAA) at the Department of Geology, Australian National University. The results are presented in Table 12.2 along with petrographic details. Samples 79 through to 76 have the high MgO, Ni and Cr and low TiO_2 typical of boninites. Silica varies considerably from 52 to 57% and is clearly mobile under the prevailing metamorphic conditions; Na, K, Ba, Rb and Sr would be expected to be mobile also. Some rocks, for example sample 91, show extensive albitization of glass or plagioclase which serves to reduce their CaO contents. Original CaO values were probably between 8 and 10%. Ti/Zr cluster between 68 and 90, with one value at 112.

The three boninite samples have distinctly LREE-depleted patterns (Fig. 12.3). Each is U-shaped, with minima in the MREE for samples 82 and 101. Sample 91, with the high Ti/Zr, has a significantly greater HREE content

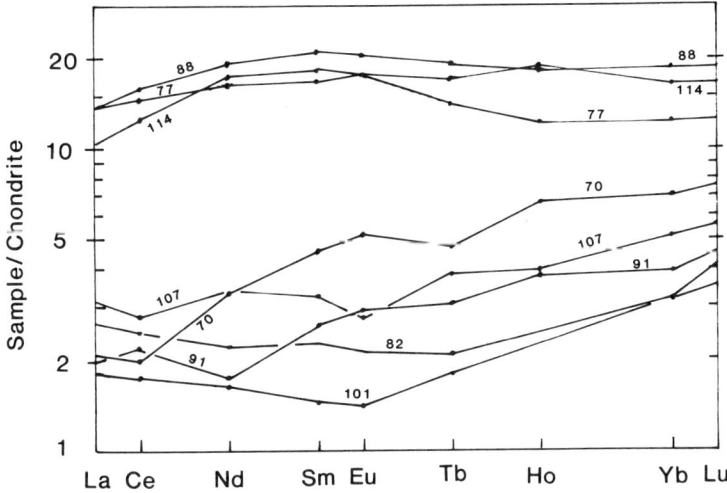

Figure 12.3 Chondrite-normalized REE patterns for Koh boninites (samples 82, 91 and 101), tholeiites (70, 77, 88 and 114) and dacite (107). Normalization values are those of Taylor and Gorton (1977).

and the minimum at Nd. These patterns fall within the spread found in Type III boninites from Cyprus (Cameron 1985).

The analysed basalts are relatively evolved, with Mg # (calculated assuming mol $Fe^{2+}/(Fe^{2+} + Fe^{3+}) = 0.10$) between 0.56 and 0.31. The Fe-enrichment trend is typically tholeiitic. Five samples have Ti/Zr between 88 and 117 and two are very much higher (152, 185). The Cr, Ni and Nb contents of these rocks are extremely low. REE patterns for three of the samples analysed are broadly similar (Fig. 12.3, samples 77, 88 and 114): all are slightly LREE-depleted and reach a maximum chondrite-normalized concentration in the MREE. Sample 70 is completely different, with LREE in the order of 2× chondritic and $(La/Yb)_N - 0.3$. Petrographically, this rock is a tholeiitic basalt, but its very low Ti, P and Zr, and $(La/Yb)_N$ are most unusual. Compositionally similar lavas are known from Guam (sample 79-6; Hickey-Vargas & Reagan 1987), from Dmitry Mendeleev Dredge Site 1438 in the forearc near the intersection of the Yap and Mariana trenches (Crawford *et al.* 1986) and sample 123-95-1 from the Lau Basin (Hawkins & Melchior 1985).

The andesite and dacite are both relatively magnesian and have high Cr and low P, Zr and Nb. The REE profile of the andesite, sample 107 (Fig. 12.3), parallels that of the two boninites 82 and 101, implying that it is derived from a boninite parent by fractionation of plagioclase feldspar (small Eu anomaly), olivine and pyroxene.

Table 12.2 Major- and trace-element analyses of metaigneous rocks from the Koh Massif, New Caledonia.

	Boninites pillow lavas								Andesite 107	Dacite 108	Tholeiites							
	79	84	101	82	80	91	76				flow 70	flow 66	dyke 77	dyke 98	pillow 114	dyke 125	pillow 88
SiO_2	50.44	52.91	49.89	54.48	55.09	53.10	58.05		62.33	68.27	52.38	53.10	50.26	50.34	55.21	54.07	49.54
TiO_2	0.08	0.14	0.15	0.16	0.15	0.28	0.16		0.23	0.16	0.40	0.82	1.50	1.90	1.64	2.25	1.99
Al_2O_3	7.14	8.54	9.50	9.59	10.00	11.38	9.36		12.76	11.89	14.79	14.77	15.28	14.42	13.27	12.63	14.80
Fe_2O_3	2.98	2.30	2.75	3.88	4.48	2.08	2.48		3.06	3.58	3.14	3.22	5.04	4.89	5.67	6.68	7.73
FeO	5.84	6.38	7.04	4.75	3.56	6.98	4.89		4.71	1.87	5.56	6.51	6.93	7.43	5.77	6.54	5.97
MnO	0.18	0.17	0.18	0.15	0.15	0.15	0.17		0.16	0.07	0.15	0.15	0.18	0.20	0.20	0.18	0.16
MgO	17.46	14.92	14.54	12.01	10.54	11.32	8.76		4.26	2.91	6.83	6.83	4.37	4.39	3.90	4.00	3.22
CaO	8.88	7.53	7.74	8.36	8.90	6.13	10.31		4.63	3.67	7.09	5.66	7.02	7.79	5.80	5.23	8.31
Na_2O	0.56	1.76	2.31	3.23	3.10	4.00	2.59		3.85	4.18	4.47	3.34	4.51	3.61	4.42	4.96	3.88
K_2O	0.07	0.17	0.05	0.23	0.07	0.07	0.01		0.05	0.22	0.45	1.86	0.62	1.31	0.18	0.31	0.49
P_2O_5	nd	nd	nd	nd	nd	nd	nd		0.02	0.02	0.02	0.11	0.25	0.33	0.13	0.19	0.14
S	nd	nd	nd	nd	nd	nd	nd		nd	nd	nd	nd	0.18	0.14	nd	nd	nd
H_2O^+	4.82	3.95	4.23	2.08	3.08	3.40	2.47		3.30	2.36	3.49	3.07	3.01	2.60	2.94	2.33	3.10
H_2O^-	1.02	0.70	1.20	0.88	0.43	0.55	0.33		0.53	0.34	0.45	0.30	0.54	0.31	0.56	0.47	0.30
CO_2	0.35	0.21	0.09	0.15	0.13	0.10	0.16		0.12	0.34	0.29	0.01	0.39	0.32	0.08	0.02	0.11
total	99.80	99.68	99.67	99.95	99.68	99.54	99.74		100.01	99.88	99.51	99.75	99.99	99.91	99.77	99.86	99.74
Ba	5	15	15	20	10	10	10		25	95	35	50	80	185	25	25	25
Rb	nd	2	1	3	nd	1	nd		0.5	6	7	21	9	22	4	6	10
Sr	9	14	29	33	20	75	10		32	52	102	75	192	247	84	72	110
Zr	7	10	10	13	11	15	13		14	26	14	52	59	130	84	136	109
Nb	nd	nd	nd	nd	nd	nd	nd		nd	1	nd	1	1	1	1	1	2
Y	2	5	4	5	4	6	4		7	9	13	25	23	31	31	41	35

Sc		42	55	47		49		42		39	38	38
V	176	193	208	229	212	240	218	264	109	233	275	383	394	372	423
Cr	1752	1217	1299	827	697	730	477	83	109	33	26	27	10	43	17
Ni	335	381	345	220	168	123	114	50	34	39	11	18	12	19	10
Cu	28	191	29	5	18	nd	67	59	1	53	53	.03	35	1	29
Zn	64	72	100	38	52	78	66	80	58	81	113	23	116	130	104
La		0.79	0.57	0.83		0.62		0.97		0.67	4.33		3.16		4.19
Ce		1.75	1.4	2.0		1.8		2.2		1.6	11.8		10.3		13.0
Nd		1.1	1.0	1.3		1.0		2.2		2.0	10.0		10.3		11.7
Sm		0.42	0.28	0.45		0.51		0.62		0.90	3.25		3.56		4.09
Eu		0.15	0.10	0.15		0.21		0.20		0.37	1.28		1.26		1.49
Tb		0.11	0.09	0.10		0.15		0.19		0.23	0.68		0.84		0.94
Ho		0.17				0.28		0.29		0.49	0.89		1.38		1.34
Yb		0.62	0.66	0.66		0.81		1.05		1.45	2.55		3.43		3.87
Lu		0.14	0.11	0.14		0.15		0.18		0.24	0.40		0.53		0.60
100Mg#	78.5	75.9	73.2	72.2	71.2	69.5	68.6	50.4	49.5	59.2	40.4	39.8	39.0	36.2	30.7
Ti/Zr	68	84	90	74	82	112	74	98	37	185	152	88	117	99	110

Petrography

phenocrysts	(Ol),	(Ol),	(Ol),	(Opx)	(Opx)	(Ol)	(Opx)	(Pl), A,	(Pl, H), Q			
	(Opx)	(Opx)	(Opx)			(Opx), (A)		(Opx)		A, (Pl)	A, (Pl)	Pl, A,	Pl, (A)	(Pl), A	(Pl), A
groundmass	A	A	A	(A)	(A, Pl)	(A)	(Opx), A					H, (Opx)	H, (Opx)		

Chemical analyses are by XRF (analyst B. W. Chappell) except for Na$_2$O by flame photometry and H$_2$O and CO$_2$ gravimetrically (analyst J. Wasik).
nd = Not detected. Detection limits for P$_2$O$_5$ 0.01%, S 0.02%, Nb 1ppm.
Mg# = mol Mg/(Mg + Fe^{2+}) where mol Fe^{3+}/(Fe^{2+} + Fe^{3+}) is set to 0.10.
Petrographic abbreviations: Ol = olivine; A = augite; Opx = orthopyroxene; Pl = plagioclase; H = hornblende; Q = quartz.
Phenocrysts are listed in order of abundance.
Pseudomorphs are in parentheses.

12.3 Formation des Basaltes

12.3.1 Geology

The basalts of the west coast of New Caledonia outcrop continuously for 200 km (Fig. 12.1) and sporadically between Bourail and Mount Dore. They are typically pillowed, metamorphosed to prehnite–pumpellyite or lower greenschist facies and are significantly less metasomatized than the sequence near Koh. Paris (1981) summarized previous work on the petrography and age of the basalts. He considered, on the basis of major elements, that they showed tholeiitic as well as calc-alkaline characteristics and ascribed them to a backarc-basin setting. A late Eocene compressional phase closed the basin and caused obduction of the harzburgitic oceanic mantle. Samples for this study were collected mainly from road cuttings or quarries (Fig. 12.4), as most surface outcrops are extensively weathered.

Boninites have so far been found only in small outcrops at or close to low-angle serpentinite thrust sheets near Nepoui. Nowhere can they be seen directly to overlie basalts, but that relationship can be inferred from the topography of the area west of the Rivière Rouge quarry. They are black, glassy rocks and occur as coarse-grained, clinoenstatite-phyric dykes or thick flows, fine-grained flows and pillow lavas. The only obvious alteration product is a yellow smectite filling vesicles, and the zeolite harmitome (identified by

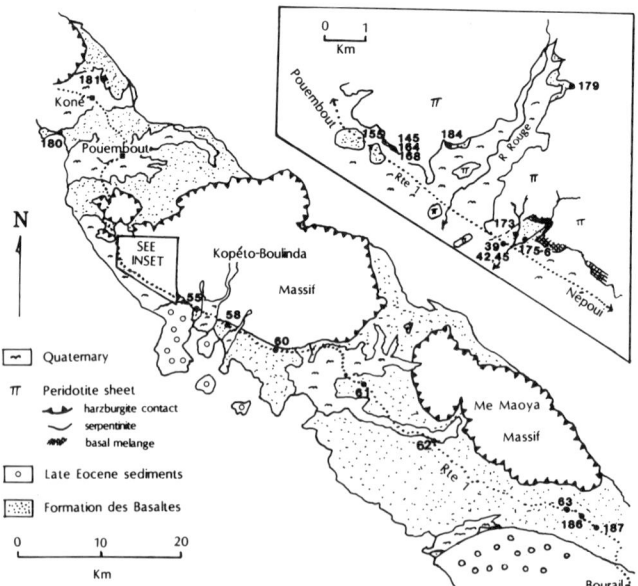

Figure 12.4 Locality map for Nepoui boninites and nearby tholeiitic basalt and gabbro samples from the Formation des Basaltes. Regional geology from Paris (1981).

XRD). Both of these minerals were avoided as far as possible in selecting the samples for chemical analysis. In the outcrops so far mentioned there is no melange at the base of the thrust sheet suggesting rehydration and later movement. East of the Rivière Blanche this melange is present and is well exposed in some of the creeks draining south. The main clasts in the melange are metasediments, porphyroblastic hornblende schists and fine-grained metabasalts.

An extraordinary feature of the boninites is that they are so fresh. Immediately below the serpentinite contacts in the quarry, in the road cutting just west of Rivière Blanche and from the location where sample 184 was taken (Fig. 12.4), the rocks are virtually unaltered, whereas a few hundred metres west of the quarry, the basalts are mainly smectite. Small outcrops of altered boninite do occur away from the contact near the road cutting; these have talc pseudomorphs after clinoenstatite and primary plagioclase feldspar (sample 45). Below the fresh harzburgite contact at locality 179, tholeiitic basalt shows the static style of low greenschist facies metamorphism which affects much of the Formation des Basaltes: augite is partially pseudomorphed by actinolite and plagioclase is still relatively unaltered (Fig. 12.5a). The only comparable metamorphism in boninites is exemplified by sample 42, which has orthopyroxene replaced by chlorite, albitized plagioclase and minor epidote along with primary interstitial quartz against the serpentinite contact. A metre further away, the boninites are fresh.

12.3.2 Petrography

The basalts are typically aphyric, with olivine (rare) and plagioclase microphenocrysts. Like the Koh tholeiites, they are augite–plagioclase rocks but show no evidence of former orthopyroxene. Augite is less commonly pseudomorphed by actinolite and plagioclase is only partially altered, in this case to clay minerals rather than albite. High-level intrusives collected nearer Bourail (samples 60 and 63) have conspicuous magnetite, augite, plagioclase and pseudomorphs after suspected pigeonite. Small amounts of primary brown hornblende and quartz are present in sample 63. General petrographic comments about the basalts are given by Rodgers (1975) and Paris (1981), and the samples analysed in this study have their petrography summarized in Table 12.3.

Boninites are of three main petrographic types. The first has olivine and orthopyroxene phenocrysts (samples 173, 168 and 176) or microphenocrysts (samples 164 and 52, the latter from Cameron *et al.* (1983)) and rare clinoenstatite phenocrysts. The most primitive (highest Mg#) rock is 173, with olivines up to 7 mm across ($Fo_{93.7-90}$, Table 12.1), much less abundant orthopyroxene phenocrysts (Mg# = 0.89–0.88) of slightly smaller size and clinoenstatite phenocrysts up to 1 mm long (Mg# = 0.90–0.89). Microphenocrysts of orthopyroxene in all five rocks range to Mg# = 0.86, and for olivine,

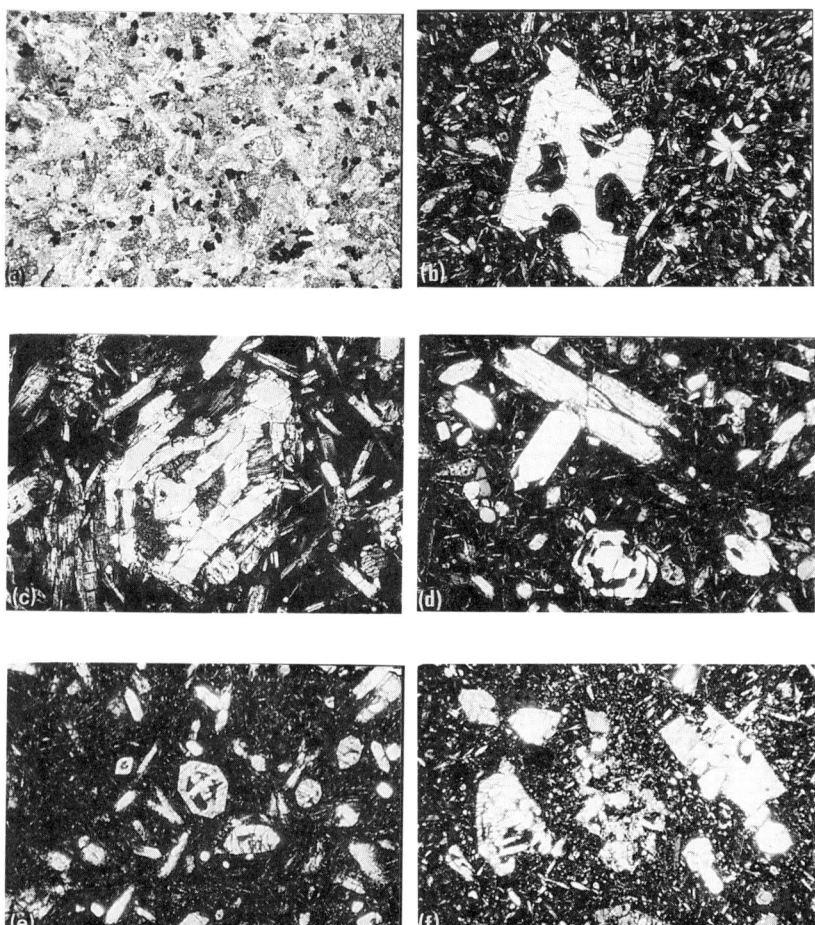

Figure 12.5 Textures in a tholeiitic basalt and boninites from the Formation des Basaltes: (a) Tholeiitic basalt, sample 179. Single olivine pseudomorph, augite partially altered to actinolite, partially altered plagioclase feldspar and relatively abundant Fe-Ti oxide. Plane-polarized light. Width of field 3.0 mm. (b) Partially resorbed olivine ($Fo_{90.5}$) in boninite sample 173. The groundmass consists of clinoenstatite, sometimes in tiny rosettes, and orthopyroxene set in clear glass (black in this photograph under crossed polars). Width of field 3.0 mm. (c) Olivine being replaced by clinoenstatite, boninite 173. The process of resorption has been crystallographically controlled, giving rise to an apparently skeletal olivine. This 'lantern' morphology is usually associated with rapid growth. Crossed polars. Width of field 0.8 mm. (d) Olivine–clinoenstatite boninite, sample 184. Olivine has partly reacted and clinoenstatite is growing (note skeletal terminations on the crystals; cf. coarser-grained rock figured on the front cover of this book). The amounts of olivine and clinoenstatite microphenocrysts are approximately equal. Crossed polars. Width of field 2.0 mm. (e) Skeletal groundmass clinoenstatite, sample 184. The 'framed cross' shape is similar to that found in skeletal orthopyroxene. Other minerals in the groundmass are olivine (fractured) and orthopyroxene (acicular grains). Crossed polars. Width of field 1.5 mm. (f) Partially disaggregated orthopyroxenite xenolith (centre), surrounded by olivine phenocrysts in various stages of resorption. Sample 173, crossed polars. Width of field 7.5 mm.

0.88. The groundmass consists of orthopyroxene and clinoenstatite laths, usually no larger than 0.1×0.3 mm, set in glass.

All olivine crystals show some evidence of resorption, either rounding effects, embayments (Fig. 12.5b) or preferential resorption along dome faces (Fig. 12.5c), resulting in skeletal 'lantern' morphologies (Fig. 12.5c). These skeletal olivines (Fo_{92-90} and essentially unzoned) do not have the delicate terminations expected from very rapid growth as is found, for example, in komatiites. Larger, partially resorbed olivines especially in samples 173 and 168 do, however, have abundant melt inclusions of similar dimensions to those found in skeletal phenocrysts of composition Fo_{92-90} in a komatiitic basalt (Fig. 5 of Nisbet et al. 1987). This generation of olivine has grown rapidly in a magma chamber and then has been slightly resorbed. It is not the most magnesian; the one phenocryst analysed with a composition more Mg-rich than Fo_{93} has three euhedral faces. Few olivines are in contact with pyroxene. Those which are have orthopyroxene or clinoenstatite moulded around part of them (Fig. 12.5c).

The second petrographic type (samples 184, 155 and 49 from Cameron et al. (1983)) has olivine and clinoenstatite microphenocrysts, typically less than 1 mm long, with groundmass orthopyroxene and clinoenstatite (Fig. 12.5d). Most of the olivine microphenocrysts ($Fo_{90.7-89.5}$) are of the 'skeletal', partially resorbed variety (Table 12.1). The proportion of olivine phenocrysts is slightly less than that of clinoenstatite (Mg# $= 0.90$–0.88) and in neither case do they exceed 10 modal% of the rock. The coexistence of olivine and clinoenstatite as discrete crystals in boninites is supported by 1 atm experiments of Tsuchiyama (1986). If the interfacial energy between olivine and protoenstatite is large and material transport fast in the magma, pyroxene will nucleate away from reacting olivines, rather than mantling them.

The third petrographic type is that described by Sameshima et al. (1983) and Campiglio et al. (1986) and shown on the cover of this book (sample 39). Their texture is dominated by coarse-grained clinoenstatite crystals (Mg# $= 0.905$–0.885) which, surprisingly, sometimes show skeletal terminations (front cover). Their general morphology is indicative of rapid growth, although not as rapid as that which gives rise to lantern-shaped groundmass clinoenstatites with compositions as Fe-rich as Mg# $= 0.865$ (Fig. 12.5e). Olivine ($Fo_{90.5-88}$) is equidimensional, rarely exceeding 0.5 mm across, and is susceptible to alteration to smectite, although not significantly in the two samples (39, 145) chosen for analysis. It forms separate crystals or, more rarely, inclusions in clinoenstatite phenocrysts. The holocrystalline rock mentioned previously, sample 45, has altered clinoenstatite. Sample 175 has a significant proportion of orthopyroxene phenocrysts.

Magnesiochromite occurs as inclusions in olivine or, rarely, as tiny groundmass euhedra. It has a relatively uniform composition across the petrographic types and lies in the range Cr# $= 0.88$–0.84, Mg# $= 0.62$–0.50 (Table 12.1), similar to chromites in the Koh boninites. Chromium contents of

Table 12.3 Major- and trace-element analyses of igneous rocks from the Formation des Basaltes, west coast of New Caledonia.

	Boninites										
	flow 173	dyke? 45	flow? 145	flow? 39	flow 184	pillow 168	pillow 164	flow 155	flow 175	flow 176	contact 42
SiO_2	55.02	56.98	56.67	57.23	57.80	57.44	57.10	57.23	58.00	58.73	62.14
TiO_2	0.21	0.21	0.22	0.23	0.28	0.26	0.26	0.20	0.25	0.30	0.28
Al_2O_3	8.65	8.90	8.92	8.96	10.03	9.48	9.88	9.99	10.60	10.54	11.39
Fe_2O_3	1.04	0.98	1.80	2.20	1.82	1.75	2.15	1.26	1.67	2.12	2.13
FeO	5.70	5.73	5.80	5.41	5.44	5.80	5.06	6.31	5.51	5.23	3.55
MnO	0.12	0.12	0.13	0.15	0.12	0.13	0.12	0.13	0.13	0.13	0.12
MgO	18.79	16.60	15.39	14.30	13.03	13.50	12.87	13.08	11.49	10.79	8.57
CaO	3.88	3.26	3.85	3.80	3.90	3.85	3.65	4.70	4.38	4.07	3.55
Na_2O	1.81	1.68	1.81	1.54	2.16	2.19	2.26	1.93	2.46	2.61	3.42
K_2O	0.59	0.43	0.50	0.53	0.95	0.84	0.70	0.61	0.59	0.57	0.23
P_2O_5	0.04	0.03	0.04	0.04	0.06	0.05	0.05	0.04	0.05	0.06	0.05
S	nd	nd	nd	nd	nd	nd	nd	nd	nd	nd	nd
H_2O^+	3.07	4.12	3.51	4.02	3.76	3.31	4.36	3.34	3.82	4.01	3.48
H_2O^-	0.29	0.61	1.01	1.53	0.55	0.52	1.89	0.70	0.98	0.59	0.88
CO_2	0.07	0.06	0.08	0.11	0.08	0.08	0.02	0.06	0.05	0.09	0.13
total	99.28	99.71	99.73	100.03	99.98	99.20	100.37	99.58	99.98	99.84	99.92
Ba	36	25	28	24	42	42	40	30	38	56	45
Rb	8.5	2	11	11	13	11	13	9	10	13	3
Sr	81	105	96	95	123	110	122	101	143	134	121
Zr	58	61	62	60	77	70	74	60	71	77	88
Nb	1.0	2	1.5	0.5	1.5	1.5	1.5	0.5	1.5	1.5	2
Y	8	8	8	9	11	9	10	7	8	11	12
Sc	18		23	23	21	22	21	25	23	22	
V	83	89	97	99	95	99	98	119	108	105	90
Cr	1350	1172	1210	1250	935	1080	1010	1040	855	675	451
Ni	680	397	355	330	310	290	291	295	300	220	165
Cu	17	1	19		18	19	16	23	20	18	18
Zn	55	62	63	67	58	60	60	64	63	61	46
La	4.92				4.22	5.85	5.20				
Ce	8.6				8.6	10.3	9.1				
Nd	4.5				4.8	5.9	4.5				
Sm	1.14				1.28	1.43	1.28				
Eu	0.37				0.41	0.48	0.41				
Gd	1.22				1.33	1.66	1.55				
Tb	0.20				0.24	0.26	0.24				
Ho	0.30				0.28	0.35	0.32				
Yb	1.09				0.98	1.30	1.25				
Lu	0.19				0.16	0.22	0.19				
100Mg #	84.9	83.2	80.4	79.3	78.5	78.4	78.4	77.7	75.1	75.0	73.6
Ti/Zr	22	21	21	23	22	22	21	20	21	23	19
Petrography											
phenocrysts	Ol, Opx	(Cen)	Cen, Ol	Cen, Ol	Cen ≈ Ol, Opx	Ol	Cen, Ol	Cen, Opx (Ol)	Ol, Opx	(Opx)	

Petrographic abbreviations: Cen = clinoenstatite; others as Table 12.2.

Table 12.3 (*continued*)

	Tholeiites										
	pillow 62	pillow 55	flow 181	pillow 58	dyke 187	flow 186	flow 61	pillow 180	flow 179	gabbro 60	gabbro 63
SiO_2	48.45	46.84	47.37	48.32	47.74	47.98	49.44	47.89	47.44	48.01	48.55
TiO_2	1.30	1.18	1.41	1.30	1.24	1.38	1.77	1.37	1.71	1.99	2.93
Al_2O_3	14.02	14.61	14.16	13.75	14.10	13.60	14.53	13.59	13.48	13.26	12.08
Fe_2O_3	3.41	2.86	3.02	4.28	3.03	3.51	6.31	4.35	4.38	4.27	6.23
FeO	7.22	7.76	7.97	7.09	8.11	7.87	3.97	6.66	8.62	9.90	10.81
MnO	0.26	0.17	0.26	0.20	0.18	0.17	0.12	0.30	0.40	0.22	0.32
MgO	8.06	7.79	8.04	7.96	7.52	7.43	6.34	6.88	7.51	5.85	4.48
CaO	10.10	10.54	10.74	9.94	11.91	9.70	10.92	9.64	8.77	9.19	7.69
Na_2O	2.49	2.05	2.58	2.44	2.43	3.81	3.12	3.44	2.86	3.37	3.18
K_2O	0.38	0.38	0.42	0.72	0.20	0.05	0.33	0.95	1.09	0.13	0.18
P_2O_5	0.09	0.11	0.12	0.07	0.09	0.11	0.16	0.12	0.14	0.18	0.27
S	nd	0.07	0.04	nd	0.10	0.08	nd	nd	0.07	0.13	0.23
H_2O^+	2.91	3.34	3.05	2.53	2.61	3.35	1.69	3.19	2.79	2.61	2.39
H_2O^-	0.85	1.94	0.40	0.92	0.38	0.48	1.90	0.84	0.85	0.49	0.35
CO_2	0.02	0.23	0.17	0.15	0.20	0.20	0.13	0.77	0.28	0.34	0.51
total	99.56	99.84	99.73	99.67	99.79	99.68	100.73	99.99	100.36	99.93	100.09
Ba	75	60	98	150	40	58	34	82	135	50	55
Rb	4	8	3	5	3	1	10	6	7	2	2
Sr	139	193	180	130	138	178	161	112	118	174	159
Zr	63	71	77	57	56	74	103	78	88	110	185
Nb	3	4	4	3	4	5	5	5	4	5	9
Y	20	21	22	18	21	23	30	23	28	34	48
Sc			44		44	45	48	45	47		
V	279	237	300	291	327	347	367	306	371	358	393
Cr	346	295	285	283	175	180	105	281	116	22	10
Ni	99	110	87	91	102	90	79	76	69	44	11
Cu	299	87	219	134	152	192	41	10	486	157	31
Zn	67	146	68	96	91	75	93	65	111	107	124
La			3.86						4.59		
Ce		12.2	11.1						12.4		
Nd		9.9	8.1						10.2		
Sm		2.79	2.99						3.58		
Eu		1.02	1.18						1.40		
Gd			4.03						4.71		
Tb		0.60	0.63						0.77		
Ho		0.99	0.88						1.02		
Yb		1.89	2.35						3.20		
Lu		0.28	0.37						0.49		
100Mg #	58.3	57.3	57.2	56.5	55.3	54.6	53.9	53.7	51.6	43.2	32.7
Ti/Zr	124	100	110	137	133	112	103	105	117	108	95

Petrography
phenocrysts   Pl, (Ol) Pl, (Ol)   Pl

olivine phenocrysts were measured at 25 keV and beam currents between 50 and 80 nA. Olivines more magnesian than Fo_{90} have Cr_2O_3 contents between 0.09 and 0.14%, but the variation is not completely systematic with $Mg\#$. Cape Vogel boninitic olivines have significantly higher Cr contents (0.27% Cr_2O_3 in Fo_{94} phenocrysts; Walker & Cameron 1983), and Fo_{94} olivines in Cambrian boninite cumulates in Victoria have up to 0.33% Cr_2O_3 (Crawford 1980).

In all Nepoui boninites, magnesian orthopyroxene phenocrysts are unzoned and show no sign of resorption. A few rocks, however, contain partially dissolved phenocrysts of more Fe-rich orthopyroxene ($Mg\# = 0.80-0.78$, Table 12.1) found clumped together as a 3 mm xenolith in sample 173 (Fig. 12.5f); these probably result from magma mixing where hot boninitic liquid is injected into the floor of a magma chamber and ultimately mixes with overlying fractionated liquid. Under these circumstances olivine may grow rapidly into the psuedoskeletal shapes seen in some of the larger phenocrysts.

The most magnesian clinoenstatite phenocryst is about 3 mol% more Fe-rich than the most magnesian olivine. Sample 41F from Cape Vogel has a similar difference in $Mg\#$ between olivine phenocrysts (Fo_{94}) and clinoenstatite ($Mg\# = 0.92$) but the latter are rimmed by clinoenstatite of composition $Mg\# = 0.94$. In Nepoui boninite 173, groundmass orthopyroxene is often rimmed by clinoenstatite and there are few clinoenstatite phenocrysts. The most primitive boninites appear to be the second petrographic type, in which olivine is only slightly resorbed and skeletal groundmass protoenstatite is forming. The third type may be typical of ponded flows in which olivine has almost completely reacted and the extensive growth of protoenstatite slowed. Orthopyroxene in olivine-orthopyroxene boninites is not sufficiently magnesian to have been in equilibrium with mantle harzburgite and may result from magma mixing.

12.3.3 Chemistry

Eleven new boninite analyses are presented in Table 12.3. When silica is plotted against MgO content, a reasonably smooth trend results for the unaltered samples (Fig. 12.6). Olivine control lines have been drawn through the points for samples 173 and 39 which reasonably represent the lower and upper bounds of silica content for a given MgO. Excess clinoenstatite does not affect the position of the points markedly, and in fact there is no petrographic evidence for it in samples 39 or 145. The SiO_2–MgO plot demonstrates the chemical homogeneity of the group in spite of petrographic differences, and the likelihood of olivine fractionation from a parental magma similar to that postulated for Cape Vogel. A few low-Ti samples from Cape Vogel plot within the field in Figure 12.6 (e.g. no. 51 of Walker and Cameron 1983), but most have 1–2% lower SiO_2 at equivalent MgO value. CaO contents are the lowest, and Na_2O and K_2O the highest of all boninite suites recorded to date. H_2O^+

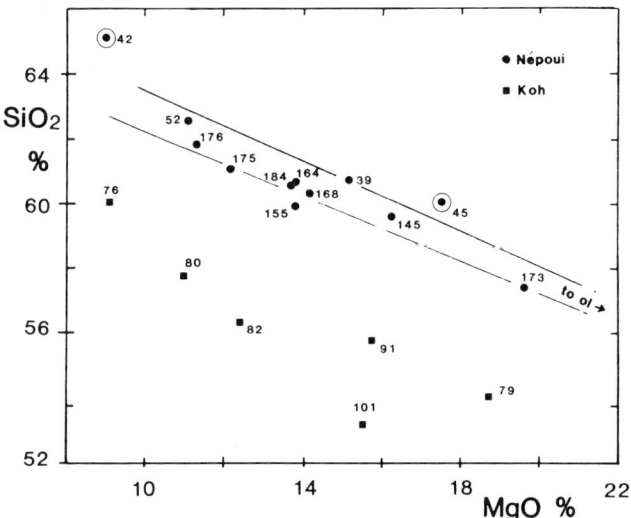

Figure 12.6 Variation of silica with MgO content (recalculated anhydrous) for the Koh and Nepoui boninites. Circled samples (42, 45) are altered. Lines define olivine control (Fo_{93}) for samples 39 and 173.

analyses are uniformly high and result from hydration of the dacitic glass (Campiglio *et al.* 1986) which remains deceptively isotropic under crossed polars while having exchanged oxygen and hydrogen with sea water (Kyser *et al.* 1986).

The trace-element geochemistry of these boninites is also unusual. They have high Zr and the lowest Sc among known boninites (Fig. 12.7). Type B Cambrian boninites from the Heathcote area and one sample from Howqua in Victoria have almost as low Sc values but much higher CaO (Crawford & Cameron 1985, Nelson *et al.* 1984). Koh boninites, on the other hand, are rich in Sc and the Cape Vogel boninites lie between the two extremes. Ti/Zr average 22 compared with 83 for Koh boninites and are distinguishable from a low-Ti, high-SiO_2 group of samples from Cape Vogel (D. A. Walker, unpubl. data) by higher overall Ti and Zr contents. Comparison with the altered samples 45 and 42 shows that Ba, Rb, Sr, Na and K are mobile, although they appear to show relatively coherent trends with MgO content in the other samples.

Nepoui boninites are LREE-enriched (Fig. 12.8). The profiles could be viewed as U-shaped, with the minimum between Tb and Yb, as is seen in some Cape Vogel samples (Jenner 1981) but their total REE abundances are much higher. Spark-source data for samples 49 and 52 agree well with INAA results reported here, except for Yb where there is a discrepancy of up to 20%.

Basalts and gabbros sampled near the boninite outcrops (Fig. 12.4) show a trend towards Fe enrichment and are obviously tholeiitic. They are more magnesian than those at Koh and contain correspondingly more Ni and Cr; Nb

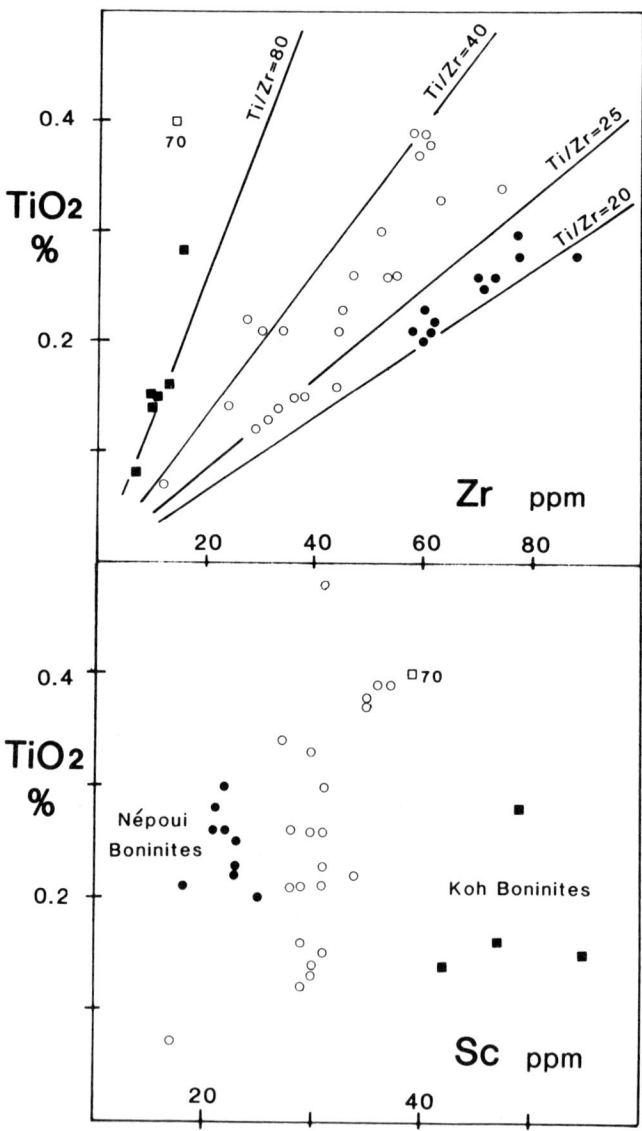

Figure 12.7 Variation of Zr and Sc with TiO$_2$ content in Nepoui and Koh boninites (symbols as for Figure 12.6) compared with Cape Vogel boninites (open circles; D. A. Walker unpubl. data).

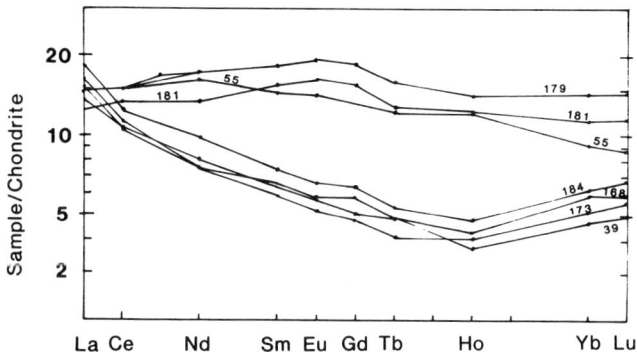

Figure 12.8 Chondrite-normalized REE patterns for Nepoui boninites (samples 39, 168, 173 and 184) and tholeiitic basalts from the Formation des Basaltes (55, 179 and 181).

concentrations are also greater. The less evolved nature of the Formation des Basaltes may be a consequence of sampling too small an area but it is significant that 20 of the 23 tholeiite analyses in Table 10a of Paris (1981) have MgO > 6%, or Mg# > 0.5. High water and Fe^{3+} contents imply extensive alteration but only K, Ba, Rb and to a certain extent Sr appear to be mobile. Ti/Zr vary from 95 to 137 and Ti/V lie in the narrow range 27–33, within that of the Koh tholeiites, whose Ti/V range from 21 to 36.

REE patterns (Fig. 12.8) are virtually identical to those from Koh tholeiites (Fig. 12.3) and those obtained from two metabasalts from the Montagne des Sources, 15 km north of Mount Dore, where cumulates overlie harzburgite tectonite (Dupuy *et al.* 1981). A third sample (176) has 0.5% TiO_2 and is similarly LREE-depleted as sample 70 from Koh. A reasonable conclusion would be that the Formation des Basaltes represents the extrusive component of the ophiolite, and forms a thrust sheet immediately below the metamorphics. The allochthonous nature of the basalts is supported by the variety of rock types with which they are in contact (1:200 000 Geology Map, Paris, 1981).

12.4 Sr–Nd isotopes and tectonic setting

12.4.1 Tholeiites

$^{143}Nd/^{144}Nd$ and $^{87}Sr/^{86}Sr$ initial ratios were determined at the Research School of Earth Sciences, ANU, for the Koh and West Coast tholeiites, assuming ages of 250 and 60 Ma respectively (Table 12.4). Normal Koh tholeiites have $\varepsilon_{Nd} \sim +8$, and sample 70 has $\varepsilon_{Nd} = +6.7$, in contrast with the Tertiary tholeiites which have ε_{Nd} lying between $+3.0$ and $+4.6$. Initial Sr values have been affected by the mobility of Rb and Sr.

Table 12.4 Sr and Nd isotopic data.

Sample	$(^{87}Sr/^{86}Sr)_i$	Sm/Nd	$(^{143}Nd/^{144}Nd)_0$	$\varepsilon_{Nd(T)}$
Koh				
boninites				
82	0.70493 ± 5	0.346	0.51207 ± 3	+4.2
91	0.70434 ± 6	0.510	0.51230 ± 3	+5.5
101	0.70516 ± 3	0.280	0.51193 ± 3	+2.7
andesite				
107	0.70541 ± 4	0.282	0.51206 ± 3	+5.2
tholeiites				
70	0.70425 ± 6	0.429	0.51228 ± 3	+6.7
77	0.70379 ± 6	0.325	0.51224 ± 3	+7.9
88	0.70364 ± 4	0.350	0.51230 ± 3	+8.6
Formation des Basaltes				
boninites				
39	0.70359 ± 6	0.267	0.51222 ± 3	+7.8
49	0.70635 ± 4	0.236	0.51223 ± 4	+8.0
52	0.70386 ± 3	0.248	0.51222 ± 3	+7.8
168	0.70338 ± 7	0.284	0.51223 ± 2	+7.9
184	0.70360 ± 5	0.242	0.51234 ± 3	+10.2
tholeiites				
55	0.70495 ± 5	0.282	0.51198 ± 3	+3.0
179	0.70449 ± 6	0.351	0.51208 ± 2	+4.6
181	0.70448 ± 3	0.369	0.51203 ± 2	+3.6

$^{143}Nd/^{144}Nd$ is normalized to $^{146}Nd/^{142}Nd = 0.636151$. Errors represent $\pm 2\sigma$ mean.
$(^{143}Nd/^{144}Nd)$ for BCR-1 was measured as 0.511833 ± 10. Other analytical details as in Nelson et al. (1984).
ε_{Nd} calculated from measured values at 250 Ma for Koh samples and 60 Ma for west coast basalts and boninites.

Most basalts from mature backarc basins (BAB) have $\varepsilon_{Nd} > +8$ (Volpe et al. 1987); lower values raise the possibility of mixing MORB sources with island-arc-type sources having ε_{Nd} values anywhere between +10 and near zero. The process could easily occur in the initial stages of splitting arcs (Crawford et al. 1981), and the Cambrian tholeiites of Victoria ($\varepsilon_{Nd} \sim +5$) have been interpreted in this light (Crawford & Keays 1987).

Wholerock Ti/V between 21 and 36 (Fig. 12.9) indicate a mid-ocean ridge or backarc-basin setting (Shervais 1982). This is supported by REE data of this study (Figs. 12.3 & 12.6) which parallel the well described patterns of N-type MORB. On a primitive mantle normalization diagram (Fig. 12.10a), the similarity in trace-element contents of the tholeiites in the two tholeiite–boninite suites is immediately apparent. Sample 70 stands out as being quite different and is discussed in the following section. Strong negative Nb spikes distinguish the tholeiites from MORB.

Tholeiitic andesites and minor basalts and dacites from the Dabi Volcanics underlying the boninites at Cape Vogel, PNG, have MgO contents from 6.5

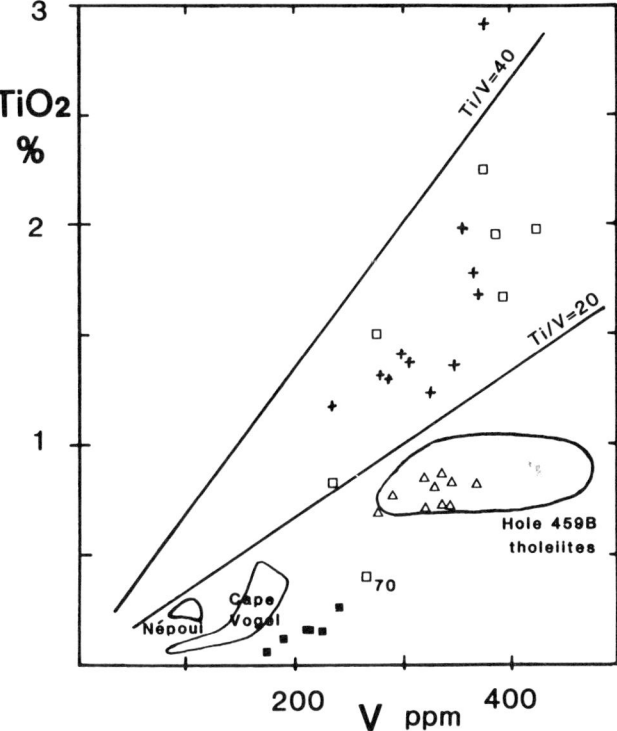

Figure 12.9 Variation of V with TiO$_2$ content in boninites from Nepoui and Koh (full square), compared with those from Cape Vogel and with tholeiites from Koh (open square), Formation des Basaltes (cross), Cape Vogel (open triangle) and the Mariana forearc (Bougault *et al.* 1981). Cape Vogel data from D. A. Walker (unpubl. data).

to 1.5%, but even in the most evolved samples TiO$_2$ is still below 0.9% and FeO* < 7% (D. A. Walker, unpubl. data). Ti/V lie between 12 and 16 (Fig. 12.9) and, on Fig. 12.10a, sample 20 with 3.6% MgO (no. 11 in Table 3 of Cameron *et al.* 1983) shows small Ti and Zr depletions, all suggestive of arc tholeiites. The tholeiitic basalts underlying 'boninite series' rocks from Hole 458 in the Mariana forearc (Hickey-Vargas 1989) have similar Nb depletions relative to La (Wood *et al.* 1981) and Ti/V = 15–19 (see Bougault *et al.* (1981) for similar tholeiites in nearby Hole 459).

Volpe *et al.* (1987) found that BAB from the Mariana trough were enriched in K, Rb, Sr, Ba and H$_2$O and depleted in the HFSE Ti, Zr and Nb relative to MORB. Arc tholeiites from the Mariana arc are further enriched and depleted in these elements respectively. It is likely (Crawford *et al.* 1981) that there is a geochemical continuum in BAB tholeiites, from those with island-arc character (Cape Vogel, Mariana forearc) to those with MORB-like characteristics from mature backarcs (e.g. Lau Basin; Hawkins & Melchior 1985). Transitional between them would be both suites of tholeiites from New Caledonia, the Victorian Cambrian tholeiites and a number of low-Ti ophiolites, including Troodos. Unfortunately K, Rb, Sr and Ba are too mobile to use as a measure of the degree of arc or MORB character in the New Caledonian rocks. It is clear though from the presence of primary amphibole in coarse-grained tholeiites from both localities that water has played a part in their petrogenesis. Both Nb and P have higher field strengths than Ti and Zr and show depletions in Fig. 12.10a in otherwise MORB-like patterns.

A spreading centre propagating into continental crust could also give rise to basaltic rocks with BAB geochemical features if the source also contained ancient arc material. Rocks with depleted Nb, P and Ti have been found at the transition from continental to young oceanic crust near the mouth of the Gulf of California (DSDP Site 475, Saunders *et al.* 1982).

12.4.2 Boninites

The two suites of New Caledonian boninites are as different chemically and isotopically as their accompanying tholeiites. In Figure 12.10b, the representative Nepoui boninite shows depletion in Ti, V and Sc relative to neighbouring

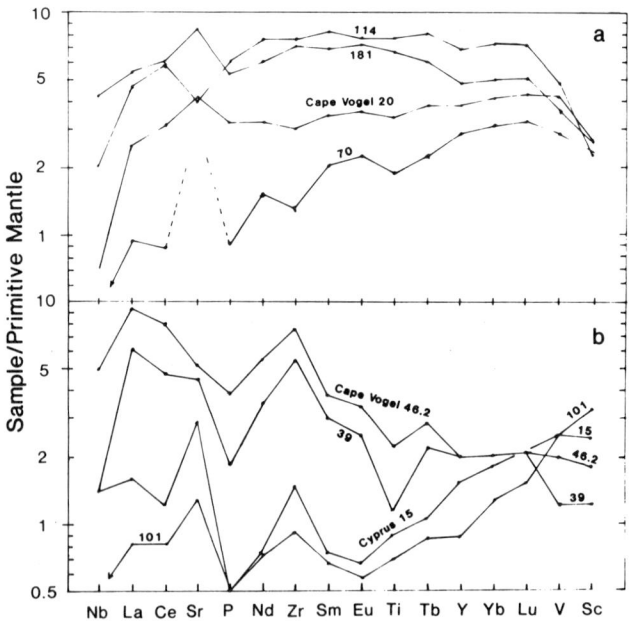

Figure 12.10 Trace-element compositions, normalized to those of primitive mantle (Sun & McDonough 1988), for representative samples of: (a) tholeiites from Koh (70, 114) and the Formation des Basaltes (181), compared with that from Cape Vogel (sample 20); and (b) boninites from Koh (101), Cyprus (sample 15 of Cameron 1985), Nepoui (39) and Cape Vogel (46.2; Walker unpubl. data). Nb and P values are uncertain below 1 ppm and 0.1% respectively.

elements whereas the Koh sample shows none of these. Both do have negative Nb and P spikes and a positive Zr spike. All other boninites except those from Cyprus have negative Ti spikes. The pattern for sample 101 parallels that of Troodos sample 15 (Cameron 1985).

Nepoui boninites are very fresh ($^{87}Sr/^{86}Sr)_I = 0.7034-0.7036$ in four samples) and have uniquely high ε_{Nd}, from $+7.8$ to $+10.2$, compared with the more typical boninite values found in the Koh boninites (ε_{Nd} $+2.7$ to $+5.5$, Table 12.4). The most silica-, LREE- and Zr-enriched boninites have the highest ε_{Nd} values, thereby dismantling any global models which attempt to integrate the chemical and isotopic peculiarities of boninites in general (e.g. Cameron et al. 1983). The very depleted nature of the source material is corroborated by extremely low Sc, V and Ca contents in the boninites. This is excellent evidence for a harzburgitic source which has undergone long-term depletion in incompatible and moderately compatible elements. High ε_{Nd} would be consistent only with a recent influx of incompatible-enriched, metasomatic fluid that had no trace of continental influences.

The Nepoui and Koh boninites are end-members of a geochemically coherent rock type. Koh (and Troodos) boninites have high Ca, Sc, Ti/Zr and are depleted in LREE and SiO_2. Nepoui and, to a lesser extent, the upper boninite unit at Cape Vogel have the reverse. The fact that boninites which appear to come from the most depleted source regions (lowest Sc, Ca) have the highest concentration of LREE, Zr and K, and vice versa, suggests that there is a genetic relationship, i.e. the more depleted the source in moderately compatible elements, the more highly metasomatized it would need to be in order that further melting can take place.

Koh boninite 91 could be derived from the source of sample 70 by a small addition of a LILE-enriched, low-ε_{Nd} fluid from the subducting slab or from an ancient subduction-related enrichment event. Sample 70 itself cannot be a second-stage melt from the tholeiite source because of its lower ε_{Nd}. Because of its strong Zr depletion, it is more likely that the source for sample 70 had some arc character. Further depletion of the source region for sample 91 by boninite extraction, followed by melting of more thoroughly metasomatized mantle wedge (lower ε_{Nd}), would result in boninites with U-shaped REE patterns with minima in the MREE (samples 82 and 101, Fig. 12.3). A similar argument applies to Dupuy et al.'s (1981) sample 17 and its relation to a suitable source for the Nepoui boninites.

It would seem as though the glassy nature of the Nepoui boninites has somehow made them resistant to weathering and to the effects of obduction of an unknown but considerable thickness of oceanic lithosphere. The complete disparity in metamorphic grade between the tholeiites and boninites could be explained if they were extruded after low-grade metamorphism of the tholeiites but before ophiolite emplacement. This presupposes that the ophiolite sheet was thin and cool enough not to induce any further regional metamorphic effects. Another possibility is that the serpentinized contacts of the ophiolite have moved since the early Eocene, covering boninites which were erupted close to the time of ophiolite emplacement.

The only later thermal event in New Caledonia resulted in the emplacement of granodiorites at 25 Ma (Paris 1981) into the southern harzburgite massif. In New Guinea, tonalites which intruded the Papuan Ultramafic Belt at 50–55 Ma have REE patterns (Jaques & Chappell 1980) tantalisingly similar to those of the Cape Vogel boninites erupted at about 59 Ma (Walker & McDougall 1982). Estimates of REE contents by XRF in the Thy Granodiorite near Noumea are approximately twice those of the Nepoui boninites (Sec. 12.3.3). $^{40}Ar/^{39}Ar$ dating of the boninites is needed to clarify what relationship they have to the Formation des Basaltes, if any. Nd isotopic evidence, based on other boninite suites, implies that associated tholeiites would have $\varepsilon_{Nd} > +10$. The Formation des Basaltes has unusually low $^{143}Nd/^{144}Nd$ ratios, especially for backarc-basin basalts, and therefore is likely to be petrogenetically and tectonically unrelated to the boninites.

Acknowledgements

I thank Tony Crawford, Malcolm McCulloch and Shen-Su Sun for interesting and helpful discussions over the years. I particularly thank Bruce Chappell for the quality of his XRF and INAA analyses, Donal Windrim for running most of the isotopic analyses and Nick Ware for his help with the microprobes. The ANU Faculties Research Fund assisted with fieldwork on two occasions. Finally, I am grateful to Jean-Pierre Paris for his advice about the field aspects of this project.

References

Avias, J. V. 1977. About some features of allochtonous ophiolitic and vulcano-sedimentary suite units and their contact zones in New Caledonia. *Int. Symp. Geodynamics SW Pacific*, Noumea, 245–64. Paris: Technip.

Bougault, H., R. C. Maury, M. El Azzouzi, J-L. Joron, J. Cotten & M. Treuil 1981. Tholeiites, basaltic andesites, and andesites from Leg 60 sites: geochemistry, mineralogy, and low partition coefficient elements. Init. Rep. DSDP Leg 60, 657–77.

Cameron, W. E. 1985. Petrology and origin of primitive lavas from the Troodos ophiolite, Cyprus. *Contrib. Mineral. Petrol.* **89**, 256–62.

Cameron, W. E., M. T. McCulloch & D. A. Walker 1983. Boninite petrogenesis: chemical and Nd–Sr isotopic constraints. *Earth Planet. Sci. Lett.* **65**, 75–89.

Campiglio, C., C. Marion & M. Vannier 1986. Etude d'une boninite à olivine de Nouvelle-Calédonie: pétrographie et chimisme des phases. *Bull. Mineral.* **109**, 423–40.

Crawford, A. J., 1980. A clinoenstatite-bearing cumulate olivine pyroxenite from Howqua, Victoria. *Contrib. Mineral. Petrol.* **75**, 353–67.

Crawford, A. J. & W. E. Cameron 1985. Petrology and geochemistry of Cambrian boninites and low-Ti andesites from Heathcote, Victoria. *Contrib. Mineral. Petrol.* **91**, 93–104.

Crawford, A. J. & R. R. Keays 1987. Petrogenesis of Victorian Cambrian tholeiites and implications for the origin of associated boninites. *J. Petrol.* **28**, 1075–109.

Crawford A. J., L. Beccaluva & G. Serri 1981. Tectono-magmatic evolution of the West Philippine–Mariana region and the origin of boninites. *Earth Planet. Sci. Lett.* **54**, 346–56.

Crawford, A. J., L. Beccaluva, G. Serri & J. Dostal 1986. Petrology, geochemistry and tectonic implications of volcanic rocks from the intersection of the Yap and Mariana trenches. *Earth Planet. Sci. Lett.* **80**, 265–80.

Dallwitz, W. B., D. H. Green & J. E. Thompson 1966. Clinoenstatite in a volcanic rock from the Cape Vogel area, Papua. *J. Petrol.* **7**, 375–403.

Dupuy, C., J. Dostal & M. Leblanc 1981. Geochemistry of an ophiolitic complex from New Caledonia. *Contrib. Mineral. Petrol.* **76**, 77–83.

Hawkins J. W. & J. T. Melchior 1985. Petrology of Mariana Trough and Lau Basin Basalts. *J. Geophys. Res.* **90**, 11431–68.

Hickey-Vargas, R. L. 1989. Boninites and tholeiites from DSDP Site 458, Mariana forearc (this volume).

Hickey-Vargas, R. & M. K. Reagan 1987. Temporal variation of isotope and rare earth element abundances in volcanic rocks from Guam: implications for the evolution of the Mariana Arc. *Contrib. Mineral. Petrol.* **97**, 497–508.

Jaques, A. L. & B. W. Chappell 1980. Petrology and trace element geochemistry of the Papuan Ultramafic Belt. *Contrib. Mineral. Petrol.* **75** 55–70.

Jenner, G. A. 1981. Geochemistry of high-Mg andesites from Cape Vogel, Papua New Guinea. *Chem. Geol.* **33**, 307–32.

Kyser, T. K., W. E. Cameron & E. G. Nisbet 1986. Boninite petrogenesis and alteration history: constraints from stable isotope compositions of boninites from Cape Vogel, New Caledonia and Cyprus. *Contrib. Mineral. Petrol.* **93**, 222–6.

Nelson, D. R., A. J. Crawford & M. T. McCulloch 1984. Nd–Sr systematics in Cambrian boninites and tholeiites from Victoria, Australia. *Contrib. Mineral. Petrol.* **88**, 169–77.

Nisbet, E. G., N. T. Arndt, M. J. Bickle, W. E. Cameron, C. Chauvel, M. Cheadle, E. Hegner, T. K. Kyser, A. Martin, R. Renner & E. Roedder 1987. Uniquely fresh 2.7 Ga komatiites from the Belingwe greenstone belt, Zimbabwe. *Geology* **15**, 1147–50.

Paris, J.-P. 1981. *Géologie de la Nouvelle-Calédonie: un essai de synthèse.* BRGM Mem. no. 113.

Rodgers, K. A. 1975. Lower Tertiary tholeiitic basalts from southern New Caledonia. *Mineral. Mag.* **40**, 25–32.

Sameshima, T., J.-P. Paris, P. M. Black & R. F. Heming 1983. Clinoenstatite-bearing lava from Nepoui, New Caledonia. *Am. Mineral.* **68**, 1076–82.

Saunders, A. D., D. J. Fornari, J.-L. Joron, J. Tarney & M. Treuil 1982. Geochemistry of basic igneous rocks, Gulf of California, DSDP Leg 64. Init Rep. DSDP Leg 64, 595–642.

Shervais, J. W. 1982. Ti–V plots and the petrogenesis of modern and ophiolitic basaltic lavas. *Earth Planet. Sci. Lett.* **59**, 101–18.

Sun, S.-S. & W. F. McDonough 1988. Chemical and isotopic systematics of oceanic basalts: implications for mantle compositions and processes. In *Magmatism in ocean basins*, A. D. Saunders & M. J. Norry (eds), Spec. Publ. 42, 313–45. London: Geological Society.

Sun, S.-S. & R. W. Nesbitt 1978. Geochemical regularities and genetic significance of ophiolitic basalts. *Geology* **6**, 689–93.

Taylor, S. R. & M. P. Gorton 1977. Geochemical application of spark source mass spectrometry. III: Element sensitivity, precision and accuracy. *Geochim. Cosmochim. Acta* **41**, 1375–80.

Tsuchiyama, A. 1986. Experimental study of olivine-melt reaction and its petrological implications. *J. Volcanol. Geotherm. Res.* **29**, 245–64.

Volpe, A. M., J. D. Macdougall & J. W. Hawkins 1987. Mariana Trough basalts: trace element and Sr–Nd isotopic evidence for mixing between MORB-like and arc-like melts. *Earth Planet. Sci. Lett.* **82**, 241–54.

Walker, D. A. & W. E. Cameron 1983. Boninite primary magmas: evidence from the Cape Vogel Peninsula, Papua New Guinea. *Contrib. Mineral. Petrol.* **83**, 150–8.

Walker, D. A. & I. McDougall 1982. $^{40}Ar/^{39}Ar$ and K–Ar dating of altered glassy volcanic rocks: the Dabi Volcanics, PNG. *Geochim. Cosmochim. Acta* **46**, 2181–90.

Wood, D. A., N. G. Marsh, J. Tarney, J.-L. Joron, P. Fryer & M. Treuil 1981. Geochemistry of igneous rocks recovered from a transect across the Mariana Trough, arc, fore-arc, and trench, Sites 453 through 461, DSDP Leg 60. *Init. Rep. DSDP Leg 60*, 611–45.

13 Boninites and tholeiites from DSDP Site 458, Mariana forearc

ROSEMARY HICKEY-VARGAS

Abstract

Pre-early Oligocene, plagioclase-free, bronzite-bearing pillow lavas (bronzite andesites) cored at DSDP Site 458 in the Mariana forearc are probably differentiation products of a boninitic magma with somewhat higher CaO contents than most other Mariana Trench and forearc boninites. They have low incompatible-element abundances ($2-6 \times$ chondritic), low Ti/Zr (54–65), and high Zr/Sm (43–48) and Hf/Sm (1.0–1.1). The lavas differ from typical boninites in having somewhat higher CaO/Al_2O_3 (av. 0.73) and lower La/Sm (1.3). Tholeiitic andesites from the lower part of the core have higher incompatible-element abundances, higher Ti/Zr (85–120) and $^{143}Nd/^{144}Nd$ (0.51305), and lower Zr/Sm (28–35) and Hf/Sm (0.8–0.9). Plagioclase-bearing bronzite andesites interbedded with tholeiites have trace-element and isotopic characteristics intermediate between those of plagioclase-free bronzite andesites and tholeiitic andesites. Pb isotope ratios in Site 458 lavas are indistinguishable from those of Pacific Ocean basalts.

Parental magmas for Site 458 bronzite andesites were probably derived by a relatively high degree of partial melting of a peridotite source more fertile (higher modal clinopyroxene content and incompatible-element abundances) than that for typical boninites. Owing to the higher source incompatible-element abundances, effects of metasomatic enrichment of Zr and LREE are less pronounced. The oceanic Pb isotope ratios, together with the absence of Zr and Hf enrichment in the tholeiites and most other arc lavas, suggest that fluids from the subducted oceanic crust are not responsible for Zr and Hf enrichment in boninites. Zr and Hf enrichment may result from a metasomatism of their lithospheric mantle sources by fluids or melts from asthenospheric mantle.

13.1 Introduction

One of the most interesting discoveries of the DSDP Leg 60 transect of the Mariana arc was the identification of boninite-like lavas in the early Oligocene igneous section of forearc Site 458 (Fig. 13.1). Boninites and boninite-like lavas of similar age have been dredged from the Mariana Trench between 12°N and 18°N (Dietrich *et al.* 1978, Bloomer 1983) and are exposed on Guam (Reagan & Meijer 1984). Three volcanic units from Site 458 (bronzite

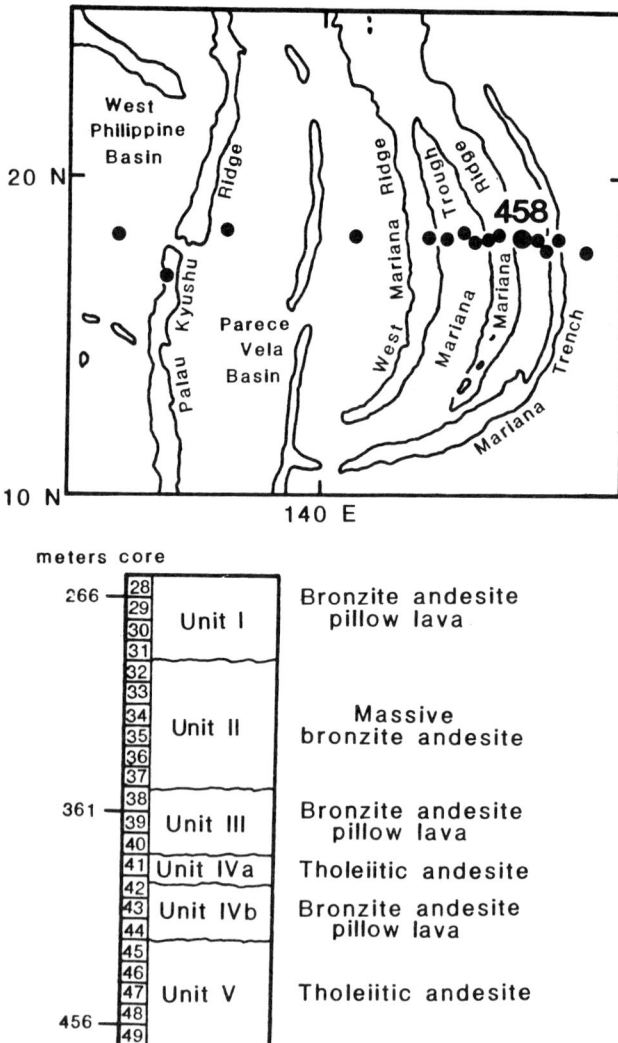

Figure 13.1 Location of DSDP Site 458 and igneous section.

340

Table 13.1 Representative major-element analyses of lavas from DSDP Hole 458, a typical Bonin Is. boninite, a Parece Vela backarc-basin basalt and an experimental melt of broadly boninitic composition.

	(1) Bronzite andesite Units I and III	(2) Bronzite andesite Bonin Islands	(3) Boninite Mariana Trench	(4) Bronzite andesite Unit IVb	(5) Tholeiitic andesite Unit V	(6) Basalt Parece Vela Basin	(7) Pyrolite melt-35% 10 kbar 1200°C
SiO_2	57.68	59.69	57.33	58.47	58.70	49.40	53.8
TiO_2	0.23	0.29	0.14	0.47	0.91	0.99	2.0
Al_2O_3	13.86	14.44	9.75	13.42	14.72	14.3	9.7
FeO	8.32	8.23	9.06	8.76	10.65	9.36	9.4
MnO	–	0.23	0.13	–	–	0.14	0.2
MgO	6.87	5.71	15.19	6.10	3.33	9.14	14.1
CaO	10.57	8.38	5.86	8.90	7.81	12.25	8.6
Na_2O	1.76	2.28	1.59	2.54	3.16	2.76	1.7
K_2O	0.33	0.51	0.93	0.39	0.36	0.46	0.4
P_2O_5	–	0.07	0.16	–	–	0.08	–
total	99.62	99.83	100.14	99.05	99.64	98.88	99.9
Mg #	0.60	0.55	0.75	0.55	0.36	0.64	0.73
CaO/Al_2O_3	0.76	0.65	0.60	0.66	0.53	0.86	0.89

Data sources: columns (1), (4) and (5), Meijer et al. (1981); (2), Kuroda and Shiraki (1975); (3), Hickey and Frey (1982); (6), Mattey et al. (1981); (7), Green (1976).

andesites) were identified as possible boninite differentiates based on their glassy texture, the presence of Mg-rich orthopyroxene (bronzite) phenocrysts and the absence of phenocryst plagioclase (Meijer et al. 1981, Natland 1981). They are high-SiO_2, high-MgO lavas, although their MgO contents are lower than in typical Mariana region boninites, due to more extensive fractionation (Table 13.1). They share many other geochemical characteristics, such as exceptionally low abundances of TiO_2 and highly incompatible elements, and low Ti/Zr (Wood et al. 1981, Bougault et al. 1981, Sharaskin 1981, Hickey & Frey 1981, Meijer et al. 1981). In this chapter, I present new Nd, Sr and Pb isotopic data for bronzite andesites and tholeiitic lavas from Site 458, and use these data, together with existing major- and trace-element data, to interpret relationships between the different volcanic units at Site 458, and between Site 458 bronzite andesites and Mariana region boninites.

13.2 Summary of section and major- and trace-element data

The Site 458 section was divided into five volcanic units (Fig. 13.1). Units I and III consist of glassy bronzite andesite pillow lavas containing less than 5% microphenocrysts of clinopyroxene and orthopyroxene. In some specimens, plagioclase occurs as minute quench crystals in spherulitic intergrowths with clinopyroxene. Unit II consists of massive flows of fine- to medium-grained bronzite andesite containing clinopyroxene, calcic plagioclase, Fe-Ti oxides and rare orthopyroxene. Lavas from Unit IVa and V are fine-grained to glassy, with microphenocrysts of plagioclase (labradorite), clinopyroxene and Fe-Ti oxides. They are distinguished from Units I–III bronzite andesites by the occurrence of plagioclase phenocrysts in glassy specimens, and the absence of orthopyroxene. These lavas were classified as tholeiitic andesites and basaltic andesites. Bronzite andesites from Unit IVb are glassy pillow lavas with microphenocrysts of clinopyroxene, orthopyroxene *and* plagioclase.

All Site 458 rocks are moderately to intensely altered and fresh material is preserved only in glassy rinds of pillow lavas. Comparisons of fresh glass and wholerock analyses indicate that loss of CaO and SiO$_2$ and addition of MgO occurred throughout the sequence (Meijer *et al.* 1981, Sharaskin 1981). Fresh glass analyses for Units I, III, IVb and V are listed in Table 13.1, along with analyses of boninites and a Parece Vela backarc-basin basalt.

Bronzite andesites from Units I and III have SiO$_2$, TiO$_2$, Al$_2$O$_3$, FeO and MgO contents comparable to a differentiated boninite from the Bonin Islands,

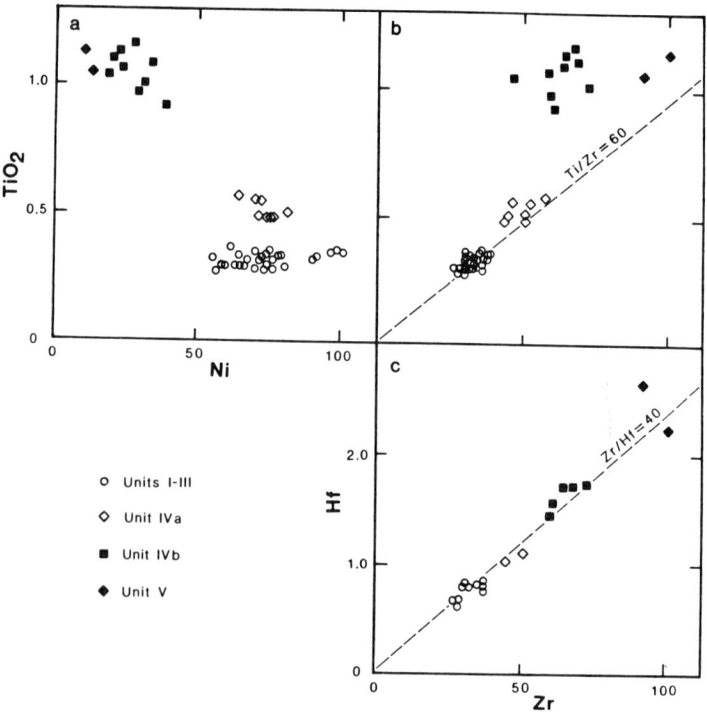

Figure 13.2 Variation diagrams for Site 458 lavas: (a) TiO$_2$ versus Ni; (b) TiO$_2$ versus Zr; and (c) Hf versus Zr. Data from Wood *et al.* (1981), Bougault *et al.* (1981) and Hickey & Frey (1981).

but higher CaO contents and CaO/Al_2O_3 (Table 13.1). Compared to MgO-rich Mariana region boninites, they have, in general, lower MgO and FeO and higher Al_2O_3 and CaO contents and distinctly higher CaO/Al_2O_3 (0.65–0.76) (Table 13.1), although plagioclase-free boninites with CaO/Al_2O_3 up to 0.68 occur in Dredge MT-50 from the Mariana forearc (Bloomer & Hawkins 1987). The higher CaO/Al_2O_3 in the Site 458 bronzite andesites is a major difference between these lavas and typical boninites (Dietrich *et al.* 1978, Sun & Nesbitt 1978, Hickey & Frey 1982). Since clinopyroxene is a phenocryst phase in the bronzite andesites and plagioclase is not, the high CaO/Al_2O_3 represent the minimum possible values for the primary magmas from which they evolved. Bronzite andesites from Unit IVb have higher TiO_2 contents and lower CaO contents and CaO/Al_2O_3 than Unit I and III lavas. Tholeiitic lavas from Unit V have similar SiO contents, lower MgO and CaO, and higher Al_2O_3, FeO and TiO_2 compared to Units I–III bronzite andesites.

Figures 13.2–13.4 summarize immobile elements Ni, Ti, Zr, Hf and REE data. Ni abundances are highest in bronzite andesites from Units I, II, III and IVb (50–101 ppm), and lowest in tholeiitic andesites from Unit IVa (10 ppm) (Fig. 13.2a). TiO_2, Zr, Hf and REE abundances in bronzite andesites from Units I–III are extremely low (2–6× chondritic abundances; Figs 13.2 &

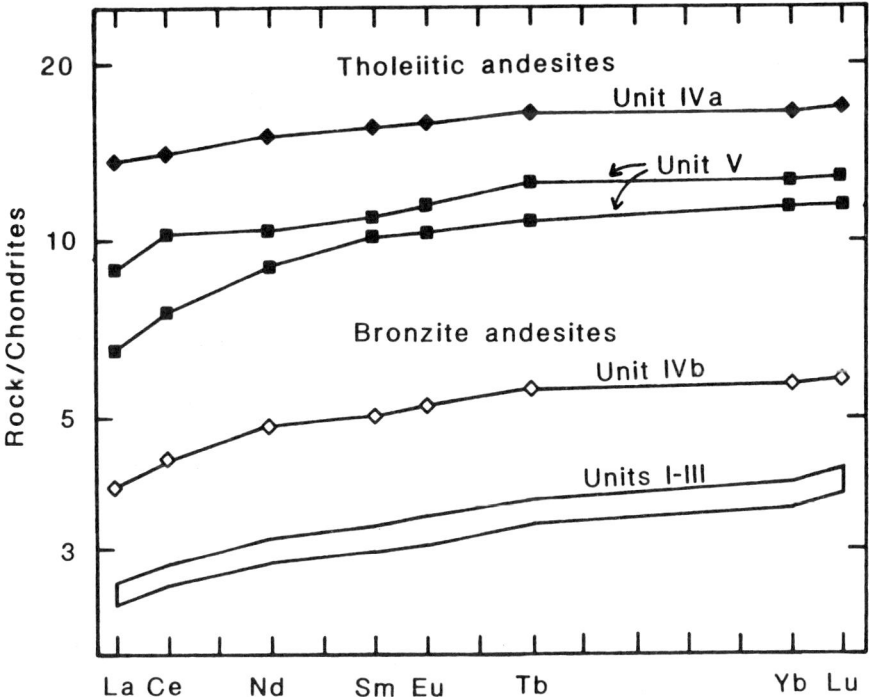

Figure 13.3 Chondrite-normalized REE abundances in igneous rocks from Site 458. Data from Hickey & Frey (1981).

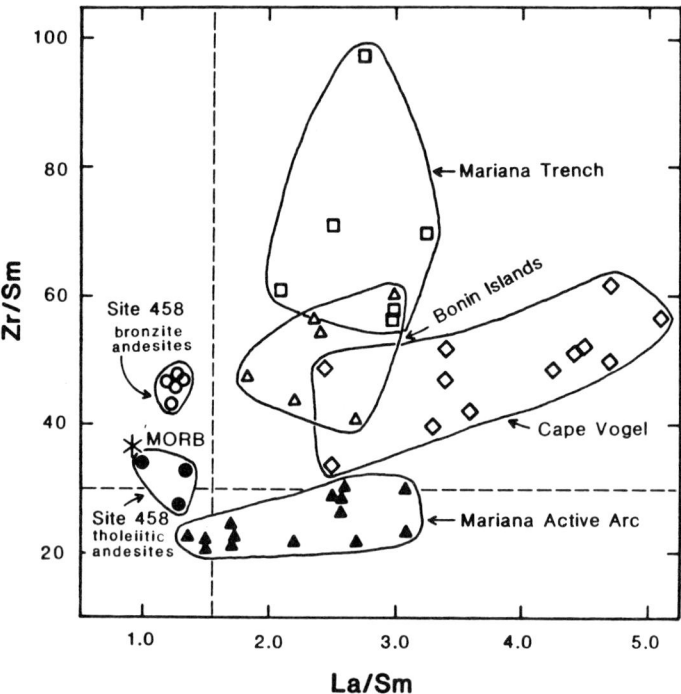

Figure 13.4 Plot of Zr/Sm versus La/Sm showing Site 458 lavas compared to boninites and Mariana active arc lavas. Boninite data from Sun & Nesbitt (1978), Dietrich et al. (1978), Jenner (1981), Hickey & Frey (1982) and Cameron et al. (1983). Mariana arc data from Dixon & Batiza (1979), Mattey et al. (1981), Meijer & Reagan (1981) and Hole et al. (1984). MORB value from Wood (1979). Broken lines are chondritic ratios.

13.3), as is typical of boninites. Bronzite andesites from Unit IVb have higher TiO_2, Zr, Hf and REE abundances, but similar Ni abundances (Figs 13.2 & 13.3). Ti/Zr in Units I–III and Unit IVb bronzite andesites (58–67) are lower than those in MORB (av. Ti/Zr = 110) and in tholeiitic andesites from Unit V (85–120; Fig. 13.2b). These low ratios appear to result from enrichment of Zr relative to Ti and REE (e.g. Zr/Sm > MORB; Fig. 13.4), which is a unique characteristic of boninites (Sun & Nesbitt 1978, Hickey & Frey 1982, Cameron et al. 1983). Unit IVa tholeiitic andesite Ti/Zr values are also low; however, this is probably caused by the crystallization and removal of Fe-Ti oxides which are abundant in these more Ti- and Fe-rich units. Unit IVa lavas do not have the high Zr/Sm shown by the overlying boninites (Fig. 13.4); Zr/Hf are similar in lavas from the different Site 458 units (35–50; Fig. 13.2c).

The relative abundances of REE in bronzite andesites from Units I–III and IVb are distinct from those of Mariana region boninites in that they have less-than-chondritic La/Sm and La/Yb, and LREE-depleted normalized abundance patterns (Figs 13.3 & 13.4). Unit IVa and Unit V tholeiitic andesites also have chondrite-normalized La/Sm and La/Yb < 1, but have flat HREE at notably higher chondrite-normalized abundances than in Units I–III and IVb lavas.

13.3 Isotopic data

Isotopic data for Site 458 are listed in Table 13.2. All Pb and Sr isotopic analyses were performed on hand-picked separates from glassy sections of the core; Nd isotope ratios were measured on these samples or wholerock powders prepared from crystalline sections. Data for a tholeiitic andesite from Mariana forearc DSDP Site 459 (Fig. 13.1) are included in Table 13.2.

Isotope ratios of Sr and Nd for Site 458 bronzite andesites and tholeiites are similar to those for modern Mariana arc volcanics (Fig. 13.5a) and also overlap the range for late Eocene–early Oligocene arc lavas from Guam (Hickey-Vargas & Reagan 1987). The $^{143}Nd/^{144}Nd$ values are lower and $^{87}Sr/^{86}Sr$ are higher in bronzite andesites from Units I–III than in tholeiitic andesites from Units IVa, V and Site 459, but $^{143}Nd/^{144}Nd$ values for the bronzite andesites are higher than those for most Mariana region boninites. Both $^{87}Sr/^{86}Sr$ and $^{143}Nd/^{144}Nd$ in bronzite andesite from Unit IVb are

Table 13.2 Isotopic data for lavas from DSDP Hole 458.

	$^{143}Nd/^{144}Nd$	Sm/Nd	$^{87}Sr/^{86}Sr$	Rb/Sr	$^{206}Pb/^{204}Pb$	$^{207}Pb/^{204}Pb$	$^{208}Pb/^{204}Pb$
Units I and III bronzite andesites							
28-1	0512963 ± 16	0.353	0.70394 ± 3	0.095	18.860	15.533	38.413
30-1	–	0.356	0.70398 ± 3	0.094	–	–	–
39-1	0.512946 ± 16	0.350	0.70389 ± 3	0.095	18.862	15.539	38.434
Unit IVa tholeiitic andesite							
41cc	0.513053 ± 18	0.334	–	–	–	–	–
Unit IVb bronzite andesite							
43-2	0513023 ± 18	0.351	0.70374 ± 3	0.066	18.857	15.546	38.454
	0.513035 ± 18	0.359					
Unit V tholeiitic andesite							
46-1	0.513059 ± 16	0.363	0.70360 ± 3	0.045	18.784	15.511	38.331
Site 459 tholeiitic andesite							
65-1	0.513052 ± 18	0.407	–	–	18.757	15.518	38.315

$^{143}Nd/^{144}Nd$ is normalized to $^{146}Nd/^{144}Nd = 0.7219$ and $^{143}Nd/^{144}Nd = 0.51263$ for USGS Rock Standard BCR-1.
$^{87}Sr/^{86}Sr$ is normalized to $^{86}Sr/^{88}Sr = 0.1194$ and $^{87}Sr/^{86}Sr = 0.7080$ for Eimer and Amend $SrCO_3$.
Pb isotope ratios are corrected for mass fractionation relative to NBS common Pb standard 981. Reproducibility is ±0.05 per mass unit or better.
Nd data for 28-1, 30-1, 43-2 and 46-1 from Hickey & Frey (1982).

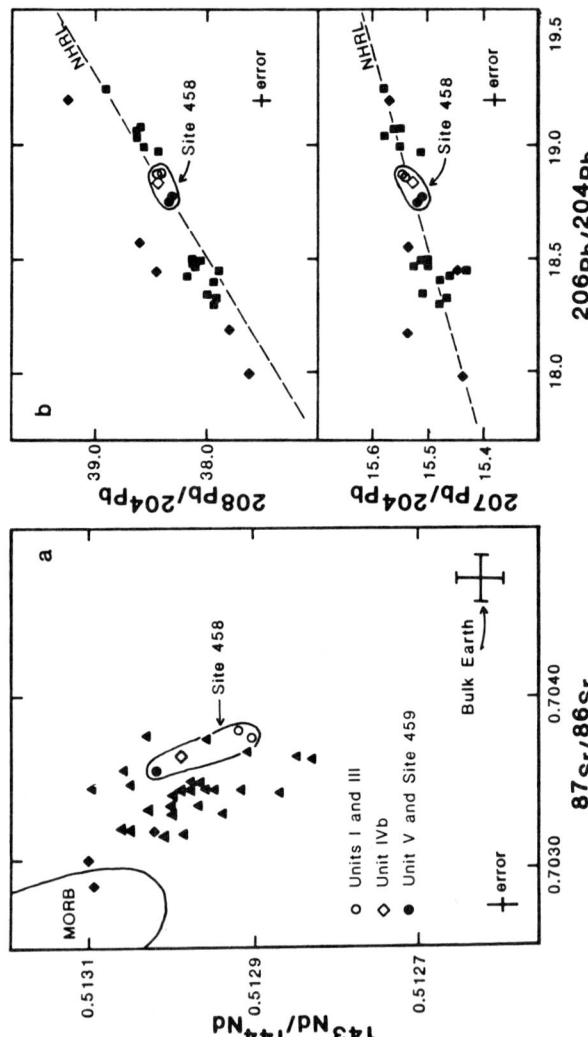

Figure 13.5 Sr, Nd and Pb isotope data for igneous rocks from Site 458. Filled triangles in (a) are data for active Mariana arc (DePaolo & Wasserburg 1977, Stern & Bibee 1980, Stern 1981, Dixon & Stern 1983, White & Patchett 1984). Filled squares in (b) are data for pre-Miocene Mariana region arc lavas (Meijer 1976, Hickey-Vargas & Reagan 1987). Filled diamonds in (a) and (b) are West Philippine Basin, Parece Vela Basin and Mariana trough basalts (Meijer 1975, Cohen & O'Nions 1982). NHRL from Hart (1984).

intermediate between values for Units I–III bronzite andesites and tholeiitic andesites from Units IVa and V.

Like other pre-Miocene Mariana region volcanics, Pb isotope ratios in Site 458 bronzite andesites and tholeiites plot along the NHRL (Northern Hemisphere regression line) and overlap with values for Pacific Ocean basalts (Fig. 13.5b). $^{206}Pb/^{204}Pb$ values are higher in bronzite andesites from Units I, III and IVb than in tholeiitic andesites from Unit V and Site 459.

13.4 Discussion

13.4.1 The importance of boninites and transitional lavas

In major-element terms, boninites are similar to experimental liquids (Table 13.1) generated by large extents of melting (>30%) of peridotite under water-saturated, low-pressure (<15 kbar) conditions (Green 1976). Although distinct from typical orogenic andesites, boninites, as near primary, subduction-related magmas, are extremely important as indicators of the compostion of the mantle beneath forearcs (Sun & Nesbitt 1978, Jenner 1981, Hickey & Frey 1982, Cameron et al. 1983). The low TiO_2, CaO and Al_2O_3 abundances of boninites, their low CaO/Al_2O_3 and Cr-rich spinels point to a harzburgitic source with low modal clinopyroxene content. Residual peridotite following boninite extraction must be exceptionally refractory. Boninites have the high Sr/Nd and Ba/La characteristic of arc rocks generally (e.g. Kay 1980), indicating that these features are characteristic of upper mantle beneath island arcs, rather than the result of partial melting and differentiation processes. Other trace-element features of boninites, such as 'U-shaped' REE patterns, and Zr and Hf enrichment (Figs 13.4 & 13.7), are not typical features of arc rocks. Several authors have suggested that LREE and Zr enrichment in boninites results from metasomatism of their harzburgitic sources by an LREE- and Zr-enriched fluid (Sun & Nesbitt 1978, Jenner 1981, Hickey & Frey 1982, Cameron et al. 1983, Nelson et al. 1984), but the source of this fluid and its relationship to the enrichment of Sr and Ba typical of arc rocks are problematic.

The relatively low-pressure, high-temperature hydrous melting conditions inferred from experimental studies (Green 1976, Tatsumi 1981, 1982) restrict boninite generation to the uppermost mantle (<45 km), whereas tholeiitic basalts may be generated at greater depths or lower P_{H_2O}. Therefore, comparison of the trace-element and isotopic compositions of boninites and tholeiitic lavas from the same area can yield information about the compositional variation within the sub-arc mantle. In this respect, lavas compositionally transitional to boninites are particularly important. Such magmas could originate by melting under temperature and pressure conditions intermediate between those appropriate for boninite and tholeiitic basalt, by melting of a relatively fertile rather than harzburgitic mantle, or by mixing of boninite and tholeiite magmas.

13.4.2 Transitional boninites from Site 458

Based on comparisons between Site 458 bronzite andesites and Mariana region boninites, as well as differences between Units I–III and Unit IVb bronzite andesites, we infer that:

(a) the parental magmas of the evolved boninitic bronzite andesite lavas from Units I–III were generated by high extents of melting of a peridotite source more fertile than that for typical Mariana region boninites, and
(b) lavas from Unit IVb were generated by mixing of bronzite andesite and tholeiite magma.

Their relatively high CaO/Al_2O_3 values are the most important evidence that the Units I–III lavas were derived from a more fertile source (higher modal clinopyroxene content) than that for typical boninites. Their CaO/Al_2O_3 and TiO_2 contents are lower than those of MORB-like backarc-basin basalts inferred to originate by melting of fertile asthenospheric mantle, implying that the source was more depleted than MORB-source mantle. The source mantle of the Hole 458 boninitic bronzite andesites may have been peridotites residual after MORB extraction. Addition of a given amount of an LREE- and Zr-enriched fluid to a moderately depleted (lherzolite) and highly depleted (harzburgite) peridotite will produce less-pronounced LREE and Zr enrichment in the moderately depleted composition. Also, the low LREE and Zr contents in the Units I–III lavas may reflect a lower degree of infiltration of the somewhat depleted source by the LILE-enriched fluids involved in boninite genesis. However, since the bronzite andesites have low absolute incompatible-element abundances similar to those of typical boninites, they must have been derived by a high extent of partial melting of this source.

Mixing of boninite magma and an LREE-depleted magma with high CaO/Al_2O_3 (e.g. a backarc-basin basalt) could also produce La/Sm, Zr/Sm and CaO/Al_2O_3 similar to those of the bronzite andesites, but absolute abundances of TiO_2 and highly incompatible elements would be significantly higher. Mixing of boninite magma and LREE-depleted primitive arc tholeiite magma might account for the low TiO_2 and incompatible-element abundances, but cannot account for the high CaO/Al_2O_3 in the bronzite andesites.

Bronzite andesite from Unit IVb has $^{143}Nd/^{144}Nd$, $^{87}Sr/^{86}Sr$ and trace-element abundances intermediate between Units I–III bronzite andesites and Units IVa and V tholeiitic lavas, suggesting an origin by magma mixing. An origin by magma mixing for this unit was also proposed by Natland (1981) based on petrographic and mineral chemical criteria. The different isotope ratios rule out generation of Unit IVb lavas by a lower extent of melting of

Table 13.3 Calculated mixture of 80% Units I–III bronzite andesite and 20% Unit IVa tholeiitic andesite compared to Unit IVb bronzite andesite.

	Average Units I–III	Unit IVa 41cc	Mixture	Unit IVb 43-2	Deviation (%)
La	0.78	4.16	1.46	1.19	+22
Nd	1.79	9.00	3.32	2.93	+10
Sm	0.60	3.03	1.09	1.00	+9
Yb	0.78	3.31	1.29	1.20	+7
TiO_2	0.29	1.31	0.46	0.50	−8
Hf	0.65	2.3	0.98	1.12	−13
Zr	28	101	43	48	−11
Sr	89	169	105	124	−15
$^{143}Nd/^{144}Nd$	0.51296	0.51305	0.51301	0.51303	−0.004
$^{87}Sr/^{86}Sr$	0.70394	0.70360[a]	0.70383	0.70374	+0.009

[a] Value for tholeiitic andesite 46-1, Unit V (Table 13.2).
Data sources: this chapter, Hickey & Frey (1981, 1982) and Wood et al. (1981).

sources for Units I–III lavas or by differentiation from the same parental magma. Source mixing, i.e. different proportions of peridotite and metasomatic fluids, is ruled out by the similar La/Sm and Ti/Zr in Units I–III and IVb bronzite andesites. Based on Figure 13.2b, Unit IVb bronzite andesites could be generated by mixing of Units I–III bronzite andesite and Unit IVa tholeiitic andesite, which has high Ti/Zr. A calculated mixture of 20% Unit IVa tholeiitic andesite and 80% Units I–III bronzite andesite matches the absolute and relative trace-element abundances of Unit IVb bronzite andesite within ±20%, and closely approaches measured $^{143}Nd/^{144}Nd$ and $^{87}Sr/^{86}Sr$ values (Table 13.3). Mixing with differentiated tholeiite magma may also contribute to the low CaO abundances and low CaO/Al_2O_3 in Unit IVb bronzite andesites (Table 13.1).

On the basis of the foregoing compositional data, a suggested origin for the Site 458 sequence is: (1) generation and ascent to a magma chamber of primary arc tholeiitic basalt magma, fractionation and eruption of differentiated Unit V; (2) generation and ascent to a magma chamber of primary bronzite andesite magma, some of which mixed with differentiated tholeiite magma and was erupted (Unit IVb); (3) eruption of differentiated arc tholeiite magma (Unit IVa); and (4) eruption of differentiated boninite magma (bronzite andesites in Units I–III). The presence of mixed rocks in Unit IVb suggests that boninite and tholeiitic magmas were generated concurrently beneath this area.

13.4.3 Source regions for Site 458 tholeiites and boninites

Based on the Al_2O_3 content of the experimental melt composition shown in Table 13.1, bronzite andesites from Units I–III could be derived from a broadly boninitic primary magma (i.e. a somewhat more Ca-rich boninite) by about 30% low-pressure crystallization of olivine and orthopyroxene. The extent of differentiation of the tholeiitic lavas is more difficult to estimate because the primary magma composition is unknown. Based on simple phenocryst addition calculations, at least 60% crystallization of olivine and clinopyroxene is required to generate these lavas from a magma in equilibrium with peridotite containing Fo_{90} olivine. According to these estimates, abundances of highly incompatible elements in parental magmas for the two lava types probably differed by less than a factor of 2 (e.g. La = 0.55 ppm for bronzite andesite and 0.8 ppm for tholeiite).

This information, together with the relatively high CaO/Al_2O_3 in the bronzite andesites and evidence that the Site 458 tholeiites and boninitic bronzite andesites were generated contemporaneously suggest that the parental magmas for the bronzite andeistes probably did not originate by melting of the peridotite residue from earlier arc tholeiite magmatism, as has been proposed for typical boninites (Crawford et al. 1981). It is more likely that the parental magmas of boninitic bronzite andesites were derived directly by melting of lithospheric mantle of the Philippine plate, residual from generation of Philippine sea-floor MORB between 85 and 44 Ma.

Experimental evidence (e.g. Tatsumi 1982) indicates that both boninites and tholeiitic basalts could be derived by melting at shallow levels (<50 km) within the lithospheric mantle under hydrous (boninite) and anhydrous (tholeiitic basalt) conditions, and such an origin is consistent with the low incompatible-element abundances inferred for parental magmas for the Site 458 bronzite andesites and tholeiites. However, temperatures sufficient to produce anhydrous melting of depleted peridotite (1350°C; Duncan & Green 1987) are unlikely to occur normally in sub-arc mantle at these levels. Therefore, it is more likely that the Site 458 tholeiites were derived by melting at greater depths than the CaO-rich transitional boninite primary magmas.

Figure 13.6 shows a possible model for the generation of magmas erupted at Site 458, based on the model of Crawford et al. (1981). These authors proposed that the hydrous, refractory peridotite appropriate as a source for typical boninites might be heated above its solidus during upwelling of anhydrous MORB-source mantle diapirs in the early stages of arc rifting and backarc-basin formation. The major difference between the Crawford et al. (1981) model and the model shown in Figure 13.6 is that boninites are derived directly from lithospheric mantle of the Philippine plate rather than from the residue of arc tholeiite generation (Fig. 13.6c) and that the arc tholeiites are generated by hydrous melting of the *asthenosphere* (Figs 13.6a, b & c). Extremely incompatible-element-poor arc tholeiites, such as those from Unit V, may be generated by high extents of melting of asthenospheric mantle hydrated by fluids from the subducted slab and heated by the rising anhydrous peridotite diapirs (Figs 13.6b & c). As slab-derived fluids are incorporated into magmas that rise to the surface, the source is depleted and asthenospheric

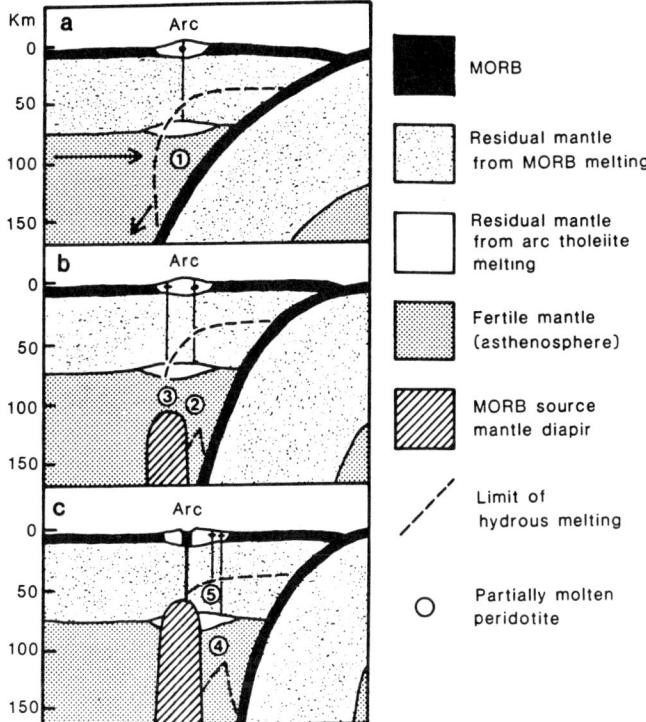

Figure 13.6 Possible model for the generation of parental magmas for bronzite andesites and tholeiitic andesites from Site 458 based on the model of Crawford *et al.* (1981): (a) Generation of arc tholeiite basalts by hydrous melting of the asthenosphere (1). (b) Beginning of diapiric upwelling. Tholeiites with extremely low incompatible-element abundances generated by high extents of melting of hydrous asthenosphere (2). Magmas with characteristics intermediate between arc and backarc-basin basalts generated by melting of asthenosphere with low H_2O content (3). (c) Continued upwelling and intrusion of diapir into lithosphere. Incompatible-element-poor tholeiites generated by hydrous melting of asthenosphere (4). Parental magmas for bronzite andesites (CaO-rich boninite) generated by melting of hydrous lithosphere (5).

melts become gradually more anhydrous and MORB-like (Fig. 13.6b). This progressive change from arc tholeiite to backarc-basin basalt magmatism may account for the elevated water, alkali-element and LREE abundances noted in Mariana backarc-basin basalts, as well as the occurrence of backarc-basin basalts with trace-element characteristics intermediate between those of arc and ocean-floor tholeiites (Wood *et al.* 1981, Fryer *et al.* 1981, Bougault *et al.* 1981).

13.5 Origin of the high Zr/Sm and Hf/Sm in boninites

An important feature of pre-Miocene Mariana region arc volcanics, including

Site 458 bronzite andesites and tholeiites, is their oceanic Pb isotope ratios (Fig. 13.5b). Among arc lavas, these are least likely to have incorporated subducted sediment into their sources (Meijer 1976). If it is assumed that most of the Sr and Pb in Site 458 lavas is derived from fluids from the subducted oceanic crust, based on the high abundances of these elements relative to REE and Ti, then a sediment-free altered ocean-floor basaltic crust is an appropriate source.

In the absence of a significant sediment component, the high Zr/Sm and Hf/Sm in Site 458 bronzite andesites are difficult to explain by the introduction of fluids from the subducted crust. Ocean-floor basalts have near-chondritic Zr/Sm and Hf/Sm (Figs 13.4 & 13.7; White & Patchett 1984) and hydrous fluids in equilibrium with altered MORB would be expected to have similar or lower ratios, because of the (assumed) greater solubility of REE compared to Zr and Hf. White & Patchett (1984) note that arc lavas have Hf/Sm similar to those of oceanic basalts and this is particularly evident for basaltic arc lavas (Fig. 13.7). If slab-derived fluids have high Zr/Sm and Hf/Sm, the appearance of Zr and Hf enrichment in boninites but not arc tholeiites could be explained

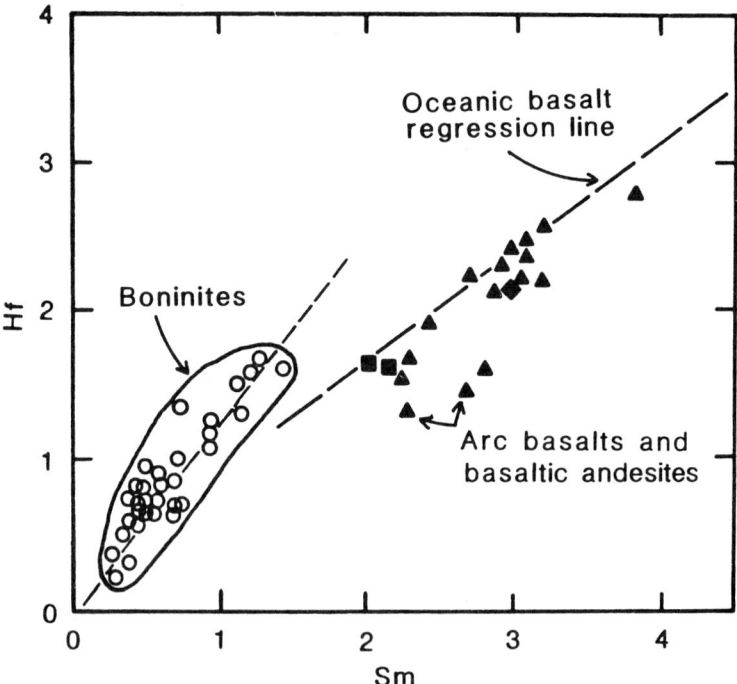

Figure 13.7 Plot of Hf versus Sm for boninites compared to arc basalts and basaltic andesites. Boninite data from sources listed for Figure 13.4. Arc data for samples with less than 56% SiO_2 and oceanic basalt regression line from White & Patchett (1984). Tholeiitic andesites from Site 458 indicated by filled squares.

if the peridotite source for tholeiites had significantly higher incompatible-element abundances than those for boninites. However, this explanation is implausible if *low* Nb/La and Ti/V, which are typical of arc volcanics, are also attributed to slab-derived fluids. Therefore, we infer that the source of Zr and Hf enrichment is unique to boninites.

If boninites are derived entirely from the lithospheric mantle and arc tholeiites mainly from the asthenosphere, we can identify this unique enrichment with mantle (not subduction-related) metasomatism, similar to the cryptic metasomatism observed in LREE-enriched harzburgite inclusions in basalts and in ophiolites (e.g. Frey 1984). Several authors have noted the similar REE patterns of these peridotites and boninites (Hickey & Frey 1982, Coish *et al.* 1982, Reagan & Meijer 1984). Metasomatic enrichment of the lithosphere could occur by introduction of LREE-, Zr- and Hf-enriched melts from the asthenosphere (e.g. phase B of Frey & Green 1974), or, in the context of a 'plum pudding' mantle (Morris & Hart 1983), Zindler *et al.* 1984), by melts of OIB (oceanic island basalt) source 'plums'. The isotopic differences between Site 458 tholeiites and boninites, higher $^{87}Sr/^{86}Sr$, lower $^{143}Nd/^{144}Nd$ *and* higher $^{206}Pb/^{204}Pb$ in the boninites are most consistent with additional OIB-like source material for these magmas.

13.6 Summary and conclusions

Units I–III bronzite andesites from DSDP Site 458 were derived by relatively advanced fractionation of a boninitic magma with higher CaO contents than other Mariana region boninites. Their parent magma may have been generated by high degrees of melting of a peridotite source with a higher modal clinopyroxene content and higher incompatible-element abundances than that for typical boninites, under similar temperature and P_{H_2O} conditions. Bronzite andesites from Unit IVb were derived by mixing of Units I–III type bronzite andesite and tholeiitic andesite magmas.

We concur with Crawford *et al.* (1981) that boninites may be generated by contact melting of lithospheric mantle during upwelling of anhydrous asthenospheric mantle into hydrous sub-arc peridotite, during the early stage of backarc-basin formation. However, because of their high CaO/Al_2O_3, it is unlikely that parental magmas for the Site 458 bronzite andesites were generated from the peridotite residue of earlier arc tholeiite magmatism. The bronzite andesites were probably generated from moderately depleted lithosphere residual from generation of Philippine sea-floor MORB, and the tholeiites generated by hydrous melting at greater depths, possibly within the asthenosphere. If this is true generally, then the unique enrichment of Zr and Hf relative to Ti and REE in boninites may result from mantle metasomatism of the lithospheric mantle by fluids or melts from the asthenosphere, whereas features shared by boninites and tholeiites in island-arc settings, such as high

Ba/La and Sr/Nd, may result from addition of fluids or melts from the subducted crust.

Acknowledgements

I thank A. Meijer, a member of the shipboard scientific party for DSDP Leg 60, for providing the original samples analysed for REE, and J. Natland for assistance in sampling the Site 458 and 459 cores. I gratefully acknowledge S. R. Hart for use of clean laboratory, mass spectrometers and mineral separation facilities at MIT.

References

Bloomer, S. H. 1983. Distribution and origin of igneous rocks from the landward slope of the Mariana Trench: implications for its structure and evolution. *J. Geophys. Res.* **88**, 7411–28.

Bloomer, S. H. & J. W. Hawkins 1987. Petrology and geochemistry of boninite series volcanics form the Mariana Trench. *Contrib. Mineral. Petrol.* **97**, 361–77.

Bougault, H., R. C. Maury, M. El Azzouzi, J.-L. Joron, J. Cotten & M. Treuil 1981. Tholeiites, basaltic andesites and andesites from Leg 60 sites: geochemistry, mineralogy and low partition coefficient elements. Init. Rep. DSDP Leg 60, 657–78.

Cameron, W. E., M. T. McCulloch & D. A. Walker 1983. Boninite petrogenesis: chemical and Nd–Sr isotopic constraints. *Earth Planet. Sci. Lett.* **65**, 75–89.

Cohen, R. S. and R. K. O'Nions 1982. Identification of recycled continental material in the mantle from Sr, Nd and Pb isotopic investigations. *Earth Planet. Sci. Lett.* **61**, 73–84.

Coish, R. A., R. L. Hickey & F. A. Frey 1982. Rare-earth element geochemistry of the Betts Cove ophiolite, Newfoundland. *Geochim. Cosmochim. Acta* **46**, 2117–34.

Crawford, A. J., L. Beccaluva & G. Serri 1981. Tectono-magmatic evolution of the West Philippine–Mariana region and the origin of boninites. *Earth Planet. Sci. Lett.* **54**, 346–56.

DePaolo, D. J. & G. J. Wasserburg 1977. The sources of island arcs as indicated by Nd and Sr isotopic studies. *Geophys. Res. Lett.* **4**, 465–8.

Dietrich, V., R. Emmerman, R. Oberhansli & H. Puchelt 1978. Geochemistry of basaltic and gabbroic rocks from the West Mariana Basin and Mariana Trench. *Earth Planet. Sci. Lett.* **39**, 127–44.

Dixon, T. H. & R. Batiza 1979. Petrology and chemistry of recent lavas in the northern Marianas: implications for the origin of island arc basalts. *Contrib. Mineral. Petrol.* **70**, 167–81.

Dixon, T. H. & R. J. Stern 1983. Petrology and chemistry of submarine volcanoes in the southern Mariana Arc. *Geol. Soc. Am. Bull.* **94**, 1159–72.

Duncan, R. A. & D. H. Green 1987. The genesis of refractory melts in the formation of the oceanic crust. *Contrib. Mineral. Petrol.* **96**, 326–42.

Frey, F. A. 1984. Rare earth element abundances in upper mantle rocks. In *Rare earth element geochemistry*, P. Henderson (ed.), 153–203. Amsterdam: Elsevier.

Frey, F. A. & Green, D. H. 1974. The mineralogy, geochemistry and origin of lherzolite inclusions in Victorian basanites. *Geochim. Cosmochim. Acta* **38**, 1023–59.

Fryer, P., J. Sinton & J. A. Philpotts 1981. Basaltic glasses from the Mariana Trough. Init. Rep. DSDP Leg 60, 601–10.

Green, D. H. 1976. Experimental testing of 'equilibrium' partial melting of peridotite under water-saturated, high pressure conditions. *Can. Mineral.* **14**, 255–68.

Hart, S. R. 1984. The DUPAL anomaly: a large scale isotopic anomaly in the southern hemisphere. *Nature* **309**, 753–6.

Hickey, R. L. & F. A. Frey 1981. Rare-earth element geochemistry of Mariana fore-arc volcanics: Deep Sea Drilling Project Site 458 and Hole 459B. Init. Rep. DSDP Leg, **60**, 735–42.

Hickey, R. L. & F. A. Frey 1981. Rare-earth element geochemistry of Mariana fore-arc volcanics: Deep Sea Drilling Project Site 458 and Hole 459B. Init. Rep. DSDP Leg 60, 735–42.

Hickey-Vargas, R. & M. K. Reagan, 1987. Temporal variation of isotope and rare earth element abundances in volcanic rocks from Guam: implications for the evolution of the Mariana Arc *Contrib. Mineral. Petrol.* **97**, 497–508.

Hole, M. J., A. D. Saunders, G. F. Marriner & J. Tarney 1984. Subduction of pelagic sediments: implications for the origin of Ce-anomalous basalts from the Mariana Islands. *J. Geol. Soc. Lond.* **141**, 453–72.

Jenner, G. A. 1981. Geochemistry of high-Mg andesites from Cape Vogel, Papua New Guinea. *Chem. Geol.* **33**, 307–32.

Kay, R. W. 1980. Volcanic arc magmas: implication of a melting–mixing model for element recycling in the crust–upper mantle. *J. Geol.* **88**, 497–522.

Kuroda, N. & K. Shiraki 1975. Boninite and related rocks of Chichi-Jima, Bonin Islands, Japan. *Rep. Fac. Sci. Shizuoka Univ.* **10**, 145–55.

Mattey, D. P., N. G. Marsh & J. Tarney 1981. The geochemistry, mineralogy and petrology of basalts from the West Philippine and Parece Vela basins and from the Palau–Kyushu and West Mariana ridges, DSDP Leg 59. Init Rep. DSDP Leg 59, 753–97.

Meijer, A. 1975. Pb and Sr isotopic studies of igneous rocks cored during Leg 31 of the Deep Sea Drilling Project. Init. Rep. DSDP Leg 31, 601–5.

Meijer, A. 1976. Pb and Sr isotopic data bearing on the origin of volcanic rocks from the Mariana arc system. *Geol. Soc. Am. Bull.* **87**, 1358–69.

Meijer, A. & M. K. Reagan 1981. Petrology and geochemistry of the island of Sarigan, Mariana Arc: calcalkaline volcanism in an oceanic setting. *Contrib. Mineral. Petrol.* **77**, 337–54.

Meijer, A., E. Anthony & M. K. Reagan 1981. Petrology of volcanic rocks from the fore-arc sites. Init. Rep. DSDP Leg 60, 709–30.

Morris, J. D. & S. R. Hart 1983. Isotopic and incompatible element constraints on the genesis of island arc volcanics from Cold Bay and Amak Island, Aleutians, and implications for mantle structure. *Geochim. Cosmochim. Acta* **47**, 2015–30.

Natland, J. H. 1981. Crystal morphologies and pyroxene compositions in boninites and tholeiitic basalts from DSDP Holes 458 and 459B in the Mariana fore-arc region. Init. Rep. DSDP Leg 60, 681–708.

Nelson, D. R., A. J. Crawford & M. T. McCulloch 1984. Nd–Sr isotope and geochemical systematics in Cambrian boninites and tholeiites from Victoria, Australia. *Contrib. Mineral. Petrol.* **88**, 164–72.

Reagan, M. K. & A. Meijer 1984. Geology and geochemistry of early arc volcanic rocks from Guam. *Geol. Soc. Am. Bull.* **95**, 701–13.

Sharaskin, A. Y. 1981. Petrology and geochemistry of basement rocks from five Leg 60 sites. Init. Rep. DSDP Leg 60, 647–56.

Stern, R. J. 1981. A common mantle source for Western Pacific island arc and 'hotspot' magmas: implications for layering in the upper mantle. *Carnegie Inst. Washington Yearb.* **80**, 455–61.

Stern, R. J. & Bibee, L. D. 1980. Esmeralda Bank: geochemistry of an active submarine volcano in the Mariana island arc and its implications for magma genesis in island arcs. *Carnegie Inst. Washington Yearb.* **79**, 465–72.

Sun, S.-S & R. W. Nesbitt 1978. Geochemical regularities and genetic significance of ophiolitic basalts. *Geology* **6**, 689–93.

Tatsumi, Y. 1981. Melting experiments on a high magnesian andesite. *Earth Planet. Sci. Lett.* **54**, 356–65.

Tatsumi, Y. 1982. Origin of high magnesian andesites in the Setouchi volcanic belt, southwest Japan, II. Melting phase relations at high pressures. *Earth Planet. Sci. Lett.* **60**, 305–17.

White, W. M. & J. Patchett 1984. Hf–Nd–Sr isotopes and incompatible element abundances in island arcs: implications for magma origins and crust–mantle evolution. *Earth Planet. Sci. Lett.* **67**, 167–85.

Wood, D. A. 1979. A variably veined suboceanic upper mantle – genetic significance for mid-oceanic ridge basalts from geochemical evidence. *Geology* **7**, 499–503.

Wood, D. A., N. G. Marsh, J. Tarney, J.-L. Joron, P. Fryer & M. Treuil 1981. Geochemistry of igneous rocks recovered from a transect across the Mariana Trough, arc, fore-arc and trench, Sites 453 through 461, DSDP Leg 60. Init. Rep. DSDP Leg 60, 611–46.

Zindler, A., H. Staudigel & R. Batiza 1984. Isotope and trace element geochemistry of young Pacific seamounts: implications for the scale of upper mantle heterogeneity. *Earth Planet. Sci. Lett.* **70**, 175–82.

14 Petrogenesis of high-Mg and associated lavas from the north Tonga Trench

TREVOR J. FALLOON, DAVID H. GREEN
& MALCOLM T. McCULLOCH

Abstract

In 1984, the *RV Natsushima* recovered a suite of primitive high-Si, high-Mg lavas from the N Tonga arc and trench. These lavas have appropriate high Mg#, Ni and Cr contents to be primary melts from upper-mantle peridotite. The mineral chemistry of phenocryst and groundmass phases combined with wholerock, glass and olivine-hosted glass inclusion major-element chemistry provide unequivocal evidence of mixing of two or more distinct parental magmas in the petrogenesis of these lavas. The major- and trace-element and isotopic (Sr–Nd) geochemistry of the lavas can be explained by partial melting at shallow depths (<10 kbar, or somewhat higher pressures if H_2O was present) of refractory mantle peridotite which had been previously 'enriched' in LILE by two distinct components. Evidence for the existence of a depleted mantle source comes from the CIPW molecular normative chemistry of the high-Mg lavas as seen in the projection from diopside onto the base of the basalt tetrahedron (JdTs–Qtz–Ol), as well as from mineral chemistry (high Cr# in Cr-spinel, and very magnesian olivines (to Fo_{94}) and orthopyroxenes) and trace-element geochemistry (low Ti/V, Ti/Sc and HFSE abundances). Evidence for the 'enriched' components comes from the Sr and Nd isotopes and from trace-element geochemical features. One 'enriched' component is required to explain the range in ε_{Nd} (+6.0 to +1.0) and $^{87}Sr/^{86}Sr$ (0.7029 to 0.7045), and is unlike that seen in other boninites or that expected from subducted pelagic sediment. This component is interpreted to be of a mantle origin, due to similarity in chondrite-normalized element abundance patterns and ratios with intraplate basalts from the Cook–Samoa–Austral Islands, which are all part of the Dupal mantle anomaly. A second 'enriched' component is required to explain high (Sr, Rb, K, Ba)/La ratios in the N Tongan lavas. This component is interpreted to be a hydrous fluid phase derived from subducted oceanic lithosphere.

14.1 Introduction

The composition of primary magmas in island-arc settings remains controversial with respect to both the nature of the source (peridotitic versus eclogitic)

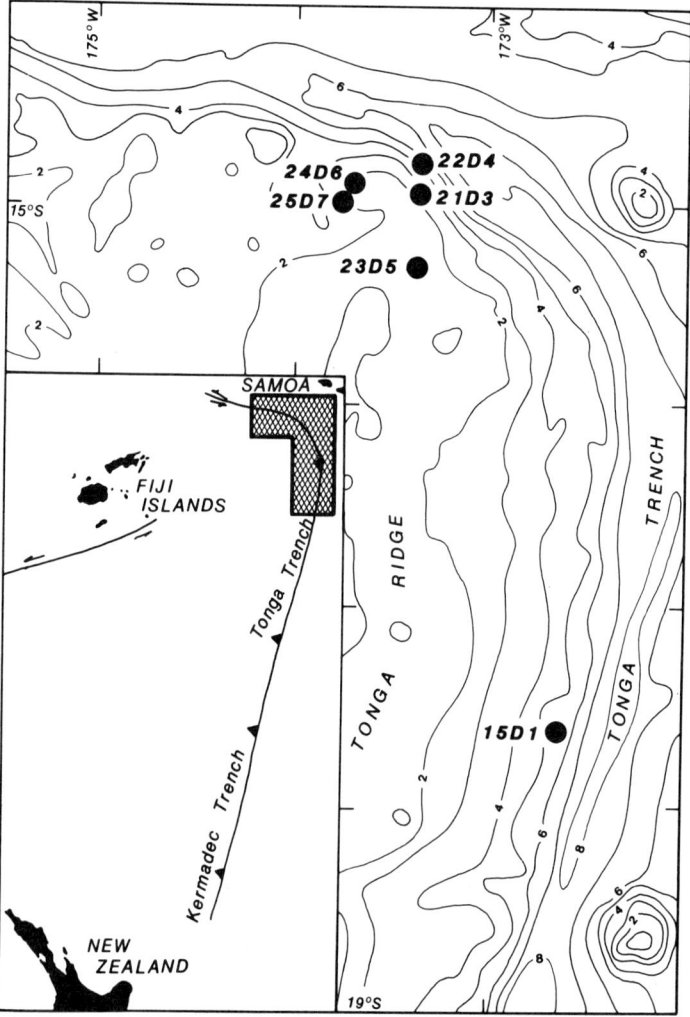

Figure 14.1 Bathymetry of the northern termination of the Tonga Trench showing dredge locations of the 1984 *RV Natsushima* cruise.

and the origin of the distinctive trace-element and isotopic enrichments observed (wedge, subducted oceanic lithosphere, subducted sediment, hydrous fluid or mantle metasomatism?). Distinctive high-Si, high-Mg lavas dredged from the N Tonga arc and trench (Fig. 14.1) during the 1984 cruise of the research vessel *RV Natsushima* are characterized by highly magnesian olivine (to Fo_{94}) and orthopyroxene (max. Mg# = 0.90) phenocrysts and glass which was high Mg# (>0.66). They are, thus, important to this debate, as they have the appropriate characteristics to be primary or near-primary liquids from partial melting of an upper-mantle peridotitic source (Falloon & Green 1986, Falloon *et al.* 1987). The high-Mg lavas are also very fresh, so that their trace-element and isotopic characteristics can be confidently used in evaluating potential source components.

Details of dredge locations, some major- and trace-element data and a brief discussion of the geochemical affinities of these high-Mg lavas are given in Falloon *et al.* (1987). Falloon & Green (1986) presented evidence from the chemistry of glass inclusions in magnesian olivine phenocrysts for the existence of extremely refractory parental magmas in the Tonga arc. In this study, we present detailed petrographic and mineral chemical data on the N Tonga high-Mg lavas, with the aim of demonstrating that magma mixing was an important process in their petrogenesis. We also present new Sr and Nd isotopic data for the dredged lavas and discuss their petrogenesis in light of existing models for the origin of island-arc magmas.

14.2 Petrography and mineral chemistry

14.2.1 Introduction

Table 14.1 summarizes the mineral chemical and petrographic features of representative samples of the dredged lavas. The mineral chemistry of pyroxene, plagioclase and Cr-spinel is summarized in Figures 14.2, 14.3, 14.4 and 14.5 and representative analyses are given in Tables 14.2, 14.3 and 14.6. The compositions of glass inclusions trapped in magnesian olivine phenocrysts are given in Table 14.5 and wholerock, groundmass and glass rind compositions of representative lavas are given in Table 14.4. The petrography and mineral chemistry of the lavas indicates disequilibrium within phenocryst assemblages and between phenocrysts and host liquid. This disequilibrium can be explained by magma mixing, a conclusion supported by compositions of glass inclusions trapped within magnesian olivine phenocrysts. The discussion below concentrates mainly on disequilibrium features of the mineral chemistry indicative of magma mixing.

The petrography and mineral chemistry define four groups of high-Mg lavas from N Tonga (Table 14.1); more evolved basaltic andesites were recovered with high-Mg lavas from Station 23 and are included in this discussion.

Table 14.1 Summary of the petrography and mineral chemistry of the dredged lavas from north Tonga.

Station no.	Group	Phenocrysts and microphenocrysts	Groundmass phases
23	Group A high-Mg lavas (sample nos. 5-24, 28)	(1) olivine (15%), Mg # 0.850–0.918, euhedral to resorbed Cr-spinel inclusions (Cr # 0.838–0.875), glass inclusions (2) orthopyroxene (18%), Mg # 0.836–0.897, zoning (R, N, U), euhedral to resorbed, glomeroporphyritic clusters common, Cr-spinel inclusions (Cr # 0.752–0.825) (3) clinopyroxene (X), Mg # 0.848, strongly resorbed (4) Cr-spinel (<1%), Cr # 0.821–0.833	fresh glass and spherulitic quench pyroxene (augite zoned to subcalcic augite (Mg # 0.848–0.723), small orthopyroxene (Mg # 0.864) and Cr-spinel (Cr # 0.767–0.874) euhedra, strongly vesicular (40%)
23	Group B high-Mg lavas (sample nos. 5-25)	(1) olivine (16%), Mg # 0.846–0.940, euhedral to resorbed Cr-spinel inclusions (Cr # 0.802–0.871), glass inclusions (2) orthopyroxene (8%), Mg # 0.865–0.811, zoning (N, R), resorbed to euhedral, glomeroporphyritic clusters common, Cr-spinel inclusions (Cr # 0.719–0.841) (3) clinopyroxene (2%), Mg # 0.902–0.829, zoning (N, R), euhedral to resorbed, Cr-spinel inclusions (Cr # 0.675) (4) Cr-spinel (<1%), Cr # 0.766–0.865	fresh glass and quench pyroxene microlites (pigeonite zoned to augite (Mg # 0.853–0.815)) Cr-spinel euhedra (Cr # 0.824–0.859), strongly vesicular (50%)

23	basaltic andesites (sample nos. 5-20, 21 23, 27)	(1) plagioclase (3–14%). An 83–96, zoning (R, O), strongly resorbed to euhedral, glomeroporphyritic clusters with pyroxene common, both having mutually subhedral grain contacts (2) clinopyroxene (1–2%), Mg# 0.839–0.765, zoning (N, R, U), euhedral to resorbed (3) orthopyroxene (1–4%), Mg# 0.827–0.773, zoning (N, R, U), euhedral to resorbed (4) olivine (X), strongly resorbed, Mg# 0.804, Cr-spinel inclusion, Cr# = 0.907	fresh hyalopilitic texture, plagioclase microlites (An 64–81), quench pyroxene laths (pigeonite zoned to augite (Mg# 0.751–0.611) or augite zoned to subcalcic augite (Mg# 0.816–0.564)), vesicular (0–40%)
24	high-Mg lavas (sample nos. 6-2, 3)	(1) olivine (14%), Mg# 0.885–0.917, euhedral to skeletal, Cr-spinel inclusions (Cr# 0.805–0.859), glass inclusions (2) orthopyroxene (4%), Mg# 0.893, euhedral (3) orthopyroxene (X), Mg# 0.863–0.898, clinopyroxene (X), Mg# 0.874, both resorbed	fresh glass and spherulitic quench pyroxene (augite zoned to subcalcic augite (Mg# 0.84–0.608)), Cr-spinel euhedra (Cr# 0.826–0.854), vesicular (30%)
25	high-Mg lavas (sample nos. 7-14, 15, 16, 18)	(1) olivine (10%), Mg# 0.874–0.921, euhedral to skeletal, Cr-spinel inclusions (Cr# 0.815–0.841)	fresh glass, spherulitic quench pyroxene (augite (Mg# 0.753–0.692)), small endiopside euhedra (Mg# 0.887) with pigeonite cores (Mg# 0.860, 3.4–4.0 wt% CaO), Cr-spinel euhdra (Cr# 0.768–0.845), vesicular (30%)

Modal percentages for phenocryst phases based on point counting of >1000 points, and have been resummed on the basis of 0% vesicles. R, N, O and U stand for reversely, normally, oscillatory and unzoned, respectively. X stands for rare (≪1%) xenocryst.

Table 14.2 Representative electron microprobe analyses of groundmass pyroxenes from north Tonga lavas.

	1	2	3	4	5	6	7	8	9	10	11	12	13	14	15	16	17	18	19	20
SiO_2	52.83	56.70	49.65	48.42	56.86	52.06	47.37	56.81	53.70	48.65	55.66	53.05	54.16	52.18	51.70	49.44	53.94	50.78	53.74	50.54
TiO_2	0.25	–	0.61	0.81	–	0.32	0.42	–	–	0.26	–	–	–	0.32	0.18	0.63	–	0.41	0.17	0.45
Al_2O_3	1.76	0.63	6.07	6.92	0.81	3.12	9.66	0.60	1.62	7.72	0.60	1.51	1.57	1.98	2.91	3.86	1.70	4.69	0.79	3.03
Cr_2O_3	1.27	0.22	–	–	0.69	–	–	0.32	0.35	–	0.30	0.28	–	–	0.29	–	–	–	–	–
FeO	5.05	8.79	8.66	11.13	7.05	6.67	14.78	8.86	5.93	11.50	10.39	7.93	15.70	16.75	8.89	18.40	15.65	14.96	17.35	16.38
MnO	–	–	–	–	–	–	–	–	–	–	–	–	–	0.25	–	0.33	0.32	0.37	–	0.26
MgO	18.71	30.24	13.88	12.91	32.94	16.75	12.88	31.51	18.27	14.85	30.00	17.41	25.68	18.06	17.49	12.93	23.76	17.23	23.19	14.17
CaO	20.13	3.42	21.13	19.81	1.65	21.08	14.88	1.90	20.13	17.02	3.05	19.82	2.95	10.46	18.54	14.40	4.63	11.55	4.46	15.17
total	99.97	100.00	100.00	100.00	100.00	100.00	99.99	100.00	100.00	100.00	100.00	100.00	100.06	100.00	100.00	99.99	100.00	99.99	99.70	100.00
Mg #	0.87	0.86	0.74	0.67	0.89	0.82	0.61	0.86	0.85	0.70	0.84	0.80	0.74	0.66	0.78	0.56	0.73	0.67	0.70	0.61
Mg #[a]	0.89	0.86	0.75	0.69	0.89	0.84	0.61	0.86	0.85	0.72	0.85	0.82	0.75	0.66	0.82	0.56	0.73	0.67	0.71	0.61
Wo	38.40	6.60	40.50	37.30	3.20	40.50	24.00	3.60	38.90	29.80	6.00	38.40	6.00	22.00	35.00	27.00	9.50	25.90	9.00	29.00
En	54.60	80.30	44.80	43.40	86.50	50.00	46.20	83.20	51.80	50.80	80.20	50.20	70.60	51.30	53.00	41.20	66.10	49.80	64.40	43.60
Fs	6.90	13.10	14.70	19.30	10.40	9.50	29.80	13.10	9.30	19.40	13.80	11.30	23.40	26.70	12.00	31.90	24.40	24.30	26.60	27.40

Dash (–) indicates below detection limit.

[a] Mg # = Mg/(Mg + Fe^{2+}) with Fe^{3+} calculated by stoichiometry.

Analyses are as follows: (1) clinopyroxene microphenocryst 7-18; (2) pigeonite core to clinopyroxene microphenocryst 7-16; (3) quench clinopyroxene microphenocryst 7-18; (5) orthopyroxene microphenocryst 6-2; (6) quench clinopyroxene 6-2; (7) quench clinopyroxene 6-2; (8) orthopyroxene microphenocryst 5-28; (9, 10) quench clinopyroxene 5-28; (11) quench pigeonite 5-25; (12) quench pigeonite 5-25; (13) quench pigeonite 5-27; (14) quench clinopyroxene 5-27; (15, 16) quench clinopyroxene 5-20; (17) quench pigeonite 5-23; (18) quench clinopyroxene 5-23; (19) quench pigeonite 5-21; (20) quench clinopyroxene 5-21.

14.2.2 Groundmass pyroxenes

The groundmasses of all four groups of high-Mg lavas consist of fresh glass, abundant laths of pyroxene and small Cr-spinel euhedra. The groundmass of the basaltic andesites from Station 23 contain plagioclase in addition to pyroxene and fresh glass; Cr-spinel is absent. Representative compositions of groundmass pyroxenes are given in Table 14.2 and shown in Figures 14.2a, b and d.

Quench pyroxenes in Station 25 high-Mg lavas are augitic and distinguished from other pyroxene compositions by their low Mg#, low SiO_2 and high Al_2O_3 contents (Table 14.2, nos. 3 and 4). Also present in the groundmass of Station 25 high-Mg lavas are small magnesian endiopside euhedra (Table 14.2, analysis 1). Microprobe analyses produced a continuous range of compositions from magnesian pigeonite to magnesian endiopside. SEM backscattered electron images, however, reveal the presence of distinct pigeonite cores to the endiopside euhedra (Fig. 14.3). Similar magnesian pigeonite cores to diopside microphenocrysts are present in some Troodos Upper Pillow Lavas (Duncan & Green 1987). Experimental studies (Duncan & Green 1987) on a Troodos parental composition (compositionally similar to Station 25 high-Mg lavas) demonstrated that these magnesian pigeonite cores started to crystallize at ~5 kbar, and were overgrown by Ca-rich clinopyroxene en route to the surface; during eruption the quench augite compositions crystallized. Similar magnesian pigeonites have been reported by Umino (1986) in his Type IV boninite lavas from Chichi-jima, Bonin Is., which have similar major-element compositions to Stations 24 and 25 high-Mg lavas. Umino (1986) showed that the Type IV boninites are unrelated to the other Chichi-jima boninites, most of which are clinoenstatite-bearing.

Quench pyroxenes in Group A high-Mg lavas from Station 23, and also those in Station 24 high-Mg lavas, range from augite to subcalcic augite (Table 14.2, nos. 6, 7, 9 and 10). Microprobe analyses indicate a range of core compositions, which zone continuously from magnesian orthopyroxene (Table 14.2, no. 5) to less magnesian quench pyroxene. SEM backscattered electron images reveal the presence of small, compositionally uniform orthopyroxene microphenocrysts with quench overgrowths. Quench pyroxenes in the Group B high-Mg lava 5-25 are zoned from pigeonite cores to augite rims. Groundmass pyroxenes in Station 23 basaltic andesites are also zoned from pigeonite cores to augite rims. Sample 5-20 is distinct in having two quench pyroxene trends (Fig. 14.2d), one extending from pigeonite cores to augite rims, the other extending from augite cores to subcalcic augite rims. Similar complex zoning of groundmass pyroxenes has been documented by Crawford et al. (1986) in primitive arc lavas from the southern Mariana forearc.

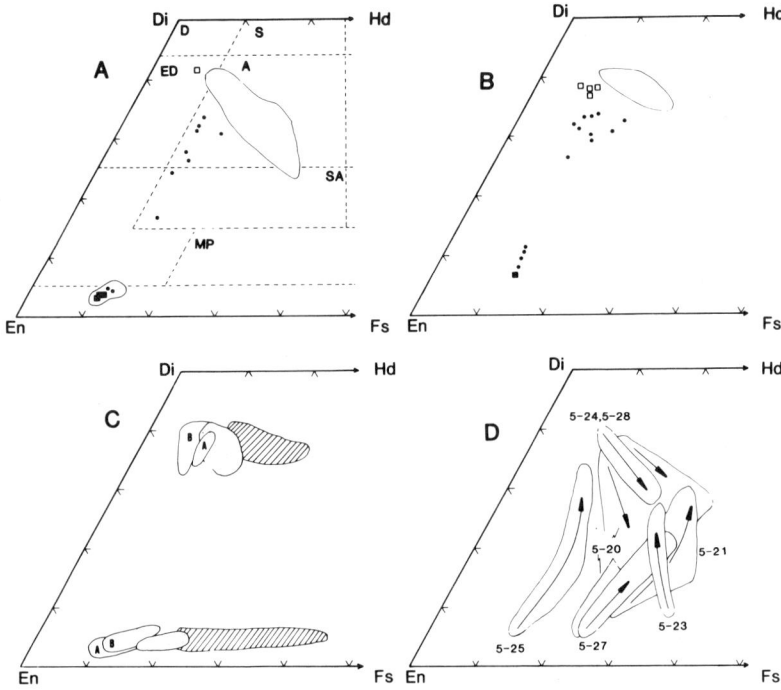

Figure 14.2 Representative compositions of pyroxene from N Tonga lavas. Pyroxene end-member solid solutions as follows: Di, diopside; Hd, hedenbergite; En, enstatite; Fs, ferrosalite. Classification of pyroxene compositions from Deer *et al.* (1966): D, diopside; S, salite; ED, endiopside; A, augite; SA, subcalcic augite; MP, magnesian pigeonite. (a) Pyroxene compositions from Station 24 high-Mg lavas (samples 6-2 and 6-3). Full squares: orthopyroxene cores to small groundmass pyroxene euhedra. Enclosing field encloses range of orthopyroxene microphenocrysts. Full circles: random microprobe spot analyses of groundmass pyroxenes indicating overlap between orthopyroxene cores and quench pyroxene rims. Field of quench clinopyroxene is indicated. Open square: xenocrystal endiopside. (b) Representative compositions from Station 25 (samples 7-14, 7-15, 7-16, and 7-18) high-Mg lavas. Full square: pigeonite core to endiopside microphenocrysts (see also Figs 14.3a & b). Open squares: endiopside microphenocrysts. Full circles: random microprobe analyses of groundmass pyroxenes, resulting in overlap between pigeonite cores to endiopside microphenocrysts and endiopside and quench augite compositions. Field encloses range of quench clinopyroxene compositions. (c) Orthopyroxene and clinopyroxene phenocrysts in N Tonga lavas compared with the range observed in Tofua magmatic arc lavas (Ewart *et al.* 1973). A corresponds to Group A high-Mg lavas from Station 23 (samples 5-24 and 5-28); B corresponds to Group B high-Mg lava from Station 23 (sample 5-25); stippled field corresponds to Station 23 basaltic andesites (samples 5-27, 5-20, 5-21 and 5-23); cross-hatched field corresponds to Tofua magmatic arc pyroxenes. (d) Groundmass pyroxene trends in dredged lavas from Station 23, N Tonga. Arrows indicate the change in pyroxene compositions from core to rim.

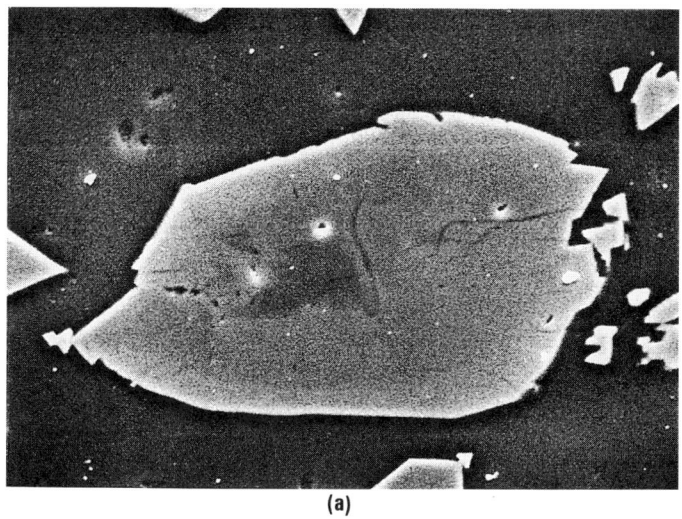

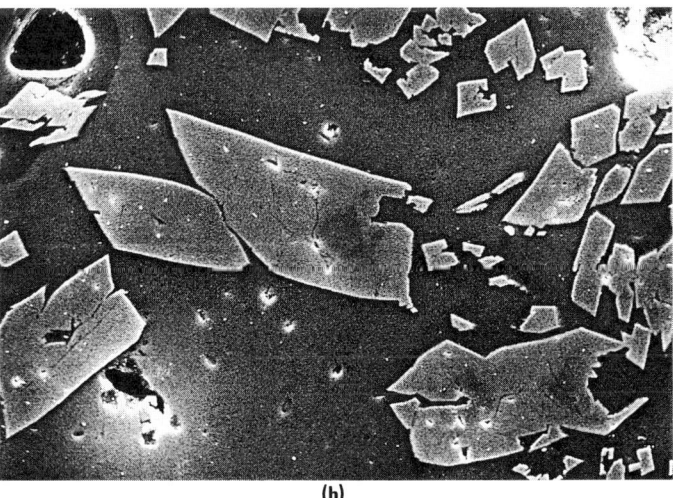

Figure 14.3 (a) SEM backscattered electron image of small endiopside microphenocryst within the groundmass of sample 7-16 (Station 25, high-Mg lava). Dark areas correspond to magnesian pigeonite cores (Table 14.2, no. 2). Scale black and white bands correspond to 10 μm. (b) Same as in (a), but now magnified; scale bars correspond to 100 μm.

Table 14.3 Representative electron microprobe analyses of pyroxene phenocrysts and microphenocrysts in north Tonga lavas.

	1		2	3	4	5		6		7		8		9		10	11		12	
	C	R	C	C	C	C	R	C	R	C	R	C	R	C	R	C	C	R	C	R
SiO_2	52.60	53.80	52.92	53.70	53.15	53.38	53.70	52.25	51.02	54.93	55.43	56.83	57.38	56.14	55.97	57.38	54.83	56.03	55.48	54.55
TiO_2	–	–	–	–	–	–	–	–	0.33	–	–	–	–	–	–	–	–	–	–	–
Al_2O_3	1.92	0.93	1.53	1.62	1.76	1.70	1.36	2.16	3.42	0.98	1.09	0.55	0.55	0.93	1.02	0.46	1.06	0.79	1.06	1.21
Cr_2O_3	0.29	0.56	0.79	0.35	–	0.42	0.65	0.61	–	0.20	–	0.51	0.35	–	0.57	0.38	0.39	0.27	–	–
FeO	7.25	6.59	5.47	5.93	5.99	7.13	6.35	8.07	10.40	13.21	10.95	8.01	7.07	9.09	8.36	6.77	13.22	11.32	11.76	13.93
MnO	–	–	–	–	–	–	–	–	–	–	–	–	–	–	–	–	0.21	–	0.27	0.30
MgO	17.52	20.52	18.62	18.27	17.75	17.27	17.84	17.99	15.91	28.74	30.46	32.55	33.39	31.75	32.32	33.31	28.21	29.66	29.55	28.09
CaO	20.42	17.60	20.67	20.13	21.35	20.10	20.10	18.92	18.92	1.94	2.07	1.55	1.26	2.09	1.76	1.70	2.08	1.93	1.87	1.92
total	100.0	100.0	100.0	100.0	100.0	100.0	100.0	100.0	100.0	100.0	100.0	100.0	100.0	100.0	100.0	100.0	100.0	100.0	100.0	100.0
Mg #	0.81	0.85	0.86	0.85	0.84	0.81	0.83	0.80	0.73	0.80	0.83	0.88	0.89	0.86	0.87	0.90	0.80	0.82	0.82	0.78
Mg # [a]	0.85	0.87	0.90	0.85	0.87	0.81	0.83	0.84	0.76	0.81	0.85	0.88	0.89	0.88	0.89	0.90	0.80	0.82	0.83	0.80
Wo	39.30	33.10	39.30	38.90	41.20	39.10	39.00	36.00	36.00	3.90	4.10	3.00	2.40	4.10	3.50	3.20	4.20	3.80	3.70	3.90
En	51.60	58.60	54.70	51.80	51.40	49.50	50.90	53.70	49.00	77.90	81.80	85.30	87.20	84.20	86.00	86.90	76.90	79.30	79.70	77.00
Fs	9.10	8.40	5.90	9.30	7.40	11.50	10.20	10.30	15.00	18.20	14.00	11.70	10.40	11.70	10.50	9.90	19.00	17.00	16.60	19.10

Dash (–) indicates below detection limit.
C = core, R = rim.
[a]Mg # = Mg/(Mg + Fe^{2+}) with Fe^{3+} calculated by stoichiometry.
Analyses are as follows: (1) reverse-zoned xenocrystal clinopyroxene phenocryst 5-25; (2) clinopyroxene phenocryst 5-25; (3) resorbed xenocrystal clinopyroxene 5-28; (4) resorbed xenocrystal clinopyroxene 6-3; (5) reverse-zoned clinopyroxene phenocryst 5-27; (6) normally zoned clinopyroxene phenocryst 5-20; (7) reverse-zoned orthopyroxene phenocryst 5-25; (8, 9) reverse-zoned orthopyroxene microphenocryst 5-24; (10) resorbed orthopyroxene core in olivine host (Fo_{90}) 6-2; (11) reverse-zoned orthopyroxene phenocryst 5-27; (12) normally zoned orthopyroxene phenocryst 5-21.

14.2.3 Phenocryst pyroxenes

Representative pyroxene phenocryst compositions are given in Table 14.3 and shown in Fig. 14.2c. Pyroxenes in the dredged N Tonga lavas are, in general, more magnesian than those so far reported from Tofua magmatic arc volcanics (Ewart *et al.* 1973). Pyroxene phenocrysts exhibit a range of zoning with respect to Mg#, from relatively unzoned, to normal- and reverse-zoned. They also show a significant range in Mg# within one individual host rock (e.g. in 5-25 pyroxene, phenocrysts range in Mg# from 0.81 to 0.90). Based on experimentally determined pyroxene K_D values for Mg–Fe partitioning between crystal and liquid (Grove *et al.* 1982, Grove & Bryan 1983), many of the less magnesian pyroxenes are not in equilibrium with their host groundmass compositions (Table 14.4). Pyroxenes in the basaltic andesites from Station 23 are less magnesian than pyroxenes in the high-Mg lavas, and overlap with Tongan magmatic arc pyroxene compositions (Fig. 14.2). Reverse zoning is also present in these pyroxenes.

14.2.4 Plagioclase

Plagioclase occurs only in the Station 23 basaltic andesites. Plagioclase compositions (Fig. 14.4 and summarized in Table 14.1) are extremely calcic, as is characteristic of Tongan arc lavas (Ewart *et al.* 1973, Falloon & Green 1986). Plagioclase phenocryst compositions, however, are significantly out of equilibrium with their host wholerock and groundmass compositions (Table 14.4) based on the empirical relationship presented by Falloon & Green (1986) relating wholerock Ca/(Ca + Na) to the expected plagioclase composition under anhydrous conditions. Plagioclase phenocrysts are hosted in rocks with CaO/Na_2O which are too low to have crystallized such calcic plagioclase. Falloon & Green (1986) argued that pressure would not have a significant effect on plagioclase composition in complex systems. The experimental work of Baker & Eggler (1987) on high-Al basalt compositions under differing H_2O contents produced no change in plagioclase compositions; rather, plagioclase compositions reflected the low CaO/Na_2O of the starting compositions chosen, confirming the conclusion of Falloon & Green (1986).

14.2.5 Olivine and olivine-hosted glass inclusions

Olivine is an abundant phenocryst phase in all N Tonga high-Mg lavas (10–16 modal%, Table 14.1). Compositions of small groundmass olivine euhedra and rims of microphenocrysts and phenocrysts are in equilibrium with host groundmass compositions; however, core compositions of microphenocrysts and phenocrysts are significantly out of equilibrium based on an olivine–liquid $K_D = 0.3$ (e.g. sample 5-25 contains olivine phenocrysts of both Fo_{90} and Fo_{94}). Compositions of olivine-hosted glass inclusions can be used to calculate

Table 14.4 Wholerock and groundmass major-element chemistry of high-Mg lavas and basaltic andesites from north Tonga.

	5-25		5-24		5-28		6-2		6-3	
	WR	GDM	WR	GDM	WR	GDM	WR	GL	WR	GL
SiO_2	53.83	57.29	52.95	56.52	53.72	56.36	55.55	58.44	56.04	58.40
TiO_2	0.36	0.31	0.15	0.21	0.14	0.22	0.30	0.33	0.31	0.29
Al_2O_3	9.63	13.00	7.92	13.07	8.72	13.17	10.10	12.53	10.60	11.92
FeO	9.71	8.90	9.70	9.22	9.65	9.13	8.37	7.79	8.46	7.94
MnO	0.19	–	0.21	–	0.20	–	0.16	–	0.17	–
MgO	16.30	8.80	20.89	9.18	19.65	9.08	15.29	8.27	13.61	9.44
CaO	8.50	9.90	7.18	10.28	7.54	10.56	8.79	10.64	9.17	10.13
Na_2O	1.27	1.50	0.85	1.13	0.69	1.13	0.96	1.35	1.14	1.26
K_2O	0.16	0.20	0.14	0.26	0.12	0.24	0.40	0.50	0.43	0.50
P_2O_5	0.04	–	0.01	–	0.02	–	0.06	–	0.07	–
Cl	–	0.10	–	0.13	–	0.11	–	0.15	–	0.13
LOI	0.24	–	0.13	–	–0.02	–	1.78	–	1.87	–
Mg #	0.77	0.66	0.79	0.66	0.80	0.66	0.78	0.68	0.76	0.70

WR = wholerock; GDM = groundmass as determined by electron microprobe broad-beam-area scans; GL = quench glass rind analysis as determined by electron microprobe broad-beam-area scans.
LOI = loss on ignition.
Major elements are in wt%.
Wholerock data determined by XRF (Falloon *et al.* 1987); all analyses are resummed to 100% volatile-free.
Mg # determined on the basis of $Fe^{2+}/(Fe^{2+} + Fe^{3+}) = 0.9$.
(–) Not determined.

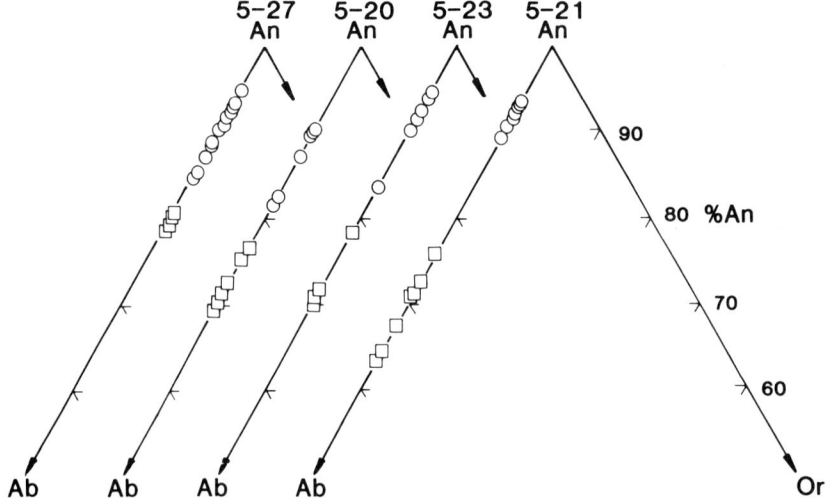

Figure 14.4 Plagioclase phenocryst and groundmass compositions in basaltic andesites from Station 23, N Tonga: open circles, cores and rims of phenocrysts and microphenocrysts; open squares, cores and rims of groundmass plagioclase.

Table 14.4 (*continued*)

	7-14		7-15		7-16		7-18		5-27		5-20	
	WR	GL	WR	GL	WR	GL	WR	GL	WR	GDM	WR	GDM
SiO_2	54.70	57.78	54.96	57.94	55.01	57.81	54.72	56.83	57.48	59.42	56.30	56.89
TiO_2	0.43	0.47	0.44	0.44	0.42	0.46	0.45	0.49	0.44	0.43	0.62	0.49
Al_2O_3	10.77	12.82	10.89	12.81	10.95	12.83	10.90	12.02	14.80	15.20	15.20	16.30
FeO	8.68	7.88	8.47	7.99	8.51	8.01	8.65	8.21	9.66	9.35	9.77	8.40
MnO	0.17	–	0.15	–	0.16	–	0.17	–	0.17	–	0.17	–
MgO	13.28	7.43	13.08	7.39	12.94	7.30	12.97	8.98	5.36	3.95	5.33	4.74
CaO	9.50	10.74	9.64	10.78	9.70	10.69	9.65	10.58	10.11	9.67	10.20	10.20
Na_2O	1.57	1.70	1.49	1.78	1.43	1.73	1.52	1.67	1.67	1.65	2.08	2.30
K_2O	0.62	0.73	0.60	0.71	0.60	0.71	0.71	0.77	0.25	0.23	0.24	0.39
P_2O_5	0.28	0.30	0.28	–	0.28	0.29	0.26	0.28	0.06	–	0.07	–
Cl	–	0.15	–	0.16	–	0.17	–	0.17	–	0.10	–	0.29
LOI	1.75	–	1.45	–	1.58	–	1.69	–	0.57	–	0.71	–
Mg #	0.75	0.65	0.75	0.65	0.75	0.64	0.75	0.66	0.50	0.45	0.52	0.53

the original parental liquid compositions which crystallized these olivines (Falloon & Green 1986, Anderson 1974, Watson 1976). Table 14.5 gives compositions of trapped glass inclusions and compositions in equilibrium with host olivines, calculated using the method outlined in Falloon & Green (1986). The glass inclusion compositions indicate that olivines in sample 5-25 (Table 14.5, nos. 3 and 4) crystallized from distinctly different liquid compositions which are not related by crystal fractionation. Glass inclusions in olivine phenocrysts from Station 24 (Table 14.5, nos. 1 and 2) do not indicate distinctly different parental compositions, but preserve evidence of a more primitive magma which can be related to the glass rind compositions by olivine–orthopyroxene fractionation.

14.2.6 *Cr-spinels*

Cr-spinel is a ubiquitous accessory phase in N Tonga high-Mg lavas, occurring as small groundmass euhedra, microphenocrysts and inclusions in all phenocryst phases. Representative Cr-spinel compositions are given in Table 14.6 and summarized in Figures 14.5a, b and c. The Cr # versus Mg # relationships of Cr-spinel compositions from dredged N Tonga lavas have been previously reported by Falloon & Green (1986) (see Fig. 14.5). Cr # values are very high, overlapping with Cr-spinels from boninites, and are distinctly more Cr-rich than Cr-spinels from the Troodos Upper Pillow Lavas and MORB (Falloon & Green 1986).

Table 14.5 Microprobe analyses of olivine-hosted glass inclusions and calculated parental magma compositions.

	1		2		3		
	a	b	a	b	a	b	4
SiO_2	58.58	55.77	59.06	57.67	57.07	52.80	54.72
TiO_2	0.35	0.29	0.39	0.36	0.39	0.29	0.15
Al_2O_3	12.42	10.39	12.36	11.46	12.71	9.53	9.57
FeO	8.25	9.04	8.05	8.49	9.80	10.88	7.22
MgO	6.25	12.89	6.68	9.53	5.65	15.71	17.15
CaO	12.36	10.34	11.79	10.93	12.75	9.56	10.50
Na_2O	1.07	0.89	1.19	1.10	1.39	1.04	0.54
K_2O	0.35	0.29	0.39	0.36	0.14	0.10	0.15
Cl	0.12	0.10	0.10	0.09	0.12	0.09	–
total	99.99	100.00	100.01	99.99	100.02	100.00	100.00
Mg #	0.60	0.74	0.62	0.69	0.53	0.74	0.82
CaO/Na_2O		11.55		9.90		9.15	19.40
CaO/Al_2O_3		0.99		0.95		1.00	1.09
CaO/TiO_2		35.31		30.23		32.69	70.00
Al_2O_3/TiO_2		35.48		31.69		32.59	64.00

(–) Not determined.
Mg # calculated on the basis of $Fe^{2+}/(Fe^{2+} + Fe^{3+}) = 0.9$.
(1a) Broad-area scan analysis of glass inclusion in Fo_{90} olivine, sample 6-2; (b) calculated parental composition in equilibrium with Fo_{90} olivine; (2a) broad-area scan of glass inclusion in Fo_{88} olivine, sample 6-2; (2b) calculated parental composition in equilibrium with Fo_{88} olivine, sample 6-2; (3a) broad-area scan analysis of glass inclusion in Fo_{90} olivine, sample 5-25; (3b) calculated parental composition in equilibrium with Fo_{90} olivine; (4) calculated parental composition in equilibrium with Fo_{94} olivine, sample 5-25 (Falloon & Green 1986).

As Cr-spinel composition is extremely sensitive to the liquid composition from which it crystallizes, the composition of Cr-spinel preserves the history of magmatic differentiation in a suite of related lavas (Dick & Bullen 1984). The composition of Cr-spinel in the high-Mg lavas from Station 23 provides evidence that mixing of relatively evolved and primitive magmas occurred in a sub-arc magma chamber. Figure 14.5a shows the variation in trivalent cation ratios $Y(Cr)$, $Y(Al)$ and $Y(Fe^{3+})$ of Cr-spinels in the high-Mg lavas. The trivalent ratios correlate well with each other, indicating that the dominant substitution is $2Cr^{3+} \Leftrightarrow Al^{3+}Fe^{3+}$ (Fig. 14.5a). Smooth trends in Figure 14.5a also indicate that the Cr-spinels can be related to a single magmatic trend. Cr-spinel inclusions in xenocrystal olivine in basaltic andesite 5-27 (Table 14.6, no. 11) plot at higher $Y(Cr)$ and lower $Y(Al)$ at a given $Y(Fe^{3+})$ than the Cr-spinels hosted in the high-Mg lavas, and therefore belong to a separate magmatic differentiation trend than Cr-spinels in the high-Mg lavas.

In Figure 14.5b, Cr # versus Mg # of Cr-spinels are plotted according to individual host rocks. Cr-spinels in high-Mg lavas from Station 23 display a

Table 14.6 Representative electron microprobe analyses of chromites from north Tonga lavas.

	1		2			3	4	5	6	7	8	9	10	11
	C	R	C	R_1	R_2									
TiO_2	0.84	0.35	–	–	–	1.61	0.42	0.30	0.20	0.30	0.32	0.23	–	–
Al_2O_3	8.54	7.19	6.79	5.81	6.75	10.62	10.72	5.95	8.37	11.46	6.69	7.90	7.04	4.22
Cr_2O_3	41.74	53.36	64.00	62.82	64.18	32.89	52.48	61.23	59.62	51.35	60.79	57.44	61.23	61.38
FeO	38.61	28.40	13.18	17.76	13.60	46.83	24.78	22.00	17.05	25.01	18.05	22.73	18.61	27.25
MgO	10.27	10.49	16.03	12.61	15.46	7.65	11.47	10.52	14.76	11.89	14.15	11.51	13.13	7.16
total	100.00	99.79	100.00	100.00	99.99	99.60	99.87	100.00	100.00	100.01	100.00	99.81	100.01	100.01
Y(Al)	0.165	0.141	0.128	0.132	0.128	0.208	0.205	0.117	0.158	0.218	0.128	0.153	0.136	0.086
Y(Cr)	0.542	0.701	0.812	0.816	0.817	0.432	0.675	0.809	0.757	0.655	0.780	0.748	0.791	0.837
Y(Fe^{3+})	0.272	0.150	0.060	0.053	0.054	0.320	0.110	0.066	0.080	0.120	0.084	0.093	0.074	0.077
Mg #	0.493	0.515	0.766	0.617	0.742	0.364	0.550	0.520	0.703	0.570	0.680	0.562	0.640	0.370
Cr #	0.766	0.833	0.863	0.861	0.864	0.675	0.770	0.873	0.827	0.750	0.859	0.829	0.854	0.907

Dash (–) indicates below detection limit.
Trivalent and divalent cation ratios calculated on the basis of stoichiometry.
C = core, R = rim.
(1) Microphenocryst 5-25; (2) microphenocryst attached to olivine (Fo_{94}), R_1 rim next to glass, R_2 rim next to olivine 5-25; (3) inclusion in clinopyroxene (Mg # 0.84) 5-25; (4) groundmass 5-28; (5) groundmass 5-24; (6) inclusion in olivine (Fo_{91}) 5-24; (7) inclusion in orthopyroxene (Mg # 0.87) 5-24; (8) inclusion in olivine (Fo_{91}) 6-3; (9) groundmass 6-2; (10) groundmass 6-2; (11) inclusion in resorbed olivine (Fo_{80}) 5-27.

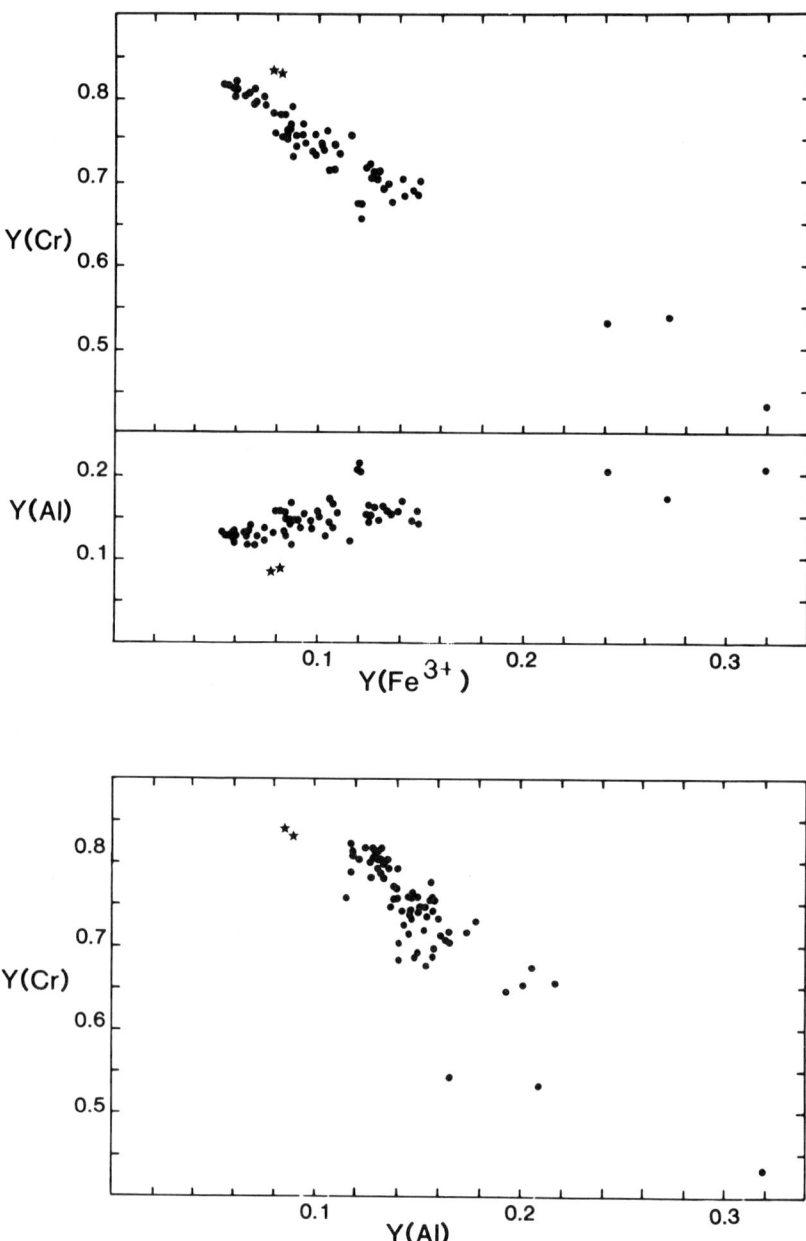

(a)

Figure 14.5 (a).

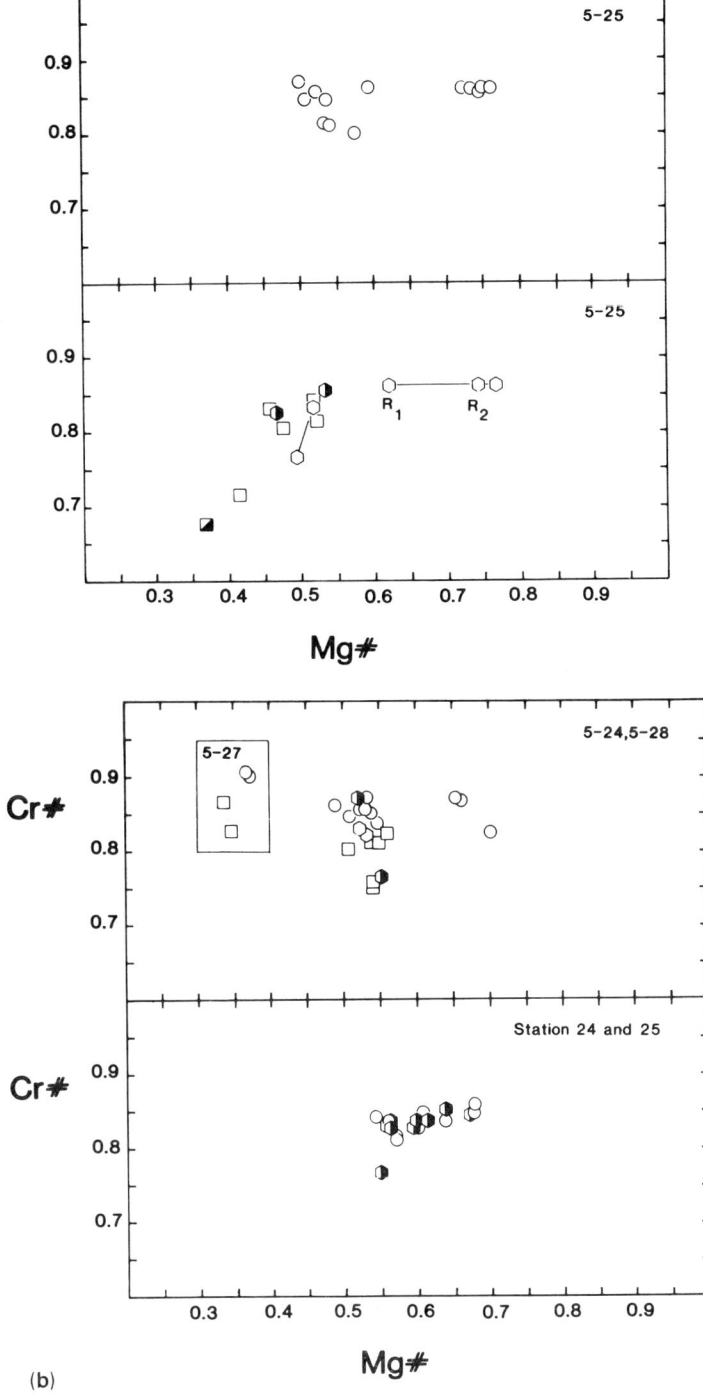

(b)

Figure 14.5 (b).

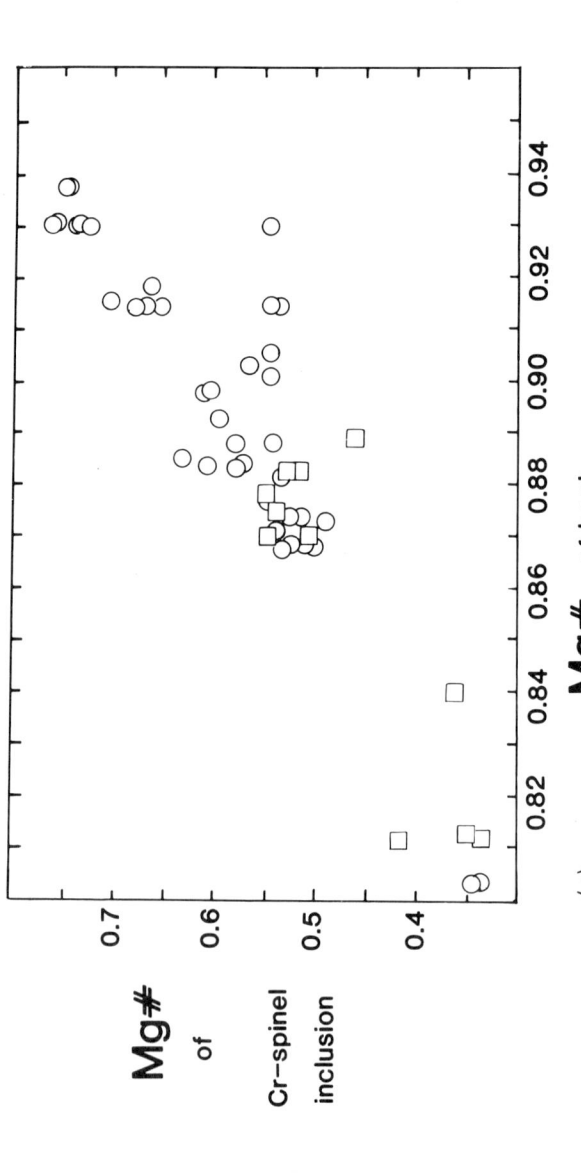

Figure 14.5 (a) Compositional relationships in Cr-spinels from N Tonga. The variation of trivalent cation ratios $Y(Cr)$, $Y(Al)$ with $Y(Fe^{3+})$ and $Y(Cr)$ versus $Y(Al)$. Asterisks are Cr-spinel inclusions in xenocrystal olivine (sample 5-27). (b) Cr # versus Mg # of Cr-spinels in N Tonga lavas. Open circles: Cr-spinel inclusions in olivine phenocrysts. Open squares: Cr-spinel inclusions in orthopyroxene phenocrysts. Half-filled square: Cr-spinel inclusions in clinopyroxene phenocrysts. Half-filled hexagons: Cr-spinel euhedra in groundmass. Open hexagons: Cr-spinel microphenocrysts. R_1 and R_2 refer to rim compositions in Table 14.6 (no. 2, also see text). Cr-spinel compositions enclosed in box are from resorbed xenocrystal olivine in 5-27: open circles, Cr-spinel inclusions in resorbed olivine xenocryst; open squares, Cr-spinel inclusions in surrounding orthopyroxene microphenocrysts. (c) Mg # of Cr-spinel inclusions versus Mg # of olivine (open circles) or orthopyroxene (open squares) host.

significant range in composition. Cr-spinels in sample 5-25 varies in Cr# from 0.67 to 0.86, suggesting mixing of relatively evolved and primitive magmas. The linear trend in Figure 14.5c, showing the relationship between Cr-spinel Mg# versus the Mg# of their host phase (olivine, orthopyroxene), suggests that the Mg# of the Cr-spinels is in equilibrium with their respective hosts.

Some Cr-spinels included in olivine have significantly lower Mg# than expected given the Mg# of their hosts. This may be due to subsolidus re-equilibration of the Cr-spinel, as the Mg# of Cr-spinel is temperature-sensitive (Dick & Bullen 1984). However, owing to the size of the olivine host, little change is observed in olivine Mg#. An example of subsolidus re-equilibration is given in Table 14.6 (no. 2) and Figure 14.5b. Here, a Cr-spinel is attached to a magnesian olivine phenocryst (Fo_{94}; the olivine host is significantly out of equilibrium with the groundmass) such that one half is enclosed by the olivine and the other half is enclosed by the groundmass. The rim of the Cr-spinel next to olivine has a much higher Mg# than the rim enclosed by the groundmass, yet the Cr# is identical. This can be explained by subsolidus re-equilibration of the Cr-spinel Mg# when in contact with the lower-temperature groundmass. The overall Cr-spinel compositional trend, taking into account the effects of subsolidus re-equilibration, is consistent with fractionation of olivine and pyroxenes depleting residual liquids in Cr and Mg. The low $Y(Fe^{3+})$ contents of the Cr-spinels (generally < 0.085) can be explained by crystallization under relatively reducing f_{O_2} conditions, significantly lower than FMQ (Barnes 1986, Murck & Campbell 1986).

14.3 Implications of mineral chemistry for magma mixing

Evidence from compositions of olivines, pyroxenes and Cr-spinels and olivine-hosted glass inclusions demonstrates that high-Mg lavas from Station 23 are the result of magma mixing of relatively evolved and primitive magmas. Compositions of olivine-hosted glass inclusions further suggest that the primitive magmas involved are distinct from each other and cannot be related by crystal fractionation. The Group B high-Mg lava sample 5-25 requires at least three different liquid compositions: liquid 1, which crystallized Fo_{94} olivine phenocrysts; liquid 2, which crystallized Fo_{90} olivine phenocrysts; and liquid 3, which crystallized less magnesian orthopyroxene and clinopyroxene. The groundmass composition of sample 5-25 (Table 14.4) is a result of some combination of these three liquids; the wholerock composition is the result of the combination of these three liquids plus their entrained phenocryst populations. Such evidence for magma mixing places severe doubts on the validity of using wholerock compositions alone to establish primary magma compositions in island-arc settings. Although the groundmass composition of sample 5-25 may be considered a parental liquid composition due to its high Mg#, it cannot be used as a constraint on the nature of primary island-arc magmas.

Compositions of olivine-hosted glass inclusions, however, may be more reliable in this regard (Falloon & Green 1986).

The plagioclase phenocryst compositions of Station 23 basaltic andesites also suggest that magma mixing has occurred. These plagioclase compositions could only have crystallized from liquids with much higher CaO/Na_2O than observed in their respective host rocks (Table 14.4). The CaO/Na_2O of glass inclusions in olivine have appropriately high CaO/Na_2O and thus provide a ready solution to the presence of such calcic plagioclase. Owing to the lower density of plagioclase relative to magnesian olivines and pyroxenes, plagioclase may float to the top of a sub-arc magma chamber after crystallizing from a primitive high-CaO/Na_2O magma at the base of a magma chamber. The plagioclase is then incorporated into a more evolved magma composition before eruption. Considerations of magma chamber dynamics allow more complex models to be proposed (O'Hara & Mathews 1981, Sparks *et al.* 1977, 1980, Huppert & Sparks 1980); these, however, are beyond the scope of this chapter.

Although the presence of resorbed xenocrystal pyroxenes (Table 14.3, Fig. 14.2) in high-Mg lavas from Stations 24 and 25 suggests that some magma mixing was involved in the petrogenesis of these lavas, other considerations indicate that this was relatively minor compared to the Station 23 lavas.

14.4 Geochemistry

Major-, trace-element and REE geochemistry of representative dredged samples from N Tonga are given in Table 14.7, and Sr and Nd isotopic compositions for representative lavas are listed in Table 14.8. Sr and Nd isotopic analyses were performed at the Research School of Earth Sciences, Canberra, following the method outlined in Cameron *et al.* (1983). Included in Tables 14.7 and 14.8 are representative samples from Stations 15, 21 and 31 (Falloon *et al.* 1987). Figure 14.6 shows a series of normalized multi-element abundance patterns for the N Tonga lavas, together with two representative patterns for sub-aerial lavas from the sub-aerial Tongan arc (Fig. 14.6a, L1 and T116, respectively).

Falloon *et al.* (1987) have presented and briefly discussed the geochemistry and affinities of the dredged lavas. The N Tongan high-Mg lavas and basaltic andesites belong to a low-Ti, low-K island-arc tholeiite series, based on their low TiO_2 and K_2O contents, lack of phenocrystal amphibole, the presence of extremely calcic plagioclase and low LILE abundances. However, with the exception of Station 15 basaltic andesites, these lavas fall in the calc-alkaline field on a FeO^*/MgO versus SiO_2 diagram (see Fig. 4 of Falloon *et al.* 1987) due to their relatively high MgO contents at a given SiO_2 level compared to most other island-arc tholeiite suites. Station 23 basaltic andesites show strong affinities to Tofua (sub-aerial N Tonga) magmatic arc (TM arc) volcanics on the basis of similar REE patterns and normalized element abundance patterns (Figs 14.6a and c). The high-Mg lavas from Stations 23, 24 and 25 show strong affinities to low-Ti ophiolitic basalts, such as Troodos Group II and III Upper Pillow Lavas (Cameron 1985, Flower & Levine 1987) and are distinctly different from clinoenstatite-bearing boninites. Below the petrogenesis of the dredged N Tonga lavas is discussed in the light of new Sr and Nd isotopic data.

Table 14.7 Representative major-, trace- and rare-earth-element compositions of dredged lavas from north Tonga and Lau Basin.

	1-38	3-44	3-24	5-20	5-24	5-25	6-3	7-18	11-3
SiO_2	53.43	54.35	59.19	56.30	52.95	53.83	56.04	54.72	49.13
TiO_2	0.96	0.20	0.25	0.62	0.15	0.36	0.31	0.45	0.71
Al_2O_3	16.90	10.67	10.96	15.20	7.92	9.63	10.60	10.90	15.07
FeO	8.47	9.41	8.83	9.77	9.70	9.71	8.46	8.65	9.70
MnO	0.16	0.19	0.18	0.17	0.21	0.19	0.17	0.17	0.18
MgO	6.08	14.99	10.80	5.33	20.89	16.30	13.61	12.97	9.18
CaO	10.84	8.66	7.78	10.20	7.18	8.50	9.17	9.65	13.13
Na_2O	2.58	1.14	1.37	2.08	0.85	1.27	1.14	1.52	2.85
K_2O	0.48	0.35	0.56	0.24	0.14	0.16	0.43	0.71	0.02
P_2O_5	0.10	0.03	0.07	0.07	0.01	0.04	0.07	0.26	0.03
LOI	1.15	0.05	0.01	0.71	0.13	0.24	1.87	1.69	−0.72
Mg#	0.56	0.76	0.71	0.52	0.80	0.77	0.76	0.75	0.65
Ba	29	106	140	55	34	41	105	230	14
Rb	9	6	10	4	2	3	8	12	2
Sr	132	151	175	126	46	68	159	348	48
Zr	59	12	19	39	5	20	32	53	21
Nb	1	<1	2	<1	<1	<1	8	16	<1
Y	29	7	7	17	5	10	8	11	22
Sc	40	50	43	44	44	44	43	44	51
V	259	247	251	298	191	225	214	223	299
Ni	52	275	117	37	501	341	189	199	133
Cr	204	1095	605	75	2027	1294	927	760	350
La	2.84	1.74	4.0	2.82	–	1.59	6.53	15.98	3.43
Ce	9.23	4.10	8.44	6.44		3.34	14.74	37.72	–
Pr	1.56	–	1.06	1.01	–	0.55	1.73	4.48	5.52
Nd	8.40	2.85	4.51	5.17	0.90	2.72	6.36	17.29	5.83
Sm	2.84	0.82	1.32	1.72	0.28	0.89	1.30	3.23	8.17
Eu	1.11	–	0.50	0.62	–	0.36	0.45	1.01	8.61
Gd	4.04	1.27	1.38	2.37	0.59	1.27	1.35	2.68	10.58
Dy	5.03	1.16	1.65	3.10	0.77	1.62	1.27	1.84	11.57
Er	3.28	0.80	1.20	2.15	0.46	1.10	0.63	0.93	11.74
Yb	3.20	0.86	1.14	2.07	0.66	1.14	0.58	0.79	11.39

LOI = loss on ignition.
All iron as FeO.
(−) Not determined.
All analyses resummed to 100% volatile-free.
Major elements in wt%, trace elements in parts per million.
Mg# determined on the basis of $Fe^{2+}/(Fe^{2+} + Fe^{3+}) = 0.9$.

Table 14.8 Sr and Nd isotopes of representative dredged lavas from north Tonga and Lau Basin.

Sample no.	Rb	Sr	Sm	Nd	Rb/Sr	Sm/Nd	$^{87}Sr/^{86}Sr$	$2\sigma \times 10^{-4}$	$^{143}Nd/^{144}Nd$	$2\sigma \times 10^{-4}$	ε_{Nd}	2σ
1-38	9	132	2.84	8.40	0.0682	0.338	0.702937	0.13	0.512958	0.10	6.01	0.19
3-24	10	175	1.32	4.51	0.0571	0.293	0.704384	0.13	0.512785	0.06	2.63	0.12
3-44 unl							0.704537	0.10				
3-44	6	151	0.82	2.85	0.0397	0.288	0.704528	0.35	0.512787	0.09	2.67	0.17
5-20 unl							0.703861	0.15				
5-20	4	126	1.72	5.17	0.0317	0.333	0.703844	0.13	0.512961	0.13	6.07	0.25
5-24	2	46	0.28	0.90	0.0435	0.311	0.704422	0.11	0.512770	0.06	2.34	0.12
5-25	3	68	0.89	2.72	0.0441	0.327	0.704047	0.25	0.512961	0.09	6.07	0.17
6-3	8	159	1.30	6.36	0.0503	0.204	0.704515	0.27	0.512708	0.08	1.13	0.17
7-18	12	348	3.23	17.29	0.0345	0.187	0.704175	0.13	0.512730	0.01	1.56	0.19
11-3	2	48	1.57	3.48	0.0417	0.451	0.703364	0.10	0.512963	0.01	6.10	0.20

unl = Unleached, all other samples leached in 6 N HCl overnight.
$\varepsilon_{Nd(0)} = 0.512650$.
$^{143}Nd/^{144}Nd$ ratio normalized to a $^{146}Nd/^{144}Nd$ ratio of 0.7219.

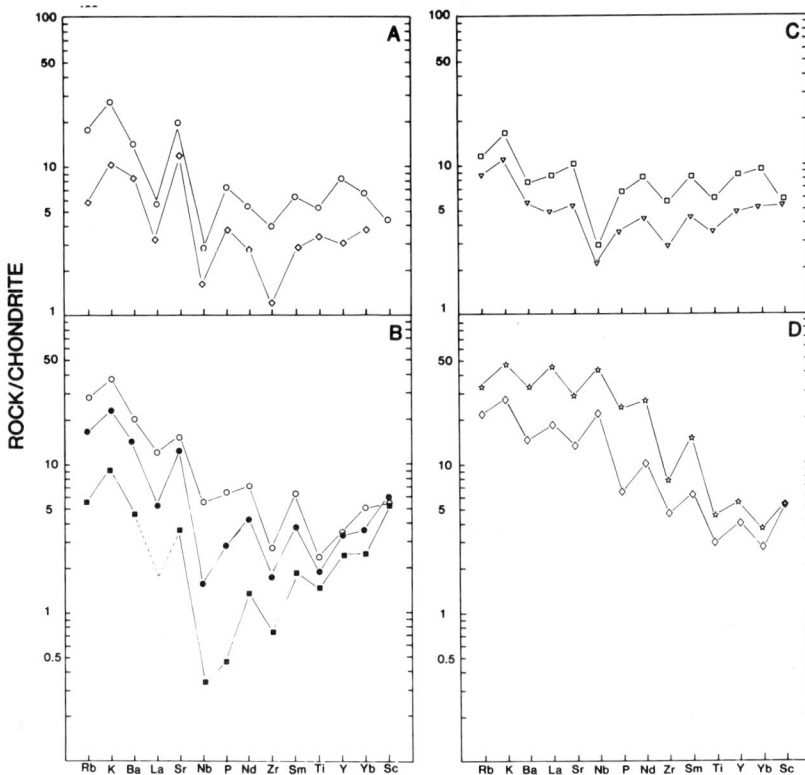

Figure 14.6 Representative chondrite-normalized abundance patterns for dredged N Tonga lavas. Normalizing values are from Thompson *et al.* (1983); normalizing value for Sc from Hickey & Frey (1982). The ordering of elements is taken from Hickey & Frey (1982), except that the order of LILE (Rb, K, Ba, Sr) is based on experimental studies of Tatsumi *et al* (1986). (a) Representative patterns of Tofua magmatic arc basaltic andesites (Ewart *et al.* 1973): open circles, L1, from the island of Late, south Tonga; open diamonds, T116, from the island of Tafahi, north Tonga. (b) Representative patterns of Station 21 lavas and Group A high-Mg lava from Station 23: open circles, 3-24, Station 21; full circles, 3-44, Station 21; full squares, 5-24, Station 23. La is estimated. (c) Representative patterns of Station 23 basaltic andesites and Group B high-Mg lava: open squares, 5-20 basaltic andesite; open inverted triangles, 5-25 high-Mg lava (Group B). (d) Representative patterns of high-Mg lavas from Stations 24 and 25: open diamonds, 6-3, Station 24; open stars, 7-18, Station 25.

14.5 Petrogenesis

14.5.1 Introduction

Petrogenesis of island-arc lavas requires mixing of two or more source components (Gill 1981, Hickey & Frey 1982, Cameron *et al.* 1983, Morris &

Hart 1984, White & Patchett 1984, Arculus & Powell 1986, von Drach *et al.* 1986, Wheller *et al.* 1987). However, controversy exists as to the identity of these source components. We believe that the petrogenesis of the N Tonga lavas is best explained by a model involving the mantle wedge plus two distinct 'enriched' components (Hickey & Frey 1982, Varne & Foden 1986, Wheller *et al.* 1987, Bloomer 1987). In this model, a depleted peridotitic mantle source (more depleted than N-MORB and OIB sources) is 'enriched' in silicate-incompatible elements by a mantle metasomatic phase, with only minor input from the subducted slab, and minimal involvement of subducted sediment. The mantle-derived 'enriching' component has similarities to the enriched component involved in petrogenesis of both ocean island basalts of the Cook–Austtral–Samoa islands, which are part of the Dupal mantle anomaly (Palacz & Saunders 1986), and also Group II ultrapotassic rocks of Foley *et al.* (1987). A second 'enriching' component is either a hydrous, slab-derived fluid phase or, alternatively, a mantle-derived fluid phase. The discussion below outlines in more detail the evidence for these three components. The approach is essentially empirical and qualitative; no quantitative numerical model is presented as such modelling at this stage would be unconstrained and essentially *ad hoc*.

14.5.2 The depleted component

In Figure 14.7, the CIPW molecular normative chemistry of the wholerock, groundmass–glass and olivine-hosted glass inclusion compositions of the N Tonga lavas are shown in the projection from diopside onto the base of the basalt tetrahedron with apices jadeite plus Ca-Tschermaks molecule (Jd + CaTs)–quartz (Qtz)–olivine (Ol). This diagram is helpful in evaluating the roles of differing source compositions, different degrees of partial melting and different depths of magma segregation on lava compositions (Green *et al.* 1987). The N Tonga compositions are also compared with the fields of Cape Vogel boninites (Jenner 1981), Troodos Upper Pillow Lavas (Cameron 1985, Duncan & Green 1987), Tofua arc volcanics (Ewart *et al.* 1973), Lau Basin backarc-basin basalts (Hawkins 1976, Hawkins & Melchior 1985) and primitive MORB glasses (references in Falloon & Green 1987).

Also shown in Figure 14.7 are olivine + orthopyroxene ± clinopyroxene + liquid cotectics defined by partial melting experiments on a MORB pyrolite composition at 10, 15, 20 and 30 kbar (Falloon & Green 1987, 1988) and Tinaquillo lherzolite at 5 kbar (Jaques & Green 1980, Falloon *et al.* 1989). An important constraint on the range of liquid compositions able to be produced from any specific source composition is an olivine control line drawn from the olivine apex through the bulk composition (Green *et al.* 1987). Liquid compositions can only be produced to the left-hand side of this control line during equilibrium batch partial melting. For example, primitive MORB glasses plot well to the left of the olivine control line through the MORB

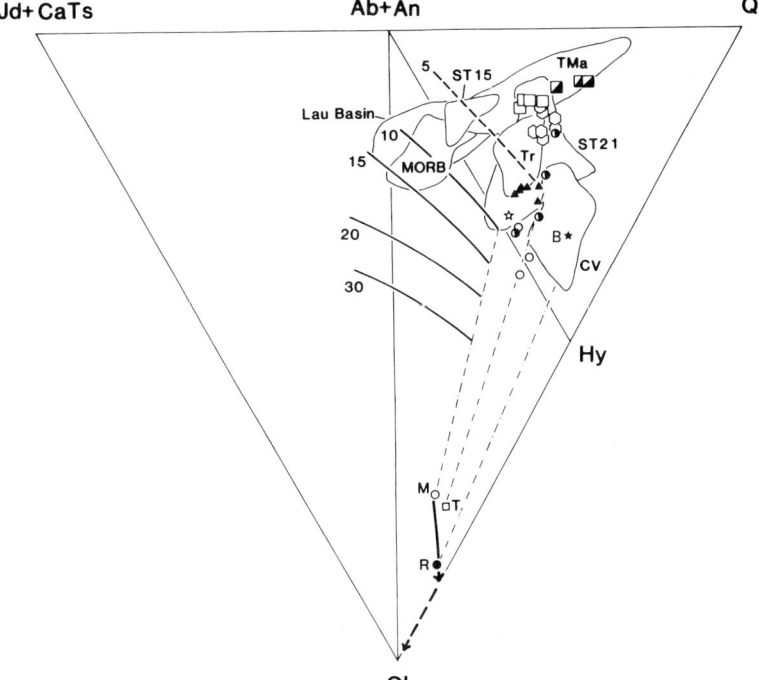

Figure 14.7 CIPW molecular normative projection from Di onto the base of the basalt tetrahedron jadeite plus Ca-Tschermaks molecule (Jd + CaTs)–olivine (Ol)–quartz (Qtz). After Green (1970), Hy, hypersthene; Ab, albite; An, anorthite.

Open circle M is a MORB pyrolite composition (Fallon & Green 1987, 1988). Open square T is Tinaquillo lherzolite (Jaques & Green 1980). Full circle R is the residual mantle composition after removal of a stage-one MORB picrite composition (Green et al. 1979). Asterisk B is the boninite parental magma composition calculated by Walker & Cameron (1983). Open star is calculated parental magma composition to Troodos Group II and III Upper Pillow Lavas (Duncan & Green 1987). Half-filled squares are groundmass compositions of Station 23 basaltic andesites (Table 14.4). Open hexagons are groundmass and glass compositions of north Tonga high-Mg lavas (Table 14.4). Half-filled circles are calculated parental magma compositions in equilibrium with magnesian olivine phenocrysts from north Tonga (Table 14.5). Open squares are wholerock compositions of Station 23 basaltic andesites (Table 14.4). Full triangles are wholerock compositions of Stations 24 and 25 high-Mg lavas (Table 14.4). Open circles are wholerock compositions of Station 23 high-Mg lavas (Table 14.4).

Field enclosing MORB are primitive MORB glasses (references in Falloon & Green 1987). TMa, Tofua magmatic arc; field encloses range in wholerock compositions (Ewart et al. 1973). CV, Cape Vogel, Papua New Guinea; field encloses wholerock compositions of boninites (Jenner 1981). Tr, Troodos; field encloses range in wholerock compositions of Group II and III Upper Pillow Lavas (Cameron 1985). ST21, Station 21; field encloses range in wholerock compositions of Station 21 lavas (Falloon et al. 1987, Falloon unpubl. data). ST15, Station 15; field encloses range in wholerock compositions of Station 15 basaltic andesites. Lau Basin field encloses range in quenched glass compositions (Hawkins & Melchoir 1985).

Full curves: locus of liquids in equilibrium with olivine + orthopyroxene ± clinopyroxene at 10, 15, 20 and 30 kbar from MORB pyrolite (Falloon & Green 1987, 1988). Broken curve: locus of liquids in equilibrium with olivine + orthopyroxene ± clinopyroxene at 5 kbar from Tinaquillo lherzolite (Falloon et al. 1989). Chain lines: locus of liquids in equilibrium with olivine only. Arrows denote change in residue compositions with progressive equilibrium batch partial melting: full arrow indicates lherzolite residues, and broken arrow indicates harzburgite residues. Dunitic residues plot at the Ol apex.

pyrolite composition, confirming that MORB pyrolite and possibly more refractory sources are capable of yielding primitive MORB glasses. However, as all primitive N Tonga compositions plot to the right of the olivine control line through MORB pyrolite, a MORB pyrolite-type source cannot partially melt to yield these lavas; a more refractory source is required.

More refractory compositions such as Tinaquillo lherzolite, or the residue composition (point R, Fig. 14.7) after extraction of a MORB picrite composition from a MORB pyrolite starting composition, are capable of producing primitive N Tonga high-Mg lavas by partial melting. The composition labelled T in Figure 14.7 is the calculated parental composition for the Troodos Upper Pillow Lavas, studied experimentally by Duncan & Green (1987). This composition is very similar to the Tongan high-Mg lavas. The Troodos composition was found to be multiply saturated in olivine and orthopyroxene at $1360°C/8$ kbar and thus may be a primary melt leaving a harzburgite residue (Duncan & Green 1987). Primary H_2O content of this Troodos parent magma was estimated by Duncan & Green (1987) to have been $<1.0\%$. The results of the experimental study by Duncan & Green (1987), combined with information gained from Figure 14.7, suggest that the high-Mg lavas from Stations 24 and 25 plus the refractory magma compositions identified from olivine-hosted glass inclusions are the result of partial melting of a refractory mantle peridotite at pressures <10 kbar, leaving a harzburgite residue. Water contents were probably $<1\%$, similar to the parent magma of the Group III Troodos Upper Pillow Lavas.

Boninites from Cape Vogel, Papua New Guinea (Jenner 1981, Dallwitz 1968) plot further to the right towards Hy (Fig. 14.7) than the N Tonga lavas and as such require an even more refractory mantle source. Composition B (Fig. 14.7) is the parental Cape Vogel boninite composition of Walker & Cameron (1983). Experimental work by Jenner (1982), Van der Laan *et al.* (1989) and Umino & Kushiro (1989) suggests boninite primary magmas are the result of hydrous partial melting ($>1\%$ H_2O) at low pressures (<8 kbar).

In summary, the CIPW molecular normative chemistry of the high-Mg lavas from N Tonga suggests that the mantle source was more refractory than MORB sources, but not as refractory as boninite sources. The CaO/Al_2O_3 of the N Tonga high-Mg lavas provide a further constraint on the nature of the mantle source. In Figure 14.8, the change in CaO/Al_2O_3 of the residue during equilibrium partial melting of MORB pyrolite is seen to increase with increasing partial melting until clinopyroxene is exhausted; CaO/Al_2O_3 then falls sharply until orthopyroxene is exhausted. This results in a large variation in CaO/Al_2O_3 with little change in bulk rock Mg#, as is observed in refractory harzburgites from Papuan ophiolites (Fig. 14.8). The CaO/Al_2O_3 of primitive N Tonga high-Mg lavas and calculated refractory magma compositions (Table 14.5) range from 0.8 to 1.10, higher than CaO/Al_2O_3 of primitive MORB glasses (0.67–0.88) and notably higher than CaO/Al_2O_3 of clinoenstatite-bearing boninite from Cape Vogel (<0.66). This is consistent with the source

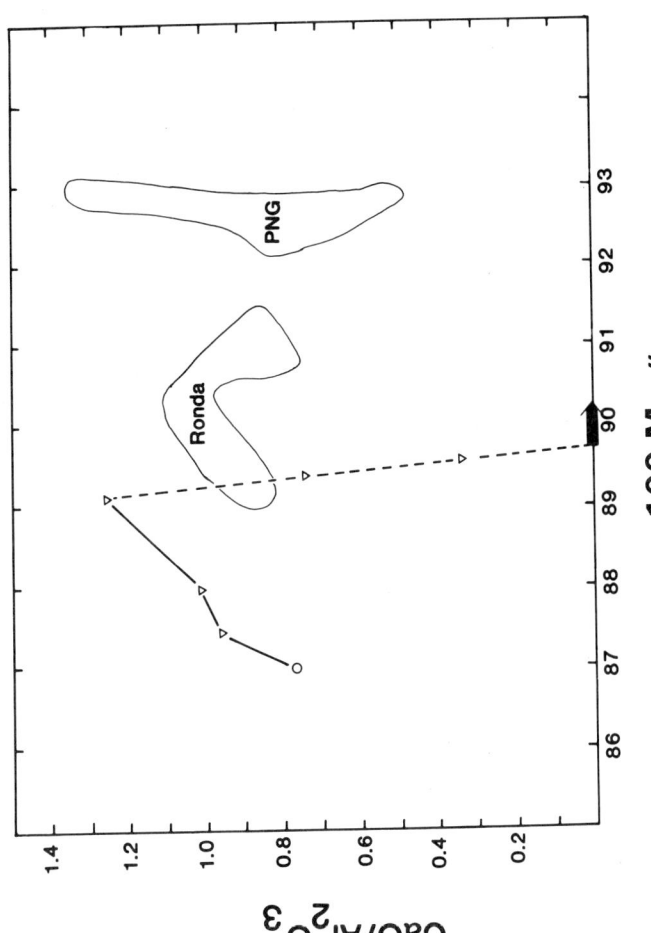

Figure 14.8 CaO/Al_2O_3 of residue compositions from equilibrium batch partial melting of a MORB pyrolite composition at 10 kbar (Falloon & Green 1987) versus 100 Mg # of the residue. Ronda encloses the field of peridotite compositions from the Ronda high-temperature peridotite intrusion (Frey et al. 1985). PNG encloses the field of harzburgite from the Papua New Guinea ultramafic belt (Jaques & Chappell 1980). Open circle is MORB pyrolite starting composition; open inverted triangles are calculated residue compositions (Falloon unpubl. data). Full lines, lherzolite residues; broken lines, harzburgite residues, bold arrow, dunitic residues.

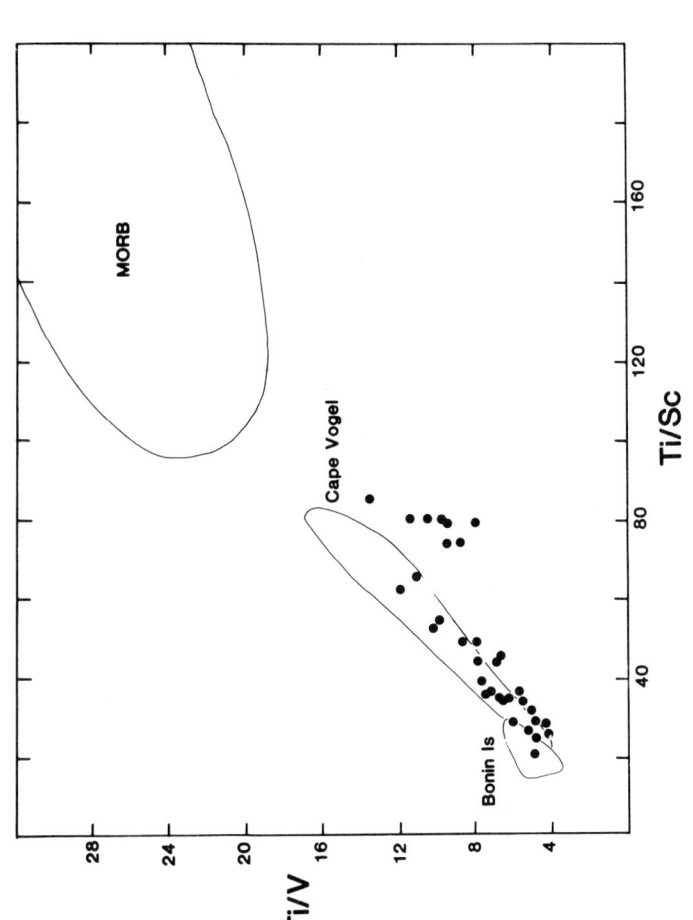

Figure 14.9 Ti/V versus Ti/Sc of dredged N Tonga lavas (this study, Falloon et al. 1987, Falloon unpubl. data). Fields for Cape Vogel, Papua New Guinea, and Bonin Islands, Japan, boninites compiled from Jenner (1981), Hickey & Frey (1982) and Cameron et al. (1983). MORB field from Hickey & Frey (1982).

for N Tonga high-Mg lavas and calculated refractory magma compositions (Table 14.5) being more refractory than MORB sources, but still containing some clinopyroxene. The clinopyroxene would be exhausted during partial melting, leaving residual harzburgite; however, the former presence of clinopyroxene is required to explain the very high CaO/Al_2O_3 in the N Tonga lavas and glass inclusions in olivine. The CaO/Al_2O_3 of boninites from Cape Vogel can probably be explained by partial melting of a more refractory harzburgite source containing no clinopyroxene.

The high CaO/Na_2O of the N Tonga lavas and olivine-hosted glass inclusions (>10) also indicate a more refractory source than MORB ($CaO/Na_2O < 9.5$). Many other trace-element characteristics of the N Tonga lavas also suggest a refractory mantle source, taking into account the effect of subsequent enriching event(s). For instance, their low Ti/V and Ti/Sc compared to MORB overlap with Bonin Islands and Cape Vogel boninites (Fig. 14.9) and are best explained by melting of a depleted mantle source (Hickey & Frey 1982, Nelson *et al.* 1984). The low TiO_2, Y and HREE abundances in the N Tonga high-Mg lavas compared to N-type and depleted MORB mantle (Figs 14.6 & 14.11a) and their relatively high Al_2O_3/TiO_2 and CaO/TiO_2 are also consistent with a mantle source more depleted in these elements than MORB source mantle (Sun & Nesbitt 1978). Finally, many petrographic and mineral chemical features of the N Tonga lavas also suggest a refractory mantle source; for example, the very magnesian olivine phenocrysts (up to Fo_{94}) and very Cr-rich spinels (up to $Cr\# = 0.87$), which are significantly more magnesian and Cr-rich respectively than observed in MORB (olivine generally $<Fo_{92}$, Cr-spinel $Cr\# < 0.60$).

14.5.3 *Enriched source components*

Figure 14.10 shows the ε_{Nd} versus $^{87}Sr/^{86}Sr$ relationships of the N Tonga lavas. Compared to Pacific MORB, the N Tonga lavas have lower ε_{Nd} and higher $^{87}Sr/^{86}Sr$, and, based on ε_{Nd} values (Table 14.8), they define two broad groups. One group, which includes basaltic andesites from Stations 15 and 23 along with the Group B high-Mg lava 5-25 and the Lau Basin backarc-basin basalt 11-3, all have ε_{Nd} of $+6.0$, despite showing a significant range in $^{87}Sr/^{86}Sr$. The range in $^{87}Sr/^{86}Sr$ is considered to be a source feature and not to be due to 'post-magmatic' sea-water alteration effects, as a comparison of leached and unleached samples shows only small differences in $^{87}Sr/^{86}Sr$. These isotopic compositions are similar to analysed lavas from the Kermadec and Tongan islands (Ewart & Hawkesworth 1987) and to rocks from the Lau Basin and young East Pacific Rise seamounts (Fig. 14.10). A similar range of $^{87}Sr/^{86}Sr$ at constant ε_{Nd} was observed in Aleutian arc lavas by von Drach *et al.* (1986), and was considered to be a result of efficient localized mantle mixing producing a homogeneous source in terms of ε_{Nd}, which was subsequently invaded by a component with a high Sr/Nd ratio. Sample 1-38 from

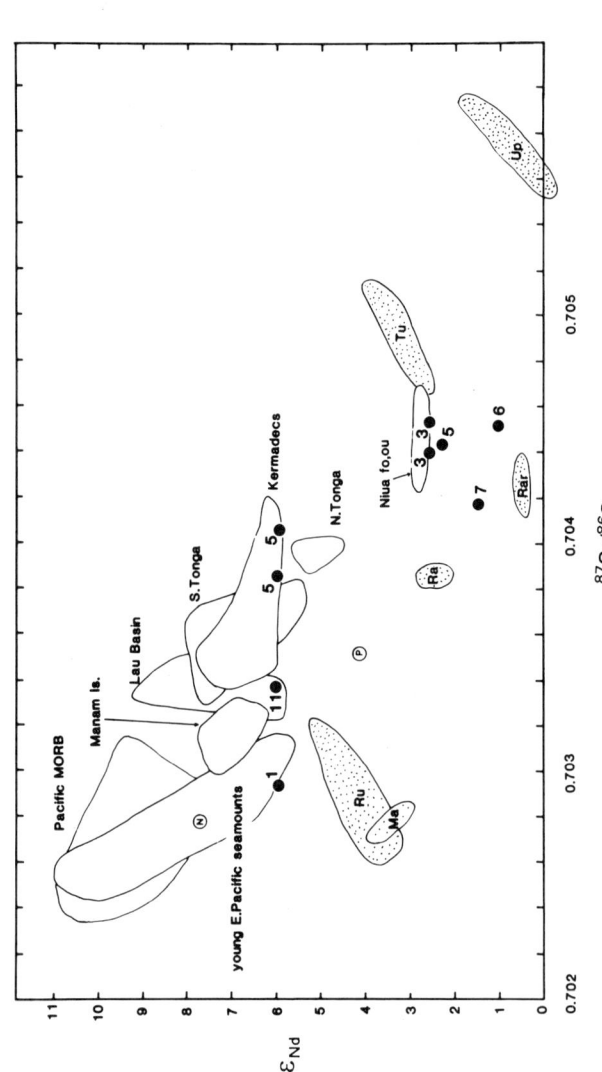

Figure 14.10 Nd versus Sr isotopic composition of dredged N Tonga lavas. Full circles: N Tonga dredged lavas, data from Table 14.8, numbers refer to dredge stations. Data for S Tonga, N Tonga, Kermadecs and Niua fo'ou from Ewart & Hawkesworth (1987). Data for Lau Basin backarc basalts from Sinton et al. (1988) and Carlson et al. (1978). Data for Manam Island from Johnson et al. (1985). Data for young East Pacific seamounts from Zindler et al. (1984). Data for Pacific MORB from MacDougall & Lugmair (1985), Newsom et al. (1986), Allegre et al. (1980), Ito et al. (1980) and White & Hofmann (1982). N and P refer to N-type and P-type MORB respectively (Le Roex et al. 1985). Stippled fields are data from ocean islands of the Samoa–Cook–Austral chain (Palacz & Saunders 1986): Ru, Rurutu; Ma, Mangaia; Ra, Rapa: Rar, Rarotonga; Tu, Tutuila; Up, Upolu.

Station 15 has significantly lower $^{87}Sr/^{86}Sr$ than other analysed Tongan lavas, confirming the conclusion of Falloon *et al.* (1987) that Station 15 basaltic andesites are unrelated to the currently active Tofua magmatic arc, but show strong affinities to Lau Basin backarc crust. The isotopic composition of basaltic andesites from Station 23 confirms their strong chemical affinities to Tofua magmatic arc volcanics (Falloon *et al.* 1987, Ewart *et al.* 1973, Ewart & Hawkesworth 1987).

The second isotopic group has lower ε_{Nd} values, between +1.0 to +2.5, and includes high-Mg lavas from Stations 24, 25 and 23 (Group A) and lavas from Station 21. The analysed Group A high-Mg lava 5-24 has similar ε_{Nd}, $^{87}Sr/^{86}Sr$ and normalized element abundance pattern (Fig. 14.6b) to Station 21 island-arc tholeiites, suggesting Station 23 Group A high-Mg lavas may be parental magmas to Station 21 lavas. The lower ε_{Nd} and higher $^{87}Sr/^{86}Sr$ of this group compared to the higher ε_{Nd} group of N Tonga lavas is consistent with their LREE-enriched REE patterns compared to the flat to LREE-depleted patterns shown by the high-ε_{Nd} group. The low-ε_{Nd} lava group is similar isotopically to lavas from Niua fo'ou Island (Lau) backarc-basin tholeiites and hotspot-associated intraplate volcanics from the Cook–Austral–Samoa Islands (Fig. 14.10) (Palacz & Saunders 1986). The normalized element abundance patterns of these N Tonga lavas (Fig. 14.6) show significant LILE enrichment relative to REE and HFSE, in common with other island-arc lavas (Arculus & Powell 1986), which is inconsistent with a depleted mantle source. Stations 24 and 25 lavas also show significant enrichment not only in LILE but also in LREE and HFSE, especially Nb and Zr. Compared to other N Tonga lavas and Tofua magmatic arc lavas. Stations 24 and 25 lavas have significantly low La/Nb and Ti/Zr and high $(La/Yb)_N$.

14.5.4 *A two-component mixing model*

The evidence for a depleted mantle source with superimposed LILE enrichments is similar in some respects to that seen in boninites (Hickey & Frey 1982, Cameron *et al.* 1983). Both Hickey & Frey (1982) and Cameron *et al.* (1983) proposed two-component mixing models which appear to explain successfully the trace-element and isotopic characteristics of boninites. Hickey & Frey (1982) defined two different end-members on the basis of trace-element and isotopic correlations with Nd. Cameron *et al.* (1983) calculated the amount of an 'enriching' component on the basis of a two-component mixing model involving REE; they found significant trace-element and isotopic correlations with the calculated amount of added enriched component.

The REE variation of the N Tonga lavas can also be explained successfully by a two-component mixing model, as significant correlations exist between trace-element contents, ratios and isotopes with the REE ratios Sm/Nd, La/Nd, or with the calculated amount of enriching component based on the method of Cameron *et al.* (1983). However, poor correlations exist between

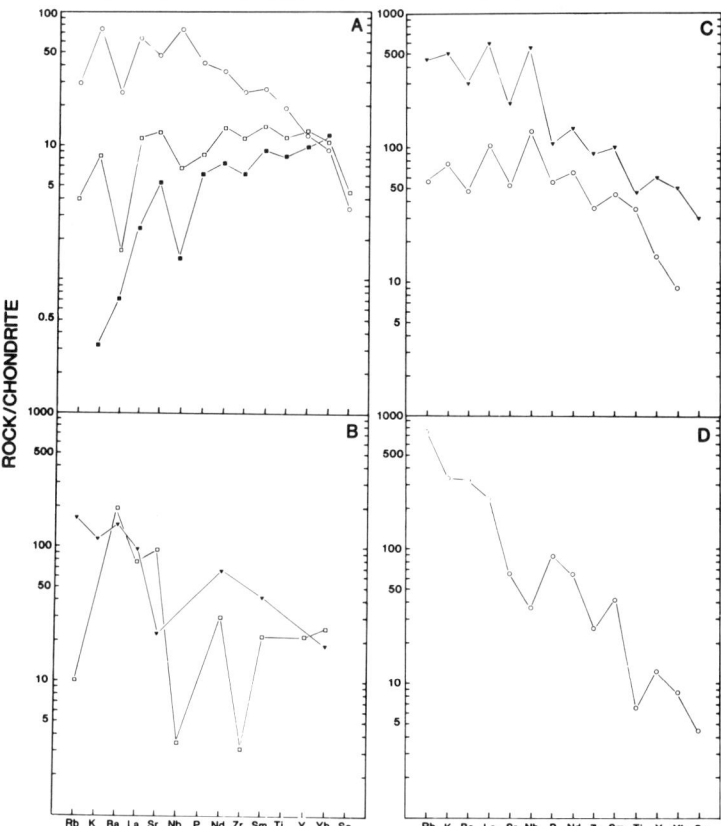

Figure 14.11 Chondrite-normalized abundance patterns of possible depleted and enriched end-member components. (a) Full squares, depleted MORB basalt from site 504B (Hole *et al.* 1984); open squares, N-type MORB V27-19 (Le Roex *et al.* 1985); open circles, P-type MORB V33-66 (Le Roex *et al.* 1985). (b) Open squares: Pacific authigenic mean weighted sediment (Hole *et al.* 1984); full inverted triangles, Tonga sediment (Jenner *et al.* 1987). (c) Full inverted triangles, Toro Ankole mafic mafurite C4802 (Foley 1986); open circles, alkali basalt R198, Rapa Island (Palacz & Saunders 1986). (d) Open circles, Batu Tara sample B4155, Indonesia (Stolz *et al.* 1988).

LILE/REE and $^{87}Sr/^{86}Sr$ (e.g. Sr/La has a relatively constant value of 45 for samples 1-38, 11-3, 5-20, 5-25 and 3-24 despite the $^{87}Sr/^{86}Sr$ varying from 0.7029 to 0.7041). Although a two-component mixing model can explain most of the trace-element and isotopic characteristics, it is not a complete explanation.

With regard to HFSE in a two-component mixing model, the low Nb and other HFSE in the Tofua magmatic arc and N Tonga lavas can be attributed to both the nature of the depleted source component and the 'enriching' components. There is no need to invoke any residual titanate phase to explain the high La/Nb in the Tonga lavas, since, for example, depleted Costa Rica MORB from Site 504B (Hole *et al.* 1984) show significant Nb depletion relative to other HFSE (Fig. 14.11a). The relatively low La/Nb in Stations 24 and 25 lavas are due to the nature of the 'enriching' component, as demonstrated by the good correlation of Nb with percentage of 'enriching' component (using the method of Cameron et al. (1983)).

14.5.5 Origin of the 'enriching' component in the two-component mixing model

The nature of the 'enriching' component in arc petrogenesis remains a subject of controversy. The two most likely 'enriching' components capable of producing the trace-element enrichments and range in Nd and Sr isotopes observed in the N Tonga lavas are either subducted oceanic sediment (dominantly pelagic), a mantle-derived phase (fluid or melt) or a combination of these components. In Figures 14.11b and c, normalized element abundance patterns of possible 'enriching' components are shown. Included are two representative pelagic sediments (Fig. 14.11b) and two examples of enrichments seen in mantle-derived rocks from Samoa (Rapa) and the East African rift (Toro Ankole) (Fig. 14.11c). For the dredged N Tongan lavas, based on a two-component mixing model, Stations 24 and 25 high-Mg lavas have the highest calculated amounts of enriched components (Cameron et al. 1983) and thus are more likely to reflect the trace-element signature of the enriched component. As can be seen by comparison of Figures 14.6d and 14.11c, there is a marked similarity between the normalized element abundance patterns from Rapa and Toro Ankole with Stations 24 and 25 high-Mg lavas. There is a pronounced dissimilarity between the normalized element abundance patterns of the sediment components, especially with regard to Nb and La, and Stations 24 and 25 high-Mg lavas (Fig. 14.11b). The similarity in normalized element abundance patterns between Stations 24 and 25 lavas and hotspot- and rift-related within-plate lavas is taken as strong evidence that the 'enriching' component is a mantle-derived phase and not subducted oceanic sediment. This suggestion is supported by Pb isotopic data for Tongan Is. lavas from Ewart & Hawkesworth (1987), which show no evidence for sediment involvement.

The enrichments seen in the N Tonga Stations 24 and 25 lavas, although of mantle derivation, are unlike mantle-derived enrichments noted in other island-arc volcanics such as those from the Indonesian arc (Wheller et al. 1987, Varne 1985, Varne & Foden 1986, Stolz et al. 1988), or those seen in boninites (Hickey & Frey 1982, Cameron et al. 1983). Varne (1985) proposed that the characteristic enrichment in K-group elements in volcanics from the eastern Sunda arc is due to a component from enriched subcontinental lithosphere. The normalized element abundance pattern for a Batu Tara leucite-bearing mafic lava (Fig. 14.11d) is typical of the enrichment present in the Indonesian arc volcanics, and is unlike patterns from Stations 24 and 25 in having a negative Nb anomaly relative to La and being much more enriched in K-group elements (Rb, Ba, K, Sr). The negative Nb anomaly is interpreted to result from equilibrium with a residual titanate phase (Varne 1985). No such residual titanate phase is required for Stations 24 and 25 high-Mg lavas. The N Tonga high-Mg lavas lack the distinctive Zr enrichment relative to Sm seen in normalized abundance patterns of boninites (Sun & Nesbitt 1978, Hickey & Frey 1982, Nelson *et al.* 1984, Falloon *et al.* 1987).

14.5.6 Evidence for a second enriched component

Although a two-component mixing model involving the REE (Cameron *et al.* 1983) can explain most trace-element and isotopic features of the N Tonga lavas, an additional component with high LILE/LREE and LILE/HFSE is also required. In Figure 14.12, the Sr/La and Ba/La of the N Tonga lavas are compared with ratios in likely depleted and enriched source components.

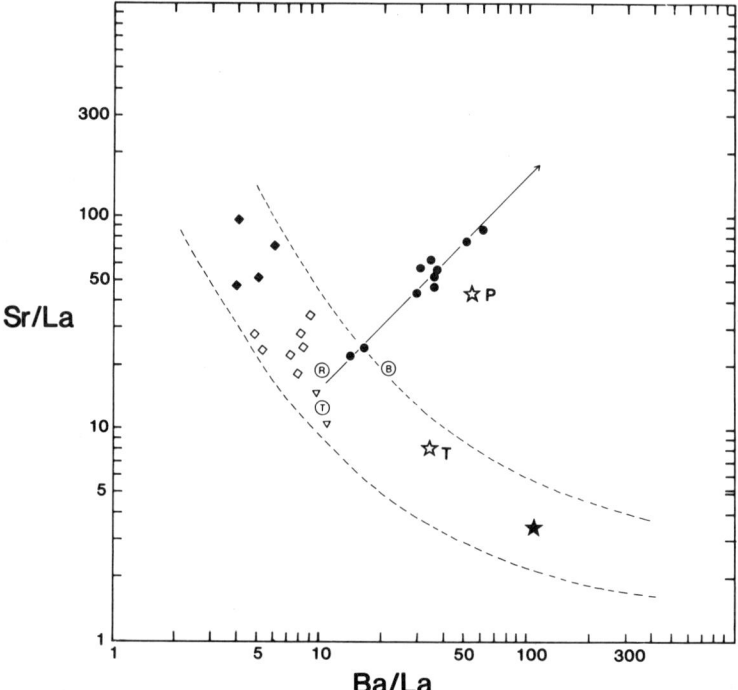

Figure 14.12 Sr/La versus Ba/La of N Tonga lavas: full diamonds, N-type MORB (Langmuir *et al.* 1977, Frey *et al.* 1974); open diamonds, E-type MORB (Sun 1980, Le Roex *et al.* 1985); circle R, R198 Rapa Island (see also Fig. 14.11); circle T, Toro Ankole C4802 (see also Fig. 14.11); circle B, Batu Tara B4155 (see also Fig. 14.11); open inverted triangle, ocean island basalts (Sun 1980); open stars T and P, Tonga and Pacific authigenic mean weighted sediment respectively (see also Fig. 14.11); asterisk, continental ultrapotassic lava EL7WAK17 (Jaques *et al.* 1984). Arrow points in direction of high Sr,Ba/La component.

Enrichments produced by mantle components fall in a broad array from a high Sr/La, low Ba/La end-member equivalent to depleted MORB to enriched end-members similar to Group I continental ultrapotassic lavas (Foley et al. 1987). The N Tonga lavas define a trend towards a component with high Sr/La and Ba/La which is difficult to identify due to the lack of any correlation between Sr/La and $^{87}Sr/^{86}Sr$ isotopic ratios. Potential sources are a hydrous fluid from subducted oceanic crust, a mantle-derived fluid component or slab-derived components in general.

Green et al. (1987) suggest an important role for reduced fluid volatiles and associated mantle degassing in the generation of basalt magma. In particular, they suggest that transform–trench intersections (such as N Tonga) are favourable sites for deep Earth degassing of reduced volatiles (dominantly CH_4). The interaction of reduced (C–H–O) fluids with oxidized subducted lithosphere produces an H_2O-rich fluid at low f_{O_2} ($\sim$ MW), which promotes melting of refractory peridotite in the mantle wedge. The presence of a degassing-derived, reduced volatile phase can explain the very low $Y(Fe^{3+})$ contents of the N Tonga Cr-spinels, which suggest very low f_{O_2} conditions (Murck & Campbell 1986, Barnes 1986). The LILE (K, Ba, Rb, Sr) would be carried from the subducted lithosphere via this hydrous fluid phase. The predicted high Sr/Nd of this fluid phase could produce a range in $^{87}Sr/^{86}Sr$ at relatively constant ε_{Nd}.

Conclusions

High-Si, high-Mg lavas recovered from the N Tonga arc and trench have appropriate compositions (in terms of Mg#, Ni and Cr contents) to be primary melts from upper-mantle peridotite. However, mineral chemical data combined with wholerock, groundmass and olivine-hosted glass inclusion data show, especially in lavas from Station 23, that magma mixing involving at least three distinct parental compositions was a major process in the petrogenesis of these lavas.

Major- and trace-element and isotope geochemistry of the lavas can be explained by involvement of both depleted and enriched source components. Major-element geochemistry can be explained by shallow (< 10 kbar or slightly higher pressures if H_2O was present) melting of a refractory mantle peridotite source containing some clinopyroxene. Trace-element and isotopic data can be explained by invoking at least two 'enriching' components. One 'enriching' component is responsible for similar enrichments to those seen in ocean island basalts from the Cook–Samoa–Austral Islands, north of Tonga. This component is most clearly seen in Stations 24 and 25 high-Mg lavas. The other 'enriching' component characterized by high LILE/La is considered to be a hydrous fluid derived from the subducting slab.

Acknowledgements

Financial support for this study was provided by a Commonwealth of Australia postgraduate award to T.J.F. and a grant from the Department of Science and Technology. Technical assistance was provided by P. Robinson and W. Jablonski. Thanks are also extended to the captain and crew of the RV *Natsushima* and to chief scientists E. Honza and K. B. Lewis and shipboard colleagues. The manuscript was improved by comments of two reviewers.

References

Allegre, C. J., O. Brevart, B. Dupre & J. F. Minster 1980. Isotope and chemical effects produced in a continuously differentiating convecting Earth mantle. *Phil. Trans. R. Soc. Lond. A* **297**, 447–77.

Anderson, A. T. 1974. Evidence for a picritic, volatile-rich magma beneath Mt Shastsa, California. *J. Petrol.* **15**, 243–67.

Arculus, R. J. & R. Powell 1986. Source component mixing in the regions of arc magma generation. *J. Geophys. Res.* **91**, 5913–26.

Baker, D. R. & D. H. Eggler 1987. Compositions of anhydrous and hydrous melts coexisting with plagioclase, augite, and olivine or low-Ca pyroxene from 1 atm to 8 kbar: application to the Aleutian volcanic center of Atka. *Am. Mineral.* **72**, 12–28.

Barnes, S. J. 1986. The distribution of chromium among orthopyroxene, spinel and silicate liquid at atmospheric pressure. *Geochim. Cosmochim. Acta* **50**, 1889–909.

Bloomer, S. H. 1987. Geochemical characteristics of boninite- and tholeiite-series volcanic rocks from the Mariana forearc and the role of an incompatible element-enriched fluid in arc petrogenesis. Geol. Soc. Am. Spec. Pap. no. 215, 151–64.

Cameron, W. E. 1985. Petrology and origin of primitive lavas from the Troodos ophiolite, Cyprus. *Contrib. Mineral. Petrol.* **89**, 239–55.

Cameron, W. E., M. T. McCulloch & D. A. Walker 1983. Boninite petrogenesis: chemical and Nd–Sr isotopic constraints. *Earth Planet. Sci. Lett.* **65**, 75–89.

Carlson, R. W., J. D. MacDougall & G. W. Lugmair 1978. Differential Nd and Sm evolution in oceanic basalts. *Geophys. Res. Lett.* **5**, 229–32.

Crawford, A. J., L. Beccaluva, G. Serri & J. Dostal 1986. Petrology, geochemistry and tectonic implications of volcanics dredged from the intersection of the Yap and Mariana trenches. *Earth Planet. Sci. Lett.* **80**, 265–80.

Dallwitz, W. B. 1968. Chemical composition and genesis of clinoenstatite-bearing volcanic rocks from Cape Vogel, Papua: a discussion. *XXIII Int. Geol. Congr.* vol. 2, 229–42.

Deer, W. A., R. A. Howie & J. Zussman 1966. *An introduction to the rock-forming minerals*. London: Longman.

Dick, H. J. B. & T. Bullen 1984. Cr-spinel as a petrogenetic indicator in oceanic environments. *Contrib. Mineral. Petrol.* **86**, 54–76.

Duncan, R. A. & D. H. Green 1987. The genesis of refractory melts in the formation of oceanic crust. *Contrib. Mineral. Petrol.* **96**, 326–42.

Ewart, A. & C. J. Hawkesworth 1987. The Pleistocene–Recent Tonga–Kermadec arc lavas. Re-evaluation of new isotopic and rare earth data in terms of a depleted mantle source model. *J. Petrol.* **28**, 495–530.

Ewart, A., W. B. Bryan & J. B. Gill 1973. Mineralogy and geochemistry of the younger volcanic islands of Tonga, SW Pacific. *J. Petrol.* **14**, 429–65.

Falloon, T. J. & D. H. Green 1986. Glass inclusions in magnesian olivine phenocrysts from Tonga:

evidence for highly refractory parental magmas in the Tonga arc. *Earth Planet. Sci. Lett.* **81**, 95–103.

Falloon, T. J. & D. H. Green 1987. Anhydrous partial melting of MORB pyrolite and other peridotite compositions at 10 kbar: implications for the origin of primitive MORB glasses. *Mineral. Petrol.* **37**, 181–219.

Falloon, T. J. & D. H. Green 1988. Anhydrous partial melting of peridotite from 8 to 35 kbars and the petrogenesis of MORB. *J. Petrol.* Special Lithosphere Issue, 379–444.

Falloon, T. J., D. H. Green & A. J. Crawford 1987. Dredged igneous rocks from the northern termination of the Tofua magmatic arc, Tonga and adjacent Lau Basin. *Austr. J. Earth Sci.* **34**, 487–506.

Falloon, T. J., D. H. Green, C. J. Hatton & K. L. Harris 1989. The anhydrous partial melting of a fertile and depleted peridotite from 2–30 kbar. *J. Petrol.* (in press).

Flower, M. J. F. & H. Levine 1987. Petrogenesis of a tholeiite–boninite sequence from Ayios Mamas, Troodos Ophiolite: evidence for splitting of a volcanic arc? *Contrib. Mineral. Petrol.* **97**, 509–24.

Foley, S. F. 1986. The origin of ultrapotassic rocks. Ph.D. Thesis, University of Tasmania.

Foley, S. F., G. Venturelli, D. H. Green & L. Toscani 1987. The ultrapotassic rocks: characteristics, classification and constraints for petrogenetic models. *Earth Sci. Rev.* **24**, 81–134.

Frey, F. A., W. B. Bryan & G. Thompson 1974. Atlantic ocean floor. Geochemistry and petrology of basalts from Legs 2 and 3 of the DSDP. *J. Geophys. Res.* **79**, 5507–27.

Frey, F. A., C. J. Suen & H. W. Stockman 1985. The Ronda high temperature peridotite: geochemistry and petrogenesis. *Geochim. Cosmochim. Acta* **49**, 2469–91.

Gill, J. B. 1981. *Orogenic andesites and plate tectonics.* New York: Springer.

Green, D. H. 1970. The origin of basaltic and nephelinitic magmas. *Trans. Leics. Lit. Phil. Soc.* **64**, 28–54.

Green, D. H., T. J. Falloon & W. R. Taylor 1987. Mantle derived magmas – role of variable source peridotite and variable C–H–O fluid compositions. In *Magmatic processes: physicochemical principles*, B. O. Mysen (ed.), 139–54. Geochem. Soc. Spec. Publ. no. 1.

Green, D. H., W. O. Hibberson & A. L. Jaques 1979. Petrogenesis of mid-ocean ridge basalts. In *The Earth: its origin, structure and evolution*, M. W. McElhinny (ed.), 265–99. London: Academic Press.

Grove, T. L. & W. B. Bryan 1983. Fractionation of pyroxene-phyric MORB at low pressure: an experimental study. *Contrib. Mineral Petrol.* **84**, 293–309.

Grove, T. L., D. C. Gerlach & T. W. Sando 1982. Origin of calc-alkaline series lavas at Medicine Lake volcano by fractionation, assimilation and mixing. *Contrib. Mineral. Petrol.* **80**, 160–82.

Hawkins, J. W. 1976. Petrology and geochemistry of basaltic rocks of the Lau Basin. *Earth Planet. Sci. Lett.* **28**, 283–97.

Hawkins, J. W. & J. T. Melchior 1985. Petrology of Mariana Trough and Lau Basin basalts. *J. Geophys. Res.* **90**, 11431–68.

Hickey, R. L. & F. A. Frey 1982. Geochemical characteristics of boninite series volcanics: implications for their source. *Geochim. Cosmochim. Acta* **46**, 2099–115.

Hole, M. J., A. D. Saunders, G. F. Marriner & J. Tarney 1984. Subduction of pelagic sediments: implications for the origin of Ce-anomalous basalts from the Mariana Islands. *J. Geol. Soc. Lond.* **141**, 453–72.

Huppert, H. E. & R. S. J. Sparks 1980. The fluid dynamics of a basaltic magma chamber replenished by influx of hot, dense ultrabasic magma. *Contrib. Mineral. Petrol.* **75**, 279–89.

Ito, E., W. M. White, V. von Drach, A. W. Hofmann & D. E. James 1980. Isotopic studies of ocean ridge basalts. *Carnegie Inst. Washington Yearb.* **80**, 465–71.

Jaques, A. L. & B. W. Chappell 1980. Petrology and trace element geochemistry of the Papuan ultramafic belt. *Contrib. Mineral. Petrol.* **75**, 55–70.

Jaques, A. L. & D. H. Green 1980. Anhydrous melting of peridotite at 0–15 kbar pressure and the genesis of tholeiitic basalts. *Contrib. Mineral. Petrol.* **73**, 287–310.

Jaques, A. L., J. D. Lewis, C. B. Smith, G. P. Gregory, J. Ferguson, B. W. Chappell & M. T. McCulloch 1984. The diamond-bearing ultrapotassic (lamproitic) rocks of the west Kimberley region, Western Australia. In *Kimberlites I: Kimberlites and related rocks*, J. Kornprobst (ed.), 225–54. Amsterdam: Elsevier.

Jenner, G. A. 1981. Geochemistry of high-Mg andesites from Cape Vogel, Papua New Guinea. *Chem. Geol.* **33**, 307–32.

Jenner, G. A. 1982. Petrogenesis of high-Mg andesites: an experimental and geochemical study with emphasis on high-Mg andesites from Cape Vogel, PNG. Ph.D. Thesis, University of Tasmania.

Jenner, G. A., P. A. Cawood, M. Rautenschlein & W. M. White 1987. Composition of back-arc volcanics, Valu Fa ridge, Lau Basin: evidence for a slab-derived component in their mantle source. *J. Volcanol. Geotherm. Res.* **32**, 209–22.

Johnson, R. W., A. L. Jaques, R. L. Hickey, C. O. McKee & B. W. Chappell 1985. Manam Island, Papua New Guinea: petrology and geochemistry of a low-TiO_2 basaltic island-arc volcano. *J. Petrol.* **26**, 283–323.

Langmuir, C. H., J. F. Bender, A. E. Bence & G. N. Hanson 1977. Petrogenesis of basalts from the Famous area: mid-Atlantic ridge. *Earth Planet. Sci. Lett.* **36**, 133–56.

Le Roex, A. P., H. J. B. Dick, A. M. Reid, F. A. Frey, A. J. Erlank & S. R. Hart 1985. Petrology and geochemistry of basalts from the American–Antarctic Ridge, Southern Ocean: implications for the westward influence of the Bouvet mantle plume. *Contrib. Mineral. Petrol.* **90**, 367–80.

MacDougall, J. D. & G. W. Lugmair 1985. Extreme isotopic homogeneity among basalts from the southern East Pacific Rise. Mantle or mixing effect. *Nature* **313**, 209–11.

Morris, J. D. & S. R. Hart 1984. Isotopic and incompatible element constraints on the genesis of island arc volcanics from Cold Bay and Amak Island, Aleutians, and implications for mantle structure. *Geochim. Cosmochim. Acta* **47**, 2015–30.

Murck, B. W. & I. H. Campbell 1986. The effects of temperature, oxygen fugacity and melt composition on the behaviour of chromium in basic and ultrabasic melts. *Geochim. Cosmochim. Acta* **50**, 1871–87.

Nelson, D. R., A. J. Crawford & M. T. McCulloch 1984. Nd–Sr isotopic and geochemical systematics in Cambrian boniniites and tholeiites from Victoria, Australia. *Contrib. Mineral. Petrol.* **88**, 164–72.

Newsom, H. E., W. M. White, K. P. Jochum & A. W. Hofmann 1986. Siderophile and chalcophile element abundances in oceanic basalts, Pb isotope evolution and growth of the Earth's core. *Earth Planet. Sci. Lett.* **80**, 299–313.

O'Hara, M. J. & R. E. Mathews 1981. Geochemical evolution in an advancing, periodically replenished, periodically tapped, continuously fractionated magma chamber. *J. Geol. Soc. Lond.* **138**, 237–77.

Palacz, Z. A. & A. D. Saunders 1986. Coupled trace element and isotope enrichment in the Cook–Austral–Samoa islands, southwest Pacific. *Earth Planet. Sci. Lett.* **79**, 270–80.

Sinton, J. M., R. C. Price, K. T. Johnson, H. Staudigel & A. Zindler 1988. Petrology and geochemistry of submarine lavas from the Lau and north Fiji back-arc basins. In *Basin formation ridge crest processes, and metallogenesis in the North Fiji Basin*, L. W. Kroenke (ed.), Circum-Pacific Council Energy Min. Resources, Earth Sci. Ser., Vol. 4, 47–84.

Sparks, R. S. J., P. Meyer & H. Sigurdsson 1980. Density variation amongst mid-ocean ridge basalts: implications for magma mixing and the scarcity of primitive lavas. *Earth Planet. Sci. Lett.* **46**, 419–30.

Sparks, R. S. J., H. Sigurdsson & L. Wilson 1977. Magma mixing: a mechanism for triggering acid explosive eruptions. *Nature* **267**, 315–18.

Stolz, A. J., R. Varne, G. E. Wheller, J. D. Foden & M. J. Abbott 1988. The geochemistry and petrogenesis of K-rich alkaline volcanics from the Batu Tara Volcano, eastern Sunda arc. *Contrib. Mineral. Petrol.* **98**, 374–89.

Sun, S.-S. 1980. Lead isotopic study of young volcanic rocks from mid-ocean ridges, ocean islands and island arcs. *Phil. Trans. R. Soc. Lond.* A **297**, 409–45.

Sun, S.-S. & R. W. Nesbitt 1978. Geochemical regularities and genetic significance of ophiolitic basalts. *Geology* **6**, 689–93.

Tatsumi, Y., D. L. Hamilton & R. W. Nesbitt 1986. Chemical characteristics of fluid phase released from a subducted lithosphere and origin of arc magmas: evidence from high-pressure experiments and natural rocks. *J. Volcanol. Geotherm. Res.* **29**, 293–309.

Thompson, R. N., M. A. Morrison, A. P. Dickin & G. L. Hendry 1983. Continental flood basalts...arachnids rule OK? In *Continental basalts and mantle xenoliths*, C. J. Hawkesworth & M. J. Norry (eds), 158–85. Nantwich: Shiva.

Umino, S. 1986. Magma mixing in boninite sequence of Chichijima, Bonin Islands. *J. Volcanol. Geotherm. Res.* **29**, 125–57.

Umino, S. & I. Kushiro 1989. Experimental studies on boninite petrogenesis (this volume).

Van der Laan, S. R., M. F. J. Flower & A. F. Koster van Groos 1989. Experimental evidence for the origin of boninites: near-liquidus phase relations to 7.5 kbar (this volume).

Varne, R. 1985. Ancient subcontinental mantle: a source for K-rich orogenic volcanics. *Geology* **13**, 405–8.

Varne, R. & J. D. Foden 1986. Geochemical and isotopic systematics of eastern Sunda arc volcanics: implications for mantle sources and mantle mixing processes. In *The origin of arcs*, F.-C. Wezel (ed.), 159–89. Amsterdam: Elsevier.

von Drach, V., B. D. Marsh & G. J. Wasserburg 1986. Nd and Sr isotopes in the Aleutians: multicomponent parenthood of island-arc magmas. *Contrib. Mineral. Petrol.* **92**, 13–34.

Walker, D. A. & W. E. Cameron 1983. Boninite primary magmas: evidence from the Cape Vogel Peninsula, PNG. *Contrib. Mineral. Petrol.* **83**, 150–8.

Watson, B. E. 1976. Glass inclusions as samples of early magmatic liquid: determinative method and application to south Atlantic basalts. *J. Volcanol. Geotherm. Res.* **1**, 73–84.

Wheller, G. E., R. Varne, J. D. Foden & M. J. Abbott 1987. Geochemistry of Quaternary volcanism in the Sunda–Banda arc, Indonesia, and three component genesis of island arc basaltic magmas. *J. Volcanol. Geotherm. Res.* **31**, 137–60.

White, W. M. & A. W. Hofmann 1982. Sr and Nd isotope geochemistry of oceanic basalts and mantle evolution. *Nature* **296**, 821–5.

White, W. M. & P. J. Patchett 1984. Hf–Nd–Sr isotopes and incompatible element abundances in island arcs: implications for origins and crust–mantle evolution. *Earth Planet. Sci. Lett.* **67**, 167–85.

Zindler, A., H. Staudigel & R. Batiza 1984. Isotope and trace element geochemistry of young Pacific seamounts: implications for the scale of upper mantle heterogeneity. *Earth Planet. Sci. Lett.* **70**, 175–95.

15 Th, U and Pb systematics of boninite series volcanic rocks from Chichi-jima, Bonin Islands, Japan

PATRICK F. DOBSON &
GEORGE R. TILTON

Abstract

Separated glasses from boninite series lavas from the Bonin Islands have Th/U near unity, and Th and U concentrations that are similar to mid-ocean-ridge basalt (MORB). While experimental work suggests that U is more incompatible than Th, U–Th isotopic disequilibrium studies and measurements of natural samples indicate that Th is more strongly partitioned into mantle melts. Magmas derived from sources probably depleted by previous melting events, such as MORB, have Th/U values that are lower than the average mantle ratio of 3.8 to 4.2. The very low concentrations of incompatible elements in boninite lavas imply that they are derived from a mantle source even more depleted than that of MORB. However, boninites are enriched in Th and U relative to high-field-strength elements when compared to MORB. The wide range of Th/U observed at a variety of boninite localities may be due to variable metasomatic addition of these elements to the boninite magma source region. Thus, the Th and U concentrations and Th/U of boninites appear to be the result of a Th- and U-enriched metasomatic fluid invading and modifying a depleted, low-Th/U source. This water-rich fluid plays a key role in boninite magma generation by lowering the liquidus sufficiently to promote melting of the mantle wedge. Pb isotopic data for the Bonin Islands lavas ($^{206}Pb/^{204}Pb = 18.546–18.613$, $^{207}Pb/^{204}Pb = 15.543–15.566$, $^{208}Pb/^{204}Pb = 38.227–38.347$) suggest that this fluid was derived from downgoing oceanic crust with a sedimentary component to account for the differences in Pb isotopic ratios for the Bonin Island lavas compared to those for MORB.

15.1 Introduction

Incompatible-element ratios have been used by geochemists as diagnostic features to help identify magma sources and petrogenetic processes (e.g. Gast 1968, Hanson 1977, Sun & Nesbitt 1977, Pearce 1983, Kay 1984). The distribution of the incompatible elements U and Th in the mantle is of interest to petrologists in formulating heat-flow models and in tracing 'enriched' and 'depleted' mantle source regions (Wakita *et al.* 1967, Tatsumoto 1978, McLennan & Taylor 1980). Although the average mantle has an estimated Th/U atomic ratio of approximately 3.8–4.2 (Stacey & Kramer 1975, Allegre *et al.* 1986), observed values for mantle-derived primary melts vary widely from 0.65–3.0 for abyssal tholeiites to 3.3–7.0 for alkali basalts (Tatsumoto 1978).

Boninite, a rare high-Mg (10–15% MgO), high-Si (55–57% SiO_2) volcanic rock found in several Pacific island arcs and in association with some ophiolites, has been the subject of numerous studies because of its unusual geochemistry (e.g. Sun & Nesbitt 1978, Cameron *et al.* 1979, Meijer 1980, Jenner 1981, Hickey & Frey 1982, Cameron *et al.* 1983). Boninite lavas have unusually high (>50) Al_2O_3/TiO_2 and CaO/TiO_2 values which demand an origin via partial melting of a previously depleted mantle source if no Ti-bearing residual phase is present (Sun & Nesbitt 1978, Duncan & Green 1980). The chemical composition of boninite melts has been interpreted to have been caused in part by addition of a fluid enriched in K, Rb, Sr, Ba and LREE to this depleted mantle source (Jenner 1981, Hickey & Frey 1982, Cameron *et al.* 1983).

The island of Chichi-jima, located in the Izu–Bonin arc system in the western Pacific Ocean about 1000 km south of Tokyo, is the type locality of boninite (Petersen 1891). The boninite series volcanics exposed on Chichi-jima include pillow lavas, submarine sheet flows and breccias ranging in composition from boninite and bronzite andesite to dacite. These rocks are overlain by quartz-bearing dacite and rhyolite lavas that form the most evolved members of the Eocene boninite suite.

As a part of a broader geological and geochemical study of boninite series volcanic rocks from Chichi-jima (Dobson 1986), five boninite and bronzite andesite wholerock samples and corresponding glass and orthopyroxene separates were analysed for Th, U and Pb concentrations and Pb isotope compositions. These samples were selected on the basis of petrographic and stable isotopic evidence (Dobson & O'Neil 1987) to ensure that the Th/U ratios are indeed primary and have not been subsequently modified by hydration and alteration. The glasses are from porphyritic, glassy pillow rims containing 5–6% phenocrysts and an additional 20% microphenocrysts. The phenocrysts in the most mafic samples, MD-94 and MD-97 (MgO $>$ 10 wt%), are mainly low-Ca pyroxene (orthopyroxene $\gg$ clinoenstatite), with lesser amounts of Ca-rich clinopyroxene, while slightly more differentiated rocks (MD-16,

MD-31 and MD-108) also contain minor amounts of plagioclase. Trace amounts of Cr-spinel are present in all samples.

15.2 Analytical techniques

Wholerock powders were prepared using a WC jawcrusher and shatterbox. The glasses were crushed, sieved and then separated using heavy liquids and a Franz magnetic separator. After washing, the 50–80 mesh size fractions were hand-picked to increase the purity of the separates. Pyroxene separates were obtained in a similar fashion, and two splits were analysed, one group of untreated samples and the other leached with 10% HF for 15 min to remove any adhering glass. Samples of approximately 50 mg glass and 100 mg pyroxene and wholerock were dissolved in Teflon capsules using a mixture of HF and $HClO_4$. A mixed $^{235}U-^{230}Th-^{208}Pb$ spike was added to wholerock and glass samples before digestion for measurement of concentrations by isotope dilution, and unspiked samples were analysed for Pb isotope compositions. For pyroxene samples, a mixed $^{235}U-^{230}Th-^{205}Pb$ spike was used for measurement of U, Th and Pb concentrations due to their very low Pb concentrations. Extraction and purification of U, Th and Pb for mass spectrometric measurements was done using the procedures of Tilton & Chen (1979). Analytical blanks were 0.2–0.6 ng for Pb, 3–6 pg for U and 6–30 pg for Th. Th and U analyses were performed at the University of California, Santa Barbara (UCSB), using an AVCO single-collector mass spectrometer. Mass spectrometric measurements for Pb were conducted at UCSB and at the US Geological Survey, Menlo Park, on a Finnegan MAT mass spectrometer. Th and U data were obtained with an electron multiplier and were corrected for mass discrimination of 0.43% per mass unit based upon replicate analyses of NBS standard U-050. All Pb data were corrected 0.12% per mass unit on the basis of replicate analyses of NBS SRM 981 Pb standard. Uncertainties are about 1–2% for U and Th concentrations and 0.7% for Pb concentrations.

15.3 Results

Four of the analysed glass separates from boninite and bronzite andesite samples have very uniform Th, U and Pb compositions, with concentrations in the ranges 0.13–0.14 ppm Th, 0.13–0.15 ppm U and 1.38–2.11 ppm Pb (Table 15.1). MD-31 has a significantly higher U content (0.22–0.43 ppm) as well as a slightly higher Th concentration. While replicate analyses of other samples prepared from different hand-pickings produced very similar results, two MD-31 glass separates showed a range of Th and U concentrations far outside analytical uncertainty; this probably reflects varying degrees of impurity or alteration in the two separates of this sample. Analyses of all wholerock samples yield a similar range of Th and Pb values (0.12–0.18 ppm Th,

Table 15.1 Th, U and Pb data from boninites, Chichi-jima, Japan.

Sample no.	Material	Th	U	Pb	Th/U	$^{206}Pb/^{204}Pb$	$^{207}Pb/^{204}Pb$	$^{208}Pb/^{204}Pb$
MD16	WR	0.13	0.07	1.70	1.86			
	glass	0.14	0.13	1.67	1.1	18.606	15.566	38.344
	Opx[a]	0.002	0.003	0.66	0.67			
	Opx	0.011	0.011		1.0			
MD31	WR	0.12	0.11	1.81	1.1			
	glass	0.18	0.43	1.86	0.42	18.606	15.566	38.334
		0.13	0.22	1.93	0.59			
	Opx[a]	0.002	0.002	0.22	1.0			
	Opx	0.013	0.012		1.1			
MD94	WR	0.13	0.09	1.55	1.4			
	glass	0.14	0.14	1.80	1.0	18.613	15.564	38.347
		0.13	0.13	1.96	1.0			
	Opx[a]	0.004	0.004	0.75	1.0			
	Opx	0.006	0.006		1.0			
MD97	WR	0.13	0.10	1.45	1.3			
	glass	0.13	0.14	1.66	0.93	18.597	15.553	38.306
		0.13	0.13	2.11	1.0			
	Opx[a]	0.008	0.003	0.18	2.7			
	Opx	0.010	0.014		0.71			
MD108	WR	0.12	0.09	1.33	1.3			
	glass	0.13	0.15	1.38	0.87	18.546	15.543	38.227

All concentrations given in ppm.
WR, wholerock samples.
[a]Leached orthopyroxene separates.

1.33–1.81 ppm Pb) but have slightly lower U concentrations (0.07–0.10 ppm) compared to those of the glass samples. Leached orthopyroxene samples contain lower abundances of Th and U (0.002–0.008 ppm Th, 0.002–0.004 ppm U) than the untreated pyroxenes (0.006–0.013 ppm Th, 0.006–0.014 ppm U). Pb contents of the leached pyroxene samples range from 0.18 to 0.75 ppm.

Glass samples were analysed for Pb isotopic compositions using both spiked and unspiked samples. Initial spike-corrected values of $^{206}Pb/^{204}Pb =$ 18.589–18.653 and $^{207}Pb/^{204}Pb = 15.572$–15.598 (Dobson & Tilton 1985) are slightly more radiogenic than data subsequently obtained for unspiked samples ($^{206}Pb/^{204}Pb = 18.546$–18.613, $^{207}Pb/^{204}Pb = 15.543$–15.566 and $^{208}Pb/^{204}Pb = 38.227$–38.347). These higher values are probably due to the use of a ^{208}Pb spike in fairly large quantities, making accurate spike corrections difficult.

Th/U ratios of boninite and bronzite andesite lavas from Chichi-jima range from 1.1 to 1.9 for wholerock samples and 0.6 to 1.1 for glass separates. These ratios are similar to values reported in previous studies of wholerock samples from the Bonin Islands, the Mariana forearc, New Zealand and the Troodos Ophiolite in Cyprus, which vary between 0.1 and 2.2 (Bougault et al. 1981,

Table 15.2 Th, U, Pb and rare-earth-element concentrations in boninites from other studies.

Sample no.	Th	U	La	Yb	Pb	Th/U	(La/Yb)$_N$	Reference
Papua New Guinea								
2984	0.40	0.12	1.47	0.809		3.3	1.23	J
47787	0.85	0.23	4.06	0.72		3.7	3.81	J
47788	0.52	0.12	2.88	0.29		4.3	6.71	J
47790	0.50	0.14	2.03	0.42		3.5	3.27	J
47805	0.77	0.17	3.25	0.41		4.5	5.36	J
47807	0.42	0.11	2.10	0.37		3.8	3.84	J
B6	0.5	0.2	2.50	0.48	1.3	2.5	3.52	C
38.2B	0.8	0.3	4.28	0.55	1.1	2.7	5.26	C
52.1A	0.3	0.1	1.75	0.73	0.8	3.0	1.62	C
Bonin Islands								
106	0.1	<0.1	0.82	0.52	1.7	>1	1.07	C
Mariana forearc								
458-33-2	0.19	2	0.9	0.8		0.10	0.76	B
458-39-3	0.15	0.1	0.58	0.86		1.5	0.46	B
458-40-1	0.13	0.16	0.52	0.80		0.81	0.44	B
458-47-2	0.20	0.19	1.57	1.56		1.1	0.68	B
458-48-1	0.22	0.10	2.26	1.44		2.2	1.06	B
New Caledonia								
49	0.4	0.2	4.40	0.75	3.4	2	3.96	C
52	0.5	0.3	4.99	1.12	2.1	1.7	3.01	C
Cyprus								
201	0.1	<0.1	0.88	1.04	1.1	>1	0.57	C
92	0.2	0.1	1.09	0.84	2.7	2	0.88	C
116	0.1	<0.1	0.56	1.11	1.3	>1	0.34	C
New Zealand								
6	<0.1	<0.1	0.48	0.76	<0.1	1(?)	0.43	C

All concentrations given in ppm.
References: J, Jenner (1981); C, Cameron *et al.* (1983); B, Bougault *et al.* (1981).

Cameron *et al.* 1983, H. Sato pers. comm. 1985). Higher Th/U, up to 4.5, were obtained for New Caledonia boninites as well as for samples from Cape Vogel in Papua New Guinea (Jenner 1981, Cameron *et al.* 1983). These data (Table 15.2) span a much larger range in concentrations than the samples measured in this study, with the ranges being Th < 0.1–0.85 ppm, U < 0.1–2 ppm and Pb < 0.1–3.4 ppm.

15.4 Secondary process affecting Th/U ratios

In order to evaluate these results, it is necessary to review the processes that control the abundances of Th and U in volcanic rocks. Because these elements can be mobilized easily by secondary processes, it is essential to ascertain whether or not these values are primary.

Post-magmatic water–rock interactions can significantly change the Th/U ratio of lavas. Because of the oxidation of U to U^{6+} in surficial environments, U and Th behave differently, with U^{6+} being much more soluble than Th^{4+} (e.g. Gabelman 1977). Rosholt et al. (1971) analysed silicic glasses and felsites for U and Th, and noted that the more oxidized crystalline samples had lost substantial amounts of U. Analyses of mid-ocean-ridge hydrothermal vent fluids and altered submarine basalts by Albarede & Michard (1986), Aumento et al. (1976) and MacDougall (1977) suggest that U is stripped from sea water by oceanic basalts during hydrothermal circulation. However, Tatsumoto (1978) found that both Th and U can be mobilized in oceanic basalts, giving both increases and decreases in their concentrations and Th/U values. It is thus critical that only unaltered samples be used in studies utilizing Th/U ratios to determine source characteristics.

Although it is difficult to estimate the effects of these alteration processes on samples used in previous work, the data obtained in this study can be carefully evaluated for possible variations due to alteration. Wholerock samples contain small amounts (2–10%) of alteration minerals, such as zeolites, palagonite mineral aggregates, calcite and quartz which occur as small veins and as vesicle fillings. Although stable isotope data for the analysed samples (Dobson & O'Neil 1987) suggest that no major water–rock interaction occurred, it is evident that some low-temperature exchange did take place. Glasses separated from these samples were cleansed of any secondary alteration, and have lower ^{18}O values than corresponding wholerock samples; they thus reflect the primary chemical signature of these lavas more accurately than the bulk wholerock samples.

Four glass and wholerock pairs (MD-16, MD-94, MD-97 and MD-108) have nearly identical Th and U concentrations (0.12–0.13 ppm WR Th, 0.13–0.14 ppm glass Th; and 0.07–0.10 ppm WR U, 0.13–0.15 ppm glass U). The MD-31 glass has significantly higher Th and U contents than its corresponding wholerock sample. A replicate analysis of this glass sample subjected to even more rigorous cleaning yielded lower, but still anomalous, values. The elevated Th and U concentrations of the MD-31 glass samples may be the product of small amounts of contamination, but no significant shifts in the Pb concentrations and isotopic compositions for these separates were observed.

Glass serves as the principal repository for Th and U in boninite lavas, and thus Th and U concentrations of glass should be somewhat higher than those of fresh bulk wholerock samples. Analysed glass separates contain a small amount of pyroxene microphenocrysts; hence they will have somewhat lower concentrations of Th and U than pure glass. Assuming the extreme case of pure glass and 25% crystals containing no Th and U, a maximum enrichment (glass/wholerock) of 1.33 is predicted. The analysed glasses contain 0 to 8%

more Th than wholerock samples, but are enriched 1.35–1.86 times the measured wholerock values in U. Assuming that the analyses of glass separates represent pristine magmatic values, this apparent overenrichment can only be attributed to U loss in the wholerock samples through leaching by circulating fluids, a process that would account for the higher Th/U found in wholerock samples. Thus, the Th/U of near unity found in fresh glass separates are interpreted as primary magmatic values and are quite different from the Th/U of 3.8–4.2 for average mantle. We now seek an explanation for these low primary values.

15.5 Primary processes affecting Th/U ratios

15.5.1 Introduction

Numerous geochemical and isotopic studies in recent years have demonstrated that the mantle is heterogeneous on large and small scales. Such heterogeneities have been attributed to processes such as partial melting and metasomatic events which mobilize incompatible elements, creating zones of enriched and depleted mantle. The effects of these processes on the distribution of Th and U in boninite magmas will be reviewed in light of our new analytical results obtained for boninite and bronzite andesite lavas from the type locality in the Bonin Islands.

15.5.2 Source characteristics and partial melting

Partial melting is an important process controlling both mantle and magma compositions. The amount and style of partial melting dramatically affect the concentrations and ratios of LILE in both melts and residual mantle. Batch melting of more than several per cent results in magmas with LILE ratios that are very similar to those of the original source. Multistage melting, especially involving small melt increments, can result in the production of magmas with LILE ratios significantly different from those of the original mantle material (Langmuir *et al.* 1977, Sun & Nesbitt 1978, Wood 1979, Duncan & Green 1980). With repeated melting events, the source is further depleted in the more-incompatible elements (MI) relative to the less-incompatible elements (LI), so that later melts will have [MI]/[LI] that are much smaller than those of earlier formed melts and the original mantle.

One characteristic of most fresh mafic igneous rocks with low Th/U is that they generally appear to have been derived from depleted mantle sources. MORB have variable but generally low Th/U (1.5–2.5), and are thought to be derived from a depleted mantle source with a Th/U of about 2 (Tatsumoto 1978). Island-arc tholeiites, another magma series with low LILE concentrations, have similarly low Th/U (Jakes & Gill 1970).

Residual mantle after MORB extraction should have even lower Th/U than their associated lavas if Th is preferentially incorporated into melts with respect to U. Unfortunately, few ophiolite assemblages have been examined for Th and U concentrations. Chen & Pallister (1981) found a large degree of heterogeneity in Pb, Th and U abundances in a study of the Samail Ophiolite. Petrographically fresh samples of layered cumulate gabbros had Th/U ranging from 0.221 to 1.88. Associated volcanic rocks from Oman have much higher concentrations, mainly as a result of fractional crystallization, but the original distribution of Th, U and Pb in these rocks has been substantially modified by subsequent alteration, as evidenced by the Pb isotopic data. Tectonized harzburgite from St Paul's Rocks in the Atlantic Ocean analysed by Morioka et al. (1971) have Th/U of 0.15 to 2.9. These ultramafic rocks may be representative of highly depleted residua remaining after successive intervals of partial melting, but it is unlikely that they have escaped modification by secondary alteration or grain-boundary infiltration.

Ultramafic xenoliths represent another 'sample' of mantle source areas, but most are associated with alkalic volcanics interpreted to be derived from enriched source regions and thereby have correspondingly high Th/U (e.g. Green et al. 1968). However, lherzolite inclusions found in the calc-alkalic lavas from Itinome-gata, Japan, have extremely low Th/U of 0.78 to 1.2 (Aoki & Fujimaki 1982, Morioka & Kigoshi 1978, Tanaka & Aoki 1981).

Accurate determination of partition coefficients for incompatible elements is a critical requirement in modelling the effects of partial melting on the concentrations of these elements. Tatsumoto (1978) determined partition coefficients (K_D values) for U and Th between clinopyroxene, the major mineral host of these elements in the mantle, and a melt of olivine tholeiite composition at 14 kbar to be 0.014 and 0.021, respectively. Additional experiments conducted at 20 kbar by Benjamin et al. (1978) using a synthetic phosphate-free haplobasaltic starting material also indicated that $K_D(U)$ (0.017) is less than $K_D(Th)$ (0.027). Recently Watson et al. (1987) measured very low partition coefficients for U between chromian diopside and a synthetic haplobasaltic liquid (0.0002–0.0005), but these experiments were conducted at 1 atm pressure and in an oxidizing environment, precluding comparison with the earlier studies.

In contrast, measurements of Th and U concentrations for natural clinopyroxene and groundmass samples (Nagasawa & Wakita 1968) indicate that Th is more strongly concentrated into the melt phase than is U. However, crystal–matrix studies may be unreliable due to heterogeneous matrix, impure crystals and equilibration problems. Goldschmidt's rule on the size of ionic radii as a control of ionic substitution suggests that U should be less incompatible than Th owing to its smaller ionic radius in the tetravalent state. The systematic decrease in incompatibility with increasing atomic number observed for the lanthanide series might apply for the actinide series as well.

Orthopyroxene is the only major crystalline phase with significant amounts of Th and U present in the boninites examined in this study. Petrogenetic models for boninite genesis require a highly depleted source with little or no clinopyroxene present (e.g. Hickey & Frey 1982), thus orthopyroxene–glass partition coefficients are important for this magma system as orthopyroxene is abundant in the depleted source. Measured partition coefficients and calculated K_D values, assuming equilibrium Rayleigh fractionation, are shown in Table 15.3. Discarding the results of MD-31, which, as stated earlier, may be attributable to alteration, the ranges in calculated K_D values for Th and U for orthopyroxene–glass are 0.017–0.070 and 0.025–0.035, respectively.

The lowest calculated K_D values in this study are fairly similar to experimentally determined partition coefficients for clinopyroxene–glass of Tatsumoto (1978) and Benjamin *et al.* (1978). Studies of peridotite nodules by Tatsumoto (1978) and Nagasawa & Wakita (1968) indicate that clinopyroxene concentrates Th and U more strongly than orthopyroxene, and thus the partition coefficients for these elements for Opx–glass should be smaller than those of Cpx–glass. The calculated partition coefficients represent maximum values, because most pyroxene grains contain microscopic glass inclusions whose U and Th concentrations are much higher than those of the parent mineral. The compositional differences between the experimental systems described above and the boninites examined may also be responsible in some way for the differences in the partition coefficients.

The partition coefficients calculated in this study are low and similar in value; hence the Th/U of the boninite depleted source material should not change when orthopyroxene is the primary host phase for Th and U. Modification of the source ratio during earlier melting events, when clinopyroxene (and perhaps an additional accessory phase with higher Th and U abundances) is present, must be responsible for producing mantle sources with low Th/U values.

^{230}Th–^{238}U disequilibrium studies have also been utilized to examine the partitioning of Th and U during partial melting. ^{230}Th is a daughter product of ^{238}U decay, with a half-life of 75 200 years. When in radioactive equilibrium, the (^{230}Th/^{238}U) activity ratio should equal unity. However, many historical basaltic lavas have (^{230}Th/^{238}U) activity ratios greater than unity. This has

Table 15.3 Calculated K_D values (Opx–glass).

Sample no.	Measured K_D(Opx–glass)			Fraction crystallized, S	Calculated K_D(Opx–glass)		
	K_A(Th)	K_A(U)	K_A(Pb)		K_D(Th)	K_D(U)	K_D(Pb)
MD16	0.014	0.023	0.40	0.27	0.017	0.027	0.43
MD31	0.015	0.009	0.12	0.26	0.018	0.011	0.13
MD94	0.030	0.030	0.40	0.28	0.035	0.035	0.44
MD97	0.062	0.022	0.10	0.24	0.070	0.025	0.11

$K_A(X) = [X^{Opx}]/[X^{glass}]$.

$K_D^{xl-liq} = 1 - \ln[1 + S(K_A^{xl-liq} - 1)]/\ln(1 - S)$.

been interpreted by some workers (e.g. Allegre & Condomines 1982) as evidence that partial melting produces magmas with Th/U greater than that of the mantle source (K_D(Th) < (K_D(U)). The (^{230}Th/^{232}Th) activity ratio may also be used to estimate the Th/U atomic ratio of the mantle source if the crustal residence time of the magma was short relative to the half-life of ^{230}Th (Oversby & Gast 1968, Condomines et al. 1981a,b, Allegre & Condomines 1982, Newman et al. 1984a). Assigning partition coefficients of K_D(Th) = 0 and K_D(U) = 0.005 and 0.0015, McKenzie (1985) demonstrated that with small fractions (< 2%) of melting using a dynamic partial melting process, significant changes in the Th/U between the melt and the mantle can be achieved. However, Krishnaswami et al. (1984) contend that (^{230}Th/^{238}U) activity ratios greater than unity are most likely due to uranium loss associated with post-emplacement alteration.

Thus, while there is some uncertainty over the relative partitioning of Th and U in mantle-derived melts, it appears that successive events of partial melting can produce mantle sources which in turn give rise to magmas with low Th/U. The nature of this mantle source, the degree of partial melting and the respective partition coefficients will constrain the Th/U and concentrations of Th and U as well as other incompatible elements in primary magmas.

15.5.3 Mantle metasomatism

Mantle metasomatism has been closely linked to the generation of alkalic lavas because they are relatively enriched in incompatible elements such as LREE and alkalis. Metasomatic fluids and melts involved in the genesis of highly alkaline magmas are believed to be CO_2-rich (e.g. Wyllie 1987), and geochemical studies of alkaline basaltic lavas and associated metasomatized peridotite inclusions (Menzies 1983, Pearce 1983) as well as experimental work (Wendlandt & Harrison 1979) indicate that these fluids can act as an important reservoir for certain incompatible elements. Alkalic lavas often have high concentrations of Th and U, and characteristically have Th/U > 3, as do their associated metasomatized peridotite inclusions (Green et al. 1968). However, carbonatite magmas can have Th/U < 1; this is interpreted by Williams et al. (1986) to arise from the partitioning of U into the CO_2-rich phase when immiscibility occurs between silicate and carbonate liquids.

Another possible source of metasomatizing fluids is dehydrating downgoing oceanic lithosphere at a convergent margin. Island-arc lavas are generally more water-rich than MORB lavas and often have higher ^{87}Sr/^{86}Sr; these features are thought to be due to the incorporation of subduction-related fluids into these melts (e.g. Gill 1981). Such fluids would probably be water-rich, and be enriched in LILE derived from oceanic sediments and altered oceanic crust. Experiments by Mysen (1979) showed that water at 5–30 kbar in equilibrium with a mantle assemblage would be enriched in REE, with an LREE enrichment pattern similar to that observed for CO_2 fluids

(Wendlandt & Harrison 1979). Dehydration experiments by Tatsumi *et al.* (1986) on synthetic serpentine samples at 12 kbar and 850°C indicate that LILE are preferentially mobilized by an aqueous fluid phase. The abundances of LILE in these fluids should be strongly dependent on the concentrations of these elements in the down-going slab. Several studies (e.g. Aumento *et al.* 1976, MacDougall 1977, Fyfe 1979) indicate that altered sea-floor basalts and oceanic sediments may serve as a sink for U, but Tatsumoto (1978) found decreases in U concentrations and increased Th/U in altered oceanic basalts. Metasomatizing fluids derived from the dehydration of such material thus could have a wide range in Th and U contents and Th/U.

^{230}Th–^{232}Th disequilibrium studies of some recent calc-alkaline volcanic rocks from the Mariana, New Britain and Aleutian arcs show that these lavas, unlike oceanic islands and MORB, have (^{230}Th/^{238}U) activity ratios <1 (Allegre & Condomines 1982, Newman *et al.* 1984b, Gill *et al.* 1987), high (^{230}Th/^{232}Th) activity ratios and correspondingly low Th/U values. These authors all propose that the low (^{230}Th/^{238}U) activity ratios observed only in island-arc magmas may be due to metasomatic U enrichment of their source regions shortly before magma genesis and eruption. Fluids derived from dehydration of subducted oceanic crust would trigger partial melting in the mantle wedge and also be responsible for the late-stage U enrichment.

15.5.4 Additional processes affecting Th/U ratios

Other possible mechanisms that might affect the Th/U of magmas include fractional crystallization and crustal assimilation. Both amphibole and clinopyroxene, the only major phases with substantial amounts of U and Th in basalts and andesites, have small ($\ll 1$) partition coefficients. Therefore, unreasonably large amounts of clinopyroxene or amphibole would have to be crystallized in order to affect the Th/U of the melt. The effects of crustal assimilation are more difficult to evaluate, because many types of country rock could be assimilated by the parent magmas. However, lavas erupted in oceanic island arcs will typically only pass through oceanic lithosphere, which has low concentrations of both Th and U. Assimilation of such material is not likely to change the Th/U of these magmas substantially.

15.6 Th/U systematics in boninite lavas

Whereas boninite lavas examined in this study have consistently low Th/U near unity (Table 15.1), samples analysed from other boninite localities (Table 15.2) (Jenner 1981, Bougault *et al.* 1981, Cameron *et al.* 1983) have a wide range of Th/U values. Some of the range in the other studies may be due in part to small degrees of alteration of the analysed samples. Isotopic and geochemical studies of boninites (Jenner 1981, Hickey & Frey 1982, Cameron *et al.* 1983) suggest

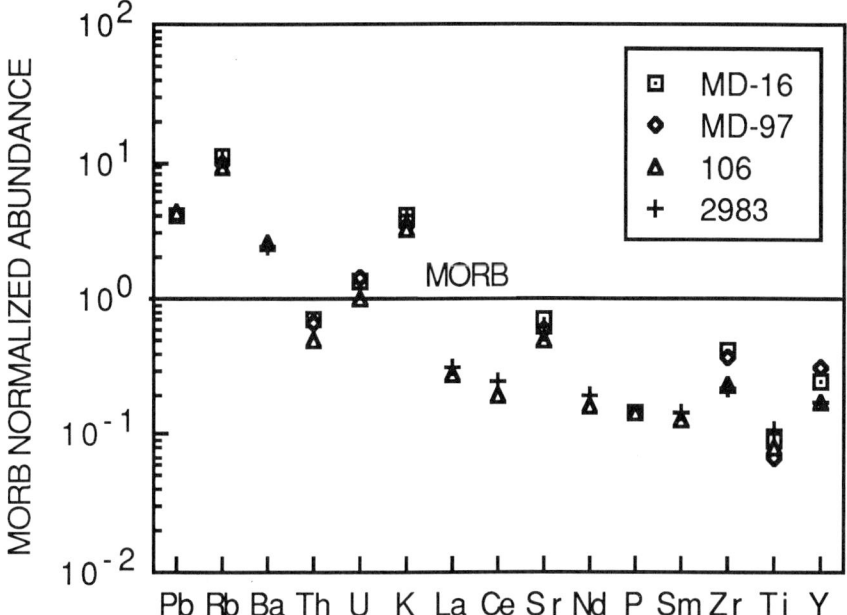

Figure 15.1 MORB-normalized concentrations of incompatible elements in boninite lavas from the Bonin Islands. MD-16 and MD-97 are from this study; 2983 is from Hickey & Frey (1982); and 106 is from Cameron *et al.* (1983). Normalizing values are from Sun (1980). Th and U glass analyses are used for MD-16 and MD-97; all other data are from wholerock analyses.

that much of the variability observed in boninite magmas is a function of differences in source compositions and the degree of metasomatic modification of the source area. The high CaO/TiO_2 and Al_2O_3/TiO_2 of boninite magmas point to previous melting events in their mantle sources (Sun & Nesbitt 1978). If Th is more incompatible than U, the mantle source region should have a low Th/U as a result of repeated depletion of the source, but later melts derived from such a source should have very low contents of incompatible elements. Whereas the concentrations of HREE and other HFSE in boninites are 50–80% lower than those of primitive MORB lavas (Fig. 15.1), U and Th abundances are not depleted with respect to MORB, and Pb concentrations (0.8–3.4 ppm) of boninites are substantially higher than those of average depleted MORB (0.4 ppm; Sun 1980). Thus, mantle metasomatism may explain the apparent enrichment of Th, U, Pb and other LILE relative to HREE.

Because their concentrations of U and Th are probably low, boninite source areas are very susceptible to shifts in the Th/U upon introduction of a metasomatizing fluid. Such shifts should be correlated with increasing Th and/or U abundances as well as increasing LREE enrichment, as predicted by the experiments of Mysen (1979) and Tatsumi *et al.* (1986). The wide range in

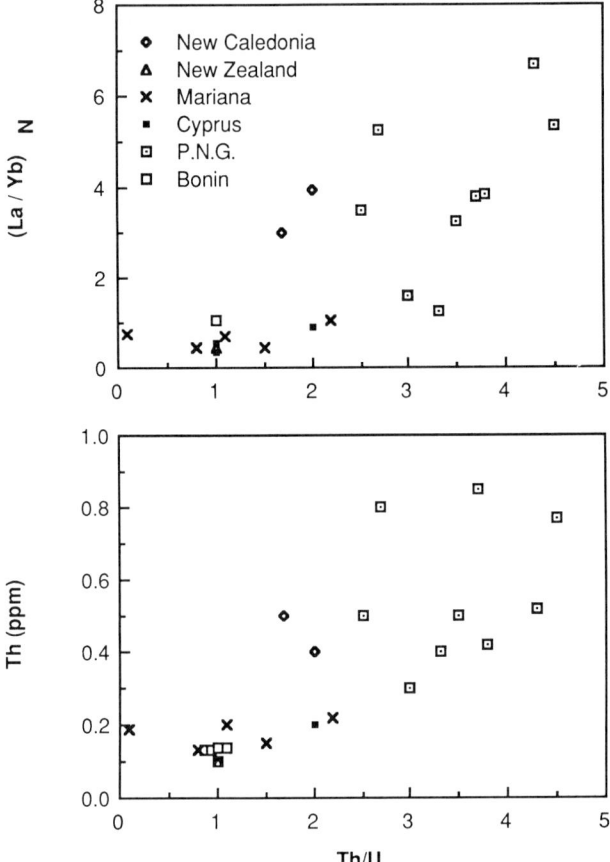

Figure 15.2 Th/U versus Th concentrations and $(La/Yb)_N$ for boninite lavas from the Bonin Islands, the Mariana forearc, New Zealand, Cyprus, New Caledonia and Papua New Guinea. Th and U glass analyses from this study; all other data (Jenner 1981, Bougault *et al.* 1981, Hickey & Frey 1982, Cameron *et al.* 1983) from wholerock analyses.

Th/U relative to rather restricted Th concentrations and LREE enrichments (Fig. 15.2) may be due in part to secondary alteration effects, but could also reflect variability in the composition of metasomatizing fluids both within and between different boninite occurrences. Boninite lavas from Cape Vogel (Jenner 1981, Cameron *et al.* 1983) have consistently high Th/U (2.5–4.5), relatively large Th abundances (0.3–0.85 ppm) and strong LREE enrichments $((La/Yb)_N = 1.23-6.71)$, while boninite magmas from Cyprus, New Zealand and the Mariana–Bonin arc system (this study, Bougault *et al.* 1981, Cameron *et al.* 1983) have much lower Th concentrations (< 0.1–0.22) and lower LREE/HREE $((La/Yb)_N = 0.43-1.07)$, suggesting that their sources underwent less metasomatic modification. Only boninite samples with flat to depleted LREE patterns and low Th contents have low Th/U. Thus, the low Th/U appear to be linked to low Th/U in the primary mantle source. Differential modification of elemental concentrations in boninite magmas by metasomatic fluids may obscure characterization of the mantle source of boninites.

15.7 Pb isotopic data: implications for sediment involvement in boninite genesis

Bonin Island lavas have radiogenic Pb isotopic ratios ($^{206}Pb/^{204}Pb = 18.546-18.613$, $^{207}Pb/^{204}Pb = 15.543-15.566$ and $^{208}Pb/^{204}Pb = 38.227-38.347$) that lie outside the field of Atlantic and Pacific MORB compositions (Fig. 15.3). Radiogenic Pb was also reported for Bonin Island boninites by Vidal et al. (1985). However, based on low $\varkappa$ (1) and μ (5–6) values calculated for analysed glass samples, non-radiogenic Pb isotopic compositions would be expected for Bonin Island lavas if these values are representative of the time-averaged composition for the mantle source.

Because of the differences in isotopic signatures between mantle and crustal reservoirs, Pb isotope studies have been utilized to measure the contribution of sediment in the genesis of convergent margin magmas (e.g. Karig & Kay 1981, Barriero 1983, White & Dupre 1986). Sediments often contain much higher Pb concentrations than those found in island-arc lavas, and thus addition of small amounts of sediment can significantly affect the isotopic composition of these magmas. While most of the Eocene to Recent Mariana arc basaltic and basaltic andesitic volcanic rocks do not show a sedimentary Pb component (Meijer 1976), some lavas from the Northern Mariana Islands (Fig. 15.3) do have anomalously high Pb isotopic ratios (Woodhead & Fraser 1985). In the Troodos Ophiolite, Pb isotopic data for volcanic glasses ($^{206}Pb/^{204}Pb = 18.60-18.70$, $^{207}Pb/^{204}Pb = 15.55-15.59$ and $^{208}Pb/^{204}Pb = 38.08-38.51$) from the Lower Pillow Lavas (Rautenschlein et al. 1985) and gabbros ($^{206}Pb/^{204}Pb = 18.23-18.82$, $^{207}Pb/^{204}Pb = 15.57-15.61$ and $^{208}Pb/^{204}Pb = 38.17-38.44$) (Hamelin et al. 1984) are very similar to those obtained for the Bonin Island samples. The Troodos results were interpreted by these workers to reflect a small sedimentary contribution ($< 1\%$) to the melt. Meijer & Hanan (1981) reported unusual Pb isotope ratios for boninite and tholeiite lavas drilled in the Mariana forearc, with lower $^{207}Pb/^{206}Pb$ ratios than those of MORB (Fig. 15.3). They interpreted these values as reflecting involvement of Pb from the subducted Pacific plate or a long-term depletion of the mantle source region.

Boninites are highly depleted in HFSE (Fig. 15.1), but partial melts from the subducted slab would be enriched in these incompatible elements and thus modify boninite sources accordingly (Tatsumi et al. 1986). Thus, it appears that Pb and other LILE are transferred by an aqueous fluid phase into the

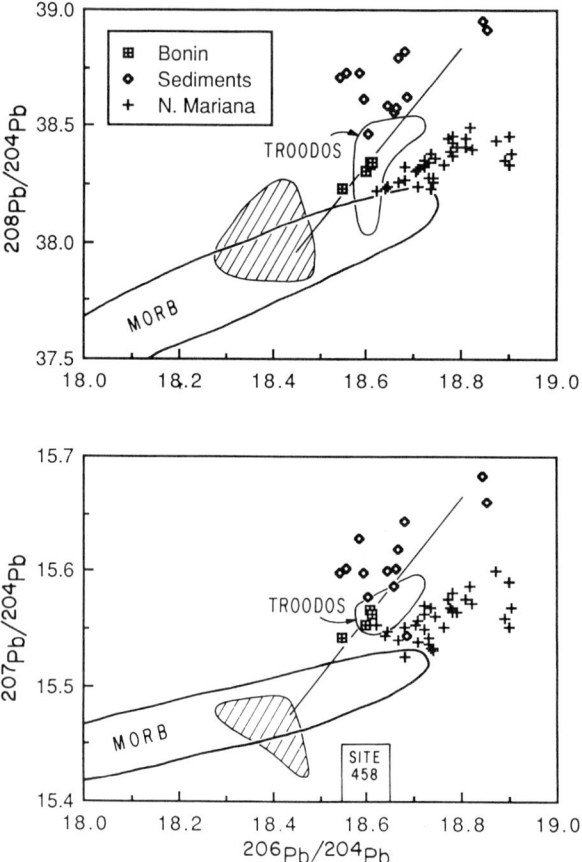

Figure 15.3 Pb isotopic compositions of boninite lavas. Atlantic and Pacific MORB fields from Sun (1980); Eocene Mariana frontal arc andesite (shaded region) and W Pacific sediment Pb data from Meijer (1976); N Mariana arc Pb data from Woodhead & Fraser (1985); Troodos field from Rautenschlein *et al.* (1985); and DSDP Site 458 boninite lavas from Meijer & Hanan (1981). Mixing lines connect the estimated compositions of the boninite source region with those of the Pacific sediments. The Bonin Islands boninite compositions lie midway between these two end-members.

source mantle for boninite magmas, and not by a partial melt of slab sediments. The Pb concentration data of boninites can be used to estimate the amount of the sedimentary contribution to boninite magma genesis. Analyses of W Pacific sediments indicate that oozes and clay-rich sediments contain 28–49 ppm Pb (Lisitzen *et al.* 1971) and range in isotopic composition from $^{206}Pb/^{204}Pb = 18.543$–$18.855$, $^{207}Pb/^{204}Pb = 15.578$–$15.661$ and $^{208}Pb/^{204}Pb = 38.464$–$38.915$ (Meijer 1976). As noted earlier, boninite magmas are strongly enriched in Pb relative to HFSE. Assuming that unmodified boninite magmas would have incompatible-element patterns similar to those of depleted MORB with concentrations 0.5 to 0.2 times those for MORB, estimated concentrations of Pb would be 0.08–0.2 ppm. Using a minimum value of 25 ppm for Pb contents of sediment, addition of 5–7% sediment component via a fluid phase would be necessary to account for the elevated (1.4–1.8 ppm) Pb contents observed in boninite lavas.

However, if Pb is less incompatible than most other magmaphile elements, as suggested by Tatsumoto & Unruh (1979), and using the Pb partition coefficients of 0.11 to 0.44 determined for the system orthopyroxene–glass (Table 15.3), then the estimate used for the unmodified Pb contents of boninite magma is too low and thus a smaller sedimentary component can account for the observed concentrations. We predict, using Wood's (1979) dynamic partial melting model, that the Pb concentrations of unmodified boninite should be of the order of 0.3–0.4 ppm. Using these higher concentrations and the upper limit of 49 ppm Pb for the sediments, an input of only 2–3% sediment component is required.

The Pb isotopic compositions of boninite lavas can also be used to estimate independently the magnitude of this sedimentary component. Although the original Pb isotope signature is obscured by the sedimentary component, an initial composition that is similar to, to slightly less radiogenic than MORB can be predicted on the basis of the depleted nature of the source. Mixing lines constructed through the observed data (Fig. 15.3) between the fields for MORB and Eocene lavas from the Mariana arc and W Pacific sediment (Meijer 1976) suggest that about 50% of the Pb is derived from the leached sedimentary component. Based on the measured Pb contents of the sediments and the boninite samples, a sedimentary contribution of 1.4–3.7% is required. The low incompatible-element concentrations of the sources of boninites make them very susceptible to isotopic modification, and thus this subducted slab contribution is perhaps also responsible for the wide range in Sr and Nd isotope ratios also observed for boninite lavas (Hickey & Frey 1982, Cameron *et al.* 1983, Karpenko *et al.* 1985).

15.8 Conclusions

The unusually low Th/U for some boninite magmas are due in part to the highly depleted nature of the boninite source as well as to late-stage metasomatic modification of the source. Previous partial melting events with clinopyroxene present in the mantle reservoir prior to boninite genesis may lower the Th/U of the source area to values of 1–2. On the basis of partition coefficients determined for the system orthopyroxene–glass for boninite lavas, further modification of the Th/U through continued melting should not occur when clinopyroxene is eliminated from the parent mantle material. Melting of this highly depleted mantle may be triggered by the introduction of a metasomatizing fluid, possibly derived from the dehydration of subducting oceanic crust. This fluid may be quite variable in chemical composition, producing enrichments in Pb, LREE and, to a lesser extent, Th and U during partial melting initiated by fluid fluxing. The range in Th/U observed between different boninite localities may be due to original mantle source differences as well as to variable metasomatic modification induced by slab dehydration fluids. Low $\varkappa$ and μ values of boninite glasses do not correspond to observed radiogenic $(^{207}Pb/^{204}Pb)/(^{206}Pb/^{204}Pb)$ and $(^{208}Pb/^{204}Pb)/(^{206}Pb/^{204}Pb)$ values. These data require late-stage fractionation and/or addition of Th, U and Pb to the source region. Pb isotope data for the Bonin Islands lavas suggest that most of the LILE content in boninite melts is derived from addition of small amounts of a leached sedimentary component that is transferred by fluids to the boninite source region and which represents 1–3% of the total magma contents. Additional trace-element and isotopic data for altered oceanic crust and sediments will help to constrain models of boninite genesis further.

Acknowledgements

We wish to thank the officials of Ogasawara National Park, Japan, for their permission to collect the samples for this study. Many thanks are due S. Maruyama and J. Blank for their assistance in the field. We gratefully acknowledge J. Wooden and J. Stacey of the USGS, who kindly provided access to their facility and assisted with laboratory and mass spectrometry procedures. Sun-Tack Kwon helped greatly with mass spectrometric measurements made at UCSB. We thank J. Wooden, H. Sato, J. G. Liou, R. G. Coleman, J. R. O'Neil, J. Gill and G. Wheller for their comments on this manuscript. Financial support for this project was provided in part by the National Science Foundation (grant EAR 84-18123 to J. G. Liou), the Geological Society of America (Harold T. Stearns Fellowship to P.F.D.), and the Stanford University Shell and McGee Funds. P.F.D. was also supported by an Exxon Teaching Fellowship during his graduate study at Stanford University.

References

Albarede, F. & A. Michard 1986. Transfer of continental Mg, S, O and U to the mantle through hydrothermal alteration of the oceanic crust. *Chem. Geol.* **57**, 1–15.

Allegre, C. J. & M. Condomines 1982. Basalt genesis and mantle structure studied through Th-isotope geochemistry. *Nature* **299**, 21–4.

Allegre, C. J., B. Dupre & E. Lewin 1986. Thorium/uranium ratio of the earth. *Chem. Geol.* **56**, 219–27.

Aoki, K. & H. Fujimaki 1982. Petrology and geochemistry of calc-alkaline andesite of presumed upper mantle origin from Itinome-gata, Japan. *Am. Mineral.* **67**, 1–13.

Aumento, F., W. S. Mitchell & M. Fratta 1976. Interaction between seawater and oceanic layer two as a function of time and depth – 1. Field evidence. *Can. Mineral.* **14**, 269–90.

Barriero, B. 1983. Lead isotopic compositions of South Sandwich Island volcanic rocks and their bearing on magma genesis in intra-oceanic island arcs. *Geochim. Cosmochim. Acta* **47**, 817–22.

Benjamin T., W. R. Heuser & D. S. Burnett 1978. Laboratory studies of actinide partitioning relevant to ^{244}Pu chronometry. *Proc. Lunar Planet. Sci. Conf.* **9**, 1393–406.

Bougault, H., R. C. Maury, M. El Azzouzi, J.-L. Joron, J. Cotten & M. Treuil 1981. Tholeiites, basaltic andesites, and andesites from Leg 60 sites: geochemistry, mineralogy, and low partition coefficient elements. *Int. Rep. DSDP Leg* 60, 657–77.

Cameron, W. E., M. T. McCulloch & D. A. Walker 1983. Boninite petrogenesis: chemical and Nd–Sr isotopic constraints. *Earth Planet. Sci. Lett.* **65**, 75–89.

Cameron, W. E., E. G. Nisbet & V. J. Dietrich 1979. Boninites, komatiites and ophiolitic basalts. *Nature* **280**, 550–3.

Chen, J. H. & J. S. Pallister 1981. Lead isotopic studies of the Samail ophiolite, Oman. *J. Geophys. Res.* **86**, 2699–708.

Condomines, M., P. Morand & C. J. Allegre 1981a. ^{230}Th–^{238}U radioactive disequilibria in tholeiites from the Famous zone (mid-Atlantic Ridge, 36°50′N): Th and Sr isotope geochemistry. *Earth Planet. Sci. Lett.* **55**, 247–56.

Condomines, M., P. Morand, C. J. Allegre & G. Sigvaldason 1981b. ^{230}Th–^{238}U disequilibria in historical lavas from Iceland. *Earth Planet. Sci. Lett.* **55**, 393–406.

Dobson, P. F. 1986. The petrogenesis of boninite: a field, petrologic, and geochemical study of the volcanic rocks of Chichi-jima, Bonin Islands, Japan. Ph.D. Thesis, Stanford University.

Dobson, P. F. & J. R. O'Neil 1987. Stable isotope compositions and water contents of boninite series volcanic rocks from Chichi-jima, Bonin Islands, Japan. *Earth Planet. Sci. Lett.* **82**, 75–86.

Dobson, P. F. & G. R. Tilton 1985. Th, U, and Pb evidence for metasomatic enrichment of a previously depleted source in boninite genesis. *Trans. Am. Geophys. Union* **66**, 1111–12.

Duncan, R. A. & D. H. Green 1980. Role of multistage melting in the formation of oceanic crust. *Geology* **8**, 22–6.

Fyfe, F. R. S. 1979. The geochemical cycle of uranium. *Phil. Trans. R. Soc. Lond. A.* **291**, 433–45.

Gabelman, J. W. 1977. *Migration of uranium and thorium – exploration significance.* Am. Assoc. Petrol. Geol., Stud. Geol. no. 3.

Gast, P. W. 1968. Trace element fractionation and the origin of tholeiitic and alkaline magma types. *Geochim. Cosmochim. Acta* **32**, 1057–86.

Gill, J. B. 1981. *Orogenic andesites and plate tectonics.* New York: Springer.

Gill, J. B., W. Johnson & J. D. Morris 1987. Excess ^{238}U, ^{226}Ra and ^{10}Be in New Britain arc magmas. *Trans. Am. Geophys. Union* **68**, 1522.

Green, D. H., J. W. Morgan & K. S. Heier 1968. Thorium, uranium and potassium abundances in peridotite inclusions and their host basalts. *Earth Planet. Sci. Lett.* **4**, 155–66.

Hamelin, B., B. Dupre & C. J. Allegre 1984. The lead isotope systematics of ophiolite complexes. *Earth Planet. Sci. Lett.* **67**, 351–66.

Hanson, G. N. 1977. Geochemical evolution of the suboceanic mantle. *J. Geol. Soc. Lond.* **134**, 235–53.

Hickey, R. L. & F. A. Frey 1982. Geochemical characteristics of boninite series volcanics: implications for their source. *Geochim. Cosmochim. Acta* **46**, 2099–115.

Jakes, P. & J. Gill 1970. Rare earth elements and the island arc tholeiite series. *Earth Planet. Sci. Lett.* **9**, 17–28.

Jenner, G. A. 1981. Geochemistry of high-Mg andesites from Cape Vogel, Papua New Guinea. *Chem. Geol.* **33**, 307–32.

Karig, D. E. & R. W. Kay 1981. Fate of sediments on the descending plate at plate margins. *Phil. Trans. R. Soc. A* **301**, 233–51.

Karpenko, S. F., A. Y. Sharaskin, Y. A. Balashov, A. V. Lyalikov & V. G. Spiridonov 1985. Isotopic and geochemical criteria for the origin of boninites. *Geochem. Int.* **22**, 1–12.

Kay, R. W. 1984. Elemental abundances relevant to identification of magma sources. *Phil. Trans. R. Soc. A* **310**, 535–47.

Krishnaswami, S., K. K. Turekian & J. T. Bennett 1984. The behavior of ^{232}Th and the ^{238}U decay chain nuclides during magma formation and volcanism. *Geochim. Cosmochim. Acta* **48**, 505–11.

Langmuir, C. H., J. F. Bender, A. E. Bence, G. N. Hanson, & S. R. Taylor 1977. Petrogenesis of basalts from the Famous area: Mid-Atlantic Ridge. *Earth Planet. Sci. Lett.* **36**, 133–56.

Lisitzen, A. D., *et al.* (16 others) 1971. Geochemical, mineralogical, and petrological studies. Init. Rep. DSDP Leg 6, 829–960.

MacDougall, J. D. 1977. Uranium in marine basalts: concentration, distribution and implications. *Earth Planet. Sci. Lett.* **35**, 65–70.

McKenzie, D. 1985. ^{230}Th–^{238}U disequilibrium and the melting processes beneath ridge axes. *Earth Planet. Sci. Lett.* **72**, 149–57.

McLennan, S. M. & S. R. Taylor 1980. Th and U in sedimentary rocks: crustal evolution and sedimentary recycling. *Nature* **285**, 621–4.

Meijer, A. 1976. Pb and Sr isotopic data bearing on the origin of volcanic rocks from the Mariana island-arc system. *Geol. Soc. Am. Bull.* **87**, 1358–69.

Meijer, A. 1980. Primitive arc volcanism and a boninite series: examples from western Pacific island arcs. In *The tectonic and geologic evolution of southeast Asian seas and islands*, D. E. Hayes (ed.), 269–82. Washington: Am. Geophys. Union.

Meijer, A. & B. Hanan 1981. Pb isotopic composition of boninite and related rocks from the Mariana and Bonin fore-arc regions. *Trans. Am. Geophys. Union* **17**, 408.

Menzies, M. 1983. Mantle ultramafic xenoliths in alkaline magmas: evidence for mantle heterogeneity modified by magmatic activity. In *Continental basalts and mantle xenoliths*, C. J. Hawkesworth & M. J. Norry (eds), 92–110. Nantwich: Shiva.

Morioka, M. & K. Kigoshi 1978. Lead isotopes in mantle derived xenoliths from Japan and South Africa. *Geochem. J.* **12**, 223–8.

Morioka, M., K. Kigoshi & S. Uyeda 1971. Uranium, thorium and potassium contents of St Paul rocks. *Geochem. J.* **5**, 1–6.

Mysen, B. O. 1979. Trace-element partitioning between garnet peridotite minerals and water-rich vapor: experimental data from 5 to 30 kbar. *Am. Mineral.* **64**, 274–87.

Nagasawa, H. & H. Wakita 1968. Partition of uranium and thorium between augite and host lavas. *Geochim. Cosmochim. Acta* **32**, 917–21.

Newman, S., R. C. Finkel & J. D. MacDougall 1984a. Comparison of ^{230}Th–^{238}U disequilibrium systematics in lavas from three hot spot regions: Hawaii, Prince Edward and Samoa. *Geochim. Cosmochim. Acta* **48**, 315–24.

Newman, S., J. D. MacDougall & R. C. Finkel 1984b. ^{230}Th–^{238}U disequilibrium in island arcs: evidence from the Aleutians and the Marianas. *Nature* **308**, 268–70.

Oversby, V. M. & P. W. Gast 1968. Lead isotope compositions and uranium decay series disequilibrium in recent volcanic rocks. *Earth. Planet. Sci. Lett.* **5**, 199–206.

Pearce, J. A. 1983. Role of the subcontinental lithosphere in magma genesis at active continental margins. In *Continental basalts and mantle xenoliths*, C. J. Hawkesworth & M. J. Norry (eds), 230–49. Nantwich: Shiva.

Petersen, J. 1891. Der boninit von Peel Island. *Jahrb. Hamburg. Wiss. Anst.* **8**, 341–9.

Rautenschlein, J., G. A. Jenner, J. Hertogen, A. W. Hofmann, R. Kerrich, H.-U Schmincke & W. M. White 1985. Isotopic and trace element composition of volcanic glasses from the Akaki Canyon, Cyprus: implications for the origin of the Troodos ophiolite. *Earth Planet. Sci. Lett.* **75**, 369–83.

Rosholt, J. N. Jr, Prijana & D. C. Nobel 1971. Mobility of uranium and thorium in glassy and crystallized silicic volcanic rocks. *Econ. Geol.* **66**, 1061–9.

Stacey, J. S. & J. D. Kramers 1975. Approximation of terrestrial lead isotope evolution by a two-stage model. *Earth Planet. Sci. Lett.* **26**, 201–21.

Sun, S. S. 1980. Lead isotopic study of young volcanic rocks from mid-ocean ridges, ocean islands and island arcs. *Phil. Trans. R. Soc. Lond. A* **297**, 409–45.

Sun, S. & R. W. Nesbitt 1977. Chemical heterogeneity of the Archaean mantle, composition of the earth and mantle evolution. *Earth Planet. Sci. Lett.* **35**, 429–48.

Sun, S. & R. W. Nesbitt 1978. Geochemical regularities and genetic significance of ophiolitic basalts. *Geology* **6**, 680–93.

Tanaka, T. & K. Aoki 1981. Petrogenetic implications of REE and Ba data on mafic and ultramafic inclusions from Itinome-gata, Japan. *J. Geol.* **89**, 369–90.

Tatsumi, Y., D. L. Hamilton & R. W. Nesbitt 1986. Chemical characteristics of fluid phase released from a subducted lithosphere and origin of arc magmas: evidence from high-pressure experiments and natural rocks. *J. Volcanol. Geotherm. Res.* **29**, 293–309.

Tatsumoto, M. 1978. Isotopic composition of lead in oceanic basalt and its implication to mantle evolution. *Earth Planet. Sci. Lett.* **38**, 63–87.

Tatsumoto, M. & D. Unruh 1979. Pb evolution in the mantle. *Trans. Am. Geophys. Union* **60**, 934.

Tilton, G. R. & J. H. Chen 1979. Lead isotope systematics of three Apollo 17 mare basalts. *Proc. Lunar Sci. Conf.* **10**, 259–76.

Vidal, P., B. Auvray, R. W. Nesbitt & C. Pin 1985. Origin of boninites from Bonin Islands: Nd, Sr, and Pb constraints. *Terra Cog.* **5**, 286.

Wakita, H., H. Nagasawa, S. Uyeda & H. Kuno 1967. Uranium, thorium, and potassium contents of possible mantle materials. *Geochem. J.* **1**, 183–98.

Watson, E. B., D. B. Othman, J.-M. Luck & A. W. Hofmann 1987. Partitioning of U, Pb, Cs, Yb, Hf, Re and Os between chromian diopsidic pyroxene and haplobasaltic liquid. *Chem. Geol.* **62**, 191–208.

Wendlandt, R. F. & W. J. Harrison 1979. Rare earth partitioning between immiscible carbonate and silicate liquids and CO_2 vapor: results and implications for the formation of light rare earth-enriched rocks. *Contrib. Mineral. Petrol.* **69**, 409–19.

White, W. M. & B. Dupre 1986. Sediment subduction and magma genesis in the Lesser Antilles: isotopic and trace element constraints. *J. Geophys. Res.* **91**, 5927–41.

Williams, R. W., J. B. Gill & K. W. Bruland 1986. Ra–Th disequilibria systematics: time scale of carbonatite magma formation at Oldoinyo Lengai volcano, Tanzania. *Geochim. Cosmochim. Acta* **50**, 1249–59.

Wood, D. A. 1979. Dynamic partial melting: its application to the petrogeneses of basalts erupted in Iceland, the Faeroe Islands, the Isle of Skye (Scotland) and the Troodos Massif (Cyprus). *Geochim. Cosmochim. Acta* **43**, 1031–46.

Woodhead, J. D. & D. G. Fraser 1985. Pb, Sr, and [10]Be isotopic studies of volcanic rocks from the Northern Mariana Islands. Implications for magma genesis and crustal recycling in the Western Pacific. *Geochim. Cosmochim. Acta* **49**, 1925–30.

Wyllie, P. J. 1987. Transfer of subcratonic carbon into kimberlites and rare earth carbonatites. In *Magmatic processes: physiochemical principles*, B. O. Mysen (ed.), 107–19. University Park, PA: Geochem. Soc.

16 Magnesian andesites from Mexico, Chile and the Aleutian Islands: implications for magmatism associated with ridge–trench collision

G. ROGERS AND A. D. SAUNDERS

Abstract

Magnesian basaltic andesites and high-Mg andesites from Baja California, Mexico, are primitive melts with Mg# ranging from 0.53 to 0.75. The lavas, erupted in a post-subduction environment, have an unusual and distinctive trace-element chemistry, with low Rb and Th and high Sr and Ba abundances. Consequently the magmas have very high K/Rb (up to 5400) and K/Th (exceeding 32 000) and very low Rb/Sr (< 0.01). Similar rocks, which we have termed 'bajaites', occur in the Aleutian Islands and southernmost Chile plate margins which, like Baja California, have experienced ridge subduction or virtual cessation of subduction during Cenozoic times.

Although not boninitic, the Baja and Aleutian lavas do show affinities with the high-Mg andesites from the Setouchi Belt, Japan. Absolute abundances of HFSE (Ti, Zr, P) are higher than in most boninites. The unusual chemistry of bajaites have been partly attributed to post-subduction breakdown of amphibole stabilized within the mantle wedge during earlier, normal subduction. Here, we develop this model further. Amphibole breakdown results from thermal relaxation and lowering of P_{H_2O} as subduction slows or ceases or as a ridge is subducted. The fluids released by amphibole breakdown ascend and interact with refractory mantle to produce high-Si, high-Mg liquids characteristic of many of the bajaite lavas. A contribution from the subsiding slab cannot be precluded although such a component cannot, alone, produce the trace-element features of the bajaite series. An extensional crustal regime may be necessary to allow the melts to rise to the surface and retain their primitive character.

16.1 Introduction

Boninites and high-Mg andesites are intimately associated with convergent plate margins. It is therefore reasonable to assume that subduction-related processes are responsible for their high-Mg, high-Si and unusual trace-element characteristics. Perhaps the most important subduction-related factor that controls the genesis of these magmas is the transfer of volatiles, especially hydrous fluids, from the subducted slab. This enables extensive wet, low-pressure melting of otherwise refractory peridotite mantle. That the mantle is indeed refractory, and had probably undergone earlier melt extraction, is indicated by the very low abundances of many of the HFSE, for example Ti, Zr and P (e.g. Meijer 1980, Wood et al. 1981, Hickey & Frey 1982).

The case for a slab-derived component is strengthened by the observation that boninites often have abundances of LILE (K, Rb, Ba, LREE) higher than would be predicted from melting of a depleted mantle source (i.e. LILE/HFSE is high). Volatiles from the slab probably include LILE, thus enriching or metasomatizing the mantle in a manner similar to that proposed for other subduction-related magmas (Saunders et al. 1980, Tarney et al. 1981, Wood et al. 1981, Hickey & Frey 1982). The apparent paradox of deriving an alkali- and LREE-enriched Mg-andesite magma from a refractory mantle peridotite can, therefore, be resolved by recourse to slab-derived metasomatism.

Boninites and related rocks are far less common in the geological record than their calc-alkaline counterparts. Whereas this may be a sampling-related problem, it would also indicate that the boninitic suite is associated with atypical subduction conditions. It has been proposed that the early Cenozoic boninites and Mg-andesites of the W Pacific region (e.g., Mariana forearc) represent magmatism associated with the *onset* of subduction in this region some 45 Ma (Meijer 1980). Injection of water-rich basaltic crust into hot, dry mantle, previously depleted during episodes of arc-magma extraction, could have initiated near-surface melting and generation of boninitic magma. Continued subduction then cools the mantle, pushes the zone of melting to greater depths, and potentially inhibits boninite magma formation, or at least results in conditions unfavourable to its arrival in an unfractionated state at the surface.

Whereas boninites (*sensu stricto*) may indeed be restricted to such an unusual tectonic environment, it is clear that Mg-andesites erupted in other arc settings owe their origin to different factors. We have recently described a suite of Mg-andesites from Baja California, Mexico (Rogers *et al.* 1985, Saunders *et al.* 1987). Similar rocks occur on Isla Cook (S Chile) and in the Aleutian Islands, which, together with Baja California, are regions of attempted or successful ridge subduction during the Cenozoic. Although significantly different from boninites (*sensu stricto*) in major-element chemistry, many of these lavas do possess some of the trace-element features of boninitic lavas (e.g. low Ti/Zr). An additional model is required to explain the origin of these rocks, and we have proposed that breakdown of amphibole during the changing conditions that accompany ridge subduction or cessation of subduction plays an important role in their genesis. Here, we develop this model further, with particular emphasis on the primitive nature of the lavas.

16.2 Tectono-magmatic setting

16.2.1 Baja California, Mexico

The tectonic setting of NW Mexico has been reviewed by several authors (Mammerickx & Klitgord 1982, Klitgord & Mammerickx 1982, Sawlan & Smith 1984, Saunders *et al.* 1987). Throughout much of the Mesozoic and early Cenozoic, the western margin of North America and Mexico was a convergent plate boundary, with extensive calc-alkaline plutonism and volcanism. Oblique collision of the Pacific–Guadalupe spreading centre with this margin 29 Ma, however, led to the development of two diverging triple junctions, the northwards-migrating Mendocino transform–trench, and the southwards-migrating Rivera ridge–transform (Atwater 1970, Dickinson & Snyder 1979) (Fig. 16.1). A transform fault system, now represented in part by the San Andreas Fault, developed between these triple junctions.

As a consequence of the southerly movement of the Rivera Triple Junction, subduction processes were progressively terminated along the continental margin, so that by 12.5 Ma subduction was only active south of 29°30′N. The spreading centre then swung anticlockwise until it lay parallel to the trench between 29°30′N and 23°30′N (Menard 1978, Mammerickx & Klitgord 1982). This portion of the ridge was abandoned as a spreading centre, but its ultimate fate is uncertain. In the northern portion, between the Molokai Fracture Zone (27°N) and an unnamed fracture zone at 29°30′N, there is evidence that the ridge was simply abandoned offshore, leaving a piece of the Guadalupe plate stranded between the ridge and the coastline. South of 27°N, it is unclear if the ridge was actually subducted or also abandoned close to the trench. Nonetheless there was essentially simultaneous cessation of subduction along much of Baja California at 12.5 Ma. North of the 29°30′N fracture zone, the ridge appears to have been subducted before 12.5 Ma, leaving a 'no-slab window' beneath much of the western margin of N America (Fig. 16.1c).

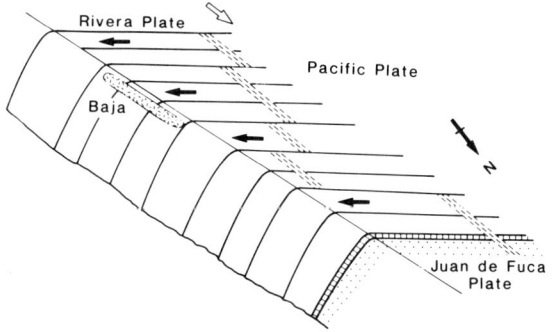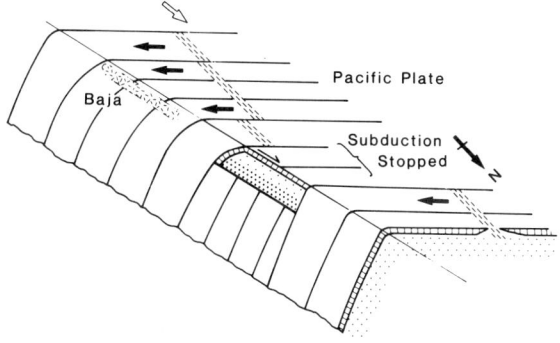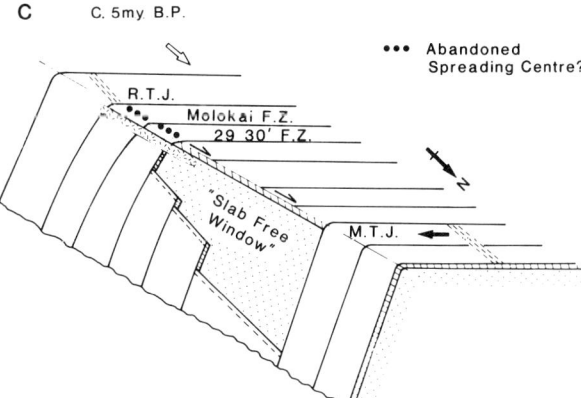

Figure 16.1 Schematic illustration of the development of a slab-free window beneath the northwestern margin of North America. The pre-Gulf location of Baja is shown for reference. Data sources: Dickinson & Snyder (1979) and Mammerickx & Klitgord (1982).

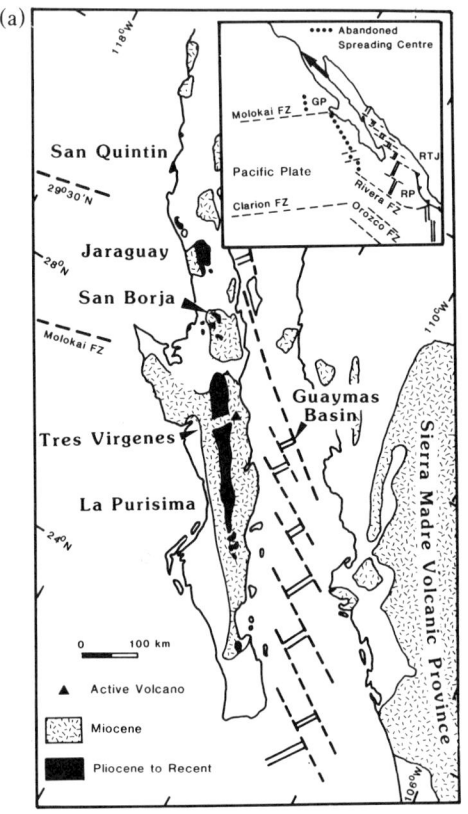

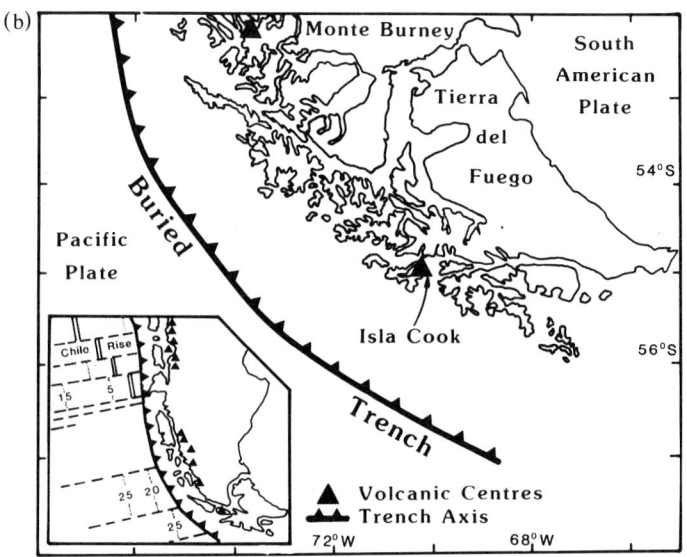

Subsequent plate reorganizations culminated at about 4.0 Ma in the northward propagation of the East Pacific Rise spreading centre, forming the Gulf of California, and coupling the Baja Peninsula to the Pacific Plate. The Baja Peninsula has since moved approximately 250 km northwestwards in the last 4 Ma as a result of spreading at the Gulf mouth (Moore 1973).

The Cenozoic to Recent magmatic history of NW Mexico was outlined by Gastil et al. (1975, 1979) and subsequently detailed by Hausback (1984), Sawlan & Smith (1984), Saunders et al. (1987) and Rogers et al. (1985, 1989). Between 32 and 21 Ma, voluminous outpourings of mainly rhyolitic ignimbrite formed the Sierra Madre Occidental volcanic field on the Mexican mainland (McDowell & Keizer 1977, Cameron et al. 1980a,b) (Fig. 16.2a). Sporadic volcanism is recorded in eastern Baja California between 28 and 10 Ma (McFall 1968, Gastil et al. 1979, Hausback 1984), and may well represent the waning, distal portions of the Sierra Madre activity. This episode of earlier, probably subduction-related, volcanism is recorded within the continental Comondu Formation, which near La Paz is known to contain tuffs as old as 25 Ma (Hausback 1984). In other areas in Baja, basaltic andesites, andesites, minor dacites and acid ignimbrites were erupted up to about 10 Ma and testify to sporadic calc-alkaline magmatism along the NW Mexican margin throughout Miocene times.

From about 12.5 Ma, when subduction ceased, there has been a variety of magmatic products. In northern Baja California Sur (BCS), a sequence of tholeiitic flood basalts, the Esperanza Basalts, was erupted at 10 Ma from fissures in the developing Proto-Gulf (Sawlan & Smith 1984). The most widespread expression of post-subduction activity, however, is the Mg-basaltic andesites and Mg-andesites, termed 'bajaites' by Rogers et al. (1985) because of their distinctive chemistry.

'Bajaites', also misleadingly termed the 'alkalic suite' by Gastil et al. (1975, 1979), range in age from 12.4 Ma to Recent (Gastil et al. 1979, Hausback 1984, 1986, Sawlan & Smith 1984, Saunders et al. 1987).

Magnesian lavas are found in discrete fields in Baja California (Fig. 16.2a). In Baja California Norte (BCN) they occur around the ranchito of Jaraguay, and to the west of Mission San Borja (Stroh 1975). To the south, in BCS, there are flows near San Ignacio (Sawlan & Smith 1984), and extensive fields between La Purisima and Comondu villages (Hausback 1984). Age data for these

---

Figure 16.2 (a) Map of Baja California indicating the locations of the main volcanic deposits. Modified after Saunders et al. (1987). Inset shows the postulated configuration of the plate boundaries and fracture zones at the present day (from Mammerickx & Klitgord 1982, Klitgord & Mammerickx 1982). GP, Guadalupe plate; RTJ, Rivera triple junction; RP, Rivera plate. (b) Map of southern South America showing the location of the Isla Cook and Monte Burney volcanic centres, and (inset) the location of the Chile Rise–Chile Trench triple junction. Ages of ocean floor (inset), and plate motion reconstructions, indicate very slow subduction along the southern Chile margin. After Stern et al. (1984), Puig et al. (1984) and Herron et al. (1977).

various fields indicate that there is no simple geographical progression of any one eruptive centre with age, and thus the magmatism cannot be attributed to the simple northwestwards migration of the Baja lithosphere over a hotspot. There is a tendency for Holocene flows to be more abundant in the La Purisima field than in the Jaraguay and San Borja fields.

The Mg-andesites tend to form blocky flows, yet have a variety of morphologies. Some are erupted from small cinder cones, others build domes and a few are passive effusions without a cone. The thickness of the flows varies considerably, from 5 m to about 80 m, and they are usually less than 3 km long, although Sawlan & Smith (1984) noted one flow in the La Purisima field which is nearly 30 km long. There is a notable absence of pyroclastics associated with the bajaitic eruptions, in contrast to the earlier volcanism of the Comondu Formation.

Sawlan & Smith (1984) recognized that the late Pliocene–Recent Mg-andesites (their 'alkalic suite') are associated with NNW-trending grabens, and that cone building and eruption have been contemporaneous with faulting. Our observations indicate that these cones exhibit a variety of erosional states and collapse features. Many flows in this area, though, were not erupted from cones, but are passive effusions, possibly associated with faulting. The faulting, if deep-seated, may have permitted rapid transit of the magmas through the crust, a potentially important factor in view of the unevolved and apparently uncontaminated nature of the lavas (see below).

Other post-subduction activity includes the late Cenozoic alkali basalt volcanism of San Quintin in northern Baja California (Bacon & Carmichael 1978, Rogers *et al.* 1985, Saunders *et al.* 1987, Storey *et al.* 1989), the active calc-alkaline stratovolcano of Tres Virgenes (Sawlan 1982), and late Cenozoic rhyolitic tuffs in northern Baja California (Gastil *et al.* 1979). The Gulf itself is floored by MORB-like tholeiites (Saunders *et al.* 1982), erupted in small pull-apart basins.

16.2.2 Southern Chile

The magmatic products erupted south of the Chile Trench–Chile Rise intersection (49°S) are compositionally distinct from their more northerly Andean counterparts, and consequently have been included within a separate Austral Volcanic Zone (AVZ) (Stern *et al.* 1984). The tectonic setting in southern Chile is analogous to that found in Baja, with a northward-migrating ridge–trench–trench triple junction resulting from collision of the Chile Rise with the Chile Trench (Fig. 16.2b).

Active, but very slow, subduction of the Antarctic plate beneath the South American plate is suggested by the low rate of convergence of the two plates (Minster *et al.* 1974), and by the negative gravity anomaly associated with the buried Chile Trench as far south as 56°S (Hayes 1966). At the latitude of Isla Cook (55°S), ocean crust of 25 Ma age abuts the continental edge, indicating that several hundred kilometres of Antarctic plate have been subducted since ridge–trench collision approximately 18 Ma (Herron *et al.* 1977).

Several volcanic centres exist within the AVZ (see Puig *et al.* 1984, Stern *et al.* 1984). These include Monte Burney, a large stratovolcano, and Isla Cook, which contains a series of Holocene andesitic domes. Data for the Isla Cook complex are sparse but indicate that the lava type is a moderately evolved Mg andesite, with low Fe/Mg and trace-element characteristics remarkably similar to the bajaites from Mexico. As with the Mexican margin, the oblique convergence at the plate boundary may well be responsible for the ease with which these magmas have ascended uncontaminated and relatively unfractionated.

16.2.3 Aleutian Islands

Two samples of Mg-andesite very similar in trace-element chemistry to those from Mexico and Chile have been found in the Aleutian Island chain (Kay 1978). One sample, from Adak Island, is undated, but a dredged sample from the Aleutian Ridge has a minimum K/Ar wholerock age of 7.9 Ma, and was recovered with 12–14 Ma diatoms. Subduction of the Kula Ridge beneath the Aleutian arc occurred about 30–35 Ma (Grow & Atwater 1970, Atwater & Molnar 1973), with an apparent hiatus of magmatic activity until about 15 Ma (Delong & McDowell 1975). If these ages are reliable, it is interesting to note that the lavas were erupted about 15–20 Ma after ridge–trench collision (cf. 18 Ma in S Chile; and up to 12 Ma after collision until the youngest activity so far, in Baja).

With such limited data, it is not possible to compare closely the tectono-magmatic setting in the Aleutians with that in Baja and S Chile. Nevertheless the occurrence of almost identical magma types in these three provinces, all of which have suffered attempted or successful ridge subduction, leads us to believe that there is a causative relationship common to all three areas.

16.3 Geochemistry

Aspects of the geochemistry of the Mg-andesite lavas from Baja, Chile and the Aleutians have been presented elsewhere (Kay 1978, Puig *et al.* 1984, Stern *et al.* 1984, Rogers *et al.* 1985, Saunders *et al.* 1987) and only a brief summary will be given here. Data for the lavas are given in Table 16.1, which includes hitherto unpublished data from the La Purisima volcanic field in Baja California Sur. All Baja lavas were erupted after the postulated ridge–trench collision; most are Holocene flows. The following summary emphasizes aspects of their chemistry of particular relevance to this volume.

Taken as a whole, the post-subduction lavas from Baja span a wide range of compositions, from 45.6% to almost 60% SiO_2 (Fig. 16.3), and vary from

Table 16.1 Selected wholerock analyses of magnesian Cenozoic lavas, Baja California, Isla Cook (Chile) and the Aleutian Islands.

Sample no. Age (Ma)	J.7.2	J.14.3	J.16.2 5.15	SB.3.2 4	SB.19.1	LP.12.3 2	LP.19.3 2.8	Isla Cook, Chile (1) 3-190	Aleutian Ridge (2) 70-B49 15?
SiO_2	57.2	56.3	53.5	47.1	55.9	55.2	58.0	59.4	58.5
TiO_2	0.65	1.18	2.22	1.96	2.15	1.78	0.79	1.02	0.94
Al_2O_3	16.5	16.3	15.7	13.6	15.6	15.7	17.4	17.7	15.2
tFe_2O_3	5.24	5.90	9.29	8.52	6.33	6.44	4.37	3.88	4.07
MnO	0.08	0.08	0.12	0.11	0.08	0.07	0.06	0.08	0.06
MgO	7.8	6.2	5.6	10.1	5.0	4.8	4.6	3.43	5.0
CaO	6.3	8.0	7.9	9.5	8.1	6.6	5.9	7.7	7.6
Na_2O	4.8	4.2	4.0	2.9	4.3	5.3	4.7	4.8	3.7
K_2O	0.88	1.84	1.02	2.49	2.26	2.31	2.62	0.57	2.2
P_2O_5	0.22	0.53	0.51	1.23	0.85	0.74	0.43	0.27	0.54
LOI	0.51	0.35	0.34	1.49	0.46	0.51	0.64	0.85	2.5
total	100.18	100.88	100.20	99.00	101.03	99.45	99.51	99.70	100.31
100 Mg #	77.7	71.1	56.6	73.5	65.0	63.6	71.3	67.4	74.2
Ni	280	150	83	260	84	80	109	29	150
Cr	322	205	161	366	160	129	128	43	–
V	128	160	244	259	195	175	106	69	–
Rb	6	9	9	14	9	3.5	7	2	17
Ba	280	804	703	1274	1050	1156	1277	147	343
Sr	1189	2068	1479	2831	2609	3039	2085	1738	2600
Th	0.60	2.09	2.14	8.32	2.88	0.60	1.11	4.24	–
Zr	63	166	134	262	257	110	133	166	170
Hf	1.69	4.13	3.53	5.84	5.38	2.98	3.15	4.16	–
Nb	2	5	8	14	10	8	5	10	–
Ta	0.09	0.29	0.46	0.74	0.54	0.51	0.20	1.02	–

La	10.4	31.6	21.7	80.6	43.4	25.8	27.1	31.3	36.3
Ce	24.1	76.1	54.1	164.3	94.7	62.0	62.1	53.6	88.5
Nd	11.4	36.3	27.4	89.0	53.7	34.4	29.0	31.0	47.0
Sm	2.02	5.79	5.15	13.97	8.47	5.63	4.41	4.92	7.73
Eu	0.63	1.68	1.69	3.34	2.52	1.74	1.30	1.54	2.32
Tb	0.20	0.45	0.53	0.91	0.61	0.37	0.33	0.40	–
Dy	–	–	–	–	–	–	–	–	2.47
Er	–	–	–	–	–	–	–	–	0.9
Tm	0.12	0.14	0.17	0.20	0.13	–	–	–	–
Yb	0.58	0.79	1.02	1.34	0.92	0.55	0.57	0.68	0.633
Lu	0.09	0.12	0.16	0.15	0.14	0.09	0.09	0.08	–
Y	6	8	14	18	11	6	5	9	–
Zn	67	73	107	126	85	102	63	–	–
K/Rb	1198	1726	951	1522	2026	5479	3246	2375	1078
Sr/Nd	104	57	54	32	49	88	−72	56	55
Ti/Zr	61.6	42.6	99.8	45.5	50.3	96.5	35.7	36.8	33.1
Ba/Nb	140	168	88	91	105	145	255	15	–
$(La/Yb)_N$	12.1	26.8	19.7	70.1	31.6	31.5	31.9	30.9	38.5
$^{87}Sr/^{86}Sr$	0.70355	0.70361	0.70368	0.70444	0.70379	–	–	(0.7027)[a]	0.7027
$^{143}Nd/^{144}Nd$	0.512913	0.512940	0.512871	0.512705	0.512898	–	–	(0.513136)[a]	–
$^{206}Pb/^{204}Pb$	18.640	18.720	18.709	18.670	18.764	–	–	(18.50)[a]	17.856
$^{207}Pb/^{204}Pb$	15.576	15.577	15.573	15.538	15.608	–	–	(15.53)[a]	15.416
$^{208}Pb/^{204}Pb$	38.35	38.46	38.41	38.34	38.51	–	–	(38.04)[a]	37.313
$\delta^{18}O$	–	–	–	–	–	–	–	(5.6)[a]	–

J.7.2 and SB.19.1 from Saunders et al. (1987); J.14.3, J.16.2, SB.3.2, LP.12.3 and LP.19.3 previously unpublished (analytical techniques described in Saunders et al. (1987)).

Isotopic data for Baja samples from Rogers et al. (1989); (1) data from Puig et al. (1984) and Stern et al. (1984) (REE, Ta, Th and Hf data previously unpublished); (2) data from Kay (1978).

[a] Average values from Stern et al. (1984).

tFe_2O_3 = Total iron as Fe_2O_3.

LOI = Loss on ignition.

$(La/Yb)_N$ = Chondrite-normalized (La/Yt) ratio.

100Mg # calculated using an assumed $Fe_2O_3/FeO = 0.2$.

Ne-normative basalts (or absarokites) to Qz-normative basaltic andesites or andesites (Fig. 16.4). The alkali basalts–absarokites were erupted as a discrete magmatic pulse between about 4 Ma and 2 Ma in both the Jaraguay and San Borja fields, but, despite their distinctive Ne-normative character, their trace-element and isotope chemistry is remarkably similar to the Hy-normative post-subduction lavas.

When compared to other subduction-related magmas, most post-subduction lavas from Baja and the Aleutians have higher MgO contents at a given SiO_2 content (Fig. 16.3). Lavas in the Jaraguay field with 57% SiO_2 attain MgO values of 7.5–8.0% although MgO contents of 6.2% at 55% SiO_2 are more common. The lavas from San Borja and La Purisima (Baja) and Isla Cook (Chile) have lower MgO/SiO_2 than those from Jaraguay, but MgO values of 4.7% at 60% SiO_2 do occur. Ni contents are also high for the SiO_2 levels found, with over 100 ppm Ni at 55% SiO_2 being common. The MgO contents of the post-subduction magmas are not as high as in boninites (*sensu stricto*), but nevertheless are higher than many calc-alkaline suites (Fig. 16.3).

The higher-than-normal MgO contents are mirrored by the high Mg# values, which range from 0.53 to 0.75, the majority being between 0.60 and 0.70. High Mg# are also seen in the Aleutian and Chile samples. The values at the higher end of the range encompass Mg# values predicted for primary magmas in equilibrium with mantle mineral assemblages, and are consistent with the high Fo contents of olivines found in the more magnesian lavas from

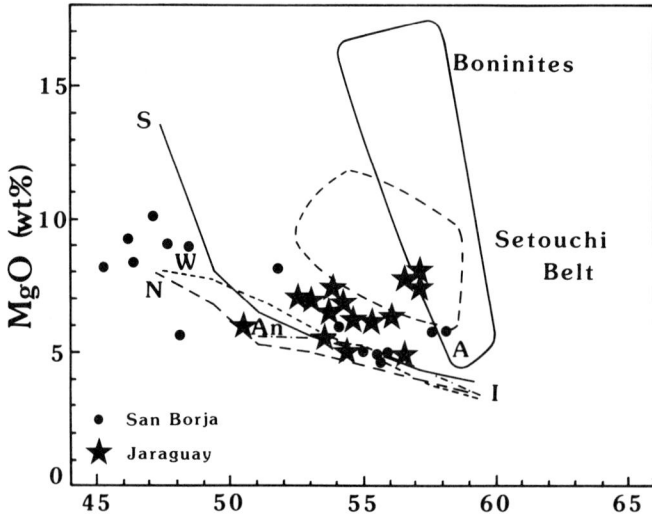

Figure 16.3 MgO versus SiO_2 for post-12.5 Ma bajaitic lavas from Baja California Norte, compared with boninitic, high-Mg, and calc-alkaline magmatic suites. A = Aleutian bajaite; I = Isla Cook bajaite. Data sources: Setouchi – Tatsumi & Ishizaka (1982a); boninites – Dietrich *et al.* (1978), Hickey & Frey (1982) and Cameron *et al.* (1983). Average trends for orogenic suites are from Ewart (1982): N, Northwest Pacific; S, Southwest Pacific; W, Western USA; A, Andes.

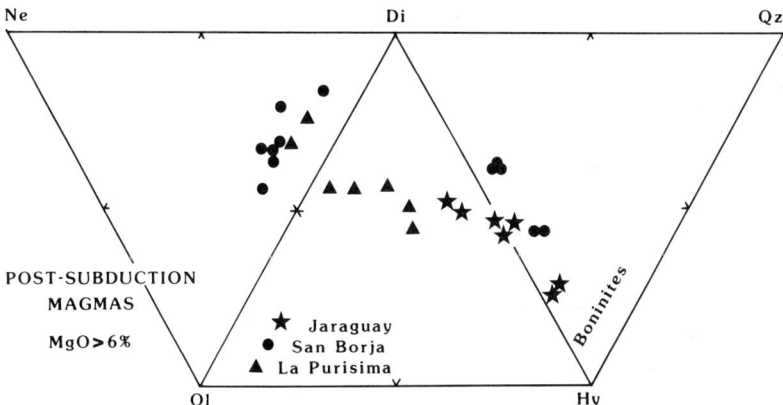

Figure 16.4 Post-12.5 Ma lavas (> 6% MgO) from Baja California plotted on the normative basalt tetrahedron.

Jaraguay (Fo_{87}–Fo_{92}). If the majority (or all) of the Baja (and Aleutian) lavas do not represent primary magmas, then this implies that the parental magmas were even more magnesian if, as seems likely, olivine was the major low-pressure fractionating phase. (Note that olivine is the dominant phenocryst phase in Baja lavas, and clinopyroxene less common; plagioclase is rare as a phenocryst phase. Many bajaites are sparsely phyric, however, with total phenocryst content rarely exceeding 15%.)

In Jaraguay, lavas with Mg# > 0.65 are found spanning almost the entire range of SiO_2 contents, suggesting that a range of near-primary magma compositions was generated from within the sub-Baja mantle. Similar, high Mg# values (~0.75) are recorded in the high-Mg andesites from Setouchi belt, Japan (Tatsumi & Ishizaka 1981, 1982a,b). The latter authors considered the high Mg# values to be indicators of primary magma compositions, although it should be noted that the MgO contents of the Setouchi lavas are higher than those recorded in Baja andesites (MgO often <9% at SiO_2 > 58%). Experimental studies support the suggestion that the Japanese magmas may be primary, although high P_{H_2O} conditions are required in the mantle (~7% H_2O at 14 kbar for high-Mg andesites from Kamiya; Kushiro & Sato 1978).

A further feature of the Baja, Chile and Aleutian lavas is their high content of alkalis, and high Na_2O/K_2O. Consequently the majority fall in the alkali field on the total alkalis–SiO_2 plot (Fig. 16.5), leading to the alkalic classification of the Baja lavas by Gastil et al. (1979) and Hausback (1984, 1986). Note that most of the lavas, with the exception of those erupted on Baja during the distinct pulse 2–4 Ma, are not Ne-normative, and should not be considered as alkaline.

The trace-element signature of the bajaites is distinctive. Superficially, the lavas resemble other subduction-related basalts and basaltic andesites, with an

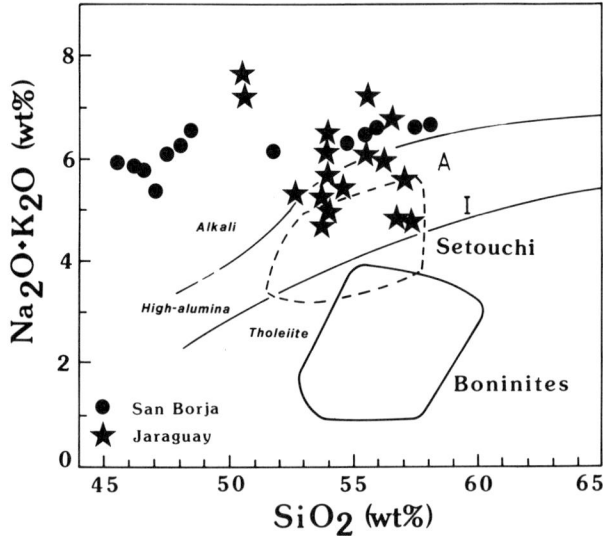

Figure 16.5 Total alkalis ($Na_2O + K_2O$) versus SiO_2 for lavas from Baja, Aleutians (A; Kay 1978) and Isla Cook, Chile (I; Puig et al. 1984). Data for boninites and Setouchi lavas as for Figure 16.3. Alkali, high-alumina and tholeiite fields from Kuno (1959). Compare this diagram with Figure 16.4, which shows that most of the lavas are Hy-normative.

enrichment of LILE relative to HFSE (Saunders et al. 1980, Gill 1981). Thus bajaites have high K/Ta and Ba/Ta relative to MORB. There are, however, several features that are distinctive. Their Rb contents are very low, generally < 10 ppm, resulting in very high K/Rb (Fig. 16.6) (>500, and typically >1500; one sample from La Purisima has a value of 5479). These values are much higher than in orogenic lavas (200–500), and are generally higher than in MORB (fresh MORB glasses have K/Rb < 2500, and usually < 1500; Sun et al. 1979, Sun 1980, Hofmann & White 1983). Th has a similar distribution to Rb, so bajaites typically have high K/Th and Ba/Th. Sr concentrations are very high, ranging from over 700 to 4000 ppm. Ba shows a similar strong enrichment, up to 3000 ppm.

K/Rb and Ti/Zr show a moderate positive correlation (Fig. 16.6), and K/Rb also correlates positively with Nb/Zr (Rogers et al. 1985). This correlation is unlikely to be due to low-pressure fractionation of a Ti-bearing phase (in the low Ti/Zr magmas) because all samples plotted on Figure 16.6 have Ni contents > 50 ppm; most have Ni > 100 ppm, and are relatively unfractionated magmas. It is most unlikely that a primitive, mantle-derived magma would precipitate a Ti-rich phase (e.g. Ti-magnetite) before olivine. Furthermore, there is a tendency for the samples with higher Ti/Zr and K/Rb to have the *lower* Ni and *higher* Sr values.

Chondrite-normalized REE profiles are illustrated on Figure 16.7. All samples are LREE-enriched, with $(La/Yb)_N = 12–49.3$, and $(Yb)_N$ usually < 4,

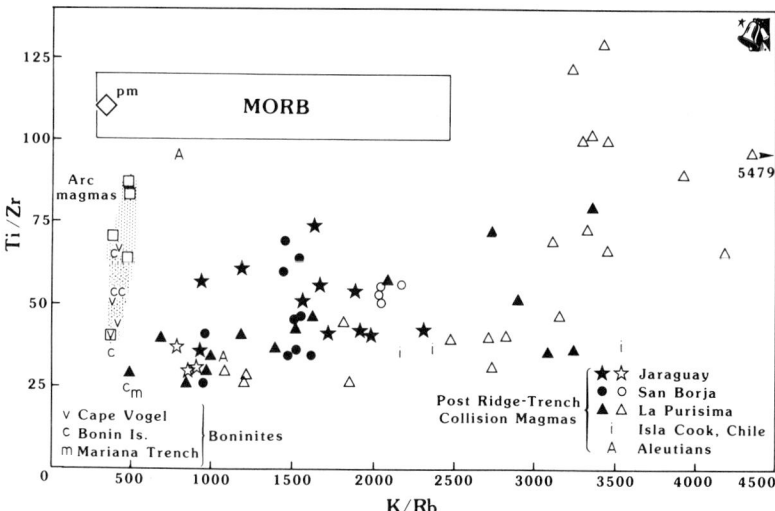

Figure 16.6 Ti/Zr versus K/Rb for Baja, Chile and Aleutian lavas, and for a selection of boninite and island-arc lava data. Data sources: Chile – Puig *et al.* (1984); Aleutians – Kay (1978); boninites – Jenner (1981) and Hickey & Frey (1982); arc magmas – Mariana basaltic andesites and andesites (open squares) – Hole *et al.* (1984); MORB – K/Rb (range) from Hofmann & White (1983) and Ti/Zr (range) from Saunders (1987); PM – primordial mantle (bulk silicate Earth) from Sun (1980). Note that symbols from Baja are coded according to the Ni content of the samples: filled symbols have >100 ppm Ni; open symbols have 50–100 ppm Ni.

and no Eu anomalies. There is a tendency for the Ne-normative basalts to have the most fractionated REE profiles. The Baja, Isla Cook and Aleutian REE patterns all have low La/Sm and high Sm/Y.

Sr, Nd and Pb isotopic data for the Baja, Chile and Aleutian samples are included in Table 16.1; the Nd and Sr data are plotted on Figure 16.8. (One $\delta^{18}O$ measurement is included for the Chile samples.) All samples fall within the 'depleted' quadrant on the $\varepsilon_{Nd}-\varepsilon_{Sr}$ plot, and there is a tendency for the Ne-normative lavas to plot towards higher ε_{Sr} and lower ε_{Nd} values. In the Baja suites, there is a temporal control of the Nd and Sr isotope data, such that ε_{Nd} increases and ε_{Sr} decreases within the younger rocks (Fig. 16.8, Rogers *et al.* 1988). This is particularly well illustrated when the syn-subduction magmas from the Comondu Formation are included on the $\varepsilon_{Nd}-\varepsilon_{Sr}$ diagram (Fig. 16.8). The majority of the analysed syn-subduction lavas plot to the right of the $\varepsilon_{Sr} = 0$ line. Following cessation of subduction, there appears to have been a progressive increase in ε_{Nd} and a decrease in ε_{Sr} with time.

Similar time-dependent relationships are observed for the trace elements (e.g. K/Rb increases in the younger suites; Saunders *et al.* 1987) and suggest that profound changes are occurring in the source of these rocks; in particular, that the magmas are tapping successively more depleted material (at least in terms of Sr, Nd systematics and Rb abundances). There are insufficient data to

Mg ANDESITES FROM MEXICO, CHILE AND THE ALEUTIANS

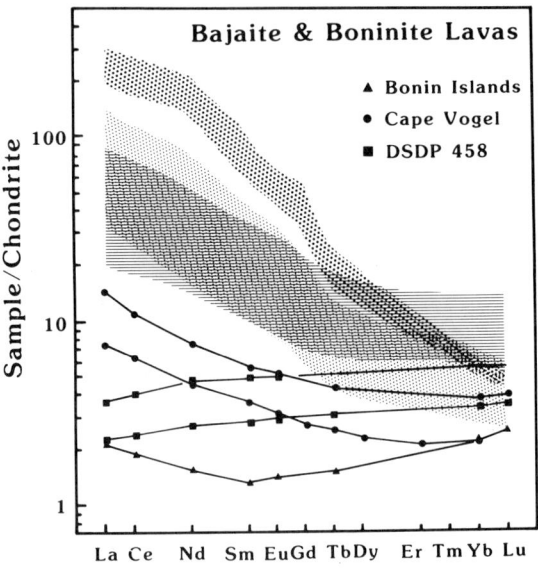

Figure 16.7 Representative chondrite-normalized REE data. Heavy stipple – range for Ne-normative bajaites from Baja California; light stipple – range for Hy-normative bajaites from Baja, Chile and Aleutians. Cross-hatched region depicts the range of values found in the high-Mg andesites of the Setouchi belt, Japan (Tatsumi & Ishizaka 1982a). Boninite data (Bonin Islands, Cape Vogel and Site 458) from Hickey & Frey (1982). Other data sources: Baja – Saunders et al. (1987), this chapter and unpublished data; Aleutians – Kay (1978); Chile – this chapter.

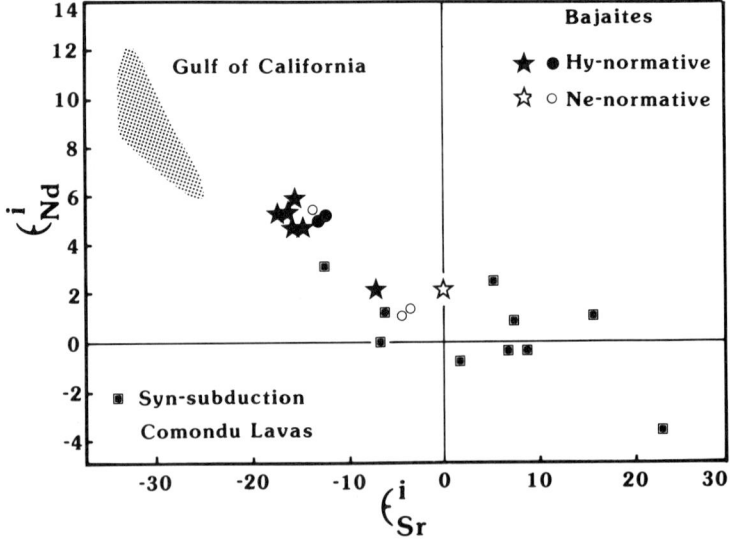

Figure 16.8 ε_{Nd} versus ε_{Sr} for lavas from Baja California (Rogers et al. 1989). Circles – Late Cenozoic samples from the San Borja volcanic field; stars – late Cenozoic lavas from the Jaraguay field. Data for the Gulf of California from Verma (1983).

monitor temporal variations in magma chemistry in the Chile and Aleutian samples, although the low $^{87}Sr/^{86}Sr$ and $\delta^{18}O$ values for the Isla Cook samples do preclude significant crustal contamination (Stern *et al.* 1984).

16.4 Comparisons with boninitic magmas

The late Cenozoic and Holocene lavas from Baja, Chile and the Aleutians are not boninites, although they do show affinities with Mg-andesites transitional to boninites. Boninites are orthopyroxene ($\pm$ olivine) phyric lavas, typically plagioclase-free. Their type locality is the Bonin Islands, but they also occur at Cape Vogel, Papua New Guinea (Jenner 1981), in the Mariana forearc (DSDP Site 458; Wood *et al.* 1981) and in the Palaeozoic Lachlan Foldbelt in SE Australia, believed to have formed by arc–continent collision (Crawford *et al.* 1984, Crawford & Cameron 1985, Crawford & Keays 1987). Boninites are, therefore, associated primarily with ensimatic regimes, within either marginal basins, forearc regions, or island arcs. Ensialic boninites erupted through and onto continental crust are, as far as we are aware, unknown, although high-Mg andesites do occur in continental settings (e.g. the Setouchi belt, Japan; Tatsumi & Ishizaki 1981, 1982a,b).

Definitions of boninites and related rocks are given elsewhere in this volume (Crawford *et al.* 1989). We would merely reiterate the consistent features of the boninite 'series' (Meijer 1980): low TiO_2 and HFSE contents (e.g. $< 0.5\%$ TiO_2), high Cr contents (> 500 ppm), and high MgO contents at intermediate SiO_2 levels. Furthermore, boninites typically have low CaO/Al_2O_3 (< 0.8) and, because of their low TiO_2 contents, they have high CaO/TiO_2 (> 15), Al_2O_3/TiO_2 (> 20) and low Ti/Zr (23–70) (see Sun & Nesbitt 1978, Hickey & Frey 1982, Crawford & Cameron 1985). Although most arc lavas have low Ti/Zr (e.g. Pearce & Norry 1979, Ewart 1982), this may be due largely to the fact that Fe-Ti oxide fractionation commences at a relatively early stage in arc magmas, and rapidly depletes Ti/Zr in residual liquids. Nevertheless, the very low Ti/Zr (often < 40) of primitive boninites requires an explanation. We shall return to this point below.

We have included representative data for boninites and other Mg-andesites on the various geochemical diagrams (Figs 16.3–16.7), wherever data are available. Several differences between the bajaites and boninites are apparent, in particular the higher TiO_2 in bajaites (Fig. 16.9), which produces lower CaO/TiO_2 (< 15) and Al_2O_3/TiO_2 (generally < 20). These values are closer to MORB. CaO/Al_2O_3 are similar to those measured in boninites (Table 16.1).

The differences between bajaites and boninites are emphasized when trace elements are considered (Fig. 16.10). Absolute abundances of REE, Sr and other incompatible elements, with the possible exception of Rb, are much lower in boninites than in bajaites. Although this may in part be due to the

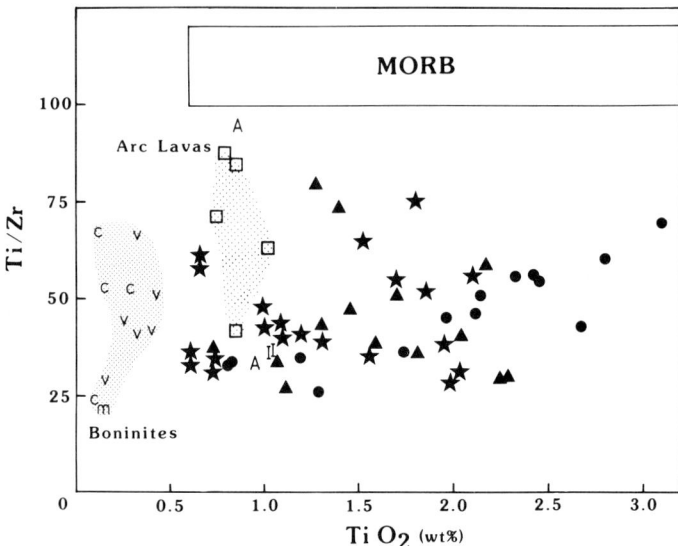

Figure 16.9 Ti/Zr versus TiO$_2$ for lavas from Baja, Chile (I), Aleutians (A), Mariana island arc (open squares and light stipple) and boninites. Symbols as in Figure 16.6. Note that the TiO$_2$ abundances in the bajaites are higher than in boninites, although this may to some extent reflect the more evolved character of the former.

effects of low-pressure fractionation (boninites being less evolved than bajaites), there are several more fundamental differences: boninites are LREE-depleted to moderately LREE-enriched ((La/Yb)$_N$ = 0.6–4.7), often with dish-shaped patterns, whereas bajaites exhibit strongly fractionated patterns ((La/Yb)$_N$ = 12–40) (Fig. 16.7). The bajaites have lower Rb/Sr and higher K/Rb. Whereas both series have broadly subduction-related element patterns with high LILE/HFSE (Fig. 16.10), it is only in the HREE and Rb that absolute concentrations are comparable. Two features that are common to both series, however, are the low Ti/Zr (Fig. 16.6) and high Sr/Nd (or Sr/Ce) (Fig. 16.10).

In several respects, the bajaite series lavas have greater affinities with the Mg-andesites from the Setouchi volcanic belt in Japan, for which only limited trace element data are available (Tatsumi & Ishizaka 1981, 1982a,b). Both series contain phenocrystal olivine and groundmass plagioclase, although the Setouchi lavas contain orthopyroxene and have higher MgO/SiO$_2$ (Fig. 16.3). Whereas the REE are mildly fractionated ((La/Yb)$_N$ = 3.5–7.4), absolute REE concentrations are lower, and Tb/Yb lower (Ishizaka & Carlson 1983), than in the bajaites (Fig. 16.7). Setouchi Sr concentrations are also markedly lower (220–427 ppm) and Rb concentrations are higher (20–139). This results in the Setouchi lavas having lower K/Rb (139–362) and higher Rb/Sr (0.08–0.51). Unfortunately, no published Rb or Zr data are available for the Setouchi lavas to compare with the bajaite series lavas; nonetheless, it is clear that the bajaites

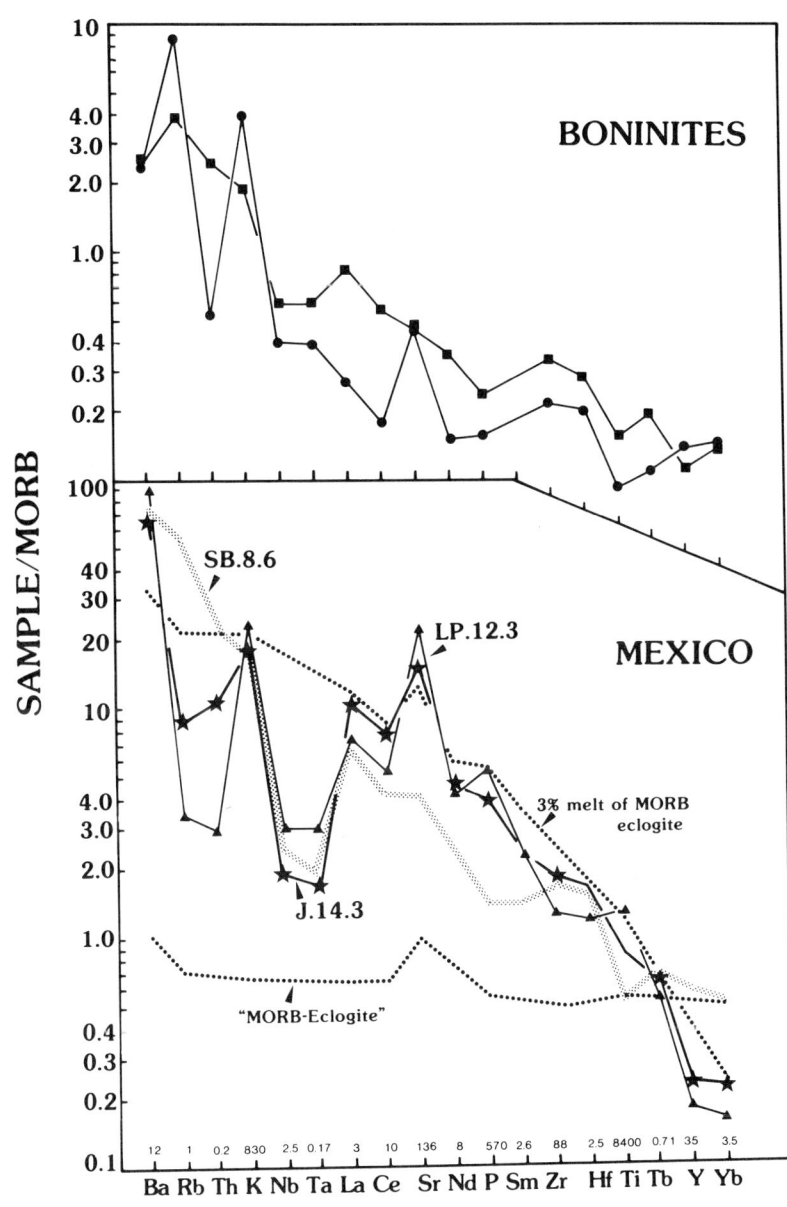

Figure 16.10 MORB-normalized diagrams for boninites (Hickey & Frey 1982, Cameron *et al.* 1983) and Mexican lavas. LP.12.3 and J.14.3 are late Cenozoic bajaites exhibiting characteristic depletion of Rb and Th, and elevated Sr abundances. Compare these patterns with a syn-subduction lava, SB.86, which shows a more typical calc-alkaline distribution of these elements. 'MORB-eclogite' represents the typical MORB values used by Kay (1978) in his models of Mg-andesite genesis via slab melting. Using a 1:1 ratio of clinopyroxene to garnet, a 3% melt has a modelled trace-element distribution shown by the upper dotted line. Note that no data for Th, Ta, Nb, Zr, Hf or Y are included in Kay's model, although Th would show a distribution similar to Rb.

do represent a compositionally distinct suite from the Setouchi Mg-andesites and basalts.

16.5 Discussion

16.5.1 Introduction

The origin of the unusual chemistry of the bajaite series lavas is uncertain. On the one hand, it has been proposed that the Chilean and Aleutian samples are derived by wet melting of subducted ocean crust under eclogite-stable conditions (Kay 1978, Stern *et al.* 1984). In the case of the Chilean samples, the presence of a slowly subducting slab lends some weight to this argument. On the other hand, Saunders *et al.* (1987) and Rogers *et al.* (1989) have argued that melting of subducted ocean crust is insufficient, alone, to produce some of the characteristics of these lavas; in particular, the high K/Rb, Sr/Nd and Ba/La ratios cannot be matched by slab-melting models. Breakdown of minor phases, probably amphibole, stabilized in the mantle wedge during earlier episodes of 'normal' subduction, appears to be necessary to fractionate these LILE. In any model, it also appears necessary for extensional tectonics (with a strong strike–slip component in the case of Chile and Mexico) to allow these magmas to rise with only modest fractionation and minimal contamination to the surface. Furthermore, any such model probably requires unusual conditions (e.g. ridge–trench collision) within the subduction zone or at the plate margin, otherwise the bajaite series would be far more common than is apparently the case. In the following sections we develop these arguments further, paying particular attention to the changing physico-thermal conditions in a subduction zone during the onset of ridge subduction.

16.5.2 Thermal conditions in the subduction environment during ridge–trench collision

Mature subduction zones, associated with calc-alkaline magmatic arcs, probably achieve a steady-state thermal regime in the subducting slab and the hanging-wall mantle wedge. Heat loss from the slab via dehydration maintains low temperatures within the subducted oceanic crust (e.g. Anderson *et al.* 1978, 1980). Decrease of temperature in the hanging wall may be retarded by hydration of the mantle peridotite, thus preserving a strong temperature contrast between the slab and the mantle (Peacock 1987).

It is likely that in mature Phanerozoic subduction zones, the slab undergoes dehydration before it can reach conditions suitable for hydrous melting (Anderson *et al.* 1978, 1980, Wyllie 1979, 1984, Hsui *et al.* 1983). This is indicated by Figure 16.11, which combines published, experimentally determined, phase relationships in basalt and peridotite with modelled geotherms. Curve I represents the modelled thermal profile for the upper surface of old (120 Ma), steeply dipping and rapidly subducting oceanic lithosphere (Anderson et al. 1980). Within this scenario, the main constraints on slab melting are that (a) sufficient water has to be retained within the ocean crust to sufficient depth for the slab geotherm to intersect the wet eclogite solidus (WES; from Wyllie 1984). This will not happen at depths less than 100 km, although it may occur at greater pressures. However, (b) amphibole within the slab is likely to break down at depths approaching 100 km, if not at higher levels. This will occur at temperatures well below the wet eclogite solidus. Similarly, (c) breakdown of serpentinite is also likely to occur well below the wet eclogite solidus, although it could be stable to greater depths than amphibole. As shown on Figure 16.11, cool, dehydrated ocean crust is unlikely to melt at any pressure, because it cannot intersect the dry eclogite solidus.

Transfer of water-soluble material, particularly K, Rb, Ba, LREE, Th, Pb, Sr and Si, from the dehydrating slab into the overlying mantle has been proposed by several workers as a means of producing the characteristic LILE/HFSE enrichment observed in arc tholeiites and calc-alkaline basalts (Jakes & Gill 1970, Hawkesworth et al. 1979, Saunders et al. 1980, Gill 1981). Reaction of such fluids with the hanging-wall peridotite is likely to stabilize a

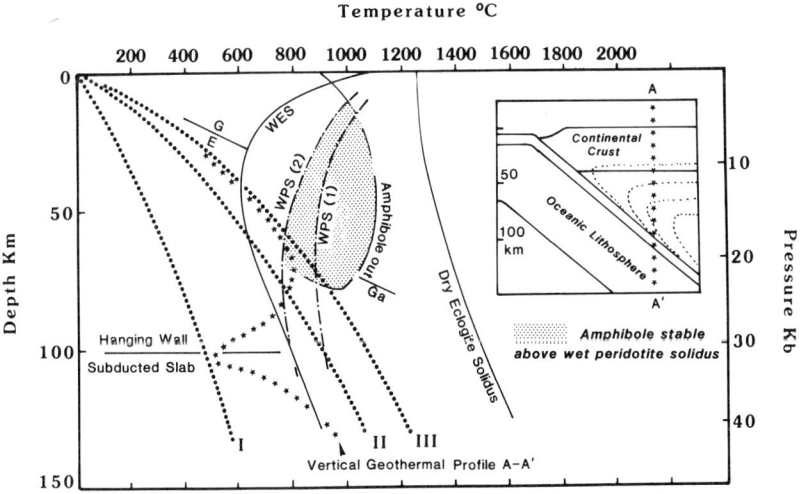

Figure 16.11 Pressure–temperature diagram illustrating the phase relationships and geotherms within mature (cold crust, cold mantle; Wyllie 1984) subduction zones (Curve I; estimated from Anderson et al. 1980). Curves II and III: shield and oceanic geotherms, respectively. Vertical thermal profile adapted from Peacock (1987) to illustrate the thermal cusp above the subducted slab, and the likely configuration of a rebound geotherm (Curves II or III, depending on the ambient geotherm). WES = wet eclogite solidus (from Wyllie 1984); WPS(1) = wet peridotite solidus, rock A of Mysen & Boettcher (1975b); WPS(2) = rock D of Mysen & Boettcher (1975b). Amphibole melting or dissociation curve from Mysen & Boettcher (1975b), rock B. Ga, garnet stable; G, E, gabbro–eclogite transition.

variety of minerals, including amphibole, particularly within the near-trench wedge (cf. Wyllie 1984). There is thus an interesting potential for generating a hydrated, 'perched' metastable mineral assemblage, formed at low-temperature, low-pressure conditions within the mantle wedge adjacent to the slab. In the next two sections, we consider these two alternative models for bajaite genesis, namely, via slab melting or melting of the metasomatized mantle wedge.

16.5.3 Production of high-Mg andesitic bajaites via slab melting

Encroachment of a spreading centre will change the configuration of the subduction zone, and the thermal regime of the slab and mantle wedge (see DeLong & Fox 1977). As progressively younger lithosphere is subducted, the angle of subduction will decrease, and the ocean crust will follow a shallower $P-T$ gradient (Cross & Pilger 1982). Actual ridge subduction, cessation of subduction, or very low subduction rates, could produce an environment suitable for slab melting, provided that the oceanic crust can cross the wet eclogite solidus (Fig. 16.11). The shallower angle of subduction for young oceanic lithosphere has the effect of swinging Curve I (the thermal profile at the surface of the slab) to a higher $P-T$ gradient. If the slab temperature can rise sufficiently before dehydration occurs, then this curve may intersect the wet eclogite solidus and induce slab melting. Furthermore, younger ocean crust is likely to be less hydrated than older crust. This will reduce the supply of volatiles to the subduction zone, and allow the slab to warm up more rapidly. Similarly, if subduction ceased, the remnant slab and wedge would approach isobaric thermal equilibrium, and the whole system would establish its own geothermal gradient.

The effect of this equilibration would also make Curve I migrate towards the shield or oceanic geotherms (Curves II and III; Fig. 16.11) In this case, the intersection of the oceanic crust with the wet eclogite solidus could occur at shallow depths (~ 60 km for a shield geotherm; less than 50 km for a sub-oceanic geotherm, which is possible if sub-ridge mantle is able to gain access to the sub-arc environment). Thus, under these conditions of subduction, or post-subduction, there is potential for slab melting.

Critical to these models is the ability of the hydrous phases to remain in the slab until the wet eclogite solidus is attained. Although a subducting slab of young oceanic crust is likely to warm up rapidly, its less hydrated state and its probable higher rate of dehydration argue against the water being held for sufficient time to allow it to reach the wet eclogite solidus. Furthermore as the slab dehydrates, the eclogite solidus temperature rises and the slab cools; there is thus a tendency for the geotherm and solidus curves to diverge, rather than to converge. In addition, shallowing of the angle of subduction may cause the zone of melting to migrate away from the trench. This is not seen in either Baja or Chile, where post-subduction magmas are erupted very close to the trench.

It thus appears that it is the restructuring of the thermal profile within the wedge rather than in the slab, during subduction of hot, dry lithosphere (Wyllie 1984), which is the major factor in bajaite genesis. To what extent do geochemical studies constrain the role of slab melting?

Kay (1978) and Stern *et al.* (1984) have proposed that slab melting, under eclogite-stable conditions, is responsible for forming the Aleutian and Chile Mg-andesites. A 3% partial melt of subducted MORB, converted to garnet and pyroxene in a 1:1 ratio, will contain high concentrations of many of the lithophile elements, because of their low distribution coefficients (see Fig. 16.10). Furthermore, melting under garnet-stable conditions will rapidly deplete the melt of HREE and Y, and could produce the REE profiles characteristic of the bajaite series lavas. The melt will also contain high concentrations of Sr, owing to the incompatible nature of this element in a clinopyroxene–garnet assemblage.

The melt produced by such a model is likely to be dacitic in composition (Ringwood 1975) and will react with the overlying mantle to form a pyroxenitic assemblage, which could then melt to produce the bajaitic magmas. However, as discussed by Saunders *et al.* (1987) and developed here, this model cannot explain all of the characteristics of the bajaitic lavas.

(a) Partial melting of MORB eclogite will not result in extensive inter-element fractionation of Rb, Th, Ba, K and LREE. Consequently, any melt from the slab will have MORB-like ratios of these elements. The K/Rb of fresh MORB is typically 1000 but may reach over 2000 in Rb-depleted samples; bajaitic lavas frequently have K/Rb in excess of 2000, and as high as 5400 (Fig. 16.6). The melt from an 'average' MORB eclogite will have slightly *reduced* K/Rb, K/Ba and Sr/Rb. The lack of fit for critical element pairs is illustrated by Figure 16.10, which includes the distribution in a modelled 3% melt of eclogite (Kay 1978). Although no data are included for Th, Nb, Ta, Zr, Hf or Y, the major discrepancy between the modelled abundances and the bajaites is clearly seen in Rb, Th and Sr. Similar arguments apply to the Chilean and Aleutian samples. We know of no mechanism whereby Rb (or Th) may be retained preferentially to K in eclogite or mantle assemblages, producing increased K/Rb, K/Ba and Sr/Rb in the melt.

(b) Similarly, the high Sr/Nd in the bajaites (34–104; Rogers *et al.* 1989) compared with MORB (~ 17) cannot be produced solely by partial melting of eclogite.

(c) The variable, but frequently low, Ti/Zr in bajaites are also difficult to explain by MORB melting unless Ti is preferentially retained (perhaps as rutile) within the slab, whereas Zr is transported from the slab. Hickey & Frey (1982) have proposed a similar model for the low Ti/Zr found in boninites. However, the more extreme bajaite lavas, with the highest K/Rb and Sr abundances, tend to have higher, MORB-like Ti/Zr

(Figs 16.6 & 16.9). Although this may be due to successively more extensive melting of the slab, with the eventual exhaustion of the Ti-rich phase, we cannot resolve the high K/Rb or the high incompatible-element abundances by such a model alone. Furthermore we cannot see how such a fractionation curve would be preserved during subsequent melting events in the mantle wedge.

In the case of the bajaites, those magmas with low Ti/Zr may reflect equilibrium of the melt with a *mantle* phase (amphibole?) with a high Ti/Zr. Those bajaites with a higher Ti/Zr (and higher K/Rb) probably represent magmas where the mantle phase has broken down and this released the high Ti/Zr (and high K/Rb) component into the melt, or where the melt has equilibrated with a MORB-type mantle with higher Ti/Zr values (see Figs 16.6 & 16.9). This progressive change of melting behaviour could occur during the changing physicochemical conditions accompanying cessation of, or slow, subduction.

16.5.4 *Ridge–trench collision, thermal rebound and the effect on the hanging-wall mineral assemblage*

Included on Figure 16.11 is an experimental amphibole breakdown curve and two solid curves for hydrated (excess water) garnet peridotite. Estimates of the geotherm in the hanging-wall mantle wedge above the slab are strongly model-dependent. A schematic illustration of a possible geotherm through a *vertical section* is included on Figure 16.11 (profile A–A'); this specific example passes through the hanging wall (HW)–subducted slab (SS) interface at 100 km. The exact profile is difficult to model, depending on the rate of subduction, the temperature of the subducted slab, the ambient geothermal gradient in the adjacent mantle, and the rate of heat and material transfer between the slab and mantle and within the mantle. Furthermore, exothermic reactions may enhance the thermal contrast between slab and hanging-wall mantle, and not only restrict the thickness of the thermal cusp (Peacock 1987) but also allow the cusp to penetrate to considerable depth along the hanging-wall–slab interface. This is illustrated qualitatively on the inset (Fig. 16.11) by the recurved isotherms.

Thermal rebound accompanying ridge subduction or cessation of subduction would result in the local mantle and ultimately the slab returning to the ambient geotherm, as described above. Under isobaric conditions, intersection of the geotherm with the garnet–lherzolite wet solidus could result, but only if sufficiently high P_{H_2O} conditions were maintained. However, it is more likely that P_{H_2O} would not be maintained, if the main source of water is from the dehydrating slab, with the consequence that, as the mantle temperature rises, the peridotite solidus temperature also rises. Intersection may not be achieved under these conditions, since we are essentially describing a dehydration front migrating through the mantle.

Critical to an understanding of the processes accompanying dehydration is the location of the water within the mantle. Most experimental data indicate that amphibole or, perhaps at greater pressures, phlogopite is the main hydrous phase within the mantle. An amphibole breakdown curve is included in Figure 16.11. It is unlikely that amphibole is stable much above pressures of 23 kbar in a spinel lherzolite, or about 26 kbar in a garnet websterite (Mysen & Boettcher 1975a,b). At shallower depths, isobaric warming of an amphibole-bearing assemblage would cause breakdown of amphibole but it would still be a stable phase above the wet peridotite solidus, if P_{H_2O} was maintained. Initial breakdown of amphibole would produce hydrous melts in equilibrium with a pyroxene + spinel (+ garnet?) + amphibole assemblage, with the eventual exhaustion of amphibole (and eventually clinopyroxene) as melting proceeded. Whether simple thermal rebound could accomplish this without a concomitant decrease in P_{H_2O} is not clear.

The melt produced by amphibole breakdown would inherit the amphibole trace-element characteristics (high K/Rb, K/Th and Ti/Zr, and possibly high Na, Sr and Ba contents) depending upon its original composition. The residual mineralogy will control the subsequent K/Rb (etc.) ratios of the melt only if the K_D values are sufficiently high; thus, if the residue is mainly olivine and othopyroxene, both of which have very low K_D values for K, Rb, Th, Ti and Zr, the element abundances and ratios in the melt will be initially determined by the proportion of amphibole contributing to the melt.

Compositional data on suitable high-pressure amphiboles are scarce, although pargasitic amphiboles within lherzolite nodules from Ataq (Varne & Graham 1971, Menzies *et al.* 1987, Menzies & Murthy 1980) do possess requisite characteristics: LREE enrichment, high Sr (600–1400 ppm), high Ba (> 1000 ppm), high K/Rb (up to 2800), low $^{87}Sr/^{86}Sr$ and high $^{143}Nd/^{144}Nd$. Isobaric conversion of the amphibole to a garnet-bearing assemblage may also produce the highly fractionated REE patterns of the bajaites, although experimental data argue against this conversion (Green 1982). Other minerals, such as those of the crichtonite–chevkinite series (Haggerty *et al.* 1986) or apatite, are also potential repositories for Sr and Ba in a metasomatized mantle, and their breakdown could release substantial amounts of these and other elements into the resultant liquid. Only amphibole, however, appears to be capable of releasing melts with the requisite high K/Rb. At this stage, we are purposefully not indicating whether all of these 'metasomatic' minerals were formed solely during prior (earlier Cenozoic or Mesozoic?) syn-subduction events or during metasomatism of the subcontinental lithosphere in a non-subduction setting, perhaps earlier in Earth history. Clearly, though, syn-subduction magmatic activity along the margins of Mexico and S Chile could have imparted a chemical and mineralogical signature on the mantle, and it is this signature which is likely to be tapped during subsequent post-magmatic activity.

16.5.5 Formation of the Mg-andesitic and Ne-normative bajaites

As mentioned earlier, the bajaite series span a compositional range from Ne-normative basalts through olivine tholeiite to quartz tholeiites. There are two problems to be addressed here. First, how does such a range of magmas originate from the mantle? Do they represent different degrees of mantle melting, or melting under different P, P_{H_2O} or P_{CO_2} conditions? Secondly, if post-subduction breakdown of mantle amphibole is contributing to the melt (as apparently required by the trace-element data), how are Qz-normative, Mg-andesite liquids produced?

Anhydrous melting experiments on lherzolites by Mysen & Kushiro (1977) and Jaques & Green (1980) indicate that, as the extent of melting increases at a given pressure, successive liquids migrate further into the olivine tholeiite field as residual orthopyroxene enters the melt. Thus, if the bajaites represent progressively higher degrees of partial melting of a common source, then there should be systematic major- and trace-element variations within the bajaite series lavas. In general, this is not seen, although Ne-normative lavas do tend to have higher abundances of many incompatible elements. Similar arguments apply to melting of a homogeneous source at different pressures, the higher-pressure regime producing the more Ne-normative melts. However, small variations in CO_2/H_2O present during melting could affect the composition of the initial melt quite profoundly (e.g. Mysen & Boettcher 1975a,b), and may be controlling the compositions of the melts generated. Available experimental data on the Setouchi high-Mg andesites (Tatsumi 1982) do not involve CO_2; however, given the analogous tectonic setting of the Baja and Setouchi lavas, it is likely that melting occurred under hydrous conditions, and such data are available.

Experimental work by Tatsumi (1982) on a clinpyroxene-bearing high-Mg andesite from the Setouchi belt, which plots close to some of the bajaites on the basalt tetrahedron (Fig. 16.4), suggested that such lavas may be produced by partial melting of lherzolite in the presence of about 7% water, at about 10 kbar. Orthopyroxene-bearing high-Mg andesites from the same area were generated by more extensive partial melting of the same source, exhausting clinopyroxene and leaving a harzburgite residue (Tatsumi 1981). The contemporaneously erupted basalts were considered to be derived from greater depths than the andesites, but with similar water contents.

The high-Mg, high-Si character of the Mg-andesitic bajaites thus appears to represent melting of mantle, perhaps under hydrous conditions, but at lower P_{H_2O} than is the case during periods of 'normal' subduction. Amphibole, stabilized under the low-temperature, low-pressure and high-P_{H_2O} conditions induced by normal subduction, breaks down as the fluids from the slab decrease, and as the geothermal gradient increases. Hydrous melts, released by amphibole breakdown, in turn flux the mantle and, with the rising temperature, cause melting within an expanded olivine stability field, giving rise to bajaite magmas. Melts from deeper in the wedge, and by implication representing lower-degree melts, presumably give rise to the Ne-normative liquids, whereas higher-level final equilibration produces the Qz-normative bajaites.

The magmas parental to the bajaites cannot have 'seen' amphibole during their last equilibration with the upper mantle, because the magmas would bear the hallmark of residual amphibole (i.e. low K/Rb). At the state of segregation of the primary magma from the mantle, amphibole must have been completely eliminated from the residual assemblage. Amphibole may have been rapidly consumed by the rising temperature and falling P_{H_2O}, so that ascending melts effectively cannibalized and scavenged the amphibole components from the mantle wedge; alternatively, the original amphibole-bearing mantle was restricted in extent, perhaps to a thermal cusp immediately overlying the subducted slab. In the latter case, fluids from the decomposing amphibole would ascend and react with the overlying, amphibole-free mantle. In either case, Ne-normative melts produced by breakdown of amphibole must at least in part have reacted with other silicate phases to produce a Qz-normative, high-Mg liquid.

The trace-element distributions are profoundly controlled by the behaviour of amphibole and any other minor phases. Initially, during subduction, these phases are stabilized in the mantle, and exert control on the abundance and ratios of critical trace elements in any liquids produced. Elements incompatible with such an assemblage (Rb, Th) are flushed through the system and end up in the crust in arc lavas and plutons (Saunders *et al.* 1987, Rogers *et al.* 1989). Other elements (K, Ba, Sr) are partially retained by the mantle assemblage. The mantle effectively acts as a giant chromatographic column, fractionating the different trace elements from ascending percolating melts (Navon & Stolper 1987). The arc environment is potentially ideal to study such a process, because it has input (slab) and output (the arc) parameters. As the physicochemical character of the mantle changes, the components within the chromatographic column, and thus the trace-element composition of the liquid, also change. Although the decomposing mineral phases cannot directly control the composition of the liquid, their contribution to the liquid may be profound. In effect, their development, and eventual breakdown, controls the initial concentrations of different elements in any partial melting equations.

The high-Mg nature of the bajaites may be related to the rapidity with which the magmas ascend through the lithosphere. Egress of the magma from the mantle and through the crust without significant contamination could result from the extensional nature of the terrain in both Baja and Chile. Nonetheless, the magmas are unlikely to be primary magmas, implying that they had more magnesian precursors. Whether or not these precursors have boninitic affinities is open to speculation. The low-CaO/Al_2O_3 and low-Ti/Zr character of the bajaites does indicate, however, that the processes involved in boninite genesis are not unique.

Acknowledgements

We thank John Tarney, Mike Norry and Mark Crawford for many stimulating discussions; Nick Marsh for assisting with XRF analyses; Nick Rogers (Open University) and Giz Marriner (Royal Holloway and Bedford New College) for the REE, Th, Hf and Ta data; Martin Menzies for providing unpublished analyses of pargasite amphiboles; and Trevor Green and Tony Crawford for their helpful reviews. Assistance in Mexico was kindly provided by Departmento de Geofisica, UNAM, Mexico DF, and in particular by Drs Surenda P. Verma and D. J. Terrell. Studies in Mexico are supported by NERC (GR3/4957).

References

Anderson, R. N., S. E. DeLong & W. M. Schwarz 1978. Thermal model for subduction with dehydration in the downgoing slab. *J. Geol.* **86**, 731–9.

Anderson, R. N., S. E. DeLong & W. M. Schwarz 1980. Dehydration, asthenospheric convection and seismicity in subduction zones. *J. Geol.* **88**, 445–51.

Atwater, T. 1970. Implications of plate tectonics for the Cenozoic tectonic evolution of western North America. *Geol. Soc. Am. Bull* **81**, 3513–36.

Atwater, T. & P. Molnar 1973. Relative plate motion of the Pacific and North American Plates deduced from sea-floor spreading in the Atlantic, Indian and Pacific Oceans. In *Proc. Conf. on Tectonic Problems of the San Andreas Fault System*, R. L. Kovach & A. Nur (eds), 136–48. Palo Alto, California: Stanford University Press.

Bacon, C. R. & I. S. E. Carmichael 1978. Stages in the $P-T$ path of ascending basaltic magma: an example from San Quintin, Baja California. *Contrib. Mineral. Petrol.* **41**, 1–22.

Cameron, K. L., M. Cameron, W. C. Bagby, E. J. Moll & R. E. Drake 1980a. Petrologic characteristics of mid-Tertiary volcanic suites, Chihuahua, Mexico. *Geology* **8**, 87–91.

Cameron, M., W. C. Bagby & K. L. Cameron 1980b. Petrogenesis of voluminous mid-Tertiary ignimbrites of the Sierra Madre Occidental, Chihuahua, Mexico. *Contrib. Mineral. Petrol.* **74**, 271–84.

Cameron, W. E., M. T. McCulloch & D. A. Walker 1983. Boninite petrogenesis: chemical and Nd–Sr isotopic constraints. *Earth Planet. Sci. Lett.* **65**, 75–89.

Crawford, A. J. & W. E. Cameron 1985. Petrology and geochemistry of Cambrian boninites and low-Ti andesites from Heathcote, Victoria. *Contrib. Mineral. Petrol.* **91**, 93–104.

Crawford, A. J. & R. R. Keays 1987. Petrogenesis of Victorian Cambrian tholeiites and implications for the origin of associated boninites. *J. Petrol.* **28**, 1075–109.

Crawford, A. J., W. E. Cameron & R. R. Keays 1984. The association boninite low-Ti andesite–tholeiite in the Heathcote Greenstone Belt, Victoria; ensimatic setting for the early Lachlan Fold Belt. *Austr. J. Earth Sci.* **31**, 161–75.

Crawford, A. J., T. J. Falloon & D. H. Green 1989. Classification, petrogenesis and tectonic setting of boninites (this volume).

Cross, T. A. & R. H. Pilger 1982. Controls of subduction geometry, location of magmatic arcs, and tectonics of arc and back-arc regions. *Geol. Soc. Am. Bull.* **93**, 545–62.

DeLong, S. E. & P. J. Fox 1977. Geological consequences of ridge subduction. In *Island arcs, deep sea trenches and back-arc basins*, M. Talwani & W. C. Pitman III (eds), 221–8. Am. Geophys. Union, Maurice Ewing Ser. 1.

DeLong, S. E. & F. W. McDowell 1975. K–Ar ages from the Near Islands, western Aleutian Islands, Alaska: indication of a mid-Oligocene thermal event. *Geology* **3**, 691.

Dickinson, W. R. & W. S. Snyder 1979. Geometry of subducted slabs related to the San Andreas transform. *J. Geol.* **87**, 609–27.

Dietrich, V., R. Emmermann, R. Oberhansli & H. Puchelt 1978. Geochemistry of basaltic and gabbroic rocks from the west Mariana Basin and the Mariana Trench. *Earth Planet. Sci. Lett.* **39**, 127–44.

Ewart, A. 1982. The mineralogy and petrology of Tertiary–Recent orogenic volcanic rocks: with special reference to the andesitic–basaltic compositional range. In *Andesites: orogenic andesites and related rocks*, R. S. Thorpe (ed.), 25–95. Chichester: Wiley.

Gastil, R. G., D. Krummenacher & J. Minch 1979. The record of Cenozoic volcanism around the Gulf of California. *Geol. Soc. Am. Bull.* **90**, 839–57.

Gastil, R. G., R. P. Phillips & E. C. Allison 1975. *Reconnaissance geology of the State of Baja California*. Geol. Soc. Am. Mem. no. 140.

Gill, J. B. 1981. *Orogenic andesites and plate tectonics*. New York: Springer.

Green, T. H. 1982. Anatexis of mafic crust and high-pressure crystallization of andesite. In *Andesites: orogenic andesites and related rocks*, R. S. Thorpe (ed.), 465–87. Chichester: Wiley.

Grow, J. A. & T. Atwater 1970. Mid-Tertiary tectonic transition in the Aleutian arc. *Geol. Soc. Am. Bull.* **81**, 3715–32.

Haggerty, S. E., A. J. Erlank & I. E. Grey 1986. Metasomatic mineral titanate complexing in the upper mantle. *Nature* **319**, 761–3.

Hausback, B. P. 1984. Cenozoic volcanic and tectonic evolution of Baja California Sur, Mexico. In *Geology of the Baja California peninsula*, V. A. Frizzell (ed.), 219–236. Pacific Section Soc. Econ. Palaeo. Mineral. no. 39.

Hausback, B. P. 1986. Gulf of California rift-related alkalic basalts in southern Baja California, Mexico. *Eos* **67**, 1281.

Hawkesworth, C. J., M. J. Norry, J. C. Roddick, P. E. Baker, P. W. Francis & R. S. Thorpe 1979. $^{143}Nd/^{144}Nd$, $^{87}Sr/^{86}Sr$ and incompatible trace element variations in calc-alkaline andesites and plateau lavas from South America. *Earth Planet. Sci. Lett.* **42**, 45–57.

Hayes, D. E. 1966. A geophysical investigation of the Peru–Chile Trench. *Marine Geol.* **4**, 309–51.

Herron, E. M., R. Bruhn, M. Winslow & L. Chauqui 1977. Post-Miocene tectonics of the margin of southern Chile. In *Island arcs, deep sea trenches and back-arc basins*, M. Talwani & W. C. Pitman III (eds), 273–84. Am. Geophys. Union, Maurice Ewing Ser. 1.

Hickey, R. L. & F. A. Frey 1982. Geochemical characteristics of boninite series volcanism: implications for their source. *Geochim. Cosmochim. Acta* **46**, 2099–115.

Hoffman, A. W. & W. M. White 1983. Ba, Rb, and Cs in the Earth's mantle. *Z. Naturforsch.* **38a**, 256–66.

Hole, M. J., A. D. Saunders, G. F. Marriner & J. Tarney 1984. Subduction of pelagic sediments: implications for the origin of Ce-anomalous basalts from the Mariana Islands. *J. Geol. Soc. Lond.* **141**, 453–72.

Hsui, A. T., B. D. Marsh & M. N. Toksoz 1983. On melting of the subducted oceanic crust: effects of subduction-induced mantle flow. *Tectonophysics* **99**, 207–20.

Ishizaka, I. & R. W. Carlson 1983. Nd–Sr systematics of the Setouchi volcanic rocks, SW Japan: a clue to the origin of orogenic magmas. *Earth Planet. Sci. Lett.* **64**, 327–40.

Jakes, P. & J. Gill 1970. Rare earth elements and the island arc tholeiite series. *Earth Planet. Sci. Lett.* **9**, 17–28.

Jaques, A. L. & D. H. Green 1980. Anhydrous melting of peridotite at 0–15 kb pressure and the genesis of tholeiitic basalt. *Contrib. Mineral. Petrol.* **73**, 287–310.

Jenner, G. A. 1981. Geochemistry of high-Mg andesites from Cape Vogel, Papua New Guinea. *Chem. Geol.* **33**, 307–32.

Kay, R. W. 1978. Aleutian magnesian andesites: melts from subducted Pacific crust. *J. Volcanol. Geotherm. Res.* **4**, 117–32.

Klitgord, K. D. & J. Mammerickx 1982. Northern East Pacific Rise: magnetic anomaly and bathymetric framework. *J. Geophys. Res.* **87**, 6725–50.

Kuno, H. 1959. Origin of Cenozoic petrographic provinces of Japan and surrounding areas. *Bull. Volcanol.* **20**, 37–76.

Kushiro, I. & H. Sato 1978. Origin of some calc-alkalic andesites in the Japanese islands. *Bull. Volcanol.* **41**, 576–85.

McDowell, F. W. & R. P. Keizer 1977. Timing of mid-Tertiary volcanism in the Sierra Madre Occidental between Durango City and Mazatlan, Mexico. *Geol. Soc. Am. Bull.* **88**, 1479–87.

McFall, C. C. 1968. *Reconnaissance geology of the Concepcion Bay area, Baja California, Mexico*. Stanford Univ. Publ. Geol. Sci. 10, no. 5.

Mammerickx, J. & K. D. Klitgord 1982. Northern East Pacific Rise: evolution from 25 my B.P. to the present. *J. Geophys. Res.*, **87**, 6751–9.

Meijer, A. 1980. Primitive arc volcanism and a boninite series: examples from western Pacific island arcs. In *The tectonic and geologic evolution of southeast Asian seas and islands*, D. E. Hayes (ed.), 269–82. Am. Geophys. Union, Monogr. no. 23.

Menard, H. W. 1978. Fragmentation of the Farallon Plate by pivoting subduction. *J. Geol.* **86**, 99–110.

Menzies, M. A. & V. R. Murthy 1980. Nd and Sr isotope geochemistry of hydrous mantle nodules and their host alkali basalts: implications for local heterogeneities in metasomatically veined mantle. *Earth Planet. Sci. Lett.* **46**, 323–34.

Menzies, M. A., N. W. Rogers, A. G. Tindle & C. J. Hawkesworth 1987. Metasomatic and enrichment processes in lithospheric peridotites, an effect of asthenosphere–lithosphere interaction. In *Mantle metasomatism*, M. A. Menzies & C. J. Hawkesworth (eds), 151–88. New York: Academic Press.

Minster, J. B., T. H. Jordan, P. Molnar & E. Haines 1974. Numerical modelling of instantaneous plate tectonics. *Geophys. J. R. Astr. Soc.* **36**, 541–55.

Moore, D. G. 1973. Plate-edge deformation and crustal growth, Gulf of California structural province. *Geol. Soc. Am. Bull.* **84**, 1883–906.

Mysen, B. O. & A. L. Boettcher 1975a. Melting of hydrous mantle, I. Phase relations of natural peridotite at high pressures and high temperatures with controlled activities of water, carbon dioxide and hydrogen. *J. Petrol.* **16**, 520–48.

Mysen, B. O. & A. L. Boettcher 1975b. Melting of hydrous mantle, II. Geochemistry of crystals and liquids formed by anatexis of mantle peridotite at high pressures and high temperatures as a function of controlled activities of water, hydrogen and carbon dioxide. *J. Petrol.* **16**, 549–93.

Mysen, B. O. & I. Kushiro 1977. Compositional variation of coexisting phases with degree of melting of peridotite in the upper mantle. *Am. Mineral.* **62**, 843–56.

Navon, O. & E. Stolper 1987. Geochemical consequences of melt percolation: the upper mantle as a chromatographic column. *J. Geol.* **95**, 285–308.

Peacock, S. M. 1987. Thermal effects of metamorphic fluids in subduction zones. *Geology* **15**, 1057–60.

Pearce, J. A. & M. J. Norry 1979. Petrogenetic implications of Ti, Zr, Y and Nb variations in volcanic rocks. *Contrib. Mineral. Petrol.* **69**, 33–47.

Puig, A., M. Herve, M. Suarez & A. D. Saunders 1984. Calc-alkaline and alkaline Miocene and calc-alkaline Recent volcanism in the southernmost Patagonian Cordillera, Chile. *J. Volcanol. Geotherm. Res.* **20**, 149–63.

Ringwood, A. E. 1975. *Composition and petrology of the Earth's mantle*. New York: McGraw-Hill.

Rogers, G., A. D. Saunders & C. J. Hawkesworth 1989. Isotope and trace element geochemistry of syn- and post-subduction magmatism in Baja California, Mexico: constraints on trace element behaviour in arc systems. *Earth Planet. Sci. Lett.* (submitted).

Rogers, G., A. D. Saunders, D. J. Terrell, S. P. Verma & G. F. Marriner 1985. Geochemistry of Holocene volcanic rocks associated with ridge subduction in Baja California, Mexico *Nature* **315**, 389–92.

Saunders, A. D. 1987. Geochemistry of basalts from Mesozoic Pacific ocean crust: Deep Sea Drilling Project Leg 91. Init. Rep. DSDP Leg 91, 483–94.

Saunders, A. D., G. Rogers, G. F. Marriner, D. J. Terrell & S. P. Verma 1987. Geochemistry of Cenozoic volcanic rocks, Baja California, Mexico: implications for the petrogenesis of post-subduction magmas. *J. Volcanol. Geotherm. Res.* **32**, 223–45.

Saunders, A. D., J. Tarney & S. D. Weaver 1980. Transverse geochemical variations across the

Antarctic Peninsula: implications for the genesis of calc-alkaline magmas. *Earth Planet. Sci. Lett.* **46**, 344–60.

Sawlan, M. G. 1982. Geochemical evolution of the Quaternary La Virgen volcano older series lavas, Baja California, Mexico. *Geol. Soc. Am. Abstr. Prog.* **14**, 231.

Sawlan, M. G. & J. G. Smith 1984. Petrologic characteristics, and tectonic setting of Neogene volcanic rocks in northern Baja California Sur, Mexico. In *Geology of the Baja California peninsula*, V. A. Frizzell (ed.), 237–51. Pacific Section Soc. Econ. Palaeo. Mineral. no. 39.

Stern, C. R., K. Futa & K. Muehlenbachs 1984, Isotope and trace element data for orogenic andesites from the Austral Andes. In *Andean magmatism: chemical and isotopic constraints*, R. S. Harmon & B. A. Barreiro (eds), 21–46. Nantwich: Shiva.

Storey, M., G. Rogers, A. D. Saunders & D. J. Terrell 1989. San Quintin volcanic field, Baja California, Mexico: 'within-plate' magmatism following ridge subduction. *J. Geophys. Res.* (in press).

Stroh, J. M. 1975. Latest Cenozoic volcanism, Baja California Norte, Mexico. Ph.D. Thesis, University of Washington.

Sun, S.-S. 1980. Lead isotopic study of young volcanic rocks from mid-ocean ridges, ocean islands and island arcs. *Phil. Trans. R. Soc. Lond. A* **297**, 409–45.

Sun, S.-S. & R. W. Nesbitt 1978. Geochemical regularities and genetic significance of ophiolitic basalts. *Geology* **6**, 689–93.

Sun, S.-S., R. W. Nesbitt & A. Ya. Sharaskin 1979. Geochemical characteristics of mid-ocean ridge basalts. *Earth Planet. Sci. Lett.* **44**, 119–38.

Tarney, J., A. D. Saunders, D. P. Mattey, D. A. Wood & N. G. Marsh 1981. Geochemical aspects of back-arc spreading in the Scotia Sea and western Pacific. *Phil. Trans. R. Soc. Lond. A* **300**, 263–85.

Tatsumi, Y. 1981. Melting experiments on a high-magnesian andesite. *Earth Planet. Sci. Lett.* **54**, 357–65.

Tatsumi, Y. 1982. Origin of high-magnesian andesites in the Setouchi volcanic belt southwest Japan. II. Melting phase relationships at high pressures. *Earth Planet. Sci. Lett.* **60**, 305–17.

Tatsumi, Y. & K. Ishizaka 1981. Existence of andesitic primary magma: an example from southwest Japan. *Earth Planet. Sci. Lett.* **53**, 124–30.

Tatsumi, Y. & K. Ishizaka 1982a. Origin of high-magnesian andesites in the Setouchi volcanic belt, southwest Japan, I. Petrological and chemical characteristics. *Earth Planet. Sci. Lett.* **60**, 293–304.

Tatsumi, Y. & K. Ishizaka 1982b. High-magnesian andesite and basalt from the Shodo-Shima Island, southwest Japan, and their bearing on the genesis of calc-alkaline andesite. *Lithos* **15**, 161–72.

Varne, R. & A. L. Graham 1971. Rare earth abundances in hornblende lherzolite xenoliths: implications for upper mantle fractionation processes. *Earth Planet. Sci. Lett.* **13**, 11–18.

Verma, S. P. 1983. Strontium and neodymium isotope geochemistry of igneous rocks from the North East Pacific and Gulf of California. *Isotope Geosci.* **1**, 339–56.

Wood, D. A., N. G. Marsh, J. Tarney, J.-L. Joron, P. Fryer & M. Treuil 1981. Geochemistry of igneous rocks recovered from a transect across the Mariana Trough, arc, fore-arc and trench, sites 453 through 461, Deep Sea Drilling Project Leg 60. *Init. Rep. DSDP Leg* 60, 611–45.

Wyllie, P. J. 1979. Magmas and volatile components. *Am. Mineral.* **64**, 469–500.

Wyllie, P. J. 1984. Constraints imposed by experimental petrology on possible and impossible magma sources and products. *Phil. Trans. R. Soc. Lond. A* **310**, 439–56.

Index

Figures in *Italic*   Sections in **Bold**

Abitibi Belt (Canada) SHMB 154
absarokite (Ne-normative basalt) 426, 429
Acadian Orogeny 282
actinolite 241, 268, 291, 317, 323
actinolite-tremolite 239
adamellite 161
adiabatic ascent, diapirs 30, 32
adiabatic melting 143
AFC processes 148, 194–5
  associated enrichment/depletion of elements 158–60
  complications involved 157–8
  komatiite as a starting material 163–4
  and SHMB 170
  supporting evidence 156–7
Akaki Canyon glass 295, 296, 303
  depletion in incompatible elements 305
  evolved nature of 301
  subduction zone signature, axis sequence lavas 306
Al_2O_3
  boninite magmas 407
  decreasing with increasing MgO 273
Al_2O_3 abundances 293, 347
Al_2O_3 contents, low, lamproites 83
albite 268, 317
Aleutian Islands 418
  tectono-magmatic setting 423
Algeria, ultrapotassic rocks 77
alkali content, high 427
alteration 169, 291, 316–17, 342
  effects on boninitic and LOTI lavas 245, *247*
  low grade, REE relatively immobile 245
  of primary minerals 239
  secondary 408
  of Th U and Pb 403
alteration minerals 401
alteration processes, and Th/U ratios 400–2
Amami Plateau 57
amphibole 58, 62, 65, 76, 77, 155, 239, 406
  in hydrated peridotite 58, 60
  as hydrous phase 439
  Na_2O and K_2O rich 77
  pargasitic 12, 439
amphibole breakdown 416, 418, 434, 435, 440–1
  melt produced 439
amphibole peridotite 108

amphibole peridotite curtain 108
amphibolite, metagabbroic 184
andesite 194, 290, 319, 421
  basaltic 342, 359, 385, 421
    magnesian 416
    Qz-normative 426
  boninite parent 319
  genesis of 115–16
  high-SiO_2 317
  orogenic, origin for 50–1
  tholeiitic 314, 332–3, 339, 342
anhydrous melting experiments 440
anorthite 115
anorthosite layers 180
Antarctic plate, subduction of 422–3
Antarctica 151
  SHMB 150
apatite 180, 439
Appalachian boninites
  and plate tectonics **10.10**
  source 280
Appalachian mountains, plate tectonic models 281–4
Appalachian ophiolites, boninitic lavas in 264–84
Appalachians (northern), tectono-stratigraphy *265*
aqueous fluids
  high temperature, work of 28
  super-heated 41
  transporting SiO_2 and Na_2 in upper mantle 23
Arakapas Fault Belt 290
arc volcanics 144
assimilation 162
assimilation-fractional crystallization processes *see* AFC processes
asthenosphere, hot, rising 68
augite 12–13, 95, 317, 323, 363
  brown hornblende rims 317
  quench 363
  sub-calcic 363
augite rims 363
Austral Volcanic Zone (AVZ) 422
Australia (SE) boninites (Cambrian) 37
Ayfon (Turkey), ultrapotassic rocks 78

backarc basins 84
backarc opening 68

INDEX

backarc spreading 66
  initiation of 63
Baja California
  Mg andesites 417–18
  tectono-magmatic setting 418–22
Baja California boninites 1, 5, 7, 13–14, 17–18, 21, 23, 26, 40, 42
  compositional variability 24
  generated from depleted mantle 43
bajaite genesis 436–7
bajaites 5, 7, 416, 417
  affinities with Setouchi Mg-andesites 432, 434
  and boninites, differences between **16.4**
  formation of 440–2
  post-subduction activity 421
  production via slab melting 436–8
  Qz-normative 441
  trace-element signature distinctive 427–9
basalt 315
  aphyric 323
  arc 302
    low-Ti 11
  backarc 34, 35, 43, 281, 283
  backarc basin 332, 351
    Lau Basin 385
    Mariana Trough 333
  high-Mg, siliceous 148–70
  intraplate 357
  komatiitic 267
  MORB type 32
  MORB-like 283
  Ne-normative 426, 429, 440
  ocean-floor affinities 234
  oceanic, mobilizing Th and U 401
  oceanic affinities 235
  ophiolitic, low-Ti 377
  porphyritic 151
  siliceous high-Mg (SHMB) 15–17
  spinifex-textured 148, 151
  subduction zone 306
  tholeiitic 54, 333
  Ti/Zr 319
  within-plate 305
  signature 234, 235
basement (W Pacific), planned drilling 41
basins, pull-apart 290
batch partial melting 228–9, 302, 380
batholiths, granitic 236
Bay of Islands ophiolite 283–4
Be, transported in hydrous fluids 57
Betic Cordillera (Spain), ultrapotassic rocks 76
Betts Cove ophiolite (Newfoundland) 39, 264, 265, 282, 283
  associated volcanic rocks 276–7
  geological setting 266–7
  tholeiites and boninites 37

Black Flag Beds 152
Black Reef Group 175
Bohemia (Central), lamprophyre dykes 78
Bohemian Massif, ultrapotassic rocks 72
Bonin (Ogasawara) Islands 50, 68
  HMA occurrences 54–7
  lavas, radiogenic isotopic ratios 409
  palaeogeography 56
Bonin (Ogasawara) Islands boninites 1, 4, 51, 134–5, 144
  classification of 91, 94
  petrography and mineralogy **4.2**
Bonin Ridge 54
Bonin Trench 54
Bonin-Mariana arc 33–4
  subduction of active spreading centre 39–40
  tensional stress 39
Bonin-Mariana arc boninites 37
boninite generation
  by contact melting of lithospheric mantle 353
  and crystallization, conditions of **4.5**
  P-T conditions 91
  restricted to upper mantle 347
  tectonic models 280–1
  tectonic setting **1.7**
    physical conditions of 29
    tectonic scenarios 29–37
  Troodos 137
boninite genesis 113, 115
  direct partial melting models 90
  key factors 10, 18
  melting and fractionation mass balances 139–42
  models 18–21
  Pb isotopic data: implications for sediment involvement **15.7**
  petrogenic models, requirements 403–4
  subduction-related origin 90
boninite melts
  buffered 139
  hydrous and LILE enriched 143
boninite nomenclature 3
boninite petrogenesis **1.6**
  enriched and depleted source components 26–9
  experimental studies 89–109
    experimental methods and results **4.3**, **4.4**
  petrography and mineralogy, selected boninites **4.2**
  models 34–7, **1.8**
  present models unsatisfactory 304
boninite segregation 18, 115
boninite source
  depleted 411
  enriched component, depleted source 192

447

INDEX

enrichment
　by subducted sediments 196–9
　selective 144
melting and metasomatism 198–9
metasomatic modification 408–9, 411
boninite source area, model for formation of *198*
boninite sources 1–2, 170
boninite suites
　major element compositional variation 24–5, **1.4**
　petrography-mineral chemistry of **1.3**
　high-Ca boninites 14–15
　low-Ca boninites 12–14
boninites 12, 289, 316, 340
　Appalachian, petrogenesis of **10.8**
　arc rock signature 347
　Archaean, Proterozoic, SHMB and modern, geochemical distinction between **6.4**
　and Archaean SHMB, similarities and contrasts 155–6
　associated with ensimatic regimes 431
　Ca/Al ratios 115
　Cambrian 281
　　mantle source 258
　Cenozoic 279–80
　　W Pacific arcs 265
　chemical and isotope similarities, island-arc basalt 155
　Chichi-ijima, Th, U and Pb data 399–400
　classification of **1.1**
　　major-element chemistry-related scheme 42
　　suites used 4–5
　consistent features 431
　as contaminated komatiites 194–6
　and convergent plate margins 417
　depleted in HFSE 409
　DSDP Site 458, source region for 349–51
　as early stage of backarc magmatism 281
　enriched in Th and U 396
　field relations **7.4**
　　boninitic marginal rocks 181, 183–4
　　quench-textured sills 183, 184
　as first magma in ocean subduction zone 281
　fractionation to bronzite andesite 41
　H$_2$O contents 113
　H$_2$O solubility in 134
　high-Ca *see* high-Ca boninites
　high-MgO and SiO$_2$ combination 189, 191
　Howqua 27, 29, 256
　interpreted as second-stage melting of refractory harzburgitic mantle 149, 150
　low Ti 198
　low-Ca *see* low-Ca boninites

　low-CaCO$_3$/Al$_2$O$_3$, in forearc settings 84
　LREE patterns 432
　LREE-depleted 318–19
　mantle sources of 170
　modern
　　intra-oceanic arcs 280–1
　　only in supra-subduction zone settings 282
　　order of crystallization 268–9
　　origin of 279–80
　　U-shaped REE patterns 274–5
　as near primary subduction-related magmas 347
　New Caledonia, Sr-Nd isotopes and tectonic setting 334–6
　origin of, experimental evidence 112–44
　　simple system analogue **5.2**
　　tectonic implications **5.7**
　Pacquet Harbour Group 267
　petrogenic scenarios 144
　petrogenic scheme *10*
　Phanerozoic 314–15
　primary 114
　radioisotopic signatures 43
　sulphide-undersaturated 170
　in supra-subduction settings 113
　Tertiary 256
　　mantle source areas, depletion and enrichment 257–8
　transitional, DSDP Site 458 348–9
　and transitional lavas, importance of 347–8
　unusual chemical features 264
　W Pacific, possible re-evaluation of **11.6**
　see also high-Mg andesites
boninitic rocks
　economic importance of 2–3
　high SiO$_2$ 246
　western Tasmania 234, 259–60
　　enriched source 268
　　major- and trace-element characteristics Table 9.2, 246–9, 256, 257
　　petrography 239
breccia 239, 397
bronzite 13, 89, 94, 95
　in boninite petrogenesis experiments 99–103, 105
bronzite andesite 4, 41, 339, 397
　boninitic 12
　Chichi-jima, Th, U and Pb data 399–400
　DSDP Site 458 340–1
　　comparison with Mariana region boninite 343
　　isotopic data **13.3**
　　like differentiated Bonin Island boninite 342–3
　　origins in magma mixing 348–9
　　parental magmas 339

INDEX

bronzitite 178
  REE patterns 188–9
Burgersfort Bulge 184
Bushveld Complex 3, 15, 149, 174
  geological setting and stratigraphy 176–80
  layered suite, evolution of 181–2
  marginal rocks
    boninitic 181
    tholeiitic 181
  and SHMB 150
Bushveld Complex (Eastern)
  Bushveld Complex 176–80
  Transvaal Sequence 175–6
Bushveld Granite 180

calc-alkaline rocks 194
calcite 268, 401
CaO abundances 293, 347
  SHMB 152, Table 6.3
CaO Contents, distinctive in orthopyroxenes, Vestfold Hills 227–8
CaO/Al_2O_3 ratios, N Tongan high-Mg lavas 382, 385
CaO/Al_2O_3 values
  high 348
  Tasmanian boninitic rocks 256
  Vestfold Hills subgroup III 223
CaO/Na_2
  of glass inclusions 376
  N Tongan lavas 385
CaO/TiO_2, boninite magmas 407
Cape Smith Foldbelt (Canada) 31
Cape Vogel (Papua, New Guinea) boninites 1, 4, 7, 11, 12, *14*, 19, 37, 40, 43, 120, 155, 156, 216, 336
  CaO/Al_2O_3 ratios 385
  coexistence of low-Ca and high-Ca boninites 25
  refractory mantle source 382
Ce enrichment 297
Celebration Dykes system, Eastern Goldfields 151
Chichi-jima boninites 4, 7, 11, 12, 19, 27, 29, 37, 42, 43, 54, 55, 91, 120, 363, 397
  coexistence of low-Ca and high-Ca boninites 24–5
  Th, U and Pb data Table 15.1
Chile Rise 422
Chile (Southern), tectono-magmatic setting 422–3
Chile Trench 422
chlorite 58, 62, 65, 239, 241, 268, 316, 317
  in hydrated peridotite 58, 60
chlorite/serpentine clots (pseudomorphs) 269
chromite 15, 25, 184, 268, 301
  cumulus 178, 186
  as inclusions 25

chromite crystallization 182
chromitite layers 178, 182
chromium content, olivine phenocrysts 325, 328
Chukotat Group 31–2
Chuniespoort Group 176
clay minerals 323
clinoenstatite 1, 91, 95, 113, 120, 121–5, 127, 134–5, 239, 269, 291, 328
  and protoenstatite crystallization 245
  quench 120
clinoenstatite crystals 325
clinopyroxene 27, 60, 76, 77, 114, 115, 161, 184, 268, 273, 280, 291, 301, 316, 317, 342, 385, 403, 406, 411
  calcic 214, 215
  absence of 228
  concentrating Th and U 404
  low-Al_2O_3 77
  spinifex-textured 151
  surrounded by phlogopite flakes 77
  zoned 13
CMAS initial melts 111–15
  liquidus phase relations *114*
Cocos plate 66
Comondu Formation 421, 429
conglomerates 236
contact melting 34–5, 36, 43
  hydrated sub-arc peridotite 34–5
continental lithosphere
  modified by subduction processes 164
  development of **6.8**
  modified melting of 169
  as SHMB magma source **6.6**
continental rifting, and boninite generation 30–2
continents, growth of 194
convection, slab-induced 144
Cowarna Rocks (Yilgarn Block) 162
  SHMB, Proterozoic dykes 150, 151
Cr abundance, LFC lavas 301
Cr concentration, high 90
Cr mineralization 202
Cr-actinolite 239
Cr-spinel 1, 76, 77, 94, 95, 101, 107, 113, 234, 236, 241, 269, 397
  Cr# vs. Mg# 370–5
  early crystallizing 77
  history of magmatic differentiation in related lavas 370
  petrography and mineral chemistry 369–75
  sub-solidus re-equilibration 375
Crimson Creek Formation 234, 235
crustal assimilation 91, 148, 406
  in AFC processes 157–8
crustal contamination 16–17, 24, 32, 78, 148, 149, 157, 167, 169
  and AFC processes 163

449

effect of 161
and SHMB (model) 150
and Vestfold Hills HMG dykes 225
crustal melting, and Ti/Zr 194
crustal recycling 169
crystal fractionation 78, 160, 162
  absence of plagioclase in early stages 273
  by growth on dyke walls 157
  estimation of, Kambalda SHMB 160–1
  layered sills and flows 157
crystal mush flows 239
crystallization, equilibrium 215
crystallization differentiation 95, 98, 107
cumulate textures 186
cumulates 188
  Bushveld complex 174–5, 180, 181
  layered 234, 236
  orthopyroxenite 174–5
  ultramafic 266
cyclic units, Bushveld Complex 178
Czechoslovakia, ultrapotassic rocks 78

Dabi Volcanics 4, 332–3
dacite 4, 41, 290, 317–18, 319, 332–3, 397, 421
  boninitic hypersthene 12
  low-Ti 199
deformation 235
  plastic 236
degassing
  at transform-trench intersections 391
  of mantle, and generation of basalt magma 391
dehydration 27, 28
  of amphibole 108
  shallow, of altered oceanic crust 34
  slab 27, 42, 51, 107, 143
  and slab heat loss 434–5
  of subducted oceanic crust 83, 108, 412
dehydration front, migration through mantle 438–9
depleted component 380–2, 385
  boninite petrogenesis 26–7
diapirism 143
  accelerated 143
  promoting partial melting 143–4
  steady-state 30, 32
  and tensional forces 144
diapirs 39, 150
  anhydrous MORB-source mantle 350
  asthenospheric, intrusion or underplating of into refractory continental lithosphere 166
  mantle, partially molten peridotite 60
  MORB-source 2, 43, 63, 108
  MORB-source mantle 30, 34, 281
    injected into hydrous mantle wedge 51
  partial melt H_2O-saturated 65

steady-state ascent 32
differentiation, tholeiitic lavas 350
diopside, chromian 403
diorite cumulates 180
disequilibrium
  and magma mixing 359
  with phenocryst assemblages 359
disequilibrium crystallization 391
dolerite 315, 316
DSDP Site 458 sequence, suggested origin for 349
Dullstroom Volcanics 175, 176, 178, 184, 199
Dundas Group 236, 238
Dundas Trough 233–4, 236
dunite 139
  layered 236
  REE patterns 188–9
dunite residue 136–7, 140, 142
Dupal mantle anomaly 357, 380
durbachite 78
dykes
  clinoenstatite-phyric 322
  diabase 268
  dolerite 291, 316
  Fe-rich 209
  high-Mg 208–29
  komatiite, prone to contamination 195–6
  lamprophyre 78
  mafic 166, 208, 209
  norite 164
  olivine gabbro 164
  picrite 164
  quartz tholeiite 164
  SHMB 151
    Proterozoic, element abundances *155*, 156
  tholeiite 151, 166
  transform sequence 291

East Antarctic Shield 229
Eastern Goldfields (Western Australia), SHMB 150, 160, 162, 170
eclogite solidus, wet 435, 436
Eldon Group 236
element abundances, SHMB 158–60
element mobilization 306
endiopside, magnesian 363
enriched component(s) 380
  boninites 27–9, 41
  Bushveld Complex micropyroxenites 192–4
  evidence for second 390–1
  Tertiary boninite 258
enriching agent origins, ultrapotassic rocks 83
enriching component(s) 391
  hydrous slab-derived fluid 380
  mantle-derived 380, 389

450

INDEX

two-component mixing model, origin of 389–91
enrichment 280
 Batu Tara leucite-bearing mafic lavas 389-90
 OIB-type 28
 Samoa (Rapa) rocks 389
 Toro Ankole rocks 389
enrichment processes, intraplate melt migration 155
enstatite 23, 236
 incongruent melting of 22, 84
epidote 241, 268, 291
equilibrium crystallization 215
equilibrium melting, progressive 115
eruptions, explosive 106
Esperanza Basalts 421
Eu anomalies
 negative 160, 170
 positive 170, 224
eutectic melting 137, 138–9
exothermic reactions 438
extension, leading to boninite generation 282–3
extensional regimes 417, 434

faults 315
 transform 63, 144, 418
  transcurrent slip 290
  transtensional 289
Fe content, changes in 125–6
Fe enrichment 329
Fe-enrichment trend 319
Fe-Ti oxides 78, 268, 342, 344
Fe/Mg ratio, Vestfold Hills HMG dykes 216
felsic crust/rocks
 Archaean and LREE enriched, as contaminant 163
 involvement in petrogenesis of SHMB 169
felsic volcanics 54
felsite 176
 high-Mg 199–200
FeO content, boninites and SHMB 154
flood basalt, tholeiitic 421
flow, laminar and turbulent in dykes 195–6
fluids
 mantle-derived 391
 subduction-related 405
  migration of 308–9
fold belts 29
Footwall Basalt (Kambalda) 164
forearc wedge, cooling of 65
Formation des Basaltes 315, **12.2**, 336
fosterite 23
fractional crystallization 16, 137, 138, 144, 301, 403, 406
 of mafic alkaline lamprophyric melts 72
fractionation 12, 138, 293

of boninitic magma 353
 crystal-liquid 3
 low-pressure 431–2
 magnetite 194
 olivine 95, 302–3
 olivine-orthopyroxene 116
 orthopyroxene 138
 polybaric, of olivine 139
 pyroxene 138, 219–20, 302–3
fractionation trends 310
fracture zone, transform 143

gabbro 180, 238, 266, 268, 315
 two-pyroxene 180
gabbronorite 180, 181
galena 168
garnet 219, 273
Gaussberg lamproite 84
geotherm, high-temperature 108
geothermal gradients
 high 50, 58, 63, 281
 increase in 441
 and order of eruption 282
geothermometer, olivine-chromite 25
Germany (East), ultrapotassic rocks 78
glass
 high Mg# 359
 principal repository for Th and U boninite lavas 401
glass inclusions
 olivine Table 14.5
 olivine-hosted, compositions of 375, 376
glomerocrysts, orthopyroxene 291
Gordon Limestone 236
granite 161
 melanocratic 78
 from partial melting of tonalite 194
 post-Bushveld 176
granodiorite 161, 336
granophyre 176
granulite terrains 209
gravity anomaly, negative 422
greenstone belts
 Archaean 2, 44, 91, 149
 Australia 39, 256
  relationships between boninite suites 25–6
 Cambrian 277
 SHMB in 169–70
groundmass pyroxene 12, 363
Guadalupe plate 418
Guam 340
Guam boninites 29

H_2O
 in boninite genesis 27
 in HMA magmas 62

451

INDEX

a key ingredient in low-Ca boninite genesis 29
migration of, normal subduction zone 59
released by decomposition of amphibole and chlorite 60
slab-derived 51
solubility of in primitive boninite melt 108
H_2O contents, and boninite variants 142
H_2O front, upward-migrating 65
H_2O saturation 106
Haha-jima basalt 54
hanging-wall mantle wedge 434
  estimates of geotherm 438
Hangingwall Basalts (Kambalda) 152, 157, 160, 164
  low initial $^{87}Sr/^{86}Sr$ 164–5
harzburgite 62, 89, 139, 280
  depleted, invasion by hydrous fluids 2
  hydrated 144
  plastic deformation 236
  residual 385
    refractory 30
  tectonized 268
  Th/U 403
harzburgite residue 115, 136–7, 140, 142, 229, 382, 440
  further melting 22, 39
harzburgite source 335, 347
  increasingly refractory 25
harzburgite-dunite 54
Hf abundances 343, 344
Hf enrichment 339, 347, 352–3, 353
Hf/Sm
  arc lavas 352
  high, origin of in boninites **13.5**
HFS element abundances *252*, 253–4, 296, 416
  low 246, 417
HFS element enrichment 387
HFS element patterns
  boninitic rocks *253*
  LOTI volcanics 249, *253*, 254
HFS element ratios, Tasmanian boninitic rocks 260
HFS elements 41, 144, 272–3
  depletion in 91
    reflecting refractory source character 113
  immobile during alteration 245
  low 90
  in two-component mixing model 388–9
  ultrapotassic rocks 82–3
high-Ca boninites 1, *8*, 11, 14–15, 21–2, 33, 37, 42, 117, 134
  coexistence with low-Ca boninites 24–6
  generated by contact melting, depleted sub-oceanic lithosphere 43
  generating mechanism 38–9
  Type IV 107

high-Mg andesite 289, 416, 426–7
  bajaites 440–1
  and convergent plate margins 417
  found in near-trench environment 51
  generation, Setouchi and Bonin Islands, petrogenic model 63–5
Greenland 91
high-pressure melting experiments 62
key tectono-magmatic features of occurrences 52
high-Mg andesite volcanism 52
high-Mg basalt, siliceous 148
  continental lithosphere as magma source **6.6**
  crustal contamination model 150
  Eastern Goldfields, enrichment factors 162
  genesis of controversial 149
  komatiitic parents and role of crustal contamination **6.5**
  LREE-enriched patterns 149
  spinifex textured 148
  sulphide-undersaturated, Eastern Goldfields 162
high-Mg dykes, Vestfold Hills 208–29
  age relations **8.3**
  petrography **8.4**
  Ree geochemistry 224–5
  subgroup I 211
    source characteristics 218–22
  subgroup II 211
    enriched in incompatible trace elements 222
    source characteristics 222
  subgroup III 211
    source characteristics 222–3
high-Mg felsite 199–200
high-pressure melting experiments 98–9
HMA *see* high-Mg andesites
hornblende 180
hotspot intraplate volcanics 387
Howqua boninites (Australia) 27, 29, 256
HREE 301
  low 90
HREE abundances, low 90, 249, 385
HREE depletion 91, 154
HREE patterns, flat 157
Hunter Fracture Zone 41
Huskisson Syncline 234
hydration
  of asthenosphere by slab-derived fluids 350–1
  of mantle peridotite 434
  near subducted slab 144
  of near-trench mantle wedge 51
hydrothermal circulation, stripping U from sea-water 401
hydrothermal mobilization and enrichment, PGE mineralization 167

452

INDEX

hydrous fluids 2, 18, 21, 22, 24, 26, 50
  in generation of low-Ca boninites 43
  K-rich 83
  LILE-enriched 42, 57
    upward migration of 63
  as LREE transporting agent 25
  slab-derived 62, 391
    migration of 304
  from subducted oceanic crust 391
  Ti and LILE enrichment of lamproite source 83
  as transport agent for enriching component(s) 27–8, 39
  from young, subducted MORB source 28
hydrous front 60
hydrous melting experiments 89
hydrous melting/melts 150, 259, 434–5, 441
hydrous minerals, crystallization of 85

igneous rocks, felsic 162
ignimbrite 421
inclusions 25, 126–7, 325
  chromite 12
  Cr-spinel 77, 214, 215
  Fe-rich spinel 214
  glass 359, 367, 369
  harzburgite 353
  lherzolite, low Th/U 403
  magnesiochromite 316, 317, 325
  in mica 77
  peridotite, metasomatized 405
  spinel lherzolite 200, 280
incompatible elements
  Bushveld marginal rocks 181
  enrichment in 82
  mantle reservoir for 169
incompatible trace-element abundances 309–10
incompatible-element enrichment 228
incongruent melting 137
  of enstatite 22, 84
  of orthopyroxene 115, 117
  of phlogopite 84
Indonesian arc lavas 85
intraplate melt migration 155
intrusions, plutons and dykes 290
ironstone 195
Isla Cook (S Chile) 418, 422–3
  boninites 40

Japan Sea 50
  backarc basin 63
Japan (SW) arc, rotation of 54
Jimberlana dyke 162

K_2O contents, ultrapotassic rocks 83
K, enrichment in 82
K-feldspar 77, 215

K-group elements, enrichment in 389
K-richterite 76
K/Rb 428
Kaapvaal craton 175
kalsilite 73
kamafugite 73
Kambalda (Yilgarn Block) lavas 15
Kambalda (Yilgarn Block) SHMB 148, 150, 151
  Archaean 148
  contaminant for 162
  estimate of chemical and isotopic composition of contaminant 161–3
  estimate of crystal fractionation 160–1
  mass-balance calculation 164
kink banding 291
komatiite 194
  Barberton-type 154, 156, 163
  basaltic 272
  contamination of 195–6
  crustal contamination 91
    extensive during dyke injection 157
  and crustal incorporation 203
  in dyke and magma chamber processes 150
  LREE-depleted 149
  PGE abundances 167
  spinifex-textured 268
  under-saturated 167
  undergoing AFC processes 161
Koudiat el Anzazza dyke (Algeria), ultrapotassic rock 77
Kula Ridge, subduction of 423

La enrichment 253
La/Nb, Scourie dykes 163–4
Lachlan Fold Belt 232
lamproites 73, 75
  Antarctica 81
  Leucite Hills (USA) 81
  LILE enrichment 78–9
  silicic 83, 84
  Spanish
    crystallization sequence 76
    and Italian, REE pattern 81
    origin of 84–5
lamproitic characteristics, from depleted mantle composition 85
lamprophyres (alkaline) (minettes) 78
latite 78
Lau Basin spreading centre 37
lavas
  alkalic, high Th and U 405
  aphyric 240, 259, 291
  arc 302
    Aleutian 385
  boninite
    multiply melted mantle source 397
    Th/U systematics **15.6**

boninitic 232, 235, 265
 alteration 245
 associated volcanic rocks **10.7**
 mineral chemistry **10.5**
 petrography of **10.4**
 source of enrichment 304
 Betts Cove 269
 Limassol Forest Complex, petrogenesis of 288–310
 W Tasmania 232, 234, 235, 237–9
dacite 397
 high-Si and high-Mg 341
 N Tonga arc and trench 359
 high-Ti 277–8, 305
 intermediate-Ti 276–7, 282
 island-arc 405
 petrogenesis of **14.5**
 komatiitic 16–17
 LFC
 crystallization sequence 291
 high CaO/TiO_2 294–6
 hyalophitic texture 291
 low-TiO_2 294–6
 primitive 301
 LOTI 232, 234, 235, 236, 237, 238, 240–1
 affinity to island-arc tholeiite 256
 formed in supra-subduction-zone environment 259
 major- and trace-element analyses 245, Table 9.3, 249, 253–4
 tholeiitic affinity 260
 low-Ti 272, 284, 305 *see also* lavas, LOTI
 mafic, Betts Cove 266–7
 magnesian 289, 421–2
 Mariana Trench 11
 MORB 144
 MORB-type 283
 N Tonga, petrography and mineral chemistry **14.2**
 olivine 291
 phyric 291
 post-subduction, Baja California, Chile and Aleutians, geochemistry of **16.3**
 quench-textured 239
 representing primary magmas 90
 rhyolite 397
 syn-subduction 429
 tholeiite, island-arc 376
 vitrophyric 14, 264
layered complexes 149
 Bushveld 3, 15, 149, 174, 176–82
 PGE mineralization in **6.7**
layered cumulates 234, 236
leaching, by circulating fluids 402
leucite 73
 Fe_2O_3-rich 77
Leucite Hills (USA) 82
 ultrapotassic rocks 72, 77–8

LFC *see* Limassol Forest Co, plex
lherzolite
 anhydrous melting experiments 440
 partial melting of 440
lherzolite source, depleted 21
LILE 41, 229
 affected by partial melting 402
 enriched relative to HFSE 428
 preferentially mobilized by aqueous fluid phase 406
 ultrapotassic rocks 82
LILE abundances 245
 increased 280
 low, N Tongan high-Mg lavas 376
LILE enrichment 2, 30, 90, 91, 136, 277, 288, 357, 387, 405
 ancient continental lithosphere 169
 from ancient crustal component, dehydrating slab 27
 hydrous source contamination 113
 Limassol Forest Complex lavas 296
 a mantle-derived feature 83
 origin of, island arc rocks 83
 Troodos lavas, and fluid interaction 308
 ultrapotassic rocks 72, 73, 78
LILE ratios, and multistage melting 402
LILE-rich fluids 348
 slab-derived 91
Limassol Forest Complex, depleted and enriched sources 301
lithosphere
 Archaean 149
 metasomatic enrichment of 353
low-Ca boninites 17, 21, 33, 42, 117
 coexistence with high-Ca boninites 24–6
 generated from a more refractory source 43
 generating mechanism 38–9
 increase in temperature for generation of 39
 petrography and mineral chemistry 12–14
 refractory sources metasomatized 42
 source of 18–19
 Type 1 1, 6, 12, 19, 21, 26, 28, 42
 and active continental margins 43
 high SiO_2 content 22–3
 Howqua (Victoria) 39
 Type 2 1, 7, *8*, 21, 23–4, 41, 42, 43
 partial melts 26
 Type 3 1, 7, *8*, 11, 12, 13, 19, 22–3, 26, 42, 43
low-temperature exchange 401
Lower Pillow Lavas (axis sequence) 290
LREE 149
LREE depletion 234, 335
 LOTI volcanics 249
LREE enrichment 25, 28, 37, 90, 149, 150, 191, 224, 249, 387, 412

INDEX

hydrous source contamination 113
LFC magmas 307
Nepoui boninites 329
and Zr 304
LREE enrichment patterns 405–6

Machadodorp Volcanics 184
Magaliesberg Formation 184
magma chamber cumulate 236
magma chamber processes 149
  open system 158
magma chambers
  axial 301
  continuous (Bushveld Complex) 178
  crustal 157
  shallow 39
  sub-arc 376
  turbulent mixing 301
magma generation
  model for DSDP Site 458 350–1
  Setouchi volcanic belt 52, 54
  under anhydrous conditions 29
magma genesis
  estimate of depth, Vestfold Hills HMG dykes **8.8**
  subduction zone, basic concepts 57–60
magma mixing 328, 348–9, 359, 391
  implications of mineral chemistry **14.3**
magma ponding 157
magma segregation 3, 17, 32, 40, 60
  depths of 65, 380, *381*
  inferred conditions 136
  upper mantle 26
magma source
  LFC lavas 306–7
  refractory, depleted 18
magmas
  arc-related 289
  arc-tholeiite 143
  bajaite, rapid ascent 441
  basalt, backarc-basin 65
  boninite 113, 348
    crystallization pressures 106
    enriched in Pb relative to HFSE 410–11
    H_2O saturation 106
    low Th 408–9
    low Th/U 411
    magmatic water in 154
    unique combination 143
    variability of 407
  boninitic 2, 17, 34, 174, 181–2
    Precambrian 44
  H_2O content and protopyroxene crystallization 105–6
  H_2O saturation 120
  high-MgO, high-SiO_2, low-TiO_2 *see* magmas, boninite
  HMA 51, 52, 68

formation of 58; in forearc wedge 62
genesis of 50
geological background Table 2.1
H_2O content 60, 62
modern analogues for eruption settings 65–8
production of 63–5
Setouchi and Bonin Islands, experimental petrology studies 60–2
komatiite
  assimilation of crust 148
  peridotic, fractionation and contamination of 44
komatiitic 194
  contamination of 157
LFC, isotopically heterogeneous 310
low Ti/Zr 438
low-Ti 269
LREE-depleted 348
MORB-type 39
noritic 166
orogenic andesite 90
parental 2, 113, 149
  bajaites 441
  Bonin Islands boninites 107
  boninitic bronzite andesite 350
  bronzite andesites 339, 348, 353
  Bushveld complex 15, 17, 174–5, **7.3**
  high-Ca boninite 27
  Limassol Forest Complex 291
  low-Ca boninites, Type 2 23
  magnesian 427
  MORB-type 18
  Negri Volcanics SHMB 164
  picritic 143
  refractory, Tongan arc 359
  sulphide under-saturated 167
  uncontaminated 169
  Upper Pillow Lavas 137, 382
  Vestfold Hills HMG dykes 219, 227
picritic 91
primary 3, 60, 90, 95
  andesitic 89
  arc volcanism 60
  boninitic 350
  composition in island arc settings 358–9
  Troodos 137
primitive 375
  arc tholeiite 348
SHMB, dry 154
silica-saturated 90
siliceous, high-Mg 30
subduction zone, H_2O-rich 58
tholeiite, island-arc 108
tholeiitic 181–2
  arc 34
  rift-related 30
magmatic activity, syn-subduction 439–40

455

INDEX

magmatism
  accompanying continental rifting 30
  alkaline 198
  arc 38–9
  associated with onset of subduction 417
  backarc, boninites as early stage 281
  boninitic, as precursor of backarc-basin spreading 34–5, 36–7
  Bushveld, implications of subducted-sediment model 199–201
  calc-alkaline 421
  convergent plate margin 289
  felsic-to-medium 63
  HMA and back-arc 54
  MORB type 39
  potassic 84
  subduction-related, and boninite generation 33–5
magnesiochromite 12, 22, 268
magnetite 121–5, 180, 184, 323
  fractionation 194
major element data, SHMB 152, Table 6.1
major- and trace-elements, Limassol Forest Complex 293–301
Manam, Papua, New Guinea 11
mantle
  anomalously enriched 85
  depleted 43, 417
    in boninite genesis 196
    mixing with subducted sediment 197–8
  LREE-depleted 200
    contamination by high-Mg felsite 200–1
  MORB-source, injection into backarc region 65
  Pb isotopic composition, preservation of 169
  'plum pudding' 353
  residual, remelting of 302–3
  sub-arc, compositional variation 347
mantle compositions, hydrous and anhydrous, melting studies 116
mantle counterflow 281
mantle differentiation 144
mantle matrix/migrating melt interaction 307–8, 310
mantle melting, hydrous conditions 440–1
mantle metasomatism 304, 353, 380, 405–6, 407
  of lithosphere 353
mantle mixing, localised, efficient 385, 387
mantle source heterogeneity 107
mantle source region, long-term depletion 409
mantle sources
  boninite and komatiite 167
  Cambrian boninites 258
  depleted 310, 357, 385
    superimposed LILE enrichment 387
  high-Mg lavas 382
  multiply melted 397
  previous melting episodes 63
  refractory 259
    N Tongan lavas 385
mantle tectonites 290
mantle wedge 380, 436
  depleted, melting of 280–1
  hanging-wall 434, 438
  hydrous 108
    melting of 108
  induced convection 58
  low temperature 108
  melting of refractory peridotite 391
  metasomatized 335, 435–6
  refractory 28
  slab-induced convection 144
  static 143
  sub-arc 289
  subforearc 62
mantle-derived phase (fluid or melt) 389
Manus Rift 67
Mariana forearc 36, 54, 339
  unusual Pb isotope ratios, boninite and tholeiite lavas 409
Mariana forearc boninites 37, 144
Mariana forearc-Guam boninites 5, 7, 13, 29
Mariana Trench 340
Mariana Trench boninites 4, 6, 7, 53
Marubewan Formation 106
Massif de Koh **12.2**
Massif de Koh boninites, Zr depletion 335
megacrysts
  calcic plagioclase 32
  othopyroxene, reverse-zoned 184, 186
Melanesian (E Papua, New Guinea) region 66, 67, 68
melilite 73
melt inclusions 32
  magnesian 32–3
melt migration 155, 288, 305–6, 307
melt percolation 307–8, 310
melting
  adiabatic 143
  anhydrous 227
    primitive Fe-rich peridotite 227
  anhydrous and hydrous conditions, at shallow levels in lithospheric mantle 350
  contact 34–5, 36, 43
    lithospheric mantle during upwelling of asthenospheric mantle 353
  eutectic 137, 138–9
  hydrous
    of asthenosphere 350
    of peridotite 89, 91, 107
  incongruent 22, 84, 137
    of orthopyroxene 115, 117

456

INDEX

low-pressure
  importance of, boninites and silicic lamproites 83
  of peridotite 216
  of refractory harzburgitic source 216
multistage 402
near-surface 417
peritectic 22–3, 84, 115, 137, 138–9, 142
progressive equilibrium 115
second-stage 143, 149
shallow, of mica-bearing mantle 84
wet 434
  low pressure of refractory peridotite mantle 417
melting zone, migration of 436–7
melts
  anhydrous 259
  asthenosphere-derived 149
  boninite 143
  hydrous 259, 441
  LREE-enriched 308
  parental
    high-SiO_2 82
    mafic alkaline lamprophyric 82
  primary 117, 136
    upper-mantle peridotite 391
  primitive 416
  quartz-tholeiitic 116
  refractory 84
  second-stage 18, 30, 31–2, 32–3
  silica rich, orthopyroxene formation 199
  siliceous 306
Mendocino transform-transform trench 418
Meredith Granite 234
Merensky Reef 178, 179–80, 182
metaboninites 315
metadiabase 268
metamorphism 169
  greenschist facies 151, 235, 245, 268, 291, 314, 316, 322
metasomatic component, Bushveld Complex 174
metasomatic fluids
  boninite-generating 83
  CO_2-rich, as incompatible element reservoir 405
  incompatible-element-enriched 335
  Th- and U- enriched 396
metasomatic modification 408–9, 411
metasomatism 21, 108, 203, 339
  of boninite source 144
  cryptic 353
  of depleted mantle 198
  of harzburgitic source 347
  hydrous 136
  mantle 304, 353, 380, 405–6, 407
  slab-derived 143, 417
metasomatized mantle wedge 435–6

metasomatizing fluids
  possible derivation 411–12
  possible sources 405–6
  shift in Th/U 407–8
Mexican Volcanic Belt 50, 66, 68
Mexico (central) 66
Mg andesite 417–18
Mg# values, high, as indicators of primary magma compositions 427
Mg-andesite 421–2
  comparisons with boninitic magmas 16.4
MgO
  high 259, 318
    ultrapotassic rocks 72, 73, 78, 82
  in quenched melts 188
mica 73, 77, 85
  TiO_2 content 77
mica-harzburgite, melting of 84
micro-orthopyroxenites, quench-textured 174, 175
microlites, clinopyroxene 94, 95
microphenocrysts 12
  augite 13
  bronzite 13, 25, 94, 95, 105
  calcic pyroxene 22, 25
  clinoenstatite 25, 95, 325
  clinopyroxene 95, 342
    zoned 15
  diopside 363
  Fe-Ti oxides 342
  labradorite 342
  olivine 94, 95, 325
    altered 95
    fractionation before eruption 106
  orthopyroxene 21, 95, 323, 342
  pigeonite 21
  plagioclase 323, 342
  pyroxene 401
micropyroxenites 214, 216
  quench-textured 183–4, 186, Table 7.1, 188, 189, 202–3
    compared with other boninites 189–92
    contamination with gneiss 195
    enriched component, depleted source 192
    low Ti/Zr 203
    petrography of 184–5
    trace-element concentrations 200
    xenocrysts 184
Middle American Trench 66
mineral compositions, natural and experimental, comparison of 135
minette, felsic 78
Molokai Fracture Zone 418
MORB 302, 305
  low Th/U 402
  Philippine sea-floor 353
  picritic 32
  subducted 28

MORB pyrolite 18
  yielding MORB glass 382
Mount Read volcanics 236
MREE enrichment 191
Mt Hunt, SHMB (Archaean) 150, 151
Mt Read volcanic belt 234
Muko-jima boninites 54
multiple saturation 60, 101, 382
multiple-saturation experiments 116
multistage melting 402

N Fijian Basin, possible site for boninitic magmatism 41–2
N Tonga arc 33
N Tonga Trench, high-Mg and associated lavas 357–91
N Tonga forearc boninites 11, 14
Na_2O
  in low-Ca boninite sources 27, 28
  solubility of 23
Na_2O contents, enhanced, Type I low-Ca boninites 19
Na_2O/K_2O high 427
Napier Complex, Enderby and Kemp Land 209, 229
  HMG dykes, undeformed 209
Navajo region (USA), ultrapotassic rocks 72, 78
Nb 220–1, 222
Nb anomaly, negative 389–90
Nb depletion 223
Nb/Zr 428
$^{143}Nd/^{144}Nd$, in LFC lavas 300–1
$^{143}Nd/^{144}Nd$ isotope composition, SHMB 152, Table 6.2
$^{143}Nd/^{144}Nd$ ratio, New Caledonia tholeiites 331–2
Nd isotope geochemistry, Tasmanian boninitic and LOTI lavas 257–9
Nd isotopes, Betts Cove and Thetford Mines ophiolite samples 275–6, Table 10.5
Nd isotopic compositions
  relationship with LOTI 258
  western Tasmania boninitic rocks and LOTI volcanics 254–5
Nd vs. Sr isotopic compositions, N Tongan lavas 385–7
Negri Volcanics (West Pilbara) 150, 151, 162
  Archaean 148
  SHMB parent magma 164
Nepoui (New Caledonia) boninites 1, 12, 29, 43, 314, 322–31
  depleted source 335
  glassy nature of 335–6
  major- and trace-element analyses Table 12.3, 328–31
  unusual compositional features 28
New Britain arc 68

Quaternary volcanoes 66
New Caledonia
  boninite-tholeiite associations 314–36
  Formation des Basaltes **12.3**, 336
  Massif de Koh **12.2**
  Sr-Nd isotopes and tectonic setting **12.4**
New Caledonia boninites 4–5, 6, 19, 40, 43
  see also Nepoui (New Caledonia) boninites
Newfoundland
  ophiolitic sequences 283–4
  Ordovician volcanics 259
Ni abundances, bronzite andesites 343, 344
Ni concentration, high 90
nickel sulphide ore 169
nickel sulphides, Kambalda 195
'no-slab' window 418, *419*
norite 184
  cumulate-textured 214
Norseman-Wiluna belt 151
North New Guinea Plate 50, *56*, 57
North Star Basalts 163
Northwestern Alps (Italy), ultrapotassic rocks 76–7

obduction 63, 65, 67, 283, 322, 335
  SW Japan arc 54
ocean-floor spreading, and boninite generation 32–3
oceanic crust
  dehydration of 108
  old, subduction of 34
  subducted 28, 91, 412
oceanic lithosphere
  dehydrating down-going 405
  downgoing, supplying water to HMA magmas 62
  hydrous, depleted 34
  subduction
    beneath hot lithosphere 51
    in normal arc-trench system 50
  young 50
  young, hot, subduction of 40, 41, 63
ocelli, felsic 15
Oki-Daito Ridge 57
olivine 1, 12, 13, 22, 58, 76, 78, 89, 94, 107, 113, 114, 115, 121–5, 126–7, 141, 161, 170, 181, 186, 214, 236, 268, 269, 288, 293, 301, 316, 427
  altered 77
  in boninite petrogenesis experiments 99–105
  coexistence with pyroxene 134
  fosteritic 22
  skeletal 151
  stability field 137, 139
  stability limit 134
  in upper-mantle peridotite 90

INDEX

zonation 127
olivine accumulation 293
olivine addition 188
olivine boninites 12
olivine control line 380, *381*
olivine crystallization 107, 113, 195
   H_2O content needed 136
olivine crystals 325
   resorption 325
olivine fractionation 95, 302–3
olivine gabbro 164
olivine and olivine-hosted glass inclusions 367
olivine orendite 77–8
olivine saturation 225
   and magma genesis 113
olivine stability 125, 136
   melt pressure and H_2O 117
olivine tholeiite 440
olivine-orthopyroxene saturation 117
ophiolite 37, 113, 315
   Appalachian, formed in early stages of subduction 282
   formation of
      near subduction zone 266
      in supra-subduction zone settings 282
      implications of occurrence of boninites 265
   low-Ti 334
Ora Banda sill 162
Orciatico (Tuscany, Italy), ultrapotassic rocks 77
orthoenstatite 1, 106
orthopyroxene 1, 58, 76, 113, 114, 115, 120, 121–5, 127, 128, 134–5, 170, 181, 269, 316, 342, 403
   in boninite petrogenesis experiments 100–5
   cumulus 180
   incongruent melting of 117
   Mg-rich 216
   poikilitic 186
   skeletal and hopper *185*
   xenocrystic 184
orthopyroxene crystallization 107, 182
orthopyroxene formation 199
orthopyroxene fractionation 138
orthopyroxene grains, quench-textured 184
orthopyroxene saturation 225
orthopyroxene-pigeonite crystals 120
orthopyroxenite 184
   nodules 214–15
orthoquartzite 235
Owen Conglomerate 236

Pacific plate, subducted 409
Pacquet Harbour Group 264, 265
   geological setting 267
   including higher-TiO_2 basalts 278

palagonite 401
Palau-Kyushu arc 54
Palau-Kyushu Ridge 33, 36, 39, 57
Papua, New Guinea region 50
Papuan ultramafic belt 336
pargasite
   Mg-rich 58
   quench 21
partial melting 2, 18, 28, 219, 280, 380, *381*
   amphibole-chlorite peridotite 60
   anhydrous 21, 229
   batch 228–9, 302, 380
   contaminated refractory wedge 165–6
   'delayed' 84
   depleted peridotic mantle 72
   depleted subforearc oceanic lithosphere 38
   differing degrees of 139
   harzburgite
      depleted, and low-Ca boninites 42
      serpentinized 143
   hydrous 22, 90
      boninite primary magmas 382
      of hot mantle wedge 51
      major-element-depleted source 301
   of hydrous column 58, 62, 65
   initiated by fluid fluxing 412
   lherzolite 440
   low-pressure 21–2
   and magnesian quartz tholeiites 32
   MORB eclogite 437
   of MORB-source mantle diapirs 30
   peritectic 22–3, 84, 115, 137, 138–9, 142
   producing Tongan high-Mg lavas 382
   refractory mantle peridotite 382
      enriched 357
   refractory peridotite source 26
   rising diapirs 108
   source characteristics and 402–5
   subducted slab 57
   successive events give rise to magmas with low Th/U 405
   variations in degree 107
partial melting models 301–4
   dynamic model 411
partial melts
   anhydrous peridotite 51
   hydrous peridotite, enriched in SiO_2 51
   silicate 28
   small-volume, migration of 305–6, 307, 310
   from subducted slab rich in HFSE 409
partion coefficients for incompatible elements 403
Pb, transferred by aqueous fluid phase into source mantle for boninite magmas 409–10
Pb concentrations, sediments 409
Pb enrichment 412

459

INDEX

Pb isotope data, Archaean and Proterozoic basalts and sulphides 167–9
Pb isotope ratios, oceanic (DSDP Site 458 bronzite andesites) 352
pelagic sediments 2, 389
Penguin Orogeny 235
Penrose stratigraphy 290
peridotite 18, 184, 302–3
  depleted 280, 308
  low pressure melting of 289
  partial melting of 279–80
  H_2O-undersaturated, melting of 42
  hydrated 58, 65
  sub-arc, contact melting of 34–5
  hydrous 63, 65
    enriched 57–8
    refractory 350
    sub-arc 353
    without serpentine 58
  hydrous and anhydrous, partial melting of 51
  hydrous column 50, 62
  hydrous melting of 89, 91, 107
  layered 266
  mantle
    hydration of 434
    unmodified melts 225
  refractory 114, 391
  a refractory residuum 27
  residual 347
    after MORB extraction as bronzite andesite source 348
  Ronda 308
  source of low-Ca boninites 19
  supra-subduction zone 2
  wallrock 229
peridotite source 219
  depleted 21
  'refractoriness' of 24
  refractory 23, 26, 28
peritectic melting 22–3, 84, 115, 137, 138–9, 142
PGE
  concentration of in magma chambers 167
  enrichment in 150
PGE deposits 3
  Bushveld Complex 179
PGE mineralization 149, 202
  Archaean and Proterozoic layered complexes **6.7**
Phalaborwa Complex, alkaline rocks 198
phase compositions, effect of bulk compositional change 125–6
phase equilibria
  in CMAS 142
  experimental, interpretation of 136–9
phase relations, near-liquidus 129–35
phenocryst pyroxenes 367

phenocrysts
  augite 13, 14, 317
  bronzite 14
  calcic clinopyroxene 215
  in chilled margins 209
  chromite 12, 25
  clinoenstatite 12, 14, 62, 232, 323, 325
    magnesian 328
  clinopyroxene, Ca-rich 397
  diopside 78
  feldspar 317
  hornblende 318
  low-Ca pyroxene 264
  olivine 12, 13–14, 15, 62, 77, 90, 118, 119, 120, 137–8, 214, 264, 291, 323, 432
    glass inclusions 359
    magnesian 212, 359
    resorbed 25
  orthopyroxene 12, 14–15, 15, *213*, 232, 323
    altered 317
    magnesian 328, 359
    Mg-rich (bronzite) 341
    reverse-zoned 196
    Vestfold subgroup I 212
    Vestfold subgroup II 214
    Vestfold subgroup III, quench-textured 214
    zoned *185*
  phlogopite 78
  plagioclase 277, 317, 342
  pseudomorphs 316
  pyroxene 277, 291, 293
    low-Ca 397
    psuedomorphed 239
  quartz 318
  sanidine 78
Philippine Plate
  melting of lithospheric mantle 350
  subduction of 54, 63
phlogopite 62, 65, 76, 77, 83, 215
  breakdown of 60
  as hydrous phase 439
  incongruent melting of 84
picrite 293
  altered 31
Piedmonte Ophiolite Nappe 76
pigeonite 95, 103, 105, 180, 214, 323
  magnesian 215, 269
pigeonite cores 363
Pilbara Block, Western Australia 148, 150, 151
pillow basalt 322
pillow lavas 13, 239, 264, 268, 317, 397
  glassy 342
  mafic 266
plagioclase 1, 13, 22, 113, 114, 141, 161,

INDEX

169, 181, 214, 223, 268, 269, 316, 323, 367, 397, 432
  albitized 241
  calcic 342, 376
  intercumulus 178
  petrography and mineral chemistry Table 14.1, 367, *368*
  in sub-arc magma chamber 376
plagioclase feldspar 317
plagioclase grains 186
plate margins
  Andean-type, subduction-related enrichment 85
  constructive 290
  convergent 37, 417, 418
    boninite generation at 33
  passive, formation of mafic-ultramafic complexes 234
  subduction zone at 434
plate tectonics, and Appalachian boninites **10.10**
platinum-group minerals 179
Platreef (S Africa) 179
plutonism, calc-alkaline 418
polysynthetic twinning 12, 135, 269
post-collisional setting, Andean-type 84
potassic rocks, Italian 85
Povungnituk Group 31–2
precious-metal enrichment 179–80
pressure, and boninite segregation 18, 21
Pretoria Group 176
priderite 73
Priestley Peak, Enderby Land 151
primary melts 113, 117, 136
primocrysts, orthopyroxene 214
protoenstatite 62, 89, 99, 101, 236, 325
  stability of 106
protopyroxene 100
  stability field of 102, 105
pseudomorphs 259, 268, 316, 317, 323
  amphibole 239
  Cen 245–6
  chlorite 239
  clinopyroxene 269
  olivine 269
  pyroxene 239
pumpellyite 317
pyroxene 95, 141, 181
  Cr-rich 215
  groundmass, petrography and mineral chemistry 363–5
  low-Ca 161, 269
  phenocryst, petrography and mineral chemistry 366–7
  xenocrystal, resorbed 376
  zoned 12
pyroxene fractionation 138, 219–20, 302–3
pyroxene variation 127–9

pyroxenite 266
pyroxenite layers 180
'pyroxenite' marker 180
pyroxenite nodules 225, 228

quartz 268, 316, 317, 401
quartz granophyre 215
quartz syenite 78
quartz tholeiite 40
quench augite 363
quench crystallization 126
quench crystals 90, *185*, 214
  orthopyroxene 186
  plagioclase 342
quench needles 12–13
quench pyroxene 363
quench textures 15, *185*, 186, 268
quench-textured rocks 188
quenched margins 209, 291

Rayner Glacier, Enderby Land 151, 163
Rb enrichment 191
reduced volatile phase 391
REE 27, 41
  low 296
  relatively immobile during low-grade alteration 245
REE abundances
  boninitic rocks *253*
  bronzite andesites 343, 344
  Nepoui boninites 329
REE concentrations 152
REE enrichment 405
  origin of, LFC lavas 305–6
REE patterns 107
  highly fractionated, bajaites 439
  LOTI volcanics 249, *253*, 254
  Spanish and Italian lamproites 81
  U-shaped 154, 266, 272, 274–5, 280, 284, 289, 305, 310, 335, 347
REE variations *278*
  N Tongan lavas 387–8
residual mantle
  melting of induced by LILE-rich fluids 91
  Th/U of 402–3
residue melting 302
resorption, olivine crystals 325
richterite 73
ridge axis crustal sequence, extensional disaggregation 290
ridge subduction 281, 418, 436
ridge-trench collision, thermal conditions in subduction environment during 434–6
Rivera plate, subduction of 66
Rivera ridge, transform 418
roll-back 283
Ronda peridotite 308
Ronda ultramafic massif 18

461

INDEX

Rooiberg Felsites 175, 176, 180, 199
Rooiberg Group 176
Rustenburg Layered Suite see Bushveld Layered Complex
rutile 155

San Andreas Fault 418
sanidine 76, 77
sanukitoids 5, 7, 51, 91, 116
Sc
   loc, Nepoui boninites 329
   moderately compatible element 219
Sc abundances 273
   LFC lavas 303–4
Scourie dykes
   geochemical and isotopic character 163–4
   norite dykes contaminated 165
   and SHMB 150
sea-floor spreading 51, 54
   constructive plate margin 290
second-stage melt addition process 155
second-stage melting/melts 18, 30, 31–2, 149
   low pressure, of diapirs 32–3
sediment subduction 169, 201, 203
   enrichment of boninite source 196–9
   implications of model for other Bushveld magmatism 199–201
sedimentary component
   ancient 27, 28
   ancient and enriched, Mediterranean ultrapotassic rocks 82
sediments 26
   in genesis of convergent margin magmas 409
   involvement in boninite genesis, implications of Pb isotopic data **15.7**
   oceanic, subducted 389
   pelagic 2, 389
   Semail Ophiolite, Pb Th and U abundances 403
serpentine 316
   decomposition of 58, 65
Serpentine Hill Ultramafic Complex 238
serpentinite bodies 315
serpentinite breakdown 435
Sesia-Lanzo zone, paraschists 76–7
Setouchi (Japan) boninites 1, 13–14, 17–18, 21, 23, 26, 40, 42
   compositional variability 24
   derivation from harzburgitic sources 27
   generated from depleted mantle 43
Setouchi lavas, MgO contents 427
Setouchi volcanic belt 5, 7, 50, 52–4, 68
   HMA occurrences 54–6
sheet flows, submarine 397
sheeted dyke complex 266, 282, 290
   relationship to LFC lavas 301
Shikoku Basin 50, 54

off-ridge volcanism 63
Shikoku Basin lithospheric plate 54
SHMB see high-Mg basalt, siliceous
silica enrichment 115
silica saturation 84
silicate partial melts 28
sills
   Bushveld complex 15, 17
   micropyroxenite, quench-textured 15
   peridotite 15
   pre-Bushveld 175
   pyroxenite 188
   quench-textured 183, 184
   syn-Bushveld 174, 175
   ultramafic 188
siltstone 235
Silverton Formation 178
SiO_2
   high 259
      ultrapotassic rocks 82
   Limassol Forest Complex lavas 296
   in low-Ca boninite sources 27, 28
   mobile during greenschist metamorphism 272
   in quenched melts 188
   in ultrapotassic rocks 73
SiO_2 solubility 23
slab dehydration 27, 42, 51, 107, 143
slab melting 435
   and production of high-Mg andesitic bajaites 436–8
   suitable environments for 436
   under eclogite-stable conditions (model) 437–8
smectite 120, 291, 317
Smithton Basin 234, 235
Snooks Arm Group 266, 277, 283
Solomon Sea plate 67
source characteristics, and partial melting 402–5
source component mixing 349, 379–80
source components
   depleted and enriched 391
   enriched 385, 387
source composition 380, *381*
South Bismark plate 67
sphene 155
spherulites 268
spinel 77, 95, 99, 114, 141, 219, 316
   Cr-rich 347
   Fe rich, Vestfold subgroup I 214
spinel fractionation, minor 220
spinel harzburgite, Ronda 19
spreading centres 32–3, 33–4, 36, 334, 418
   active, subduction of 2, 38, 39, 43
   back-arc basins 33
   backarc, migration of 281
   East Pacific Rise 421

INDEX

effects of encroachment on subduction zone 436
marginal basin 284
'resurrection' of 39
subducted 39, 90
Lau Basin 37
North New Guinea plate 63
West Philippine Sea 36
spreading ridges, oceanic 281
spreading systems, active, Melanesian region 66
$^{87}Sr/^{86}Sr$
  in LFC lavas 300–1
  Mediterranean ultrapotassic rocks 82
$^{87}Sr/^{86}Sr$ ratio, New Caledonia tholeiites 331–2
Sr abundances, SHMB 152, Table 6.3
Sr depletion 211, 212
Sr/Nd ratio, bajaites 437
Stillwater Complex 3, 149
  parental magmas 180–1
  SHMB parent magma 165
subcontinental lithosphere, fertile 63
subducted sediment 304
  melting of 200
subducted slab 2, 155, 434
  dehydration of 107
  metasomatic halo 34
subduction 34, 66, 84, 165–6, 281–2, 283
  of active spreading centres 38, 43
  of active/dying ocean ridge system 144
  of hot, dry lithosphere 437
  of hot lithosphere 65
  of oceanic crust 65
  of oceanic lithosphere 50, 51
  of sediments 169, 196–9, 199–201, 203
  Shikoku Basin plate 54
  young oceanic lithosphere 63
  hot 40, 41, 51
subduction processes, termination of 418
subduction zone magma genesis, basic concepts 57–60
subduction zone setting, for Bushveld Complex 201
subduction zones 50, 264
  at plate margin 434
  effects of encroachment of spreading centre 436
  mature, slab melting 434–5
subduction-related processes 417
  Andean-type 72
Success Creek Group 235
sulphide droplets, magmatic 151
sulphide globules 196–7
sulphide saturation 149
  and precipitation 167
  through magma mixing 167
sulphide undersaturation 162

sulphides 168, 184
  massive 195
  nickel 165, 195
  PGE/sulphur ratio high in depleted mantle 197
  plantiniferous 182
  retaining PGE in system 196–7
sulphur saturation 196
supra-subduction zone environments 259, 260
supra-subduction zone mantle wedge, reactivated 84
supra-subduction zone setting 310, 315
supra-subduction zone volcanics 256–7
  La enrichment 253
supra-subduction zones 5, 143

T-MORB 30, 31
Ta abundance 305
Taconic Orogeny 282
Tasmania (western)
  boninite and low-Ti (LOTI) lavas 232–60
  mineral chemistry and geochemistry 241–55
  relationship between 259
Tauride foldbelt 78
tectonic scenarios, boninite generation 29–37
  continental rifting 30–2
  ocean-floor spreading 32–3
  subduction-related magmatism 33–7
tectonism
  transtensional 288
  strike-slip 290
tectonites, residual harzburgite 291
temperature
  high
    for generation of HMA magmas 51
    in upper mantle wedge 33
  increase required for low-Ca boninite generation 39
  upper-mantle, raising of 33–4
tensional regime 84
tensional stress, in forearcs 39
$^{230}Th-^{232}Th$ disequilibrium studies, calc-alkaline volcanic rocks 406
$^{230}Th-^{238}U$ disequilibrium studies 404–5
$^{230}Th/^{232}Th$ activity ratio 405, 406
$^{230}Th/^{238}U$ activity ratio 404–5, 406
Th, mobilized in oceanic basalt 401
Th/U low, derivation from depleted mantle sources 402
Th/U ratios
  additional processes 406
  primary processes **15.5**
  secondary processes **15.4**
Th/U systematics, boninite lavas **15.6**
thermal conditions in subduction environment during ridge-trench collision 434–6

INDEX

thermal erosion 195
thermal gradient, high 62, 63, 68
thermal rebound 438–9
thermal relaxation 416
Thetford Mines ophiolite 264, 265, 282
  associated volcanic rocks 277–8
  geological setting 268
tholeiite 333–4
  aluminous 181
  arc 333, 350
    Mariana arc 333
  backarc-basin 2, 387
  Bushveld Complex 201
  DSDP Site 458, source region for 349–51
  flat REE patterns 166
  generated by hydrous melting 353
  island-arc 278, 280, 281, 282
  low Th/U 402
  magnesian quartz 17, 30, *31*, 32
  Massif de Koh 317
  New Caledonia, Sr-Nd isotopes and tectonic setting 331–4
  North Star Basalts 163
  silica-saturated 90
thrust sheets 4
  serpentine 322
Ti 220–1
  as a compatible element 194
  enrichment in 235
  fractionation of from Zr 192–3
  incompatible in olivine 195
Ti vs. Zr, Appalachian samples 274, *276*
Ti/Zr 197, 428
  bajaites 437–8
  low 28, 203, 280, 341
    Bushveld marginal rocks 181
    Cape Vogel boninites 221
  primitive mantle 192
  residual mantle 198–9
  Tasmanian boninitic rocks 256, 260
  Vestfold Hills subgroup I 221
  western Tasmania boninitic rocks and LOTI volcanics 249, 254
Ti/Zr values 344
Tinaquillo lherzolite 382
  anhydrous partial melts 21
Tinaquillo peridotite 18
TiO_2
  low 309, 318
    in Appalachian samples 273–4
    Bushveld marginal rocks 181
TiO_2 abundances 293, *295*, 303, 343, 344, 347
  low 289, 341, 385
TiO_2 contents, ultrapotassic rocks 79, 82
TiO_2 enrichment of depleted source 83
titanate phase, residual 390
Tofua magmatic arc 387, 388

tonalite 161, 194, 336
  contamination by 170
Tonga arc, tensional stress 39
Tonga Trench 37
trace-element abundances, variation in, LFC lavas 296–300
trace-element data, SHMB 152, Table 6.1
trace-element distribution, controlled by amphibole behaviour 441
trace-element enrichment
  imposed on depleted mantle source 304
  styles, LFC 304
trachyte 78
transform fault 144
transform sequence extrusives 290
transform tectonized zone 290
transtensional tectonism 288
Transvaal Sequence 178
  sills and dykes 15
trench roll-back 281
triple junctions 418, 422
Troodos Massif, Cyprus 289–90
Troodos Ophiolite 1, 5, 119, 289
  Pb isotopic data 409
Turkey, ultrapotassic rocks 72
twinning, polysynthetic 12, 135, 269
two-component mixing model 387–9

  mobilized in oceanic basalt 401
  oxidation of 401
  sink for 406
ultramafic belt, Papuan 4
ultramafic rocks 235–6
ultrapotassic rocks 72–85, 380
  defined 72
  derived by differentiation from upper mantle partial melts 72–3
  geological setting and mineralogical features 75–8
  island arc 79, *80*, 81
  major groups 73
  petrogenesis **3.4**
  wholerock chemistry 78–82
Upper Hangingwall Basalt 162
Upper Pillow Lavas, Troodos Ophiolite 1, 5, 11, 14, 16, 22, 26, 42, 43, 214, 216, 290, 363

vanadium 219
varioles 268
Venterdorp lavas 175
Vermont Formation 184
Vestfold Hills, Antarctica 11, 151, 163
  high-Mg dykes 208–29
Victoria (Australia) ophiolites, tholeiites and boninites 37, 43
volatile supply to subduction zone 436

INDEX

volatile transfer 417
volcanic arcs 113
volcanism
  calc-alkaline 418
  high-Mg andesite 52
  HMA 52
  island-arc 281, 282
  off-axis 290
  off-ridge 54, 63
  primitive mafic 290
  slab-induced, migration of 283
  subduction-related 421
    absence of 63
volcanoes, New Britain arc 66

W Mediterranean
  regional plate dynamics 84
  ultrapotassic rocks 72
    emplacement in post-collisional environment 83
wadeite 73
wallrock peridotite 229
wallrock reaction 228–9
water-rock interactions, post-magmatic, changing Th/U ratio of lava 401
weathering 245
wedge hydration 51
wehrlite 141
West Philippine Basin 50, 55, 56, 63
West Philippine Sea 37
  spreading centre 33–4, 36
wet melting, of subducted oceanic crust 434
Whim Creek belt 151
Wilson cycle 29
  ocean closing stage 32
  ocean-floor spreading phase 32–3
  rift-to-ocean stage 30–1

Witwatersrand Sequence 175
Wolkberg Group 175
wollastonite 127
Woodlark Ridge 67
Woodlark Rift 67–8

xenocrysts 291
  calcic plagioclase 32
  Cr-Al-spinel 76
  olivine 76, 293, 301
  orthopyroxene 76
  zircon 157, 162
xenoliths 328
  peridotite 197
  ultramafic 76
    high Th/U 403
    PGE abundances 167

Y 301
Y abundances, low 385
Y concentration, LFC lavas 298, 300
Y enrichment 235
Yilgarn Block (Australia) 148, 150, 151
  lavas 15

zeolites 401
zone refining 228
zoning 367
Zr 28, 220–1, 304
  as an incompatible element 194
  high, Nepoui boninites 329
  incompatible in olivine 195
Zr abundances 343, 344
  low 249
Zr enrichment 150, 155, 156, 221, 235, 297, 339, 344, 347, 352–3, 353
Zr/Sm, high, origin of in boninites **13.5**

DEC 0 4 1989